Math Instruction for Students with Learning Difficulties

This richly updated third edition of *Math Instruction for Students with Learning Difficulties* presents a research-based approach to mathematics instruction designed to build confidence and competence in preservice and inservice PreK-12 teachers. Referencing benchmarks of both the National Council of Teachers of Mathematics and Common Core State Standards for Mathematics, this essential text addresses teacher and student attitudes towards mathematics as well as language issues, specific mathematics disabilities, prior experiences, and cognitive and metacognitive factors. Chapters on assessment and instruction precede strands that focus on critical concepts. Replete with suggestions for class activities and field extensions, the new edition features current research across topics and an innovative thread throughout chapters and strands: multi-tiered systems of support as they apply to mathematics instruction.

Susan Perry Gurganus is Professor Emeritus of Special Education at the College of Charleston, USA.

Math Instruction for Students with Learning Difficulties

3rd Edition

Susan Perry Gurganus

Routledge
Taylor & Francis Group

NEW YORK AND LONDON

Third edition published 2022
by Routledge
605 Third Avenue, New York, NY 10158

and by Routledge
2 Park Square, Milton Park, Abingdon, Oxon OX14 4RN

Routledge is an imprint of the Taylor & Francis Group, an informa business

First edition published by Pearson 2006
Second edition published by Routledge 2017

Library of Congress Cataloging-in-Publication Data
A catalog record has been requested for this book

ISBN: 9780367561871 (hbk)
ISBN: 9780367559588 (pbk)
ISBN: 9781003096733 (ebk)

DOI: 10.4324/9781003096733

Typeset in Bembo
by Newgen Publishing UK

To Al, for forty-six years of dinner talks on teaching and writing

Contents

Preface

This book is intended for a range of educators: teachers in training, elementary and secondary mathematics teachers with struggling students, special education and other support teachers devising supplementary instruction and interventions in mathematics, administrators seeking to further STEM learning in their schools, and parents assisting their children. The third edition was written during the pandemic of 2020–21, when many teachers floundered to engage their students through a computer screen, where the lack of immediacy proved perilously stultifying. Precisely this circumstance informed specific strategies presented here. Some of you for whom this text is assigned reading may not be particularly keen on the topic, as many of my students at the College of Charleston were not at first, but I trust you can summon focus so that your PreK-12 students might become more involved, persistent, and successful learners.

Unquestionably, mathematics figure prominently in our lives. Virtually everything we do involves some aspect of calculation. We estimate time and space while driving our cars, negotiating traffic, filling up the gas tank. We conduct minute computations when tallying class absences, scoring tests, recording grades for the term. Even pursuits as mundane as baking cupcakes, checking the calendar, hanging a picture are mathematical in nature. The study of mathematics is in and of itself a dynamic, social, cognitively enhancing endeavor. Increasingly, mathematics literacy bespeaks a skill-set essential to higher function in our rapidly progressing Information Age. For our students to be better suited to the challenge, we must be more effective teachers.

Since the first and second editions of this text in 2007 and 2017, there has been a gratifying explosion of research on mathematics teaching and learning for students who struggle. Research has expanded across mathematics domains and grade levels, including experiences that ready young children for school entry and prepare high-school students for post-secondary education. This edition updates that research, which clearly evinces that these students can achieve grade-level standards if provided appropriate, research-based instruction and interventions.

The third edition includes a new thread across all chapters and strands: the multi-tiered systems of support (MTSS) framework. The framework is introduced in Chapter 1, with a full description of the three-tier model. Chapter 2 discusses MTSS for young children and disability identification, whereas Chapter 3 includes a section on assessment for each tier. Chapters 4, 5, and 6 include research for instruction and intervention within MTSS models. Each strand includes a special box illustrating an intensive intervention for the content focus. This edition also includes many familiar features: pedagogical chapters, *Try This* opportunities, and content-focused strands. Magnifying glass icons denote terms defined in the glossary. Scenarios involving teachers and students, adapted from my interviews with students on their mathematics understanding, introduce each strand and are embedded in some chapters.

My own enthusiasm for mathematics began with excellent teachers in the public schools of Raleigh, North Carolina. My interest in teaching students with difficulties began at Mars Hill College, where I served as a departmental tutor and engaged in student-teaching practica.

I sought to establish what students did understand and built on those concepts and skills. That approach served me well in teaching elementary, middle, high-school, and college classes over five decades. My passion for art is inextricably linked to mathematics. Most illustrations in the text are my own. It is my fervent hope that this book will elucidate strategies to inspire lifelong wonder for mathematics in students who contend mightily for mastery.

Acknowledgments

Once again, I recognize my former students at the College of Charleston, most now special education teachers, administrators, or professors. They field-tested drafts of the first edition, have sent teaching anecdotes over the years, and are named in word problems throughout the text. Thank you also to colleagues across the country for encouragement, feedback, and text adoptions. I acknowledge the work of a growing number of researchers in the field of mathematics for students with learning difficulties; their names are in the text and references. Research during the pandemic was severely restricted, including work in libraries. A special thanks to the staffs of Addlestone Library, College of Charleston, and North Carolina State University's D. H. Hill Library, for their virtual access and interlibrary loan programs.

Thank you to Al Gurganus, Professor Emeritus of The Citadel, for his translations of German, Dutch, and Norwegian text. Thanks to Michelle Harrell, Director of Education, and Caroline Rocheleau, Egyptologist and Curator of Ancient Art, of the North Carolina Museum of Art for their contributions about art and mathematics. I appreciate the critical review of the algebra chapter by Dorothy Miner, retired high-school mathematics teacher, Wake County Public Schools. Special thanks to the scholars and organizations providing reprint and adaptation permissions for this edition: Keith Lenz (University of Central Arkansas), Asha Jitendra (University of California, Riverside), Jon R. Star (Harvard University), National Council of Teachers of Mathematics, Council for Exceptional Children, and the National Association for the Education of Young Children; as well as the publishers Cambridge University Press, SAGE, PRO-ED, Taylor & Francis, and hand2mind.

Thanks to the second-edition reviewers for their insightful and valuable comments. Finally, I am grateful for the encouragement and advice of my editor at Routledge, Misha Kydd, and the careful work by my editorial assistant, Olivia Powers, and my copyeditor, Martin Noble.

1 Mathematics in Today's Schools

The Context for Learning Challenges

Chapter Questions

1. How are teachers' and students' attitudes toward mathematics related to learning?
2. What mathematics reforms have influenced current mathematics standards?
3. What national standards have an effect on state and local mathematics curricula?
4. How do the mathematical practices relate to the content standards?
5. What mathematics topics are emphasized in PreK through Grade 12 classrooms?
6. How are multi-tiered systems of support (MTSS) models creating a new context for mathematics teaching and learning?

Schools in the 21st century are experiencing many forces for change in order to prepare students for college and career demands and to meet the needs of increasingly diverse students. National and state-wide curriculum standards have been adopted widely, with their high-stakes assessments. Technology offers new and challenging resources for teaching and learning. School-district initiatives offer students more options for study and concentration areas. Students with learning difficulties—those with disabilities, language differences, impoverished backgrounds, and others—find themselves in classrooms with high demands that offer varying degrees of learning support. This text addresses the variables that can assist students with learning difficulties in being successful in today's mathematics learning settings: instructional, assessment, resource, interpersonal, and intrapersonal contexts. This first chapter sets the stage with discussions of teacher and student dispositions about mathematics, mathematics reform movements, current curricular standards, multi-tiered systems of support, and the need for collaboration among professionals.

> Angela Smith was hired immediately after graduation as a special education teacher at Balsam Middle School. The principal informed her that she is to support regular mathematics classes for three periods of the day and will teach special pull-out mathematics classes two periods each day. Leaving the principal's office, Angela feels panic as she struggles to recall the mathematics programs reviewed over only a two-week period in her special education methods course.
>
> Joseph Lopez met with the assistant principal for instruction at Pine Road High School and was informed that a total of 20 students with learning, emotional, and communication disabilities will be in his Geometry, Algebra I, and Mathematics I classes. Joseph recalls accommodations for students with physical, visual, and hearing disabilities but cannot imagine what he should plan for students with other learning difficulties.
>
> Chris Johnson feels fortunate to have a teaching position at Pinetops Elementary School after relocating to Pine Grove to be closer to her family. But the meeting with the principal has left her puzzled. What is a mathematics support teacher?

DOI: 10.4324/9781003096733-1

In a meeting of new teachers in the district, Angela expresses her anxiety about teaching a subject she is not very strong in herself. Joseph talks about his love of everything mathematical but uncertainty regarding individual student needs. Chris feels confident that her five years of elementary teaching will be beneficial for instructional planning but is not sure about how to work effectively with so many other teachers.

Angela, Joseph, and Chris are facing the challenges of new teaching positions that will involve mathematics instruction or the support of that instruction for students who are struggling to learn skills and concepts. In a fairly short period of time they must understand their instructional roles, get to know students, review the district mathematics curriculum, and develop long- and short-range instructional plans in cooperation with other teachers. All three teachers have doubts about their abilities to meet the needs of students with learning difficulties within the mathematics curriculum. However, they all have many strengths and demonstrate professional attitudes about developing new skills.

What teacher characteristics are most closely associated with student achievement in mathematics, especially for students who struggle to achieve? Multiple studies cite strong understanding of concepts in conjunction with skill in specific pedagogies for mathematics as requisite for student achievement (Baumert et al., 2010; Campbell et al., 2014; Hill et al., 2019). Mathematics knowledge alone is not enough. General pedagogical skill is not enough. Deep understanding of mathematics concepts and how to teach those concepts is critical for student learning. Additionally, teachers' knowledge of students—knowing what their students know and don't know—is related to student outcomes in mathematics (Hill & Chin, 2018). Teachers of students who struggle with mathematics must be skillful with a broader range of content, be able to trace learning pathways for individual students, demonstrate a wider array of pedagogical tools, and have expectations of students that will help build knowledge and positive attitudes towards mathematics. Unfortunately, in many states and districts, disadvantaged students tend to be served by less qualified teachers in mathematics (Goldhaber et al., 2016; Hill et al., 2019; Max & Glazerman, 2014).

Two of the most critical factors for increasing student achievement are the student's self-efficacy beliefs in mathematics and the teacher's high expectations for students (Campbell et al., 2014; Clark et al., 2014; Friedrich et al., 2015; Klehm, 2014; Tosto et al., 2016). These dispositions of students and teachers are among the strongest predictors of student achievement. The following sections introduce teacher and student dispositions toward mathematics, with strategies for building positive views, expectations, and aspirations.

Mathematical Dispositions

Many teachers and teacher candidates, especially in elementary and special education, are anxious about their responsibility for teaching mathematics. They may have had poor experiences themselves (Bekdemir, 2010; Gresham, 2009; Humphrey & Hourcade, 2010). They may have taken only one mathematics-education course, unlike their peers teaching middle- or high-school coursework. Special educators, often certified to teach K–12, may be dealing with higher levels of content than they feel prepared to teach. General educators may have had an exceptional children or diverse students overview course, but not a course specific to pedagogies for working with struggling students.

When asked about memories of their own math classes, teachers and teacher candidates cited indelible memories:

- My teacher lacked mathematics skills and could teach a lesson only by following the textbook and checking the solutions (Bekdemir, 2010, p. 320).

- My teacher asked me [to explain something in class and I said] I couldn't do division and then all my classmates laughed at me (Bekdemir, 2010, p. 320).
- I took a computer class (in college) instead of a math course. I wouldn't have passed a math course…because I was always doing badly in it (Humphrey & Hourcade, 2010, p. 28).
- I've never been a math person, but I can't have my students ever know that. I'm really good at finding someone to explain the problem to me before I try and help my kids (Humphrey & Hourcade, 2010, p. 28).
- I had the best time in math class. I remember getting up every day during elementary school and couldn't wait to get to school and do math (Gresham, 2009, p. 35).
- Ever since I was in elementary school I have hated math! I remember going home crying, not wanting to go to school, screaming at my mother for making me do my homework, … begging her to let me quit school,… all by fifth grade! (Gresham, 2009, p. 35).
- Math was always easy, fun, and something I always wanted to do. I would play math games instead of playing with dolls (Gresham, 2009, p. 35).
- As soon as the test was over, I forgot everything about math that day. I feel so bad for my students who are like me (Gresham, 2009, p. 28).

Teachers' beliefs about their personal capabilities to help students learn is termed *teacher (or instructional) self-efficacy* (Klassen et al., 2011). There is evidence that teachers' self-efficacy beliefs about teaching mathematics is strongly correlated with their past experiences in mathematics (Voss et al., 2013). Further, there appears to be a reciprocal relationship between teacher self-efficacy and the quality of their instruction (Holzberger et al., 2013). Other influences on these self-efficacy beliefs are level of mathematics anxiety (also correlated with past experiences), experiences in preservice coursework, and experiences in internships engaged in teaching mathematics. Field experiences, especially, give preservice teachers feedback on their performance, an important contributor to self-efficacy beliefs (Gresham, 2009). Higher levels of teachers' self-efficacy and lower anxiety about teaching mathematics are also related to greater student achievement (Hadley & Dorward, 2011).

Teachers and teacher candidates should reflect on their own experiences with mathematics and beliefs about teaching mathematics if they are to be effective teachers. Since past experiences in school, which may not have represented the most effective teaching methods, have such a strong influence on future teachers, these candidates must understand their own beliefs about teaching mathematics and the sources of those beliefs (Clark et al., 2014).

To begin this exploration, locate the closest statement to your own on the following scale of views:

1. I avoid teaching math at all costs. I was not a good math student myself and I'd hate to teach my students the wrong way.
2. I really dislike math but if you give me a good textbook with lots of practice problems, I can muddle through.
3. Math is just another subject in school. I do (plan to do) the lessons and assign the work. I will spend extra time drilling what I know will be on the end-of-year tests.
4. I wish I knew more about teaching math. If I knew more effective ways of teaching it I wouldn't feel so intimidated. I'm not sure I'll be able to be an effective mathematics teacher at this point.
5. I'm OK in math and try to learn more by going to workshops, consulting colleagues, and reading journals. I want to learn more.
6. I'm a whiz at math but I'm not sure how to get the concepts across to students.

7. I enjoy (look forward to) teaching math and understand the concepts fairly well. I use (plan to use) a lot of hands-on and problem-solving activities in my classes. I still want to learn more.
8. I feel confident in teaching math and would feel comfortable serving as a mentor to new teachers. I take every opportunity to increase my knowledge and skills.

What experiences led to your current views? What potential experiences could enhance your views about teaching mathematics to diverse students? One strategy for teachers or teacher candidates to get into a more mathematical frame of mind is to engage in games and other activities based on mathematical concepts. Accessible games and activities are also encouraged for beginning the school year with students. Begin with interest, engagement, and success! Appendix I provides several individual and small-group activities that can create interest in mathematics. Try at least one of these with a colleague or with a group of students. Other activities can be found in mathematics journals, professional books, and on websites for teachers.

What teacher dispositions are critical for promoting student learning of mathematics? Some are certainly global, important for all teachers: a keen personal interest in learning and growing; reflective teaching; a view that all students can learn but each student is an individual learner; high but realistic expectations of learners; flexibility in adapting instruction, materials, and assessment to varying contexts; the ability to create a positive yet goal-oriented classroom environment. Global dispositions about student learning are not sufficient. Teachers should seek to develop domain-specific dispositions (Jacobson & Kilpatrick, 2015). For teaching mathematics these include:

*   genuine interest in mathematical concepts and connections;
*   a view of one's mathematics knowledge as emerging, and that growth is enhanced by interactions with students and colleagues (Davis & Renert, 2013);
*   the ability to notice a student's thinking about mathematics and adapt instruction accordingly (Thomas et al., 2016);
*   the belief that students who struggle in mathematics can learn and achieve higher-level thinking (Klehm, 2014);
*   a persistence with finding solutions to problems;
*   the willingness to consider multiple processes or multiple solutions to the same problem, to allow for the messiness of student exploration; and
*   an appreciation for mathematics-related applications such as those in music, art, architecture, science, geography, demographics, or technology.

Teachers are faced with many conflicting views, messages that tend to lead to dispositions that will not be beneficial for their students in the long run. Early-career teachers are particularly susceptible to these forces. They include views such as: you've got to cover the material so there's no time for exploratory activities, don't try to be creative in mathematics, mathematics is a static knowledge base so just cover the topics in the textbook, only standardized testing shows what a student knows, and you've got to keep the top students interested so you can't do group work in your class (Nolan, 2012). How can teacher candidates, early career teachers, and teacher mentors disrupt these messages? Through more opportunities for reflection and dialog, for risk-taking, for recognizing the structures and regulators at play in one's practices. Through continued professional development that involves true engagement with peers and an exploration of concepts and pedagogy.

Student Dispositions

What about the mathematics dispositions of students? There is compelling evidence that a student's academic self-concept in mathematics, or *self-efficacy*, is one of the strongest factors in achievement

(Wang et al., 2012). A student's self-efficacy is the "belief in one's capabilities to organize and execute the courses of action required to produce given attainments" (Bandura, 1997, p. 3). Self-efficacy influences motivation, self-regulation, courses of action, level of effort, perseverance, and, ultimately, level of accomplishment. The self-efficacy and performance relationship appears to be reciprocal, with performance results feeding back into students' self-efficacy beliefs (Arens et al., 2017). Self-confidence from efficacy beliefs may be a key factor that combines with new learning to enable students to manage difficult tasks. The opposite disposition, learned helplessness, results when students view their failures as insurmountable and out of their control. Helplessness is accompanied by passivity, loss of motivation, and declining performance.

The most influential sources of efficacy information are *mastery experience, vicarious experience, social persuasion,* and *physiological indexes* (Bandura, 1997). Mastery experiences are performance outcomes, the positive and negative experiences students have when participating in class, completing assignments, and taking tests. Students who experience success in their mathematics endeavors begin building positive beliefs in their abilities. However, easy successes are not helpful; experience with success in challenging tasks that require perseverance and even involve setbacks along the way lead to stronger efficacy beliefs. Vicarious experiences, or viewing models by which to compare one's capabilities, can positively or negatively influence self-efficacy beliefs. Teachers should be cautious in using comparative models so that students focus on the instructive elements for self-improvement and not simply make an evaluative comparison. Social persuasion, such as verbal feedback on performance, can also promote or undermine self-efficacy beliefs, depending on its use. Good persuaders must cultivate students' beliefs in their capabilities, structure activities for success, and encourage students to engage in self-evaluation, not merely voice positive encouragers. Finally, students acquire information from their physiological and emotional states such as anxiety and stress. For example, negative thoughts and fears about a mathematics test could result in lower self-efficacy and therefore lower performance on the test. Information gained from these four sources does not automatically affect self-efficacy; the student must interpret—or form *attributions* about—the results of their efforts (Schunk & DiBenedetto, 2016).

Mathematics-education researchers have proposed broader views of student dispositions that have a critical role in achievement. The *productive mathematical disposition* refers to students viewing mathematics as useful and worthwhile, that it can be understood, and that they can be effective learners (Kilpatrick et al., 2001). Clark et al. (2014) discussed the *mathematics identity* of students, a construct that includes students' perceptions of their mathematics ability and how perceptions influence performance, engagement in mathematics activities, and motivations to achieve at a high level.

Teachers providing mathematics instruction should be aware of student dispositions and actively promote positive student dispositions toward learning. Desirable student dispositions include:

- seeing the world mathematically;
- willingness to take risks and explore multiple problem solutions;
- persistence with challenging problems;
- taking responsibility for reflecting on one's own work;
- an appreciation for the communicative power of mathematical language;
- willingness to question and probe one another's thinking about ideas;
- willingness to try different tools for exploring mathematical concepts;
- having confidence in one's abilities;
- perceiving problems as challenges.

(National Council of Teachers of Mathematics, 2000; National Governors Association Center for Best Practices & Council of State School Officers, 2010)

Teachers know that students' attitudes towards mathematics can range from enthusiasm, interest, and confidence to dislike, rigidity of thought, avoidance, anxiety, and even phobia. What makes the difference? In addition to success with difficult tasks and positive performance feedback, as described previously, there is evidence that teachers' beliefs about their instructional efficacy predict student levels of academic achievement, regardless of level of student ability (Campbell et al., 2014). Teachers with a high sense of instructional efficacy tend to view all students as teachable, believe they can overcome negating community influences through effective teaching, devote more classroom time to academics, maintain an orderly classroom, encourage struggling students, hold high expectations, and encourage student self-direction (Bandura, 1997).

Encouraging Struggling Students

Students who struggle with learning mathematics, as reflected on the 2019 *Nation's Report Card*, include 19% of 4th graders, 31% of 8th graders, and 38% of 12th graders (National Center for Education Statistics, NCES, 2019). Fifty percent of fourth-grade and 68% of eighth-grade students with disabilities scored below basic. For English-language learners (EL), 41% of fourth graders and 72% of eighth graders scored below basic. *Basic level* indicates that students demonstrated some evidence of understanding mathematical concepts and procedures in five content areas. It indicates students have partial mastery of prerequisite knowledge and skills that are fundamental for the grade level. *Below basic* indicates a significant lack of understanding. For example, a fourth grader could not compare other simple fractions to $\frac{1}{2}$, or an eighth grader couldn't mentally form a new shape from six choices.

Approximately 7% of elementary and secondary students have a specific learning disability in mathematics, although estimates range from 3% and 7% (e.g., Geary, 2013; Morsanyi et al., 2018; Soares et al., 2018). Students with other disabilities as identified under the Individuals with Disabilities Education Improvement Act (IDEA, 2004) include those with cognitive or developmental disabilities, emotional disabilities, communication disorders, traumatic brain injuries, and some severe health impairments such as attention deficit hyperactivity disorder (ADHD). Students who are learning English as a second or other language (EL) may have related mathematics learning difficulties. Other students, such as those with cultural differences, scarce resources at home, highly mobile families, or poor past instruction in mathematics may also struggle with mathematics.

For these students with challenges learning mathematics, positive dispositions are a critical foundation for achievement. Concrete methods for promoting these dispositions towards mathematics include:

- Seek out student interests and plan activities that make connections with those interests. Connect mathematics lessons to real-world applications.
- Personalize math lessons by using student interests, names, real events, and student-created problems. Some teachers name classroom "discoveries" after students: The Sally Brown Proof.
- Allocate two or three minutes at the beginning of each mathematics class for warm-up activities with familiar material. Begin with success!
- Create classroom procedures that allow students to take risks and make mistakes without punishment or humiliation. Students will learn a lot from mistakes if those are viewed as learning opportunities, not failures.
- Encourage students to set personal short- and long-term goals in mathematics and keep track of their progress through individual portfolios or graphs.

- Check for student understanding when introducing new concepts and adjust explanations and examples until students demonstrate strong understanding. Check understanding by watching students work and by listening to their explanations.
- Communicate clearly with students the *why* of mathematics for the year. What new learning will they accomplish? Why will it seem they are working on some of the same topics as last year? How will this learning be beneficial in the long run? Listen to students' explanations of their views of mathematics.
- When students have accomplishments, guide them in making explicit connections with their efforts.
- For students who struggle, identify strengths also. What do students already understand? What is a potential *learning trajectory* from those concepts to new concepts?
- Introduce tools and manipulatives as a means for developing deeper understanding of mathematics concepts.
- Model positive dispositions—about mathematics and about working collaboratively with other teachers for student learning.

Most importantly, teachers should know their students—what they understand and can do in mathematics, what they are struggling with, and their individual resources (Hill & Chin, 2018). This knowledge is important for instructional planning but also for setting expectations for achievement. There is evidence that teachers' expectations of their students in mathematics can be inaccurate (Gentrup et al., 2020). Inaccurately high expectations— higher than achievement, cognitive abilities, background, and motivation would suggest— tend to be related to greater achievement gains. Inaccurately low expectations can be related to lower gains than should be achieved. Wang et al. (2020) found, across ten middle schools, that teachers tended to expect more of high achievers and unjustifiably less from low achievers than their achievement indicated. In fact, in a review of teacher expectation studies over 30 years, Wang et al. (2018) found strong and consistent evidence that teachers typically hold lower expectations for low socio-economic status students and for students with learning disabilities. Teacher expectations should be slightly higher than students' actual skills. Teachers should assist students in setting learning goals, provide high-quality feedback, and be aware of their own biases that might affect expectations (Rubie-Davies et al., 2015). The combination of positive teacher expectations and student dispositions towards mathematics learning will provide a critical component for success.

Mathematics Reform

Most educators will point to the 4 October 1957 launching of the Soviet satellite *Sputnik 1* as the flash point for reform initiatives in mathematics and science education in the United States. However, reform initiatives have been proposed throughout the history of education in this country—whenever the purpose of public education and the results of educators' efforts have been debated. The 1878 call for pragmatism by Charles Sander Peirce, the 1893 Committee of Ten's high-school curriculum proposals, early 20th-century John Dewey's progressive education initiatives, and the visionary writings of James Bryant Conant in the 1950s and 1960s on school reform, national goals, professionalizing teachers, and comprehensive high schools were all major reform efforts that had an impact on mathematics instruction (Parker, 1993).

It has been only since the 1980s, however, that reform efforts have been truly national in scope with intense public scrutiny. The 1983 report *A Nation at Risk: The Imperative for Educational Reform* promoted national educational reform (NCEE, 1983). Citing dismal performance data of students in the United States as compared with those of the rest of the industrialized world, the report called for more rigorous high-school studies. It was the catalyst for higher standards for college admission, a nationwide system of standardized achievement testing, more homework, longer school days and years, career ladders and other incentives to attract better qualified teachers, and more state and local financing for school reforms. Dozens of other national reports on various aspects of education followed in the next few years. Most called for increased standards for students, better teacher preparation, and accountability to the community.

The reform initiatives of the mid-1980s represented a convergence of attention on standards-based curriculum development and requests by colleges and employers. For educational evaluation, these initiatives represented a shift from the traditional focus on input measures, such as per-pupil spending and teacher–student ratio, to outcome measures of student achievement. They spurred the 1989 and 2000 National Council of Teachers of Mathematics curriculum standards, the Goals 2000 Educate America Act (1994), the Trends in International Mathematics and Science Studies (NCES, since 1995), and the No Child Left Behind Act of 2001 (NCLB, ESEA, 2001).

Although these reform documents mentioned students with disabilities only in general terms, the impact on these students has been significant. The 1997 reauthorization of the Individuals with Disabilities Education Act (IDEA), with its 1999 regulations, made mandatory the inclusion of students with disabilities in state and district assessments and related standards-based curricula. The 2001 NCLB Act and 2004 IDEA required special education teachers to be highly qualified in the content areas, such as mathematics, that they teach. The NCLB Act required states to test students in mathematics (and reading) annually in Grades 3 through 8 and once in Grades 10 through 12, and that all schools make adequate yearly progress or face interventions or restructuring.

National Mathematics Standards

Leadership in reorienting reform efforts to focus on curriculum, instruction, and assessment standards for mathematics was provided by the National Council of Teachers of Mathematics (NCTM) in 1986 when it established the Commission on Standards for School Mathematics and involved constituent groups in the development of national standards (Romberg, 1993). The *Curriculum and Evaluation Standards for School Mathematics* were published in 1989, the *Professional Teaching Standards* in 1991, and *Assessment Standards for School Mathematics* in 1995. The driving vision statement for the standards was "All students need to learn more, and often different, mathematics and … instruction in mathematics must be significantly revised" (NCTM, 1989, p. 1).

Criticism of the standards and their development process was immediate. The standards were developed primarily through expert opinion and consensus, rather than research review. The product, therefore, was plagued by vague constructs, pedagogical dogma, and idealistic goals. The curriculum standards were criticized for being, on the one hand, an idealistic vision for promoting conversations about mathematics education while, on the other, attempting to establish clear expectations for student achievement by the end of each grade level span (K–4, 5–8, 9–12). They were criticized for promoting constructivist methods and eliminating standard algorithms and basic fluency, an extreme paradigm shift that led to "math wars" in California, Wisconsin, Massachusetts, Texas, New York, and other states. Since California, Texas, and New York are the most populous, their standards influenced textbook and assessment development nationwide.

Special educators cited the complete absence of references to students with disabilities, especially egregious given the increasing diversity of the K-12 student population in the 1990s (Hofmeister, 1993; Mercer et al., 1993). They also questioned the fundamental process of directing change by standard setting rather than through validated, replicable, and affordable educational interventions that have been demonstrated to work with specific students. Also of concern was the emphasis on broad-based thinking skills rather than domain-specific ones. Further, they challenged the rigid adherence to an extreme constructivist paradigm where students invent their own knowledge and spend little time practicing routine skills, where teachers pose open-ended problems and provide opportunities to explore and converse, but don't directly instruct.

Because of the intense criticism, The Commission on the Future of the Standards revised the NCTM standards, a process that resulted in the 2000 *Principles and Standards for School Mathematics* (NCTM, 2000). The writing group appointed by the commission circulated a discussion draft of the revised standards in 1998 and equity was one of 19 issues raised. The issue concerned whether either curriculum or instruction should vary to meet the differing needs of various groups of students. *Should we be concerned with meeting individual needs?* In response, the equity section of the 2000 standards document (pp. 12–14) stated that equal access refers to "reasonable and appropriate accommodations … made as needed to promote access and attainment for all students," not identical instruction. "Some students may need further assistance to meet high mathematics expectations," such as "increased time to complete assignments or oral rather than written assignments." They "may need additional resources such as after-school programs, peer mentoring, or cross-age tutoring." What was not stated is revealing: These students have access rights to the curriculum, and many are served with individualized education programs (IEPs) or *504 plans* that require accommodations and collaboration among educators.

In response to criticism of the earlier standards, another document, *A Research Companion to NCTM's Standards*, was published in 2003 (Kilpatrick et al., 2003). This support document mentioned students who struggle in mathematics in only a few paragraphs in its entire 413 pages. In the chapter on implications of cognitive research, Siegler included a section on individual differences where he described three types of mathematics students: good, not-so-good, and perfectionists. The "not-so-good" type is slower, less accurate, uses less-advanced strategies, and performs poorer on tests. In the last two paragraphs Siegler cited Geary's term *mathematical disabilities* as describing about 6% of students. This group was described as similar to the not-so-good group, but their problems are a result of a combination of limited background knowledge, limited processing capacity (working memory), and limited conceptual understanding. According to Siegler, these difficulties "need to be addressed," but he cited none of the research on instruction for students with disabilities. Other authors in this edited work briefly addressed specific concerns for students with disabilities such as an over-reliance on prescriptive methods rather than formative assessment, equity issues for large-scale assessment (Wilson & Kenney, 2003), and providing an opportunity to learn that takes into account prior knowledge and student-engagement needs (Hiebert, 2003).

The instructional focus for the 2000 standards appeared to present a better balance of teacher guidance and student discovery, between concept understanding and skill mastery. Further, the consistency of the content strands across grade-level spans was an improvement. However, a confusing dichotomy of purpose remained. Was this a document of idealized visions to promote discussion among mathematics educators or a requisite set of student expectations?

The 2000 document began with these sentences:

> Imagine a classroom, a school, or a school district where all students have access to high-quality, engaging mathematics instruction. There are ambitious expectations for all, with accommodation for those who need it. Knowledgeable teachers have adequate resources to support their work and are continually growing as professionals. The curriculum is

mathematically rich, offering students opportunities to learn important mathematical concepts and procedures with understanding.

(NCTM, 2000, p. 3)

This vision was compelling and challenging, especially for teachers of students with learning difficulties or specific mathematics disabilities. While the standards did not address special needs learners' instructional considerations, the emphasis on *all students* should be inclusive. The equity principle came closest to assuring access and high expectations for all students:

> Excellence in mathematics education requires equity—high expectations and strong support for all students.... Low expectations are especially problematic because students who live in poverty, students who are not native speakers of English, students with disabilities, females, and many nonwhite students have traditionally been far more likely than their counterparts in other demographic groups to be the victims of low expectations.... mathematics must be learned by *all* students.

(NCTM, 2000, pp. 12–13)

NCTM published the *Curriculum Focal Points for Prekindergarten through Grade 8 Mathematics* in fall 2006 because so many mathematics curricula were still packed with content at each grade level, leaving teachers and parents overwhelmed. NCTM cited a study by Reys et al. (2005) that reviewed the standards of ten states and found between 26 and 89 topics for fourth grade. The fourth-grade curriculum in those states was packed with so many topics that teachers raced to cover everything at a surface level. The *focal points* are the "most important mathematical topics for each grade level," to promote a deeper understanding of fewer topics (NCTM, 2006, p. 5). NCTM identified three focal points per grade level, along with connections to the other standards at that grade level and the critical processes. For example, fourth graders should develop:

> [1] quick recall of multiplication facts and related division facts and fluency with whole-number multiplication; [2] an understanding of decimals, including the connections between fractions and decimals; and [3] an understanding of area and determining the areas of two-dimensional shapes.

(NCTM, 2006, p. 16)

Related to these focal points were connections in algebra (patterns and rules), geometry (properties of two-dimensional shapes), measurement (angle measurement and classification), data analysis (tool use for tables and graphs), and number and operations (place-value extensions, multidigit division, equivalent fractions). Fourth grade should have a focus on three primary topics with seven or eight connecting topics. The three focus topics should be developed deeply over the year, with support from the secondary topics. Less is more.

By 2005, 49 states had adopted state-wide mathematics standards (Palacios, 2005). While the NCTM 2000 content standards were by grade-level band (PreK-2, 3–5, 6–8, and high school), states typically adopted grade-specific curriculum standards based on a combination of traditional, NCTM, and international standards. There was a great deal of inconsistency across states (and school districts) regarding which mathematical topics would be introduced and developed at which grade levels. This was a critical concern because of the increasing use of large-scale textbook adoptions and assessments under the NCLB Act. It was also a problem for college admissions standards and mobile families.

To address those and other state-level issues, the Council of Chief State School Officers and the Center for Best Practices of the National Governors Association initiated an effort in 2009 to develop a national set of standards in mathematics (and English language arts). The writing team included mathematics content experts; educators and supervisors: assessment staff from ACT (American College Testing), the College Board, and Achieve; and experienced teachers (Dossey et al., 2012). *The Common Core State Standards for Mathematics* (CCSSM, NGA & CCSSO) were released in 2010 and by 2013 were adopted by 45 states. The CCSSM were a more cohesive set of standards and balanced concept understanding with procedures and fluency. The standards were organized by grade levels and 13 domains.

The US Department of Education quickly endorsed the standards, which led, in part, to political controversies. The "math wars" part two ensued. Most of the issues raised by politicians were myths or misrepresentations. Many politicians claimed that the standards were created by the federal government and imposed on the states, when the states were the leaders. They claimed that the CCSSM required assessments, when those were actually required by the 2001 NCLB Act. They claimed that CCSSM homework, curriculum, and books were confusing, when the standards included none of those elements.

There were valid criticisms, however. The standards were rushed into implementation by many states, without adequate training for teachers or parent education. Small-scale or sequenced adoptions may have been better, and the acknowledgment that systems change takes several years, with curriculum and textbook development, assessment design and validation, and a more gradual implementation. Some educators pointed out that grade-level expectations were uneven (e.g., too much at the 6th grade level), that some important content was missing (e.g., data in Grades 1–4, discrete mathematics, and deductive systems), some content was introduced too late (e.g., decimal notation for fractions), and what was expected for ALL students wasn't differentiated from what could be expected of some students (Usiskin, 2014).

Others raised concerns about students with disabilities (Jitendra, 2013; Powell et al., 2013). Should all students be required to learn the same content? Will students with mathematics learning disabilities have the necessary time and instruction to develop deep mathematical understanding? Will students with working memory, linguistic, or basic-skills difficulties be able to participate in group work requiring generating ideas, discussing multiple approaches, and defending their reasoning? What will happen with students who cannot meet one or more grade-level standards, even with interventions? For students with the most challenging cognitive disabilities, are the common standards really the most appropriate goals for their IEPs?

The Common Core State Standards Initiative offered a two-page document, *Application to Students with Disabilities*, that stated, "These common standards present an historic opportunity to improve access to rigorous academic content standards for students with disabilities" (NGA & CCSSO, 2010, *Frequently Asked Questions*). The document stressed that *how* the standards are taught and assessed is critical. For students with disabilities to meet the standards, their instruction must incorporate: supports and related services; IEP goals aligned with grade-level standards; teachers and other personnel who are qualified to deliver high-quality, evidence-based instruction; and instructional supports, accommodations, and assistive technology that assure access but do not change the standards. The authors asserted that substantial supports and accommodations would be needed for students with the most significant cognitive disabilities.

Some states opted to pull out of the Common Core and developed their own standards, under political pressure, so they would be state controlled. However, many of those states' standards borrowed heavily from the CCSSM because mathematics education has evolved toward those learning goals and the standards, at least in the upper grades, reflect what colleges expect (Wilson, 2014). Today's students need to compete with students internationally in mathematics and related fields. Today's students must have skills in reasoning, problem solving, justifying their reasoning, and communicating and collaborating with others. Most state standards, whether based on the NCTM, CCSSM, or international standards, will reflect those requirements. Most standards will continue to evolve as the demands of college and careers change and as research continues to inform about best practices in curriculum, learning, and instruction.

Special education and support teachers should be familiar with their state's mathematics curriculum and assessment standards across *all* grade levels. They need to know the content addressed in regular classrooms, to assess their students' gaps in learning, and to reference general education content for individual learning goals on IEPs. Students with mild to moderate disabilities are most often engaged with mathematics instruction in general education classes and must take all state- and district-mandated tests. Even students receiving mathematics instruction in separate special education settings must have access to the general curriculum with appropriate accommodations and support.

Core mathematics teachers should also be familiar with the full K-12 span of the standards, regardless of their grade-level teaching assignments. The eighth-grade teacher initiating a unit on statistics must understand how the concepts were developed from the earliest grade levels, recognize gaps or advances in learning among students, and prepare students for future concept development.

TRY THIS

Locate your state mathematics standards and compare them with the NCTM and/or CCSSM standards.

Scope of the Mathematics Curriculum

This section provides an overview of the *Common Core State Standards for Mathematics* (CCSSM, NGA, & CCSSO, 2010), with references to the *Principles and Standards for School Mathematics* (NCTM, 2000). These standards, in original or revised form, are used by most states in the United States. The standards for mathematical practice, which are applied across all grade levels, are discussed first. These standards inform teachers about practices to integrate within mathematics classes for students to be successful with the content.

A brief introduction to the content standards follows. These are developed in more detail in the *Content Strands* in the second part of this book. The content standards are articulated here by domain, rather than by grade level. Educators first studying the content standards may feel overwhelmed with the amount of content addressed. However, a longitudinal view will show how the same topics are developed over several years in a spiral and interconnected pattern, or trajectories for learning. For example, adding and subtracting (composing and decomposing) are introduced in kindergarten for numbers up to 19. In first grade, students should be fluent for addition and subtraction within 10 and second graders within 20. Second graders also use addition and subtraction within 100 and apply those to measurement, data, and place-value concepts. Third, fourth, and fifth graders continue to use addition and subtraction to solve word problems and apply these concepts to multidigit whole numbers and other rational numbers. Even in middle school, students continue to use addition and subtraction with algebraic expressions and equations, data analysis, and geometry.

Mathematical Practices

One goal for students in mathematics is to develop the habits of mind and practices that are used by mathematicians and successful problem solvers. Both the NCTM standards (2000) and the CCSSM (2010) included practices in which students should be engaged while doing mathematics. These practices support student understanding of the content standards and are essential for teachers to understand and promote.

Sometimes educators and parents overlook the practice standards as they seek out grade-specific content standards. But practice standards provide a means for students to develop necessary mathematics skills and dispositions for mathematics class as well as to apply mathematics concepts in other situations. These practices help students connect mathematics content and become more proficient. The teacher's role is to provide settings, models, and guidance for these processes to develop and to assess student proficiency in using the practices.

The NCTM identified five process standards: problem solving, reasoning and proof, communication, connections, and representation. They address ways of acquiring and using knowledge and are developed across the entire mathematics curriculum. They can also be applied across other content areas and real-world problems. The CCSSM outlined eight practices that "mathematics educators at all levels should seek to develop in their students" (NGA & CCSSO, 2010). These practices were informed by the NCTM process standards and the National Research Council's report *Adding It Up: Helping Children Learn Mathematics* (Kilpatrick et al., 2001). Although eight specific practices were listed, they are interrelated. Problem solving is the first standard for NCTM and CCSSM and the other standards contribute to problem-solving practices. Figure 1.1 provides a comparison of NCTM and CCSSM processes and

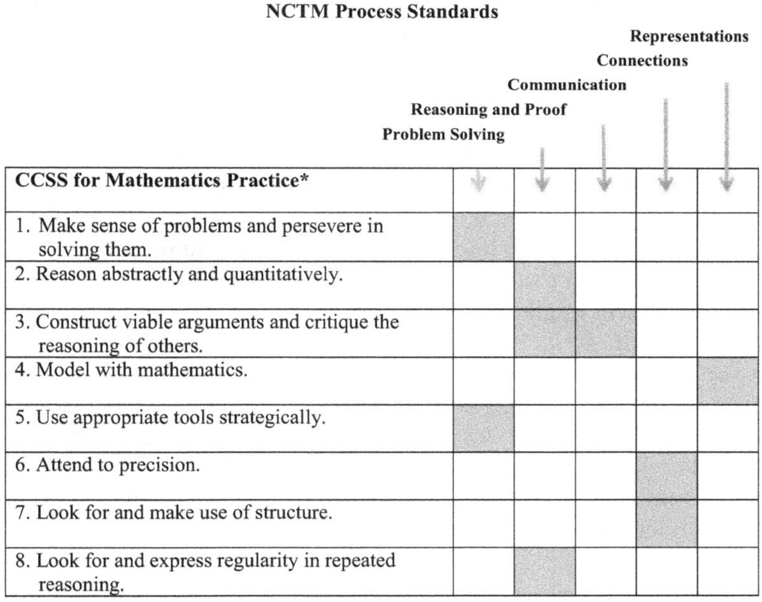

NCTM Process Standards

CCSS for Mathematics Practice*	Problem Solving	Reasoning and Proof	Communication	Connections	Representations
1. Make sense of problems and persevere in solving them.	■				
2. Reason abstractly and quantitatively.		■			
3. Construct viable arguments and critique the reasoning of others.		■	■		
4. Model with mathematics.					■
5. Use appropriate tools strategically.	■				
6. Attend to precision.				■	
7. Look for and make use of structure.				■	
8. Look for and express regularity in repeated reasoning.		■			

*Also influenced by strands of mathematical proficiency from Kilpatrick et al., (2001): adaptive reasoning, strategic competence, conceptual understanding, procedural fluency, and productive disposition.

Figure 1.1 Comparison of NCTM and CCSSM Process Standards

practices, with the strongest areas of overlap noted. Each of the CCSSM standards for mathematical practice is discussed, with illustrations, in the section that follows.

1. *Make sense of problems and persevere in solving them.* Problem solving is a major focus of the curriculum; engaging in mathematics IS problem solving. Problem solving is what one does when a solution is not immediate. Students should build mathematical knowledge through problem solving, develop abilities in formulating and representing problems in various ways, apply a wide variety of problem-solving strategies, and monitor their thinking in solving problems. Problems become the context in which students develop mathematical understandings, apply skills, and generalize learning. Students frequently solve problems within collaborative groups and even create their own problems. Younger students might use a diagram or concrete objects to solve a problem. More advanced students can draw on past experience solving problems, create equations, and use tools such as graphing calculators to solve problems. Good problem solvers ask themselves if their solutions make sense and seek to understand the approaches of others. Problem solving is addressed throughout this book, with Chapter 5 focused on specific types of mathematics problems and their strategies.

2. *Reason abstractly and quantitatively.* When solving problems, students should make sense of the parts of a problem and their relationships. Students should learn to decontextualize (to represent a situation more abstractly, as with symbols) and to contextualize (to provide referents for symbols encountered). For example, a student may be presented with the problem, *When will a train arrive at a station 100 miles away if it departs at 9:15 am and travels an average of 78 miles per hour?* The student would decontextualize the problem by using symbols for distance, time, and rate. The unknown is time. Time = distance/rate. In this case $t = 100/78$ or 1.282 hours after departing. Now the student may need to contextualize, recalling that one hour is 60 minutes (and time is not a decimal system), therefore 1.282 hours would be one hour and just over a quarter of an hour. How to convert 0.282 of an hour into minutes? Decontextualizing again, the student sets up a proportion: 0.282 is to 1 as x is to 60. The equation would be: $.282/1 = x/60$ and solving for x finds that 0.282 represents 16.92 minutes. Adding an hour and 17 minutes to 9:15 am would result in 10:32 am (or 10:31:55 am depending on how precise the answer should be). Another student might have worked solely in the abstract, simply manipulating formulas that were familiar.

A younger student begins to use the names for numbers, a more abstract form than the concrete. The student is shown 5 blocks and the teacher asks her how many she will have if the teacher removes 2 blocks. The student decontextualizes if she says, "Well, 5 take away 2 equals 3. I'll have 3 left." She doesn't need to move the objects to find the answer. The ability to decontextualize and contextualize provides a powerful tool for reasoning and problem solving. Teachers should model these reasoning processes aloud.

3. *Construct viable arguments and critique the reasoning of others.* Most fields of study require students to explore new situations, reason, and make arguments about a point of view or conclusion. Mathematics uses specific methods and terminology for this process. Students who are proficient are expected to use previously established results, assumptions, and definitions to make conjectures and arguments about new situations. They are expected to communicate their reasoning to others and defend their conclusions as well as respond to and question the arguments of others.

For example, one third-grade student is asked how many fence posts will be needed for a fence 50 feet long if there is a post every 5 feet. He divides 50 by 5 and states that 10 posts will be needed. Another student challenges that answer by stating that this situation is not like a number line with a zero in the beginning position. "You will need a fence post at the beginning point so there will be 11 posts." The first student doesn't understand so the second student draws a picture of the fence with the 11 posts. "The equation would be $50/5 + 1 = p$," she explains.

Eighth-grade students may encounter two sets of *data* that are conjectured to be related, such as the shoe sizes of all members of the class and their hand lengths. One group of students constructs a two-way table and uses relative frequencies to describe the association between the two variables. Another group plots the data onto a coordinate plane and describes the patterns of those points. A third group creates an equation of a linear model and tests the data using that equation. Each of the groups defends its arguments for the others while the teacher guides the students to see why each procedure works and is related to the others. Teachers should prompt students to use specific language for their defenses and challenges, for example, "Our group concluded that.... We began with the assumption that...." or "I don't understand your finding at this point in your argument."

4. *Model with mathematics.* Critically related to the previous practices, modeling is the skill in which students use the mathematics they know and apply it to other situations and problems. They generalize their knowledge to novel situations. Mathematics is now a tool in their problem-solving endeavors. For example, a middle-school student is making cupcakes. He wants to increase the number of cupcakes specified in the recipe from 12 to 40. Proportional modeling becomes a tool as he converts cups and teaspoons to greater amounts. A high-school student might encounter a situation where growth in a variable seems to be exponential, such as the number of hits on a YouTube page. She applies a logarithm to predict how many views her page will receive in a week.

Mathematic modeling is related to the NCTM process standard of *representations*. It also incorporates problem-solving, reasoning, and communication skills. We want students to develop a repertoire of modeling tools—such as data collection and representation, diagrams, tables, equations, three-dimensional models, and graphing skills—to represent new situations and find solutions. Our goal is for students to apply these skills in other subject areas, in everyday life, and eventually in the workplace.

5. *Use appropriate tools strategically.* The tools specified in this practice standard are those that are more hands-on, such as pencil and paper, dice, blocks, protractor, yardstick, calculator, spreadsheet, statistical software, and interactive websites. These are the technologies that assist students engaged in mathematics, some low-tech and some higher-tech. Students should become familiar with a range of tools and their best uses in varying situations. For example, a second grader uses a ruler to measure lengths in inches and centimeters. She develops the concepts of length and units of measure, and that centimeter units are smaller than those of inches. She can use a ruler later in third grade to compare sides of rectangles and compute perimeter and area. Her ruler may also be a good tool in fifth-grade volume problems or at home when she is measuring her desk drawer for an organizer.

Teachers at each grade level should introduce new tools while assuring students maintain and extend their use of previous tools. Teachers should model tool use for students and make them available for students to use as needed.

6. *Attend to precision.* Proficient students use precise mathematics language when communicating with others, use symbols accurately and consistently, and label units or diagrams clearly. Over the grade levels, students develop an understanding of situations where more or less precise results are needed. For example, we can estimate the amount of paint needed to paint a room, but a more precise measure is needed for administering liquid medication.

Another aspect of precision is in making calculations—students should calculate accurately and efficiently. They should choose the most efficient means for calculation and be able to explain clearly their process and results. For example a student is faced with solving this problem for x: $\dfrac{3}{x} = \dfrac{39}{12}$. The student must select the most efficient and accurate method. She might (a) cross-multiply to achieve $39x = 36$, with $x = \dfrac{36}{39}$ and then $x = \dfrac{12}{13}$. Or (b) she might recognize immediately that in $\dfrac{39}{12}$, both terms are divisible by 3 leading to $\dfrac{13}{4}$. Therefore, cross-multiplying yields $13x = 12$ and $x = \dfrac{12}{13}$. Or (c) she might divide 39 by 12 first, resulting in $3\dfrac{1}{4}$. Next, dividing $3\dfrac{1}{4}$ by $3 = \dfrac{3}{1} \times \dfrac{4}{13}$ resulting in $\dfrac{12}{13}$. Method (a) is what most students would likely attempt. Method (b) is what fluent students would use as a shortcut. Method (c) is fraught with potential for miscalculation.

The mathematical practices encourage precise use of language from the earliest grades. Rather than telling students, "When you add, the answer will be larger" (which does not always apply), say instead, "When you add (or combine) the numbers, your result is the sum." When students use language such as: "I minus-ed these two things here and got that number for an answer," encourage more precise language: "I subtracted the number 17 from 86 and the difference is 69."

7. *Look for and make use of structure.* Understanding and using mathematics requires us to recognize patterns and structures. As our understanding of concepts grow, we base new learning on previous recognitions. For example, a student working on multiplication fluency always stumbles on 7×8. He readily recalls that $7 \times 7 = 49$. So 7×8 will be 7 more than 49: one more is 50 and 6 more is 56. He thinks this will be an efficient strategy.

A middle-school student encounters a problem asking her to create an equation to find the age of Annie if she is only half the age of Billy. Billy is three times as old as Carla, who is 6. The student starts writing out equations using information from the problem. Billy is three times as old as Carla: $B = 3C$ (because Billy is older). Annie is half the age of Billy: $A = \dfrac{B}{2}$ (because Annie is younger). The student knows that elements in these equations can be substituted with known information. $C = 6$ and $B = 3 \times 6 = 18$, therefore it follows that $A = \dfrac{18}{2}$, leading to Annie's age of 9. Do these ages make sense in the original problem? Yes. The student recalls that the structure of $B = 3C$ can be confused in this word problem; one needs to remember that B would be larger than C (Billy is older than Carla). Another use of structure in this example is the substitution of equivalent terms.

Teachers can promote structure recognition by pointing out the known elements in new problem situations. "Does anyone recognize something familiar in this graph? When you look at this long equation, do you see parts you recognize?"

8. *Look for and express regularity in repeated reasoning.* Students should be encouraged to keep the big picture in mind while attending to details of problems. They can keep attention on the overall problem, checking for sense making, while performing a series of steps to solve the problem. In calculations, students should look for repetitions and determine whether shortcuts are viable. Some teachers call these break-through or "ah-ha" moments when students can recognize and take advantage of repetitions within mathematics.

For example, a first-grade classroom is working with numbers up to 100 using a 1 to 100 numbers chart (see Appendix II). One student notices that all the numbers in the first column end with the *digit* 1. Another student adds that all the numbers in the second column end with a 2. A third student shouts out, "…and rows start with the same number!" The teacher smiles while saying, "Why don't we try to read the numbers down the first column?"

A third grader decides she really, really dislikes problems such as 5000 − 2436. She voiced her frustration aloud one day in her work group. Another student explained that she was frustrated until she was working the problem 100 − 86 last week in the context of taking 86 cents from a dollar to figure the change needed. She realized that 99 − 85 would yield the same result because basically she took a penny from both numbers before subtracting. Others in the group exclaimed, "Let's see if it works for your problem! Let's try it both ways!" They were all delighted to find an easier way to solve those pesky "zeros subtractions." "Would it work for 5009 − 3467?" another student asked.

Teachers can promote repeated reasoning by selecting problem situations that encourage these insights. They should guide students to notice patterns and repetitions and ask questions such as: Will this work on all examples? Can we extend this to another situation? Have you seen this pattern before?

All of the mathematical practices are related to problem solving, seeking solutions. All require students to reason and select from their toolbox of prior knowledge and skills. And all these practices challenge students to higher-order thinking for their ability levels. Make sense, persevere, reason, critique, construct, model, select, use, attend, and express are all verbs—mental actions that will engage students making connections within the content.

TRY THIS

Select items from one state's mathematics assessment instrument. Evaluate the process standards required. For example, the following test item requires all five process skills—problem solving, communication, connections, reasoning, and representation:

Three days ago the mold on a piece of bread was 2 square inches. Today the mold covers 4 square inches. Find the rate of change in the area covered by the mold.

Content Standards

The mathematical practices discussed in the previous section are critical for the development of each of the content standards. The NCTM outlined five content standards and these are applied from PreK through Grade 12. The CCSSM used the term *domain* to describe how content standards are organized. There are 13 content domains, but many of these are continuations of others. For example, the CCSSM classify fractions as a 3–5 topic when fractions are actually rational numbers, also explored in ratios and proportions (6–7), expressions and equations (6–8), and algebra (high school). While the CCSSM do not specifically address fractions in Grades 1 and 2, students are expected to use halves, fourths, and thirds for partitioning shapes within the geometry and measurement domains.

In the sections that follow, mathematics content is organized into six major domains, incorporating the terminology from both NCTM and CCSSM. These six content areas are described in more detail in the *Content Strands* in the second part of this book, with recommendations for instruction with students with mathematics difficulties. Teachers should note that both the

Table 1.1 Comparison of NCTM and CCSSM Content Standard Domains by Grade Span

NCTM	Number and Operations	Algebra	Geometry	Measurement	Data Analysis and Probability
CCSSM					
K	Counting and Cardinality Number and Operations in Base Ten	Operations and Algebraic Thinking	Geometry	Measurement and Data	Measurement and Data
1–2	Number and Operations in Base Ten	Operations and Algebraic Thinking	Geometry	Measurement and Data	Measurement and Data
3–5	Number and Operations in Base Ten Number and Operations in Fractions	Operations and Algebraic Thinking	Geometry	Measurement and Data	Measurement and Data
6–7	The Number System Ratios and Proportional Relationships	Expressions and Equations	Geometry		Statistics and Probability
8	The Number System	Expressions and Equations Functions	Geometry		Statistics and Probability
High School	Number and Quantity	Algebra Functions	Geometry		Statistics and Probability

Sources: NCTM content standards reprinted with permission from *Principles and Standards for School Mathematics*, copyright 2000, by the National Council of Teachers of Mathematics. All rights reserved. CCSSM content standards from *Common Core State Standards for Mathematics*. © Copyright 2010 National Governors Association Center for Best Practices and Council of Chief State School Officers. All rights reserved.

NCTM and CCSSM emphasized two to four critical topics (or focal points) for each grade level, through Grade 8. These critical areas of focus are developed in the *Content Strands*. See Table 1.1 for CCSSM and NCTM content domain comparisons.

Number Sense, Place Value, and Number Systems

From counting objects and saying the names for numbers in preschool through studying complex numbers of high school that include an imaginary component, numbers are a tool for doing mathematics and communicating about mathematics. Numbers are symbols, words, or figures that represent a quantity or value. The term *number* is difficult to define because it is an idea, a human creation, a way to order and organize elements of our environment. Numbers are often organized by their systems: natural numbers, integers, rational numbers, and so forth (see Figure A.1 in Strand A). Numbers also can be classified by their use: labeling, ordering, coding, measuring, counting, and computing. Understanding and working with numbers is essential for success in the other content domains. Work with the other mathematics content areas, in turn, enhances number understanding.

Number sense is a complex concept dealing with our innate ability to individualize objects and extract numerosity of sets. It is the intuition about numbers (estimations, comparisons, simple addition, and subtraction), understanding of the meaning of different types of numbers and how they are related and represented, and the effects of operating with numbers. Number sense is an ability that is further developed through experience. Students in early grades formalize early number sense through learning number symbols, working with larger numbers, and becoming fluent in basic number facts. As they progress to higher levels, students develop abilities in estimation, representation, analyzing relationships, and working with more abstract number systems.

The base-10 number system allows for the manipulation of numbers of all sizes and types by using only 10 number symbols. Without a *place-value system*, we would be forced to memorize the name and symbol for each possible unit. Place value means that the symbol for a number, for example "4," has the value or meaning of 4 ones in the first place, 40 or 4 tens in the second place, and 400 or 4 hundreds if positioned in the third place. It would mean $\frac{4}{10}$ if placed immediately to the right of a decimal point. Place value is difficult for children for several reasons. It requires good spatial perception, new language, and multistep cognitive manipulation. In addition, it requires an understanding of multiplicative properties of number (multiples of 10) usually before multiplication has been introduced.

Number systems in middle school include rational numbers, numbers that can be expressed as fractions, the division of two integers with the denominator not equal to zero. Rational numbers are applied in measurement, geometry, and data analysis. Middle-school students are also introduced to negative numbers, exponents, and *irrational numbers.* In high-school mathematics, regardless of how that may be organized into courses, number systems are extended to include complex numbers, vectors, and matrices. Students calculate using these numbers, applying the properties of these systems. They solve problems in geometry, algebra, trigonometry, and statistics using a wider range of number-system tools and a wider array of units of measure.

Operations and Properties of Number

The four simplest operations—addition, subtraction, multiplication, and division—are interconnected forms of calculations with real numbers. These operations are used with integers, fractions, decimals, and within algebraic equations. In the simplest sense they are ways of composing and decomposing numbers. Addition is composing two or more numbers into a new number. Subtraction is breaking a number apart, decomposing a number into two or more parts. Subtraction of integers is the same as adding the opposite. Multiplication is composing groups or repeated addition, and division is partitioning, or repeated subtraction. As numbers increase in size or complexity, various computational algorithms, with multiple operations, are applied to numbers.

Other mathematical operations include exponentiation (applying an exponent), taking roots, negation, union and intersection (of sets), and composition (of *functions*). The term *operation* means to perform an act on something. Applying the operation of addition to the values of 2, 4, and 7 results in the sum 13. Applying the square root to the value 25 yields the root 5. There are many more operations specific to fields such as statistics, calculus, and computer science.

What is critical knowledge about operations? Students should understand the effect of each operation on different types of numbers. They should become computationally fluent, able to use efficient and accurate methods for each operation. An important related skill is estimation—both for solving problems where exactness is not required and for determining the reasonableness of exact answers. Students should be able to employ a variety of tools and strategies in performing computations and explain the processes used. Students who simply memorize algorithms by rote, rather than understand the concepts and connections, will be less able to apply computations and adjust strategies in problem-solving situations.

Operations also have special properties, such as the identity, associative, commutative, and distributive properties. For example, a student who understands the identity property of addition realizes that adding a zero will not change the sum. In multiplication the identity property is the fact that any number multiplied by one results in the original number. For addition and multiplication, understanding the commutative property supports students' fact fluency, as they can be sure that 2 + 5 will yield the same sum as 5 + 2 and 4 × 6 is equivalent to 6 × 4. Students should have experiences with many forms of problems to form a deep understanding of these properties.

Rational Numbers

Rational numbers are all numbers that can be expressed as a ratio ($\frac{a}{b}$ where $b \neq 0$). All rational numbers are included within number and operations for the NCTM standards. The CCSSM separate fractions from the decimal (base-10) representation of rational numbers through grade five, while the middle-school focus is on ratios and proportions. Both sets of standards integrate rational numbers within other domains of geometry, measurement, algebra, and data analysis.

Fractions and their operations and applications are challenging for students who struggle with mathematics. Sometimes mathematics learning disabilities are not evident until fourth grade when more conceptually complex problem solving is required. For students comfortable working primarily with whole numbers, the conceptual differences can seem counterintuitive and difficult to visualize or manipulate at a concrete level. For example, many students understand that multiplying positive whole numbers yields a product that is a larger number than the two *factors*. But multiplying by a simple fraction will result in a smaller product: $12 \times \frac{1}{3} = 4$. Other students may be confused with the morphing of math symbols into new applications, as with the changing signs for division and multiplication.

Kindergarteners are introduced to fractions with halves, fourths, quarters, and wholes describing geometric shapes and their partitions (shares). Second graders work with coins and their values in decimal form. Formal study of fractions as numbers that can be operated on, compared, and used in measurement begins in third grade. By fifth grade, students are applying all four basic operations and their properties to fractions and decimals and converting fluently between fraction, decimal, and *percent* forms of number. Middle-school students work with ratios and proportions to solve problems, with extensions to geometry, measurement, and algebra. They are making the shift to representing division as a ratio. They are fluent with operations involving the range of rational numbers, including those with negative numbers and absolute value.

Geometry and Spatial Sense

Concerned with properties of space and objects in space, geometry is one of the most appealing topics for students. The world of space and objects becomes a playground for exploration as geometry has so many real-world applications. For example, students at Pinetops Elementary School are going to assist in the design of the new playground. They want to have a softball diamond and a soccer field. They must research the regulation sizes for each and decide the best position for the rectangular soccer field related to the softball diamond and its outfield. One problem solved leads to three new problems.

Geometry is fundamentally based on three undefined terms: point, line, and plane. An understanding of these terms is necessary for understanding other terms and concepts: angle, parallel, congruence, polygons, circles, and solids. Measurement, proportion, functions, and algebraic concepts are also important for the study of geometry.

Geometric ideas are useful for representing and solving problems in mathematics and other fields (science, architecture, geography, engineering, sports, the arts, and social sciences). Geometric experiences involve analyzing and manipulating the characteristics and properties of two- and three-dimensional objects and using different representational systems, methods, and tools such as transformations, symmetry, visualization, spatial reasoning, graphing, and computer animations to solve problems. As with the study of algebra and the number system, geometry involves analyzing patterns, functions, and connections and developing and using rules (theorems or axioms) within the system to solve problems or develop more complex relationships.

Students in Grades K through 2 study properties of two- and three-dimensional shapes and explore relative positions, directions, and distances using these shapes. Grade 3 through 5 students begin using coordinate systems, transformations, and other means for analyzing the properties of shapes. They use geometric models to solve problems involving area, perimeter, angles, and symmetry. Middle-school students create and critique inductive and deductive arguments involving geometric concepts and use coordinate geometry to examine properties of shapes and objects. They use geometric models and tools to extend number and algebraic understandings. They use formulas to solve problems. By high school, students are testing conjectures, analyzing trigonometric relationships, expressing geometric properties with equations, and applying geometric models to solve problems in other disciplines.

Algebra, Expressions, Equations, and Functions

Algebraic thinking is a precursor to formal algebra, emphasized in Grades K through 5. It involves recognizing patterns, making generalizations about arithmetic operations, and operating on unknown quantities. Students who demonstrate grade-level mathematical practices (e.g., problem solving, reasoning) have tools for thinking algebraically. They think about situations, or problems, where there is an unknown and consider the operations and properties that will help solve for the unknown. For example, Betty is in a long line waiting to get into the cafeteria. She notices that the teacher by the door is letting in eight students at a time. If Betty is the fortieth student in line, how many groups will the teacher let in the door before Betty's group? What is the unknown? How can you set up the problem using arithmetic operations? A student engaged in solving this problem is applying algebraic thinking.

Algebraic thinking prepares students to move from arithmetic to algebraic expressions and equations. Algebra is the study of abstract mathematical structures involving finite quantities. It involves the symbolic representation of quantitative relationships and the subsequent manipulation of various aspects of the representation. The earliest experiences children have with algebra are with open sentences and missing numbers. For example, *If we have 10 books on this shelf and 2 are yours, how many belong to me?* translates into: $10 - 2 = ?$ or $2 + ? = 10$. From their earliest exposure to the = sign, in kindergarten, students are taught that it means *equivalence* (not here comes the answer). For example $2 + 3 = 5$ is the same as $5 = 2 + 3$. The combined values of 2 and 3 are equivalent to the value of 5. An equation is a mathematical sentence that shows two equivalent expressions.

In the middle grades, students begin using mathematical expressions that include a more advanced notation system. Parentheses (and brackets) are used for multiplication, as in the expression $3(2 + y)$. The diagonal division slash (/) or horizontal line is used for division instead of the obelus ($\div$) symbol. These symbols will evolve and have different meanings in different mathematical circumstances. Middle-grade students also use positive and negative integers as well as exponents within algebraic expressions. They evaluate expressions and rewrite them, which leads to solving equations (equalities and *inequalities*).

Students' work with number, operation, property, patterns, functions, and geometry compliments algebraic understanding. Elementary students develop fluency in working with symbols, numbers, operations, and simple graphing. Middle-school students develop concepts of linear functions, geometric representations, and polynomials. And high-school students explore other types of functions including rational, exponential, quadratic, and trigonometric. Extensions of working with patterns, *functions* include variables that have a dynamic relationship: changes in one will cause a change in the other(s). Functions can be depicted with equations, tables, spreadsheets, graphs, and geometric representations. Deeper understanding of the algebraic characteristics of our number system allows students to explore structures and patterns, pose and solve problems in a number of ways, and develop foundations for the next level of mathematics study.

Measurement, Data, and Statistics

Imminently hands-on yet elusively abstract, measurement skills and concepts can be engaging but challenging to teach. What can be measured? Time, energy, space, and matter. Each of these physical aspects of our world has its measurable aspects, their respective measurement tools, and units of measure. Some textbooks and curriculum frameworks classify money as a measurement topic; however, money is not measured but counted (unless all quantification is considered a form of measurement). Measurement is a critical topic for other mathematics applications and is related to many other topics outside mathematics, such as measuring opinions on a scale and using statistics to study trends in the economy.

Students in kindergarten and first grade use informal measurement systems, such as comparing heights as taller or shorter or measuring the length of an object using unnamed units. Standard units are introduced in the second grade with US Customary System and metric systems taught simultaneously. The USCS system reinforces fraction concepts and the metric system reinforces the base-10 place-value system and decimal concepts.

Young children develop the concept that objects have various attributes, some of which can be measured. They develop the language to express measurement ideas such as *longer* or *more*. They begin to associate specific attributes with units and tools of measurement and make simple measurements fairly accurately. Elementary students gain experience with a variety of tools and measurement concepts, in both metric and customary systems. They work with formulas for perimeter, area, and volume of various shapes. By the secondary grades, students gain experience with derived attributes (ratios of measurements), conversions, formulas, precision, and error concepts.

Data are quantifications of aspects of the world. It seems that everything is quantified: sports scores, weather records, income levels, test scores, stock-market trends, population patterns, and political views. Data analysis and the application of statistical methods are used across the curriculum. Students are taught how to collect and record data, to represent data in various forms, and to interpret and use data. Students are taught to describe their data collections with frequency charts, measures of central tendency, and various graphs and charts. Higher-level concepts include variability, significance, correlation, sampling, and transformations.

The study of *probability* assists us in making more accurate estimations with problems involving uncertainty. Probability helps answer the question, "How likely is some event?" Applications range from educational assessment, business, politics, and medicine to scientific phenomena. Students engaged in probability activities will use their knowledge of number and operations, variables and algebraic equations, problem-solving skills, measurement and graphing, and logical reasoning. Ultimately, skill with probability and statistics should enable students to make more informed decisions in all aspects of their lives.

TRY THIS

Select a test item from a state or district mathematics test and analyze the content knowledge required. For example, the following item requires facility with number and operation, geometry, and measurement concepts.

Draw a rectangle and a triangle that have the same area. Label the dimensions. Show that the areas are the same.

Multi-Tiered Systems of Support and Collaboration

Public schools in the United States are expected to effectively educate students with disabilities in general education classrooms, to the maximum extent appropriate, to increase their access to the core curriculum (IDEA, 2004). From the time of the original law, this concept has been termed *least restrictive environment* (EHA, 1975). Over the years the concept has been refreshed with characterizations such as the "inclusive education movement" and the "regular education initiative." Between 1989 and 2017, the proportion of students with disabilities in general education classrooms (for at least 80% of the school day) increased from 31.7% to 63.4% (NCES, 2018). By 2017, 71.4% of students identified with specific learning disabilities were served in general education classrooms.

It is difficult to compare inclusive efforts of other countries due to differences in terminology, but the United Nations Convention on the Rights of Persons with Disabilities (2006) asserted the right to inclusive education for all students in Article 24:

> States Parties recognize the right of persons with disabilities to education. With a view to realizing this right without discrimination and on the basis of equal opportunity, States Parties shall ensure an inclusive education system at all levels and life-long learning.... States Parties shall ensure that a) Persons with disabilities are not excluded from the general education system on the basis of disability, and that children with disabilities are not excluded from free and compulsory primary education, or from secondary education, on the basis of disability; b) Persons with disabilities can access an inclusive, quality and free primary education and secondary education on an equal basis with others in the communities in which they live; c) Reasonable accommodation of the individual's requirements is provided; d) Persons with disabilities receive the support required, within the general education system, to facilitate their effective education; e) Effective individualized support measures are provided in environments that maximize academic and social development, consistent with the goal of full inclusion. (United Nations, 2006, Article 24, 1–2)

Educators and researchers in the US have implemented various models of screening, pre-referral intervention, levels or tiers of intervention, progress monitoring, and school-wide models of interventions since the passage of P.L. 94–142 (EHA, 1975). School-wide systems with intervention levels for behaviors (e.g., Colvin et al., 1993) and reading (e.g., O'Connor, 2000) have a much longer history than those for mathematics and other areas. In the 1997 reauthorization of IDEA, Congress recognized the potential of whole-school approaches and interventions within the general education setting:

> Over 20 years of research and experience has demonstrated that the education of children with disabilities can be made more effective by.... providing incentives for whole-school

approaches and pre-referral intervention to reduce the need to label children as disabled in order to address their learning needs.

(IDEA, 1997, 20 USC 1400 (c)(5)(F))

These amendments to IDEA required that IEP teams consider positive behavioral interventions, a precursor to the school-wide *Positive Behavior Interventions and Supports* (PBIS, Sugai & Horner, 2002) model that includes functional behavioral assessments, team-based decision-making, universal screening, continuous monitoring and data collection, and a three-tiered system of interventions and supports. The regulations required evaluation teams to consider classroom assessments as part of the initial evaluation process, but did not require IEP teams to consider academic systems of intervention and support.

The 2001 Elementary and Secondary Education Act Amendments (No Child Left Behind Act) provided incentives for school-wide programs and comprehensive school reform that were to promote academic achievement of all students in core academic subjects using scientifically based research as a basis for programs. However, grant-funded programs under this legislation were tied to other requirements, such as private matching funds or state-wide curriculum standards, which soon became politicized. Further, many of the instructional programs and assessment tools recognized as high quality for these school reforms had not been adequately field tested with students with disabilities.

In 2002 the President's Commission on Excellence in Special Education issued a report calling for a model of prevention rather than a model of failure, waiting for children to fail before interventions could be provided (USDOE, OSERS). The report also urged educators to consider students with disabilities as general education children first and emphasized that special education and general education have shared responsibilities, therefore should not be treated as separate systems for funding or services. The Commission recommended refocusing on results for students over process compliance. Even IEPs had become compliance documents rather than tools for individualized educational services. The Commission recommended the incorporation of research-based early intervention programs and response-to-intervention (RTI) models for identification of all high-incidence disabilities and progress monitoring throughout IEP implementation. However, the primary rationale articulated in the document was reducing the explosive growth in Other Health Impairments (i.e., ADHD) and Specific Learning Disabilities (SLD) categories.

In the 2004 reauthorization of IDEA, Congress reemphasized the importance of whole-school approaches and added the need for scientifically based early reading programs, positive behavioral interventions and supports, and early intervening services. Curiously, the only place in the law that referred to an intervention system for academic concerns was within the procedures for determining whether a child has a specific learning disability: "a local educational agency may use a process that determines if the child responds to scientific, research-based intervention as a part of the evaluation procedures" (IDEA, 2004, 20 USC § 1414 (b)(6)(B)). This process is called *Response to Intervention* (RTI; or responsiveness-to-intervention or pyramid of interventions) in the professional literature (e.g., DuFour et al., 2004; Vaughn & Fuchs, 2003), in subsequent US Office of Special Education Program (OSEP) research initiatives, and by the National Association of State Directors of Special Education (2005).

In the past few years, school-wide models have been developed that place more emphasis on equitable supports and services for all students who may be struggling within the general education curriculum than on physical classroom placement or identification before intervention (Choi et al., 2020). The Every Student Succeeds Act (ESSA, 2015) defined multi-tiered systems of support as "… a comprehensive continuum of evidence-based, systemic practices to support a rapid response to students' needs, with regular observation to facilitate data-based instructional decision making" (ESAA, 2015, § 8002 (33)). Section 2013 (local use of funds), stated that:

> the programs and activities … shall address the learning needs of all students, and … may include developing programs and activities that increase the ability of teachers to effectively teach children with disabilities, including children with significant cognitive disabilities, and English learners, which may include the use of multi-tier systems of support and positive behavioral intervention and supports, so that such children with disabilities and English learners can meet the challenging State academic standards.
>
> (ESAA, 2015, § 2013 (b)(3)(F))

Most states are using a multi-tiered systems of support (MTSS) or response-to-intervention (RTI) framework for addressing the needs of students at risk for poor learning outcomes (Schiller et al., 2020). According to the National Center on Response to Intervention (funded by OSEP), essential components of MTSS include: universal screening, progress monitoring, a multilevel prevention system with increasingly intense levels of instruction and intervention, and data-based decision making (NCRTI, 2010). Most MTSS and RTI frameworks include three levels or tiers of prevention or intervention, as depicted in Figure 1.2.

- *Tier 1* includes a research-based core curriculum within general education, universal screening at least twice a year, differentiated learning, problem solving to identify interventions, and appropriate accommodations.
- *Tier 2* is typically small-group instruction with evidence-based interventions and progress monitoring at more frequent intervals, over 10 to 15 weeks of short (20–40 minutes) sessions 3 or 4 times a week.
- *Tier 3* is the most intensive, individualized instruction typically within special education settings or one-on-one instruction with at least weekly progress monitoring.

(NCRTI, 2010)

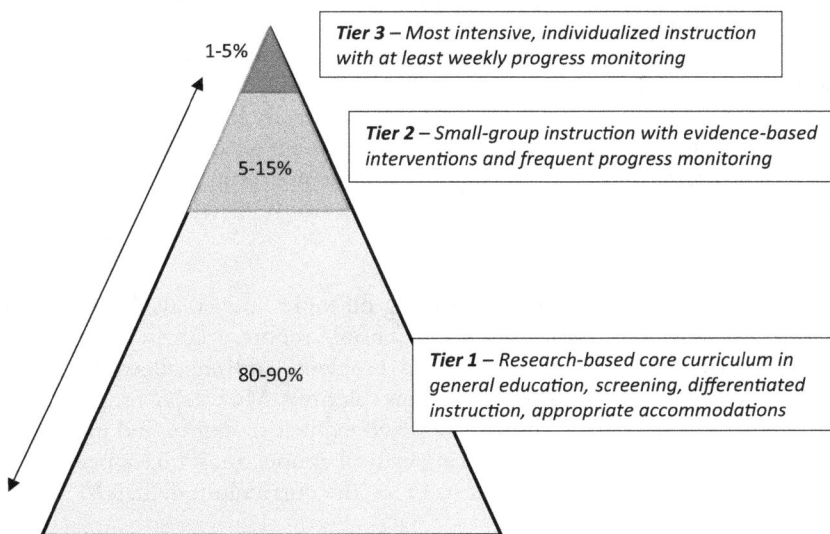

Figure 1.2 Multi-Tiered Systems of Support Model

Tiers 2 and 3 are sometimes offered as supplements to Tier 1, to provide increasingly intensive instruction for targeted content (Pullen et al., 2018). Special education services may be integrated within any of the tiers but are more common as interventions in Tiers 2 and 3. Models of MTSS and RTI vary considerably with some districts implementing four, five, or more levels of prevention and intervention (D. Fuchs et al., 2012). Some states implemented combined academic and behavioral systems under the MTSS framework, while others separated RTI and PBIS interventions.

MTSS models have generated hundreds of studies into efficacy, impact, and concerns of these school-wide approaches. Benefits of a school-wide model for interventions include the reduction of unnecessary referrals to special education, increasing the use of research-supported interventions with accompanying progress monitoring, involving all educators in a school in a problem-solving process to address student needs, increasing differentiated instruction and formative assessment within general education settings, allowing more efficient use of resources, and providing the most intense interventions to those students with the most needs (Johnson & Smith, 2008; Werts et al., 2014). Some concerns relate to system administration responsibilities, scheduling issues, time for interventions, integrating academic and behavioral systems, teacher training for assessment and instruction requirements, roles of various professionals in the school, parental involvement, possibility of delays in special education services, lack of quality assessments and interventions for some topics, and implementation of models with fidelity (Hale et al., 2010; Johnson & Smith, 2008; Werts et al., 2014). To implement successful models, Mason et al. (2019) recommended allowing teams a school year to plan schedules, professional development sessions, interventions, and systems before implementing MTSS models. Further, they cited coaching, with content-expert coaches, as critical for long-term and sustainable systems change.

Most early research on MTSS and RTI was focused on early reading and math skills, not higher-level skills, writing, or other content areas (Hughes & Dexter, 2011). However, more research into a range of mathematics interventions within multilevel frameworks and how to make decisions about student progress has been published recently. Research on assessments and instructional interventions are discussed in Chapters 3 through 6. The *Content Strands* include boxes with example intensive interventions for each domain.

During a professional development session for first-year teachers in the district, Angela Smith, Joseph Lopez, and Chris Johnson met again to compare experiences. They all agreed that mathematics classes today are not like their own experiences in school. "We had to sit in rows, not speak, and solve long algorithms exactly like the teacher modeled," shared Chris. "Our third-grade teacher took away free time if we could not pass the timed one-minute fluency tests," complained Angela. "I remember going to the blackboard to show proofs in front of the whole class and feeling superior if I had a solution and no one else understood it," recalled Joseph.

"Our elementary mathematics classes are working on topics such as algebra and statistics that we weren't introduced to until late middle school," reported Chris. "These students are working in small groups on teacher-posed problems and are allowed to use any method that works as long as they can justify their solutions. My role as a support teacher has been to build up gaps in prior knowledge, teach explicit strategies, and provide alternative examples when students are not making critical connections. I have been working closely with all the school's teachers to get to know the curriculum demands."

"At the middle school I have actually been a co-teacher with three of the mathematics teachers," Angela shares. "I had no idea how advanced the concepts had become—algebra, statistics, and integrating geometry with everything. The students love the hands-on work and working in small groups and, surprisingly, they are not off task. Perhaps that's because they have two teachers but also because the math they're working on is interesting to them—we use sports, current events, music, space travel—whatever is interesting for these students has powerful mathematics applications. Our goal is concept understanding and making connections, not just working problems."

"I have been surprised how much I enjoy working with the students who struggle with mathematics," exclaims Joseph. "It has been a challenge for me to figure out what they already understand and how I should present the next concept in these college-prep classes. The special educators have been so helpful in showing me how to design informal assessments of student learning and multiple examples for new concepts. I was worried I would have a group of students working out of middle-school workbooks, but we've been able to work on the same concepts with the extra time for interventions in the mathematics support period. These students are actually going home and challenging their parents with interesting mathematics problems. Imagine that!"

In the professional-development session, Angela, Chris, and Joseph will be introduced to the MTSS model that will begin in their schools next term. Chris and Angela will support mathematics teachers as they implement Tier 1 with all students. They will assist with screening and progress monitoring as well as develop appropriate accommodations for individual students. Both teachers, along with other special education and language support teachers will implement Tiers 2 and 3 as individual students are identified as needing more intensive instruction. They will plan various small groupings for specific mathematics interventions. Joseph will have more assistance in designing accommodations, assessments, and interventions within his class (Tier 1) and he will continue to have the support from other professionals for Tiers 1 through 3.

Today's mathematics instruction is evolving to meet the needs of more diverse students in mixed groups: students with disabilities, language and cultural differences, varying socio-economic backgrounds, and a range of abilities. Schools and teachers are under increasing pressure to demonstrate that all their students are performing to grade-level standards. The mathematics curriculum is more standardized across classrooms because of national and state standards and state-level testing programs. Today's mathematics teachers must provide high-quality instruction to very diverse student groups in a climate of reform and accountability.

The roles of educators responsible for mathematics instruction are changing. General education mathematics teachers serve more diverse students and work with more support personnel than ever before. Special educators provide a broader array of support services including full co-teaching within core mathematics classes and special interventions in pull-out programs. Other teachers specialize in mathematics to offer a range of pull out, co-teaching, coaching, and after-school services, working with the total mathematics program in the school. For districts with decision making at the school-building level, program models may be extremely varied and creative as they are planned by partnerships among educators, support staff, administrators, parents, and university colleagues.

References

Arens, A. K., Marsh, H. W., Pekrun, R., Lichtenfeld, S., Murayama, K., & vom Hofe, R. (2017). Math self-concept, grades, and achievement test scores: Long-term reciprocal effects across five waves and three achievement tracks. *Journal of Educational Psychology*, *109*(5), 621–634. doi: 10.1037/edu0000163

Bandura, A. (1997). *Self-efficacy: The exercise of control*. W. H. Freeman.

Baumert, J., Kunter, M., Blum, W., Brunner, M., Voss, T., Jordan, A., Klusmann, U., Krauss, S., Neubrand, M., & Tsai, Y.-M. (2010). Teachers' mathematical knowledge, cognitive activation in the classroom, and student progress. *American Educational Research Journal, 47*(1), 133–180. doi: 10.3102/0002831209345157

Bekdemir, M. (2010). The pre-service teachers' mathematics anxiety related to depth of negative experiences in mathematics classrooms while they were students. *Educational Studies in Mathematics, 75*, 311–328. doi: 10.1007/s10649-010-9260-7

Campbell, P. F., Nishio, M., Smith, T. M., Clark, L. M., Conant, D. L. Rust, A. H., DePiper, J. N., Frank, T. J., Griffin, M. J., & Choi, Y. (2014). The relationship between teachers' mathematical content and pedagogical knowledge, teachers' perceptions, and student achievement. *Journal for Research in Mathematics Education, 45*(4), 419–459. doi: 10.5951/jresematheduc.45.4.0419

Choi, J. H., McCart, A. B., & Sailor, W. (2020). Achievement of students with IEPs and associated relationships with an inclusive MTSS framework. *The Journal of Special Education, 54*(3), 157–168. doi: 10.1177/0022466919897408

Clark, L. M., DePiper, J. N., Frank, T. J., Nishio, M., Campbell, P. F., Smith, T. M., Griffin, M. J., Rust, A. H., Conant, D. L., & Choi, Y. (2014). Teacher characteristics associated with mathematics teachers' beliefs and awareness of their students' mathematical dispositions. *Journal for Research in Mathematics Education, 45*(2), 246–284. doi: 10.5951/jresematheduc.45.2.0246

Colvin, G., Kame'enui, E., & Sugai, G. (1993). Reconceptualizing behavior management and schoolwide discipline in general education. *Education and Treatment of Children, 16*, 361–381.

Davis, B. & Renert, M. (2013). Profound understanding of emergent mathematics: Broadening the construct of teachers' disciplinary knowledge. *Educational Studies in Mathematics, 82*, 245–265. doi: 10.1007/s10649-012-9424-8

Dossey, J. A., Halvorsen, K. T., & McCrone, S. S. (2012). *Mathematics education in the United States 2012.* NCTM. www.nctm.org/uploadedFiles/About/MathEdInUS2012.pdf

DuFour, R., DuFour, R., Eaker, R., & Karhanek, G. (2004). *Whatever it takes: How professional learning communities respond when kids don't learn.* Solution Tree.

Education of All Handicapped Children Act (EHA), Pub. L. No. 94–142, 89 STAT. 773 (1975). www.govinfo.gov/content/pkg/STATUTE-89/pdf/STATUTE-89-Pg773.pdf

Elementary and Secondary Education Act (The No Child Left Behind Act) of 2001, Pub. L. No. 107–110, § 101 Stat. 1425 (2001). www.congress.gov/107/plaws/publ110/PLAW-107publ110.htm

Every Student Succeeds Act, 20 USC § 6301 (2015). www.congress.gov/114/plaws/publ95/PLAW-114publ95.pdf

Friedrich, A., Flunger, B., Nagengast, B., Jonkmann, K., & Trautwein, U. (2015). Pygmalion effects in the classroom: Teacher expectancy effects on students' math achievement. *Contemporary Educational Psychology, 41*, 1–12. doi: 10.1016/j.cedpsych.2014.10.006

Fuchs, D., Fuchs, L. S., & Compton, D. L. (2012). Smart RTI: A next-generation approach to multilevel prevention. *Exceptional Children, 78*(3), 263–279. doi: 10.1177/001440291207800301

Geary, D. C. (2013). Learning disabilities in mathematics: Recent advances. In H. L. Swanson, K. Harris, & S. Graham (Eds.), *Handbook of learning disabilities* (2nd ed., pp. 239–255). Guilford Press.

Gentrup, S., Lorenz, G., Kristen, C., & Kogan, I. (2020). Self-fulfilling prophecies in the classroom: Teacher expectations, teacher feedback and student achievement. *Learning and Instruction, 66*. doi: 10.1016/j.learninstruc.2019.101296

Goals 2000: Educate America Act, Pub. L. No. 103–227 (1994). www2.ed.gov/legislation/GOALS2000/TheAct/index.html

Goldhaber, D., Quince, V., & Theobald, R. (2016). *Reconciling different estimates of teacher quality gaps based on value added.* National Center for Analysis of Longitudinal Data in Education Research. www.caldercenter.org/sites/default/files/Full%20Text_0.pdf

Gresham, G. (2009). An examination of mathematics teacher efficacy and mathematics anxiety in elementary pre-service teachers. *The Journal of Classroom Interaction, 44*(2), 22–38.

Hadley, K. M., & Dorward, J. (2011). Investigating the relationship between elementary teacher mathematics anxiety, mathematics instructional practices, and student mathematics achievement. *Journal of Curriculum and Instruction, 5*(2), 27–44. doi: 10.3776/joci.2011.v5n2p27-44

Hale, J., Alfonso, V., Berninger, V., Bracken, B., Christo, C., Clark, M., Cohen, A., Davis, S., Decker, M., Denckla, R., Dumont, C., Elliott, S., Feifer, C. Fiorello, D., Flanagan, E., Fletcher-Janzen, D., Geary, M., Gerber, M., Goldstein, N., … Yalof, J. (2010). Critical issues in response-to-intervention, comprehensive evaluation, and specific learning disabilities identification and intervention: An expert white paper consensus. *Learning Disability Quarterly*, *33*(3), 223–236. doi: 10.1177/07319487100 3300310

Heibert, J. (2003). What research says about the NCTM standards. In J. Kilpatrick, W. G. Martin, & D. Schifter (Eds.), *A research companion to principles and standards for school mathematics* (pp. 5–23). NCTM.

Hill, H. C., Charalambous, C. Y., & Chin, M. J. (2019). Teacher characteristics and student learning in mathematics: A comprehensive assessment. *Educational Policy*, *33*(7), 1103–1134. doi: 10.1177/ 0895904818755468

Hill, H. C., & Chin, M. J. (2018). Connections between teachers' knowledge of students, instruction, and achievement outcomes. *American Educational Research Journal*, *55*(5), 1076–1112. doi: 10.3102/ 0002831218769614

Hofmeister, A. M. (1993). Elitism and reform in school mathematics. *Remedial and Special Education*, *14*(6), 8–13. doi: 10.1177/074193259301400603

Holzberger, D., Philipp, A., & Kunter, M. (2013). How teachers' self-efficacy is related to instructional quality: A longitudinal analysis. *Journal of Educational Psychology*, *105*(3), 774–786. doi: 10.1037/ a0032198

Hughes, C. A., & Dexter, D. D. (2011). Response to intervention: A research-based summary. *Theory into Practice*, *50*(1), 4–11. doi: 10.1080/00405841.2011.534909

Humphrey, M. & Hourcade, J. J. (2010). Special educators and mathematics phobia: An initial qualitative investigation. *The Clearing House: A Journal of Educational Strategies, Issues, and Ideas*, *83*(1), 26–30. doi: 10.1080/00098650903267743

Individuals with Disabilities Education Act, 20 USC § 1400 *et seq.* (1997). www2.ed.gov/policy/ speced/leg/idea/idea.pdf

Individuals with Disabilities Education Improvement Act, USC 20 §1400 *et seq.* (2004). www.congress.gov/108/plaws/publ446/PLAW-108publ446.pdf

Jacobson, E., & Kilpatrick, J. (2015). Understanding teacher affect, knowledge, and instruction over time: An agenda for research on productive dispositions for teaching mathematics. *Journal of Mathematics Teacher Education*, *18*(5), 401–406. doi: 10.1007/s10857-015-9316-9

Jitendra, A. K. (2013). Understanding and accessing standards-based mathematics for students with mathematics difficulties. *Learning Disability Quarterly*, *36*(1), 4–8. doi: 10.1177/0731948712455337

Johnson, E. S., & Smith, J. (2008). Implementation of response to intervention at middle school: Challenges and potential benefits. *TEACHING Exceptional Children*, *40*(3), 46–52. doi: 10.1177/00400599080 4000305

Kilpatrick, J., Martin, W. G., & Schifter, D. (Eds.). (2003). *A research companion to principles and standards for school mathematics*. NCTM. www.nctm.org

Kilpatrick, J., Swafford, J., & Findell, B. (2001). *Adding it up: Helping children learn mathematics*. National Academy Press. doi: 10.17226/9822

Klassen, R. M., Tze, V. M. C., Betts, S. M., & Gordon, K. A. (2011). Teacher efficacy research 1998–2009: Signs of progress or unfulfilled promise? *Educational Psychology Review*, *23*, 21–43. doi: 10.1007/ s10648-010-9141-8

Klehm, M. (2014). The effects of teacher beliefs on teaching practices and achievement of students with disabilities. *Teacher Education and Special Education*, *37*(3), 216–240. doi: 10.1177/0888406414525050

Mason, E. N., Benz, S. A., Lembke, E. S., Burns, M. K., & Powell, S. R. (2019). From professional development to implementation: A district's experience implementing mathematics tiered systems of support. *Learning Disabilities Research & Practice*, *34*(4), 207–214. doi: 10.1111/ldrp.12206

Max, J., & Glazerman, S. (2014). *Do disadvantaged students get less effective teaching? Key findings from recent Institute of Education Sciences studies*. National Center for Education Evaluation and Regional Assistance. https://ies.ed.gov/ncee/pubs/20144010/pdf/20144010.pdf

Mercer, C. D., Harris, C. A., & Miller, S. P. (1993). Reforming reforms in mathematics. *Remedial and Special Education*, *14*(6), 14–19. doi: 10.1177/074193259301400604

Morsanyi, K., van Bers, B., Mccormack, T., & McGourty, J. (2018). The prevalence of specific learning disorder in mathematics and comorbidity with other developmental disorders in primary school-age children. *British Journal of Psychology*, *109*(5). doi: 10.1111/bjop.12322

National Association of State Directors of Special Education (2005). *Response to intervention: Policy considerations and implementation*. www.rtinetwork.org/rti-marketplace-list/entry/1/138

National Center for Education Statistics (since 1995). *Trends in International Mathematics and Science Studies*. http://nces.ed.gov/timss/index.asp

National Center for Education Statistics (2018). *Digest of Education Statistics* (Table 204.60). https://nces.ed.gov/programs/digest/d18/tables/dt18_204.60.asp

National Center for Education Statistics, National Assessment of Educational Progress (2019). *NAEP report card: Mathematics*. www.nationsreportcard.gov

National Center on Response to Intervention (2010). *Essential components of RTI—A closer look at response to intervention*. www.rti4success.org/resource/essential-components-rti-closer-look-response-intervention

National Commission on Excellence in Education (1983). *A nation at risk: The imperative for educational reform*. www2.ed.gov/pubs/NatAtRisk/risk.html

National Council of Teachers of Mathematics (1989, 2000). *Principles and standards for school mathematics*. www.nctm.org/Standards-and-Positions/Principles-and-Standards/

National Council of Teachers of Mathematics (2006). *Curriculum focal points for prekindergarten through grade 8*. www.nctm.org/curriculumfocalpoints/

National Governors Association Center for Best Practices & Council of Chief State School Officers (2010). *Common core state standards for mathematics*. www.corestandards.org/Math/

National Governors Association Center for Best Practices & Council of Chief State School Officers (2010). *Common core state standards: Frequently asked questions*. www.corestandards.org/about-the-standards/frequently-asked-questions/

Nolan, K. (2012). Dispositions in the field: Viewing mathematics teacher education through the lens of Bourdieu's social field theory. *Educational Studies in Mathematics*, *80*, 201–215. doi: 10.1007/s10649-011-9355-9

O'Connor, R. E. (2000). Increasing the intensity of intervention in kindergarten and first grade. *Learning Disabilities Research & Practice*, *15*(1), 43–54. www.tandfonline.com/doi/abs/10.1207/SLDRP1501_5

Palacios, L. (2005). *Critical issue: Mathematics education in the era of NCLB—Principles and standards*. North Central Regional Educational Laboratory.

Parker, F. (1993). *Turning points: Books and reports that reflected and shaped U. S. education, 1749–1990s*. (ED369695). ERIC. https://files.eric.ed.gov/fulltext/ED369695.pdf

Powell, S. R., Fuchs, L. S., & Fuchs, D. (2013). Reaching the mountaintop: Addressing the common core standards in mathematics for students with mathematics difficulties. *Learning Disabilities Research & Practice*, *28*(1), 38–48. doi: 10.1111/ldrp.12001

Pullen, P. C., van Dijk, W., Gonsalves, V. E., Lane, H. B., & Asheworth, K. E. (2018). RTI and MTSS: Response to intervention and multi-tiered systems of support. In P. C. Pullen & M. J. Kennedy (Eds.), *Handbook of response to intervention and multi-tiered systems of support* (pp. 5–10). Routledge. doi: 10.4324/9780203102954

Reys, B. J., Dingman, S., Sutter, A., & Teuscher, D. (2005). *Development of state-level mathematics curriculum documents: Report of a survey*. University of Missouri, Center for the Study of Mathematics Curriculum. https://mospace.umsystem.edu/xmlui/handle/10355/2242

Romberg, T. A. (1993). NCTM's standards: A rallying flag for mathematics teachers. *Educational Leadership*, *50*(5), 36–41.

Rubie-Davies, C. M., Peterson, E. R., Sibley, C. G., & Rosenthal, R. (2015). A teacher expectation intervention: Modeling the practices of high expectation teachers. *Contemporary Educational Psychology*, *40*, 72–85. doi: 10.1016/j.cedpsych.2014.03.003

Schiller, E., Chow, K., Thayer, S., Nakamura, J., Wilkerson, S. B., & Puma, M. (2020). *What tools have states developed or adapted to assess schools' implementation of a multi-tiered system of supports/response to intervention framework?* (REL 2020–017). Institute of Education Sciences, National Center for Education Evaluation and Regional Assistance, Regional Educational Laboratory Appalachia. https://ies.ed.gov/ncee/edlabs/projects/project.asp?projectID=4580

Schunk, D. H., & DiBenedetto, M. K. (2016). Self-efficacy theory in education. In K. R. Wetzel & D. B. Miele (Eds.), *Handbook of motivation at school* (2nd ed., pp. 34–54). Routledge. doi: 10.4324/9781315773384

Soares, N., Evans, T., & Patel, D. R. (2018). Specific learning disability in mathematics: A comprehensive review. *Translational Pediatrics, 7*(1), 48–62. doi: 10.21037/tp.2017.08.03

Sugai, G., & Horner, R. H. (2002). The evolution of discipline practices: School-wide positive behavior supports. *Child and Family Behavior Therapy, 24*(1–2), 23–50. doi: 10.1300/J019v24n01_03

Thomas, K., Huffman, D., & Flake, M. (2016). Pre-service elementary teacher dispositions and responsive pedagogical patterns in mathematics. In A. G. Welch & S. Areepattamannil (Eds.), *Dispositions in teacher education: A global perspective* (pp. 31–56). Sense Publishers. doi: 10.1007/978-94-6300-552-4_2

Tosto, M. G., Asbury, K., Mazzocco, M. M. M., Petrill, S. A., & Kovas, Y. (2016). From classroom environment to mathematics achievement: The mediating role of self-perceived ability and subject interest. *Learning and Individual Differences, 50*, 260–269. doi: 10.1016/j.lindif.2016.07.009

United Nations (2006). *Convention on Rights of People with Disabilities.* www.un.org/development/desa/disabilities/convention-on-the-rights-of-persons-with-disabilities.html

US Department of Education, Office of Special Education and Rehabilitative Services. (2002). *A new era: Revitalizing special education for children and their families* (ED473830). ERIC. https://files.eric.ed.gov/fulltext/ED473830.pdf

Usiskin, Z. (2014, 10 April). *What changes should be made for the next edition of the Common Core Standards?* Article presentation at the annual meeting of the National Council of Teachers of Mathematics, New Orleans, LA, United States. http://ucsmp.uchicago.edu/resources/conferences/2014-04-10/

Vaughn, S., & Fuchs, L. S. (Eds.). (2003). Redefining learning disabilities as inadequate response to instruction [Special issue]. *Learning Disabilities Research & Practice, 18*(3). doi: 10.1111/1540-5826.00070

Voss, T., Kleickmann, T., Kunter, M., & Hachfeld, A. (2013). Mathematics teachers' beliefs. In M. Kunger, J. Baumert, W. Blum, U. Klusmann, S. Krass, & M. Neubrand (Eds.), *Cognitive activation in the mathematics classroom and professional competence of teachers* (Vol. 8, pp. 249–271). Springer. doi: 10.1007/978-1-4614-5149-5_12

Wang, S., Rubie-Davies, C. M., & Meissel, K. (2018). A systematic review of the teacher expectation literature over the past 30 years. *Educational Research and Evaluation, 24*(3–5), 124–179. doi: 10.1080/13803611.2018.1548798

Wang, S., Rubie-Davies, C. M., & Meissel, K. (2020). The stability and trajectories of teacher expectations: Student achievement level as moderator. *Learning and Individual Differences, 78*, 101819. doi: 10.1016/j.lindif.2019.101819

Wang, Z., Osterlind, S. J., & Bergin, D. A. (2012). Building mathematics achievement models in four countries using TIMSS 2003. *International Journal of Science and Mathematics Education, 10*, 1215–1242. doi: 10.1007/s10763-011-9328-6

Werts, M. G., Carpenter, E. S., & Fewell, C. (2014). Barriers and benefits to response to intervention: Perceptions of special education teachers. *Rural Special Education Quarterly, 33*(2), 3–11. doi: 10.1177/875687051403300202

Wilson, L. D., & Kennedy, P. A. (2003). Classroom and large-scale assessment. In J. Kilpatrick, W. G. Martin, & D. Schifter (Eds.), *A research companion to principles and standards for school mathematics* (pp. 53–67). NCTM.

Wilson, R. (2014, 25 March). Indiana first state to pull out of common core. *The Washington Post.* www.washingtonpost.com/blogs/govbeat/wp/2014/03/25/indiana-first-state-to-pull-out-of-common-core/

2   Foundations of Mathematics Learning

Chapter Questions

1. What skills and concepts are precursors for mathematics learning?
2. What developmentally appropriate practices with young children lead to mathematics concept development?
3. Identify some early signs of difficulties with learning mathematics concepts.
4. How are students with learning disabilities in mathematics identified?
5. What are the general characteristics of students with learning difficulties that will affect mathematics teaching and learning?

Zachary reaches out and selects a red block from among a group of brightly colored wooden blocks. He pulls it in to his chest for a better grasp. Next he moves it up with both hands to his mouth and gnaws on a corner before dropping it. That block loses Zachary's attention when he notices a mobile of animals circling above his head. He reaches up and tugs at the lion. The mobile dances away from his grasp, swinging back and forth over Zachary's head.

Kristy is outside in the backyard with three of her friends from preschool. They are stacking plastic blocks into walls with a gap for the door. Kristy decides the walls need a roof so that they can hide inside. She finds some fallen palm branches and creates a roof. Only two of the children can fit inside so they decide to make the house larger. Kristy's friend suggests that they make windows and use those blocks for more walls.

David loves helping his mom bake cookies. He measures the flour into a big bowl. She adds soft butter and sugar. He asks his mom how many chocolate chips they should add. "Do we count them all first?" After they drop the dough onto a cookie sheet, his mom says, "The oven's ready." David asks, "How does it know how hot to get for our cookies?"

Anyone who claims that mathematics is simply a subject in school or is the formal study of numbers and operations should watch Zachary, Kristy, and David. Children are born cognitively programmed to be curious about their world. Zachary is analyzing three-dimensional objects in space, studying their properties, and considering a theory of pendular movement. Kristy and her friends are studying volume and surface area. David is formulating new rules about when to count or measure and when to estimate. All three children are observing and interacting with elements in their world and adding to their growing understanding about them. These are mathematics and problem-solving pursuits. These activities are critical

DOI: 10.4324/9781003096733-2

for normal cognitive development that will lead to more formal mathematics and problem-solving experiences in school.

Precursors of Mathematics Learning

What concepts and skills are developed during early childhood that form the foundation for mathematics learning in school? How are these concepts and skills developed? Read the following descriptions that illustrate milestones in the first five years in the life of a child named Jessica. Think about the concepts and skills Jessica is developing, how they are developed, and how they may be related to later, more complex, mathematics.

At three months Jessica is lying on a carpet in the den of her home surrounded by blocks and toys. When her dad picks up a ball she follows it with her eyes and smiles. She is starting to reach for objects close by.

At six months Jessica is rolling over and picking up her blocks. Of course they go right into her mouth. Her mom sits her up and gives her a green block. Jessica can hold it with either hand. When her older brother reads a book with Jessica, she looks at the pictures and makes vocalizations.

When Jessica turns one year old she can pick up her new toy animal with her thumb and one finger. She can put one block on top of another and when her brother hides a block behind himself, Jessica points for it, vocalizing in words, "there" and "mine." Jessica is walking with a bit of help.

At 18 months Jessica can bring toys to others when asked for them by name. She loves exploring new things like the cardboard boxes that came with the new dishes. She also enjoys scribbling with her crayons on large sheets of paper. When her brother reads books, Jessica can name things in the pictures using words like "baby" and "doggy."

Two-year-old Jessica enjoys turning the pages in a book when her dad reads and can describe the pictures using short phrases such as, "boy threw it" and "dog running." She can recall events in the past such as playing with a friend or visiting a relative. She can run over to her toy box and find each toy when her mom asks her for it by name. She loves to listen to music and beat on an old pan. She can lace large beads onto a string. Jessica is aware of general times of the day such as naptime and dinnertime.

At 30 months Jessica can use crayons more deliberately, making lines and shapes to cut out with scissors. She can put a simple puzzle together and take it apart again. Jessica also understands the concepts of today and tomorrow, and the prepositions in and under. She loves to hear the same story over and over.

Shortly after her third birthday, Jessica begins describing objects: red car, fuzzy bear, big chair. She understands many size and location words like big, small, over, and behind. Many of Jessica's sentences begin with question words: why and how. She can manipulate clay and blocks to create representations of real things. When asked to count eight crayons, Jessica holds out her fingers and counts while touching each crayon—1, 2, 3, 4, 5, 1, 2, 3.

Between the ages of 2 and 5 Jessica is very quickly learning language—new words and the syntax and morphemes that are common. Most everything Jessica learns is through her experience with her world. She explores in her room—toys, puzzles, books, games—pretending she is teaching her stuffed animals and telling them stories. She explores in her back yard—swinging, playing in the sandbox, and kicking a ball with her older brother.

By age five Jessica is able to classify objects by color or shape, she can understand that one thing might have more than one name (dog, terrier), and she is beginning to learn symbols such as numbers and letters in her world. Jessica is cutting and pasting and drawing. She is counting out crayons for her friends when they visit. And she loves to make patterns with her blocks on the floor, making up rules as she plays. Jessica uses over 4000 words in five- to six-word sentences, but sometimes makes mistakes with plurals and past tense. She knows left from right, can identify coins, and is beginning to relate special times of day to times on the clock.

TRY THIS

Which of the descriptions involving Jessica have direct relationships to later formal mathematics learning?

In the first two years of life, children like Jessica are in what Jean Piaget termed the *sensorimotor stage* (1952). The child's world is explored through the senses and motor skills, through personal interaction. When objects cannot be sensed, they do not exist because these children cannot represent objects mentally. Between the ages of two and six or seven, Jessica is in what Piaget called the *preoperational stage*. In this stage children can think and talk about things outside their immediate experience. They can make mental representations and use memory to recall events in the past. More recent psychologists posit that children in this age range understand more about their own mental processes and those of other people than previously understood (Sophian, 2013). Young children often learn more advanced concepts with guided experiences. Conceptually engaging activities mediated by the parent or teacher have an impact on the development of more formal thinking at earlier ages (Fleer & Hedegaard, 2010). Preschool children may not be performing at the preoperational level in all tasks or domains. Table 2.1 provides an overview of typical math-related behaviors at two age points for each of five major mathematics domains.

The next sections explore the cognitive development of mathematics abilities and the related concepts and skills developed during early childhood.

Early Development of Cognitive Structures

At birth the infant's brain weighs about 27% of the adult brain. (Society for Neuroscience [SfN], 2018). After birth the brain grows rapidly, with the number of neurons in the cortex increasing by 23%–30% in the first three months. By the time a child is five years old, her brain is 90% of its adult size. The number of connections between neurons (synapses) increases rapidly during the first two years of life—more than one million new connections per second (Center on the Developing Child, 2016), resulting in 50% more synapses than the adult brain. Synaptic connections are pathways for signal transmission within the brain and between the brain and other parts of the body such as muscles and sense organs. This period of explosive synaptic overproduction is followed by a process termed *synaptic pruning*. This reduction process (and *myelination* ⟍) is experience-dependent and the pruning of excess connections leads to more precise and organized neural systems (SfN, 2018). The production-reduction process occurs in different time spans for different brain regions, with some maturing during the preschool years (e.g., visual cortex) and others not until late adolescence (e.g., prefrontal cortex), when the number of synaptic connections is approximately that of the adult. Brain networks continue to be formed through adulthood, especially in the prefrontal cortex where higher-level cognitive abilities are processed. Some researchers assert that new neurons continue to be formed in adulthood and migrate to specific locations in the hippocampal dentate gyrus (input to *hippocampus* for learning and memory processes ⟍) and in the olfactory bulb (smell; SfN, 2018).

Table 2.1 Mathematics Concept Development in Young Children

Mathematic Domain	Concepts	Three Years	Six Years
Number & Operations	Counting	Counts 1 to 10. Counts 1 to 4 objects with one-to-one correspondence. Recognizes numerals for number words and quantity 0 to 9.	Counts to 100, by 1s, 2s, 10s. (To 200 by end of year.) Names number before & after to 99. Estimates up to 100.
	Comparison	Compares small sets visually. Uses directional and relational words (bigger, longer). Using ordinals first and last.	Uses counting to compare collections (one to 20), uses ordinal terms (1st, 2nd, etc.) Uses number lines for representations and comparisons.
	Addition and Subtraction	Recognizes difference in adding 1 or taking 1 away. Mentally determines sums and differences within 4 or 5.	Solves verbal word problems using counting-based strategies, sums to 18. Informally solves compare-type subtraction word problems (less than 18).
Fractions & Decimals	Part-whole	Understands that part is less than a whole. Puts together and takes apart puzzles with up to 15 pieces.	Can divide up for sharing up to 20. Uses halves, thirds, fourths as labels.
Geometry & Spatial	Shapes	Recognizes simple geometric shapes in environment. Begins to classify by one characteristic.	Sorts shapes into classes by variety of attributes. Identifies shapes by attributes and name. Makes a picture by combining shapes.
	Spatial Orientation	Begins to use relationship words (under, behind, between).	Draws simple map of immediate area. Uses representational words for positions in space (under, front).
Patterns & Algebraic Thinking	Patterns	Notices patterns in daily events, especially in actions and objects.	Notices patterns in math (e.g., adding 1). Understands concepts of even and odd, adding two, adding five.
	Reasoning	Demonstrates beginnings of deductive reasoning in everyday problems (e.g., which crayon is missing from what is present).	Understands that evidence is needed to justify ideas, may use multiple examples. Increasingly able to justify outcomes through examples, rules, characteristics.
	Equivalence	Applies concepts of same and different to describe items.	Recognizes properties such as $(3 + 6 = 6 + 3)$. Beginning to understand that = means the same value as.

(continued)

Table 2.1 Cont.

Mathematic Domain	Concepts	Three Years	Six Years
Data & Measurement	Represent	Uses pictures to represent features on a map or graph. Beginning to collect relevant data for solving problems.	Represents data on bar graphs (symbols). Uses data to answer a question.
	Units	Orders from smallest to largest.	Compares length of objects. Measures with copies of a unit.
	Time	Understands daily time concepts (afternoon, evening, tomorrow).	Knows when events take place. Beginning to learn to tell time.

New synaptic connections are formed, and unused connections pruned, throughout life (SfN, 2018). The connections that form in early childhood—weak or strong—are a foundation for those that form later. This type of brain *plasticity* ⟍ allows learning from experience, formal schooling, and continuous development. Alterations in the brain that occur during learning result in more efficient nerve cells and connections. Recent brain-imaging techniques show that learning specific tasks creates localized changes in the brain related to the tasks—changes in white matter such as axion growth, thickness of myelin, internodal distance, new patterns of connectivity—optimizing speed and efficiency of impulse transmission (National Academies of Sciences, Engineering, and Medicine, 2018; Zatorre et al., 2012). Interaction with other people and the environment is essential for normal brain development. Impairments in sensory (vision, hearing, touch) and cognitive systems such as language development and memory can cause long-lasting effects on brain functioning, especially impairments that occur during critical developmental periods (Bruer, 1998). Exposure to adverse childhood experiences—such as living in poverty, maltreatment, or even low parental education—has been associated with structural and functional changes in the brain (Merz & Noble, 2017). Adverse experiences in early childhood have been linked to differences in language development, executive functions, and social-emotional processing that may lead to persistent mental health problems and lower academic achievement. However, it is important to note that children who begin school behind their peers are not doomed to be school failures. Early interventions can make significant contributions to children's academic skills. Language development and cognition (early learning capabilities) appear to be relatively resilient processes (National Research Council [NRC], 2000). There are less-resilient aspects of cognition including vocabulary, language proficiency, understanding number concepts, and executive functions that are very susceptible to early influences of poverty. Early interventions can improve these deficits upon school entry, but subsequent school environments will determine whether the improvements are maintained.

During the first five years of a child's life, general cognitive processes begin contributing to mathematics development and other areas. These *domain-general abilities* continue to develop into adulthood and contribute at varying rates with differences in the individual's age, level of knowledge, and mathematics content (Geary et al., 2019). The most significant domain-general abilities for mathematics achievement are fluid intelligence, executive functions (e.g., working memory, attention, inhibition control, cognitive flexibility), other memory functions, and language ability.

Fluid intelligence refers to the capacity to think logically and to solve problems in novel situations independent of acquired knowledge (Cattell, 1986). It influences the way complex numerical, spatial, or conceptual relations tasks are processed and acted upon through experience. Fluid intelligence is closely related to mathematics ability because mathematics is an "evolutionary novel domain" with demands on reasoning, problem solving, and assimilating new concepts (Geary et al., 2019, p. 10). In a longitudinal study across prekindergarten and kindergarten (ages 3–5), Chu et al. (2016) found that fluid intelligence was significantly related to early knowledge of the cardinal value of number words (*cardinal principle knowledge*, CPK), discriminating values of sets, and competence with nonverbal computations. However, while CPK predicted end-of-kindergarten mathematics achievement, the role of intelligence declined, demonstrating the need for fluid intelligence for novel learning, but not as much for applying previous learning. Bornemann et al. (2010) found, in a study of 11th graders, that fluid intelligence was a leading indicator of mathematics performance in geometry and algebra. They found that fluid intelligence is especially important for higher-order mathematics achievement and that higher-achieving students also allocate more cognitive resources (focused attention and exploration of several possible strategies) to novel tasks than their lower-achieving peers.

Some of the resources that students need to draw on for mathematics tasks are *executive functions*. Executive function (EF) is an umbrella term for the processes needed for purposeful, goal-directed control of thoughts, behaviors, and emotions (Best & Miller, 2010). It is a higher-level cognitive system, a management system, that integrates and controls skills from lower levels such as processing speed and memory span. The preschool years are considered a sensitive period for executive function development because of rapid changes in the prefrontal cortex from ages two to five (Purpura, Schmitt, & Ganley, 2017). It has been theorized that the key developmental transitions during early childhood include increases in cognitive and behavioral control through the acquisition of more complex rule systems, thereby allowing for the development of more complex EF skills and the use of these skills for the development of other areas (Purpura, Schmitt, & Ganley, 2017). Important EF skills during early childhood include *inhibition* (suppressing irrelevant information), *shifting* (flexibility in moving between mental tasks), and *updating* (monitoring information processing by adding or deleting information), all functions within working memory (WM); Miyake et al., 2000; Purpura, Schmitt, & Ganley, 2017). In a study of five-year-olds, Simanowski and Krajewski (2019) found that these basic executive functions influenced children's early quantity-number competencies, arithmetic fluency, and predicted first- and second-grade mathematics achievement. Updating significantly affected knowledge of the number-word sequence and basic numerical skill development while shifting and inhibition significantly influenced quantity to number-word connections and conceptual understanding of number from kindergarten through Grade 1. More discussion on the components of EF and their effects related to mathematics learning difficulties (MD) is found in a later section of this chapter.

Executive functions work with long-term memory (LTM) for storage and retrieval of information. Young children begin using their memory abilities, building their knowledge base, as they interact with the environment and recall those experiences. Infants will respond to familiar faces and music. Children enjoy retelling stories and singing songs over and over again. As they begin noticing environmental print, children begin to understand the role of letters and numbers as abstract representations for familiar things. Names of streets, stores, candy, and numbers on houses and roads begin to take on meaning. New concepts should be connected to previous learning and real-life experiences of children so that cognitive structures are formed in LTM.

Language development is critically integrated with mathematics development. Children use language to express relationships, assign labels, manipulate concepts, and communicate understandings with others. Research has demonstrated that basic language skills and early numeracy are interrelated and influence each other reciprocally (Peng et al., 2020; Toll &

Van Luit, 2014). In a meta-analysis of 344 studies on the relation between mathematics and language, Peng and colleagues (2020) found that language serves as both a *medium* (to communicate, represent, and retrieve math knowledge) and a *thinking function* for mathematics (using language to think about abstract concepts and their relations). Language may bolster the connection between cognitive processes and mathematics learning. Teachers in early childhood settings should use specific and consistent math language, provide meaningful practice in everyday situations, and continually verify children's understanding of number concepts.

If young children are English language learners (EL), they may demonstrate adequate mathematics abilities in preschool but fall behind by the end of first grade (Martin et al., 2019). It has been established that heterogenous grouping and sufficient time allocated for mathematics impact the mathematics achievement of kindergarteners who are EL (Garrett & Hong, 2016). Teachers should attend to the systematic development of specific mathematics language—oral and written—in students who are simultaneously learning English, not just to their general language skills. Often teachers focus primarily on vocabulary for English learning in mathematics class, but that has the effect of simplifying and even handicapping these students (Moschkovich, 2018). Teachers should support talking to learn mathematics with understanding, viewing students' first language, everyday language, and gestures as resources. They should revoice student contributions more formally and ask for clarifications to understand student thinking about conceptual understanding and mathematics reasoning. They should plan tasks that provide students with opportunities for "talking to learn with understanding" and use their teacher talk to support and scaffold student understanding (Moschkovich, 2018, p. 29). Students learning English can thrive in mathematics if they are able to participate in rich mathematical communication in classrooms (de Araujou et al., 2018).

Other general cognitive abilities related to mathematics achievement are often overlooked: spatial skills, patterning, statistical learning, and causal learning. Spatial skills and patterning are related in their cognitive structures as each involve similar executive processes and some language abilities. Spatial skills include spatial visualization, form perception, and visual-spatial working memory (Mix et al., 2016). Mix and Cheng (2012) asserted that "the connection between space and math may be one of the most robust and well-established findings in cognitive psychology," in a literature review of the topic (p. 198). The authors concluded that spatial skills as early as preschool are predictive of later spatial skills and that connections between spatial and mathematical skills exist, emerge quite early, and may be causally related such that improved spatial skills lead to improved mathematics skills. Related to spatial skills, *patterning* is a non-numerical skill that focuses on repeating patterns (e.g., red, red, blue, red, red, blue…). By the end of preschool, most children are able to follow and complete missing items in patterns, create duplicate patterns, and extend patterns beyond models (Rittle-Johnson et al., 2019). Patterning skills are related to spatial skills and each are unique predictors of mathematics knowledge (Rittle-Johnson et al., 2017, 2019). Refer to Strand D for a more detailed discussion of these spatial skills and their development.

Another emerging area of cognitive research related to mathematics learning is *statistical learning*, the range of ways in which children (even infants) are implicitly sensitive to the statistical regularities in their environments (NRC, 2015). It is "incidental learning of structured patterns encountered in the environment" (Conway, 2020, p. 280). For example, 11-month-old infants demonstrated the use of regularities in the environment to draw inferences and make predictions (Xu & Denison, 2009). The infants were surprised, after shown a box of many red balls and a few white balls, when the box was poured out all white balls appeared. Conway, in reviewing previous research, summarized that statistical learning is multifaceted and is operationalized by two brain systems: cortical plasticity (associative and perceptual learning) and the executive system (attention and working memory). Early learning seems to be dominated by associative learning (through about age four), while attention and other cognitive control mechanisms progressively become more influential over time.

Related to statistical learning, infants and young children also have an intuitive understanding of *causal inference*; they experience observations and learning that allow them to conclude that one variable causes (or prevents) an effect. Children learn these relationships by using their senses to explore the environment, to observe actions and reactions, and to actually cause reactions. For example, pushing a ball across the floor may cause it to hit the wall and come rolling back. Crying causes someone to pick the child up. Smiling causes other people to smile back. Causal inference is a critical skill for formal mathematics. Children as young as 16–24 months can observe patterns of statistical contingency between causes and effects, learn causal properties of objects, and intervene to generate desired effects (Gweon & Schulz, 2011). Three- to five-year olds can learn from difference-making evidence, generalizations from previous causal experiences, to predict variables in new situations that would likely cause a new effect (Goddu & Gopnik, 2020).

Mathematics-Specific Concepts and Skills Developed during Early Childhood

Other concepts and skills that begin development during early childhood are specifically related to mathematics development; they are *domain-specific skills*. One of the earliest mathematics-specific abilities is children's innate *approximate number system* (ANS; Geary et al., 2019; Xenidou-Dervou et al., 2018). This sense of quantity, without the use of symbols, allows children to discriminate between amounts (e.g., group of three blocks compared with group of two). The ANS allows children (and adults) to have a rapid and intuitive sense of numbers and their relations. For young children, the size and the ratio of sets (2:3 is harder than 2:5) allows them to be quickly compared without the need for counting (Odic, 2018). The ANS also supports mental transformations, such as arithmetic operations, ordinal relationships, and proportional reasoning, without the use of symbols. Development of the ANS in young children has been correlated with development of symbolic math ability. However, it is not clear whether children map number words to ANS representations before or after they understand the cardinal principle, or whether there is a reciprocal developmental relationship (Chu et al., 2018). Even as children develop symbolic understanding, their ANS may continue to assist them in judging reasonableness of problem solutions. More discussion of the ANS and the development of number sense can be found in Strand A.

Early numeracy has been hypothesized to develop in three overlapping phases or levels (Krajewski, 2008). In the first level—basic numerical skills—infants have the capacity to discriminate quantities, although they cannot perceive different numbers of objects as precisely different. They can recognize small sets of objects without counting (*subitizing* using ANS abilities) and compare sets with small but different amounts. With the acquisition of language, children can compare sets with words such as *more* and *less*. They begin to learn to count or recite number words (counting sequence), but still do not connect these words with quantities.

In level two, young children begin linking number words to quantity. First they develop an imprecise conception of the attribution of number words to quantities and assign number words, such as *much* and *a little bit* to rough quantities (Krajewski & Schneider, 2009). Then children distinguish between adjacent numbers and link exact quantities with number words. They apply the counting sequence to fixed sets (*one-to-one counting*) and can name the total in a set (*cardinality*), about ages three to four. Children independently develop, through experiences, the realization that a quantity can be divided into parts or pieces that can again be combined into the original, about ages four to five.

Level three is characterized by concepts of number relationships. Children can represent part–whole relations between quantities with precise number words. They can decompose or combine quantities into new quantities with new number words without having to use physical objects. Preschool children are beginning to use written symbolic-based skills as a foundation for more advanced skills such as addition and subtraction with Arabic numerals (Purpura & Lonigan, 2013).

Krajewski and Schneider (2009) noted that it is important to understand that children may have reached level three with small numbers while still operating in level two with larger numbers. Also, children may not pass through levels at the same time for verbal number words and Arabic numbers. Levels one and two represent competencies that are mathematics precursor skills while level three reflects true mathematical understanding. In a longitudinal study, the researchers found that 25% of the variance in mathematics achievement at the end of fourth grade was predicted by children achieving level two in kindergarten. Low-performing fourth graders demonstrated large deficits in level one and two competencies in kindergarten, so early interventions in quantity-number competencies are important.

As the first mathematical concept, the achievement of *cardinal principle knowledge* (CPK) is a critical point in children's development and predicts their readiness for mathematical learning upon entry to school (Geary et al., 2018). CPK is understanding that the last counting word stated refers to the total number of items in a set (O'Rear & McNeil, 2019). Geary and van Marle (2016) found that children who were CPKs at the beginning of preschool had higher intelligence scores, knew more numerals, could employ count words, and demonstrated more accurate ANS. Children with delayed CPK showed deficits in executive function and letter recognition (general-domain deficits). The early CPKs most likely had more exposure to counting and number activities before entering preschool. The delayed CPKs demonstrated across-the-board quantitative deficits in preschool but also executive-function deficits. Enumeration, the ability to count collections of objects with number words up to three of four objects, may be critical for the transition to CPK for at-risk children. The researchers recommended assessing quantitative knowledge at the beginning of preschool and the gains in that knowledge throughout preschool as a better assessment for school readiness (Geary et al., 2018). An overview of early number skills with examples is displayed in Table 2.2.

"Young children are more capable of learning mathematics than was previously believed. They can do much more than just 'macaroni math,' gluing pieces of macaroni to paper" (NRC, 2012). Many researchers and educators are calling for more attention to early mathematics and related cognitive skills, asserting that children are born ready to learn and early learning environments that support cognitive skills related to mathematics are essential for successful formal learning in school. In fact, early mathematics skills have been repeatedly shown to predict later mathematics *and* reading achievement (Litowski et al., 2020). Willoughby et al. (2019) found, in surveying early childhood teachers for a study on the relationship between EF and academic achievement, that most teachers spent an average of 90 minutes or more per day on reading and language arts but only 23% of kindergarten teachers, 31% of first-grade teachers, and 36% of second-grade teachers devoted the same amount of time to mathematics. Further, there is compelling evidence that kindergarten mathematics instruction is limited and focuses on material below children's ability levels, spending too much time on content already mastered (Claessens et al., 2014). Young children's development of early mathematics skills may be hindered by teachers and parents underestimating children's abilities for their age, teachers' anxiety about teaching mathematics, and a lack of teacher training in these early mathematics competencies. Klibanoff et al. (2006) found, in a study of preschool and daycare teachers' mathematical speech, that the amount of specific "math talk" that these teachers used was significantly related to growth in children's skills. Increasing talk about mathematics in preschool settings, especially conversations that include rich quantitative information, has the

Table 2.2 Early Number Skills with Example Tasks

Early Number Skills	Example Tasks
Precursor Skills	
Subitizing	Child is shown a box with 2 balls and a box with 3 balls and is asked, "Which is more?" (without counting).
Using Compare Words	Child is asked which hand has more pennies. "Can you say that in your own words? Can you tell me about the other hand?" (more, fewer; taller, shorter)
Verbal Counting	Child is asked, "Can you count starting at the number 1? See how high you can count."
Arabic Numeral Recognition (early symbolic)	Shown numerals on flash cards, "Can you say the names of these numerals?" (starting with 0–5, then to 9, then 10 if the child is succeeding)
One-to-One Counting	Child is asked, "Can you count out the red beads? While you are counting please say the number word aloud for each bead."
Advanced Counting	
Cardinality	Child is shown a group of 5 crayons. "Can you tell me the total number of crayons? It is OK to count them, but then tell me the total."
Give-a-Number	"Here are some toy cars. Can you hand me two?" (1–4 first, then add 5)
Forward and Backward	"Can you count starting from zero up to 5? Can you start at 5 and count back to zero?" (other starting points)
Mental-Verbal Tasks without Symbols	
Estimation	"Here is a jar filled with cubes. Can you estimate the number inside? We will count them to check."
Non-symbolic Compose/ Decompose	"Here are five books on this shelf. If I put two more on the shelf, how many will be there? What if I remove three from the shelf? Without using books, can you tell me what four added to two equals? What about six subtract three?"
Mental Number Line	"You have done well with your numbers! Now I want you to think of the number four. Is it before or after the number six? Can you tell me a number that comes between two and eight on the number line?"
Symbolic	
Symbolic Mapping	"Here are some cards with the numbers 1 to 5 written on them. Can you match those with the circles of dots here?" (1–5, then to 9)
Symbolic Comparing Magnitudes	"Here are some cards with numbers on them. Can you put them in order from smallest to largest?" (mixture of 0, 2, 3, 4, 5, 8, 9, 10)
Symbolic Addition and Subtraction (without objects)	"If we have 4 (show number 4) and add 3 more (show number 3), how many will we have? I have written here for you 7 − ____ = 5. Can you write in the missing number?"

Note: These number skills are not necessarily in order. Children will have success with smaller numbers (1–4) before achieving cardinality (4–5). Numbers up to 10, then up to 20+ will take longer to develop. For example, children may have reached verbal counting to 10 before Give-a-Number is firm with numbers 1 to 5.

potential to increase children's preparedness for school. In addition to increasing time and talk related to mathematics, early childhood professionals should understand each child's abilities and skills and build on those using developmentally appropriate yet challenging practices.

Importance of Informal Mathematics Knowledge

Ginsburg (1977) termed the knowledge that children develop in everyday settings prior to attending formal schooling *informal knowledge*. Most preschoolers arrive at school with important mathematical competencies, such as a sense for numbers and counting, that are foundational for formal mathematics learning if understood by educators. Even with older children, the everyday, informal knowledge that is developed through experience can be tapped for enhancing formal mathematics learning. Purpura, Baroody, and Lonigan (2013) found that the relation between informal and formal mathematics is fully mediated by number knowledge. Informal knowledge does not directly impact formal, rather children must map their informal knowledge onto numeral knowledge, and then onto formal knowledge. Depth of numeral knowledge, understanding related procedures and concepts, is necessary for formal knowledge.

Seo and Ginsburg conducted an interesting study of the types of informal mathematical activities in which four- and five-year-old children were engaged in natural settings (2003). The researchers classified observable activities by their characteristics:

- *Classification activities* involved sorting, grouping, or categorizing objects.
- *Magnitude activities* were statements made about global magnitude of objects, direct or side-by-side comparisons, or judgments without quantification.
- *Enumeration activities* involved saying number words, counting, subitizing, and even reading and writing numbers.
- *Dynamics* involved putting things together, taking them apart, or making other transformations such as turning and flipping.
- *Pattern and shape activities* included identifying or creating patterns or shapes and exploring the properties and relationships of shapes. (Seo & Ginsburg, 2003, pp. 93–94)

After coding 15-minute videotaped segments of 90 children, the researchers concluded that most children (88%) engaged in mathematical activities naturally and that about 43% of the time observed was spent in math-like activities. Very significant in their findings was the conclusion that there were no income level or gender differences in these activity levels. In general, children engaged in pattern and shape activities the most, and classification the least, and were capable of achieving quite complex levels of performance. For example, some children demonstrated estimating the number in a set without counting and transforming a rhombus shape into a trapezoid. Other studies of preschool children have concluded that, in the early stages of numerical development, there is gender equality on the performance of a range of symbolic and nonsymbolic tasks (e.g., Bakker et al., 2019). Boys and girls *should* be equally competent in acquiring formal mathematics concepts and skills.

Young children's informal knowledge related to mathematics is predictive of later formal mathematics achievement. Shanley et al. (2017) studied early number skills—both informal and formal—in children across kindergarten and Grade 1. Informal early number skills (devoid of numerals including counting, quantity comparisons, and number words) were significantly and positively associated with first-grade mathematics summative achievement for an intervention group, but not for an at-risk control group. Gains in formal early-number skills (number identification and sequencing; use of numbers to describe, compare, and combine quantities) were significantly associated with mathematics achievement gains at the end of K and Grade 1. The researchers concluded that interventions that target informal and formal early number skills in kindergarten can lead to mathematics achievement gains at the end of first grade. In a study of informal numeracy skills of 781 preschool children and their later fifth-grade mathematics achievement, Nguyen et al. (2016) found that counting and numeracy skills, especially advanced counting, were most predictive of later achievement over other early competencies such as spatial-geometric, measurement, data analysis, and patterning. Advanced counting was defined as counting with cardinality, counting forward or back from a given number, and conceptual subitizing.

TRY THIS

Visit a child-care center or preschool and observe children ages three to five. Using the Seo-Ginsburg categories, note the informal mathematics activities in which children are engaged.

Informal mathematics knowledge is also important for older students. It may be harder to extract informal from formal learning and some misconceptions may be more rigidly held by older students, but teachers need to assess prior knowledge in whatever form for better connections to new mathematics learning. Misconceptions of older students, sometimes termed *biases*, are often caused by inadequate concept development through a wide enough range of examples during formal instruction that connect with prior learning or by limited informal experiences. For example, if a child had only cube-shaped blocks to play with, he could not compare other three-dimensional shapes and their properties. Viewing a diagram of a pyramid would be confusing for this child.

Signs of Learning Difficulties in Young Children

When should parents, teachers, and other caregivers be concerned about the development of young children in areas related to later mathematics learning? While much attention has been focused on identifying at-risk factors in young children for language and literacy issues, only recently have efforts targeted mathematics concepts and related cognitive abilities. Some of the same at-risk factors for language and literacy development are also relevant for mathematics. For children younger than six years old in the United States, 19% are in families with incomes below the poverty level and more than half (52%) of children ages three and four do not attend preschool (Annie E. Casey Foundation, 2019). Students from impoverished backgrounds also have more difficulty overcoming disadvantage than wealthier students throughout the school years. Early childhood educators must be able to identify and address developmental discrepancies and environmental disadvantages to provide early interventions related to concepts foundational for mathematics learning.

With the caveat that young children develop at individual rates but with the reality of early experiences in cognitive development being significant and predictive of later achievement, educators should be concerned when observing these characteristics in three- and four-year-olds:

- *Language*: problems with naming objects, following simple directions, rhyming, recalling number words 0 to 10, or using language to express needs or thoughts.
- *Social/emotional*: not making choices, following simple rules, engaging in play with other children, or sticking with a task or activity; easily frustrated or angry; overly egocentric as compared with peers.
- *Sensory/motor*: avoiding hands-on tasks; clumsy and immature large motor skills as compared with peers; excessively disorganized; over- or under-reacts to environmental stimuli; awkward pencil, crayon, or scissors grip for age.
- *Cognitive*: cannot count objects to 4 or 5, cannot name colors and simple shapes, does not recall simple words or directions, cannot sort objects by one attribute, cannot make simple comparisons, cannot offer a simple reason for an action.

Quite a bit of research in recent years has focused on early predictors of later mathematics achievement. In a longitudinal study—315 children were assessed one year prior to school entry and again in Grades 3 and 6—Träff et al. (2020) found that skills and knowledge at lower levels were necessary for students to succeed in higher-level mathematics. Basic number abilities in preschool accounted for 4% variance in third grade and third grade arithmetic skills accounted for almost 18% variance in sixth grade. The importance of basic number skills decreased over time, while the importance of general cognitive abilities (working memory, language, rapid automatic naming) remained about the same.

In an eight-year longitudinal study, Geary et al. (2017) found that intelligence, central executive, and reading achievement were the domain-general abilities that made the most significant contributions to mathematics achievement from 2nd to 8th grade and those effects were constant across grades with greater effect than domain-specific abilities in the lower grades. Domain-specific abilities, especially after third grade, increasingly affected mathematics achievement. Teaching mathematics-specific competencies has the potential to reduce individual differences in achievement across grade levels—number knowledge and basic arithmetic in the early grades and fractions knowledge in later grades. In examining possible visuospatial predictors, Li and Geary (2013) found that children with the largest gains (not initial ability) in visuospatial (VS) memory from 1st to 5th grade had higher end-of-fifth-grade achievement than those with smaller gains. Also predicting later elementary mathematics achievement were speed of numeral processing and the domain-general abilities of intelligence, attention, and inhibitory control.

It has been established that mathematics experiences of children ages three to five influences readiness for kindergarten (Murray & Harrison, 2011). Mathematics performance in the early elementary grades predicts later elementary and middle-school performance (Nelson & Powell, 2018). Elementary-grade performance predicts middle and high-school performance (Watts et al., 2014), and so forth. Formal mathematics is a curriculum domain with an interconnected hierarchy of concepts and skills for which learning at a specific age and grade level is a foundation for subsequent learning. Young children who do not develop strong foundational mathematics skills, especially in counting (stable order, one-to-one correspondence, cardinality, number word use), comparing (symbolic magnitudes, estimations, mental number line), and combining (composing and decomposing verbally and with number symbols), are at risk for a poor beginning in formal mathematics instruction and achievement across the grade levels.

Do multi-tiered systems of support (MTSS) apply to children ages three to five? Yes! These systems are especially important for young children identified as developmentally delayed, English language learner, or at-risk for learning and/or behavioral difficulties, but should be available for all children. Preschool MTSS should be integrated within school-wide programs when possible. These support systems should include the same elements as those for school programs: high-quality curriculum for all children (at a developmentally appropriate level); initial screening and on-going assessment with continuous progress monitoring; collaborative problem solving among team members; and multiple tiers for increasingly individualized, intense, targeted, and frequent intervention. Screening and assessment for early mathematics skills of young children are discussed in the next chapter.

Developmentally Appropriate Mathematics for Young Children

The term *developmentally appropriate practice*, or DAP, was applied to the concept of planning children's activities in preschool settings according to age appropriateness by the first National Association for the Education of Young Children (NAEYC) monograph on early childhood programming (Bredekamp, 1987). DAP relates to the developmental appropriateness of activities based on age, individual growth patterns, and cultural factors. Individual development involves aspects of cognitive, physical, social, and emotional development as well as familial and other environmental influences. NAEYC revised the definition of DAP in 2020:"…methods that promote each child's optimal development and learning through a strengths-based, play-based approach to joyful, engaged learning." The 2020 revisions emphasized the importance of social, cultural, and historical contexts not only for the child but also for the adults involved in any aspect of early childhood education (NAEYC, 2020, p. 5).

Activities in preschool and early school settings (homes, child care centers, and schools for ages birth through seven) can range along a continuum from developmentally inappropriate to developmentally appropriate (Hart et al., 1997). Practices on the developmentally inappropriate end would include teaching through lecture; requiring rote memorization of isolated facts; forcing paper and pencil tasks too early; designing curriculum into isolated formal school topics of math, science, and social studies; administering standardized tests as a means of measuring learning; applying whole-group learning goals; and requiring children to sit at individual desks for most learning experiences. In contrast, DAP builds on the child as the source for learning activities. Children have options in the learning environment, activities integrate curriculum areas, the environment includes hands-on explorations, and teachers are able to guide children in flexible ways depending on developmental needs.

Developmentally appropriate practice does not preclude important mathematics learning, however. Young children can develop concepts and attitudes that will have a significant impact on their formal mathematics learning. The NAEYC and the National Council of Teachers of Mathematics (NCTM) developed a joint position statement about appropriate mathematics experiences for children ages three to six titled *Position Statement on Early Childhood Mathematics: Promoting Good Beginnings* (NAEYC, Adopted in 2002, Updated in 2010). Their ten recommendations are illustrated here with elaborated examples. The ten practices are in italics, copyright © 2010 NAEYC®. All rights reserved. Reprinted with permission.

1. *Enhance children's natural interest in mathematics and their disposition to use it to make sense of their physical and social worlds.* For example, observe children playing with objects for exploration. Children naturally sort, classify, and compare objects such as blocks, balls, and other toys. Their curiosity in the world around them should be encouraged through enriched environments and encouragement from others. Early explorations in supportive environments can help children develop confidence in their abilities from an early age. Early positive experiences with mathematics provide a foundation for life-long dispositions about learning such as curiosity, persistence, interest in mathematical concepts, willingness to try different tools for exploring concepts, and perceiving problems as challenges.

2. *Build on children's experience and knowledge, including their family, linguistic, cultural, and community backgrounds; their individual approaches to learning; and their informal knowledge.* Before age two, most children develop the general concepts of shape and number. Between two and six or seven they should have experiences that develop classification, comparison, counting, locating parts in wholes, ordering, using initial mathematics language, measuring informally, making simple graphs, and using simple mathematics symbols. Number systems and mathematics language may be influenced by linguistic differences. For example, Asian-language children, because of linguistic differences, are more likely to show larger numbers in a base-10 system and interpret part-whole and calendar concepts sooner than non-Asian children, although cultural values for mathematics education also contribute to achievement differences (Ng & Rao, 2010). Asian languages represent numbers in exact base-10 references, whereas non-Asian languages do not always emphasize the number system structure and may actually reverse the order of spoken and written language (e.g., fourteen for 14 in English, four and ten in German, but ten-four in Chinese). In Asian languages, geometric figures are typically

named by a combination of everyday words (e.g., triangle would be three-corner-shape), and the days of the week and months of the year are referred to by number.

Experience with counting, ordering, and classifying is influenced by opportunities to engage with a wide variety of materials and activities. The four-year-old who helps her father with cooking family meals and setting the table may develop concepts of number, spatial relationships, and measurement earlier than her peers. A five-year-old child who travels with his mother on the subway develops concepts of value of money, time, distance, and spatial navigation. Early childhood teachers should find out what children already understand and develop experiences for children to relate their informal understanding to the vocabulary and concepts of mathematics.

3. *Base mathematics curriculum and teaching practices on knowledge of young children's cognitive, linguistic, physical, and social-emotional development.* Although generalizations about child development can assist teachers in planning experiences, exact timelines based on these generalizations should not be imposed. One child may be using a pencil or crayon to draw representations of objects while another child's fine motor skills are not as developed so she uses larger implements and cut-outs. One child perceives a drawing as representing real objects while another child the same age cannot make that connection yet. New learning experiences should be based on each child's current developmental level but work to gradually scaffold the child's understanding to a higher level. With common curriculum standards beginning in kindergarten, it is critical that teachers not narrow educational experiences to teaching a set of skills by a certain date, that they maintain focus on individual children and their holistic development. Highly specific timelines for skill mastery can lead to poor concept understanding and even anxiety and trauma for some children.

4. *Use curriculum and teaching practices that strengthen children's problem-solving and reasoning processes as well as representing, communicating, and connecting mathematics ideas.* The mathematical processes and practices described in Chapter 1 also apply to young children. Using developmentally appropriate activities, teachers should ask questions, set up learning experiences, and scaffold understanding so that children can form the foundations for these processes. Consider the following preschool setting.

> Mr. Sanchez is sitting at a low table with three children—Amy, Jennie, and Bart. His goal for the interaction is to have the children count buttons and make representations of their findings. Mr. Sanchez asks the children how many buttons they are wearing today. Amy counts five, Jennie three, and Bart finds eight on his clothing. Mr. Sanchez asks comparison questions such as, "Who has the most buttons? Who has the smallest number?" This is to encourage the children to use their mathematics vocabulary. Bart exclaims that he has the same as Amy and Jennie together but this notion confuses the girls. His buttons are not like theirs at all. Mr. Sanchez explains to the children that they are going to create a graph showing the number of buttons using the chart paper and crayons. Each child draws his or her buttons in a column over his or her name while the other children help by counting the buttons drawn. The children are eager to add to their graph so they ask Mr. Sanchez if they can each question another child outside the group about buttons and draw those on the graph.

What mathematics practices were involved in this activity? Was the activity developmentally appropriate for preschool children? Did the activity allow for differences in individual development?

5. *Ensure that the curriculum is coherent and compatible with known relationships and sequences of important mathematics ideas.* The national curriculum standards (NCTM, 2000; NGA & CSSO, 2010) encouraged developing the *big ideas* of mathematics rather than teaching isolated facts and skills. An example of concept development from ages three and six was illustrated in Table 2.1. The close connections among the mathematics content areas are especially evident in this age span. Mathematics educators and researchers, concerned about the development of big ideas over time, across grade levels, have begun to study learning progressions, or *trajectories*. "Learning trajectories are empirically supported descriptions of the likely obstacles and landmarks students encounter as they move (toward) more sophisticated ideas" (Confrey et al., 2014, p. xvii). Learning trajectories allow for partial ideas and digressions when needed. They provide a language for describing student proficiencies and learning gaps. They are not linear; rather, they interact with prior learning and other current learning.

Learning trajectories include a hierarchical sequence that anticipates fall back to earlier levels. The higher levels require increasingly abstract reasoning. Trajectories are developed in connection with a sequence of instructional activities. For example, a trajectory for young children on length might have the following levels (PreK through Grade 1; Sarama & Clements, 2009; Sarama et al., 2011):

*   *Pre-length quantity recognition:* Ask students to compare objects by different attributes. Some students compare size generally but without dimensions.
*   *Length quantity recognition:* Challenge students to compare tall people for the tallest. Students identify length and distance, but cannot compare length.
*   *Length direct comparison:* Ask students to organize themselves by height. Have students compare sticks to find the longest. Students align objects to compare and use terms long, longer, and longest.
*   *Indirect length comparison:* Challenge students to compare objects not adjacent. Students begin to form mental images and create a counting scheme for units of space. This level of the trajectory may actually be concurrent with the next level.
*   *End-to-end accumulation:* Ask students to compare lengths when there are not enough units for comparison. Students make longer comparisons by the repetition of shorter units.
*   *Unit repeating and relating:* Ask students to use a simplified ruler to check measures. Students use rulers with some success, but one student aligns one end with the one unit mark instead of zero. Students demonstrate use of mental imagery and perceptual information to attempt measuring tasks.

Additional levels follow across months and years with students demonstrating increasing concept understanding and more complex and sophisticated coordination of schemes, moving back to more foundational concepts when obstacles or uncertainties arise. Activities may differ and students may skip ahead or need more time, depending on their conceptual understanding. Additionally, vocabulary may be critical for some children to construct meaning at higher levels. Learning trajectories also can be connected with curriculum standards to inform teachers about what to teach in relation to other topics.

6. *Provide for children's deep and sustained interaction with key mathematical ideas.* Rather than occasional and haphazard mathematics-related events or planned formal mathematics lessons covering dozens of topics, the standards call for a balanced approach with young children. Present concepts in a logical sequence and provide in-depth experiences including extensions by families or caregivers. This approach requires planning and focusing on big ideas, especially in the areas of number, measurement, and geometry.

An example of a shallow mathematics event would be for the teacher to ask students to count the new leaves on a classroom plant. This counting experience is not connected to deep mathematics concepts and does not promote the processes. A stronger task could be comparing the leaf growth on two or three plants, recording data, and making predictions. Extensions of this number-sense activity include a graphing activity or another pictorial representation.

One misconception of many early childhood teachers about mathematics is that "mathematics should not be taught as stand-alone subject matter" (Lee & Ginsburg, 2009, p. 41). There appears to be an avoidance of saying that it's time for math, as if teachers are afraid their students will be anxious and fearful. We don't hesitate to say it's time for reading. Young children's interest in and abilities for mathematics are often underestimated. Time should be set aside specifically for mathematics so that children develop the deep concept understanding, vocabulary, and dispositions they will need entering school.

7. *Integrate mathematics with other activities and other activities with mathematics.* Activities with young children lend themselves to concept integration. In addition to planned time for mathematics study, concepts, vocabulary, and skills should be integrated across the curriculum. Concept integration leads to stronger understanding and generalization. Children's literature, music, games, science explorations, social skills, social studies, and even classroom routines have many opportunities for the integration of mathematics concepts if planned at the appropriate level and with appropriate depth, not just random activities.

One first-grade teacher plans weekly themes around which she constructs activities that support the entire curriculum. The first week of February was "All about Snowflakes" and activities included reading *The Snowy Day* by Ezra Jack Keats (1962), singing a song created by the class called *I'm a Little Snowflake*, and studying the shapes and symmetry of cut-out snowflake designs. For science concepts the class predicted the freezing and melting rates of ice and for social studies they read about the first person to photograph snowflakes, Wilson Bentley. For writing they created a group book called *The Lonely Snowflake* using new words such as cold, dark, white, and night (and one student wanted to learn to spell the word hexagon). These activities, with expert teacher guidance, promoted problem solving, reasoning, communication skills, and persistence that could be individualized, as well as deeper concept development across the curriculum.

The mathematics that is integrated within other activities should follow logical sequences and allow for depth and focus, not a random grab bag of activities. Again, purposeful planning and careful attention to individual prior knowledge and learning pathways is critical for rich mathematical experiences for younger children. In the activities described previously, some children were still working with the whole numbers 1 to 5, while others wondered about larger numbers and how to record those. Some children could use triangles to create patterns with more complex attributes while others were still working with basic rectangles.

8. *Provide ample time, materials, and teacher support for children to engage in play, a context in which they explore and manipulate mathematical ideas with keen interest.* With young children, play can be the best context for assessment, problem solving, and concept development. Play has been called the leading activity or motivator during preschool years (Fleer & Hedegaard, 2010). The teacher's responsibility is to provide an environment with space and materials for exploration, to carefully observe and listen to children during these play activities, and to mediate and scaffold concept and skill learning. Teachers should ask questions that will stimulate children in making connections, building extensions, and developing new understandings. For example, a child is playing with blocks of different colors, sizes, and shapes. The teacher asks the child to sort the blocks by color, then by shape. He asks the child to line up the blocks from smallest

to largest and to count the blocks. The teacher is scaffolding attribute understanding, mathematics vocabulary, and sequential counting.

9. *Actively introduce mathematical concepts, methods, and language through a range of appropriate experiences and teaching strategies.* Teachers need to plan specific learning experiences for young children with mathematics objectives in mind. Typically these activities also have motor, language, and social-skills objectives for individual children. For example, a mathematics objective appropriate for most five-year-old children is: Children will count by 5s from 0 to 20 (for some children to 50). The planned activities for this objective could involve music, art, games, money, or time. The teacher should plan explicit activities and assessments for the objective, make deliberate plans for new vocabulary and skill development, and assure that each child is learning something new, extending or revising previous understandings of number. But the activities are still individualized and developmentally appropriate, not paper and pencil tasks at the symbolic level. Appropriate mathematics learning at this level cannot be left to chance. Further, one activity on this objective will not be sufficient. Children need a variety of activities with a range of examples, materials, and discussions to develop deep concept understanding.

10. *Support children's learning by thoughtfully and continually assessing all children's mathematical knowledge, skills, and strategies.* Assessment strategies at this level must be individually relevant, frequent, and developmentally appropriate. The most authentic and informative assessments are careful observation, children's products over time, open-ended questions, and performance assessments in real contexts such as play. For the example of counting by fives, the teacher should listen to each child count by five in another time and setting, using nickels, analog clocks, or objects. The assessment should take place in a natural setting, not formalized into a testing format. Authentic and natural assessments can, therefore, occur continually as they are embedded within day-to-day activities. These informal assessment strategies are still planned and documented, they are just more developmentally appropriate and valid for young children.

A panel of experts in mathematics instruction for young children reviewed high-quality research and affirmed the need for developmentally appropriate practice in their five recommendations for practices and strategies for teaching math to young children (Frye et al., 2013, pp. 1–2):

1. Teach number and operations using a developmental progression.
2. Teach geometry, patterns, measurement, and data analysis using a developmental progression.
3. Use progress monitoring to ensure that math instruction builds on what each child knows.
4. Teach children to view and describe their world mathematically.
5. Dedicate time each day to teaching math, and integrate math instruction throughout the school day.

The authors provided research support and detailed examples for each of these recommendations. For example, to encourage children to notice patterns in the world around them, show children a calendar and look for repetitive patterns, such as days of the week and months of each season. Portions of this guide could also be a resource for parents who want to support the mathematics development of their children.

When Zachary, Kristy, David, and Jessica begin formal mathematics instruction, will their teachers tap their informal mathematics knowledge to make linkages to formal learning? Will they have had the rich, developmentally appropriate experiences in preschool settings that provide strong concept understanding for more formal mathematics learning? Will any of these children have difficulties learning mathematics concepts and skills?

Mathematics Learning Difficulties

According to large-scale national reports, anywhere from 19% to 66% of students are not meeting expected achievement goals in mathematics. For example, the *Nation's Report Card* (National Center for Education Statistics, 2019) reported that 59% of fourth graders and 66% of eighth graders scored below proficient on mathematics achievement measures, with 19% of fourth graders and 31% of eighth graders scoring below basic levels. At a basic level, students should show some evidence of procedural knowledge and concept understanding in the five content areas at their grade level. To be proficient, students are required to reason, make conjectures and inferences, and support their reasoning consistently across mathematics content areas.

Clearly, not all of the 19% to 31% of students scoring below basic on these large-scale measures had identified disabilities in mathematics. Approximately 6% to 7% of K to 12 students have an identified specific learning disability in mathematics (MLD, Geary, 2015; Morsanyi et al., 2018; Shalev, 2007). Another 10% to 15% have been characterized as low achievers (LA), despite average abilities, demonstrating mild but persistent learning difficulties in mathematics (Berch & Mazzocco, 2007; Murphy et al., 2007). While research studies are not consistent in how they define MLD and LA, some researchers propose cut-off scores on achievement measures of the 10th *percentile* \ for MLD and 11th to 25th percentiles for LA (or mathematics difficulties, MD), across more than one academic year for the purposes of studying student characteristics and the impact of interventions (Geary, 2015; Swanson et al., 2018; Zhang et al., 2020). Some students with persistent MD may not have had access to quality instruction, of the type being measured that requires standards-based problem-solving and reasoning skills. Other students are impoverished, lacking the supports of home or community that are required to get an early start in learning and continue to make gains during the school years. An increasing number of students are struggling to learn English while learning mathematics and other subjects in school.

The sections that follow explore specific characteristics of students with identified MLD, including associated cognitive processes, and general characteristics of students who struggle with mathematics as observed in classrooms.

Theoretical Perspectives of Mathematics Learning Disabilities

The Individuals with Disabilities Education Improvement Act (IDEA, 2004) definition for specific learning disabilities (SLD) included the phrase: "…a disorder in one or more of the basic psychological processes involved in understanding or using language, spoken or written, that may manifest itself in the imperfect ability to listen, speak, read, write, spell, or do mathematical calculations…" (Part A, §602.30). Students who are identified with specific learning disabilities in mathematics must meet criteria for deficits in mathematics calculation and/or problem solving. The fifth edition of the *Diagnostic and Statistical Manual of Mental Disorders* (APA, 2013) included *specific learning disorder* under neurodevelopmental disorders, indicating the cognitive basis and life-long nature of SLD, and bringing the APA criteria closer to those used in schools under IDEA. This edition eliminated the specific but separate reading, mathematics, and written expression disorders of previous editions because those areas often overlap and difficulties in those academic areas are actually symptoms of underlying neuropsychological disorders.

Current definitions of SLD include the manifestation of difficulties with specific academic demands in school. Students are typically identified with unexpected low achievement in one or more areas such as mathematics or reading, despite average ability and targeted interventions. Underlying these academic manifestations are one or more deficits in cognitive processes that are much more difficult to assess and vary considerably from student to student. Cognitive-processing deficits most commonly cited in research involving students with MLD are problems with working memory, including verbal and visual representations, attention, processing speed, and executive function (Geary et al., 2019; Johnson et al., 2010) and related language deficits (Purpura, Logan, et al., 2017).

A prevalent model for identifying students with specific learning disabilities is the *response-to-intervention* (RTI) process, considered part of multi-tiered systems of support (MTSS) in many school districts. IDEA 2004 required states to permit the use of RTI and prohibited them from requiring the discrepancy model, with little guidance on how to implement RTI for identification purposes. RTI-like systems may date back to 1982 when the validity of student identification for SLD and resulting overidentification were addressed by a National Research Council study (Pullen et al., 2019). Pre-referral interventions were initiated in many states for the collection of student progress information during specially designed interventions prior to a formal referral, but were often viewed as not informative for instruction and lacking validity. The RTI model for identification was developed to add validity to a pre-referral identification system, with the student's response to research-supported interventions a measure of the need for special education. Fuchs et al. (2003) outlined the process as progress-monitoring during instruction by the regular classroom teacher followed by additional or different instruction for students who did not respond, again with progress monitoring. Those students who still failed to respond qualified for special education placement (or a more comprehensive evaluation). Because of underlying deficits in domain-general cognitive processes (e.g., working memory, language) and their changing profiles across ages or content domains, specific learning disabilities are difficult to diagnose and describe for individual students.

Deficit or Difference?

Other researchers challenged the concept of learning disability altogether (e.g., Banks, 2014; Baum et al., 2014; Lewis, 2014). These challenges have risen from frustrations with the lack of an operational and widely recognized definition of SLD, the complexity of learning difficulties, a lack of consensus about identification methods that are reliable and valid, focus on deficits rather than abilities, and weak evidence for linking specific deficits with effective instructional approaches, especially in mathematics, beyond basic elementary topics. Teachers and parents are often frustrated with the "wait to fail" model of SLD identification—the simplistic discrepancy-between-scores method of identification—and the lack of appropriate and individualized intervention recommendations from the diagnostic process.

Those who promote reconceptualizing MLD as a difference rather than a deficit propose that students who struggle in mathematics and have average or above-average abilities may actually have different pathways of development and understand mathematics concepts and representations differently than their peers (Lewis, 2014). The problem with characterizing a MLD as a deficit, Lewis argued, is that the diagnosis offers no differentiation from other struggling students and does not contribute to a remedial intervention. A cognitive deficit also implies an unchangeable condition. A cognitive difference urges us to examine what students understand, and build on those understandings. Lewis and her research subject Dylan published a report on how they took an "emancipatory approach," a collaboration between researcher and subject, to share Dylan's difficulties as a student with persistent and severe mathematics difficulties (Lewis & Lynn, 2018). They discussed the tools and compensatory strategies Dylan used to succeed in upper-level mathematics courses in college and have a career as a statistician. They challenged myths of students with deficits in mathematics as not being able to learn higher mathematics, the requirement for mastering the basics, and the need for speed and efficiency. Understanding what a student is able to do, to encourage the student to achieve, and to assist them in finding supports and strategies needed is an important lesson for teachers from this single-subject research.

In fact, expert views on MLD identification and intervention are not so dissimilar. Most educators and psychologists with extensive experience with MLD recommend a robust evaluation process that not only identifies (or rules out) the mathematics learning disability and its nature, but provides enough information to guide the development of an individualized program of remediation (NJCLD, 2010). A robust evaluation includes empirical evidence that the student's low achievement is not due to other factors (poor teaching, environmental causes, other disabilities, or emotional issues; Fletcher et al., 2007). A robust evaluation illuminates what students understand as well as their misconceptions within the range of complex mathematics topics and processes, not simply number facts and calculation (Lewis, 2014). A robust evaluation includes measures of cognitive-processing ability including executive functions, processing speed, and receptive and expressive language (Johnson et. al., 2010). A robust evaluation includes interventions designed to facilitate learning and instructional planning, not delay services (Fuchs et al., 2012). A robust evaluation informs the development of a student's individualized education program (IEP). The identification decision is not the end of a robust evaluation process; rather, dynamic evaluation activities continue to measure student progress within instructional settings. The challenge is practical implementation, with fidelity to validated protocols by highly trained educators and other professionals, as is discussed in Chapter 3.

Cognitive Processes Involved in Mathematics Disabilities

Modern concepts of cognitive processes and executive control involved in learning date back to the 1950s. When studying individual differences in mathematics learning, researchers have taken a range of approaches including longitudinal studies of the development of various abilities that are predictors of later achievement; comparison studies examining differences between groups (e.g., MLD vs. typically achieving, MLD vs. younger students); and profile studies of relative strengths and weaknesses on specific mathematics tasks. Most studies of MLD have focused on deficits in cognitive processes, specifically those in executive functions, and how those relate to domain-specific learning.

As discussed in a previous section of this chapter, executive functions (EF) are cognitive processes in the prefrontal areas of the frontal lobes of the brain and connected structures. They include working memory, attention, inhibition, self-monitoring, goal setting, and initiation activities. Executive functions are complex, multiple, and interrelated; they cue and direct mental functioning that varies depending on setting and task; and they follow a developmental trajectory from infancy through adulthood (Otero & Barker, 2014).

Many researchers in special education and psychology characterize working memory as the central executive function that directs other cognitive processes (e.g., Baddeley, 2006; Geary, 2011; Swanson, 2012). "Working memory (WM) is the limited-capacity system that combines mental operations and temporary information storage in the support of goal-directed cognition and behavior" (Hitch et al., 2018, p. 64). WM is where incoming (new information) and prior knowledge is retrieved and stored temporarily for integration with new learning, as is illustrated in Figure 2.1. Working memory includes focused attention and active processing of a limited amount of information at a time. The ability to retain information in

External
Stimulation

Information Loss

Sensory
Registry

Long-term Memory

Declarative
(Explicit)

Episodic

Semantic

Phonological
Loop

**Working
Memory**

Attention

inhibition
updating
switching

Episodic
Buffer

Nondeclarative
(Implicit)

Automatic habits,
skills, procedures

Visual-Spatial
Sketchpad

priming, encoding,
consolidation,
retention, retrieval

Verbal,
Motor
Output

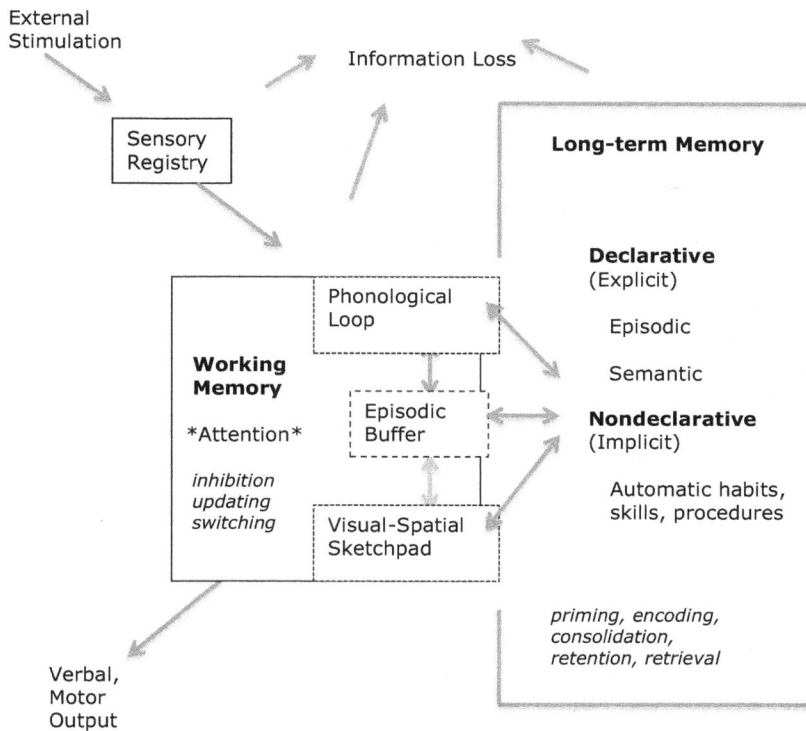

Figure 2.1 Information-Processing Model of Learning

WM increases throughout childhood and adolescence. Learning requires working memory, long-term memory (LTM), and motivation. Motivation plays a large role in effectively and efficiently allocating capacity to working memory, according to some theories (e.g., Atkinson et al., 2019; Hayes, 2000). Motivation can assist the student in directing and maintaining attention for important and prioritized information in WM.

The central executive is the control center for working memory, the most complex component (Baddeley, 1986, 2018). It is a concept based on functions, not a specific location in the brain. Functions include focusing attention on relevant information (from external sources or from LTM), dividing attention between stimulus streams, switching between tasks (which might not be a unitary function), and the capacity to interface with LTM. The central executive controls attention during learning and coordinates with LTM. The central executive works much like a boss, making decisions about priorities and collecting information, but delegating tasks to be completed to employees, or subsystems.

Essential and interactive subsystems (dependent systems) of the central executive are the phonological loop and visuospatial sketchpad (Baddeley, 1986). The phonological loop deals with language-based information, spoken or written (lip-read and signed material as well). It stores information read or heard in an articulated code that continues to loop within WM while being used. This is where vocabulary, procedures, math facts, math symbols, and other information presented in written or oral form (from external sources or LTM) is processed and held during processing. The visuospatial sketchpad holds and processes visual and spatial information. Information such as graphs, number lines, charts, objects in space, and drawings (from external sources or LTM) are held and processed here. Visualizing (creating mental images) from other information is also active in the sketchpad. Baddeley proposed the episodic buffer in 2001, which forms an interface between the WM subsystems and LTM and serves as a back-up store for WM. It may hold perceptual or linguistic information as bound representations (e.g., phrases into sentences, objects into scenes). Others argued that WM buffers are actually activated portions of LTM (Coolidge & Wynn, 2005) or that there is a focus-of-attention component that is a resource for retaining items in WM (Cowan, 2001).

Long-term memory has been classified as declarative (explicit) and nondeclarative (implicit). Declarative LTM is further separated into episodic (events in one's past experience) and semantic (facts, concepts, language) types. Nondeclarative LMT doesn't require conscious awareness and includes procedures, skills, and habits that are automatic. Some processes that occur in LTM include priming, encoding, consolidation, retention, and retrieval. A person must categorize and organize new information within WM in order to send it to a retrievable place in LTM.

If we consider the mathematical tasks presented in school, the cognitive processes that are required can be illustrated using an information-processing model as in Figure 2.1, much like a computer-processing analogy. Which processes in WM are employed in the following scenarios? How is LTM involved?

Consider four-year-old Jo Lynn attempting to combine five objects with two objects. Her senses of vision and touch provide information about the objects. Her WM attends to a collection of five and another collection of two. Reaching into LTM, Jo Lynn retrieves memories of similar objects (VS sketchpad) and their names (phonological loop), as well as the counting numbers (phonological loop). Holding "five" and "two" in WM, she goes back to LMT for a procedure of combining that she's used in the past. Recalling the "count-on" procedure, Jo Lynn actives LMT again (phonological loop) to retrieve "six" and "seven." She articulates, "I have seven."

Consider a student, Zack, presented with drawings of triangles of different types and asked to describe the differences. There are isosceles, equilateral, scalene, and right triangles. Zack attends to the shapes and receives visual information. The verbal information from his teacher is, "Do you recall these four types of triangles? What are the similarities and differences?" His WM is activated and seeks out past experiences with triangles in LTM. Both the VS sketchpad and phonological loop are engaged as Zack mentally matches the triangles in front of him with those retrieved from LTM now in WM. He thinks of terms to describe comparisons: length of sides and size of angles (phonological loop and VS sketchpad working together) and verbalizes those using the names of the triangles (output).

Consider Elaine, whose teacher is explaining that the equation $y = 2x$ is represented by a line on a coordinate plane displayed on the white board. Elaine's attention turns to the diagram on the white board (activating WM processes) and her WM connects with LTM where she recalls the coordinate plane and the x- and y-axes (involving both phonological loop and VS sketchpad). She sees a straight line and highlighted points on the line: (-2, -4); (-1, -2); (0, 0); (1, 2); (2, 4). The teacher asks, "If x is 10, what would y be?" Elaine recalls that she can use the linear equation $y = 2x$, substituting 10 for x, and find that y would be 20 (drawing from LTM and phonological loop portion of WM).

When can cognitive-processing activities break down or be insufficient for the task? When:

- accurate and appropriate information is not readily retrieved from LTM;
- extraneous or irrelevant information competes for space in WM;
- the capacity of WM is not large enough to manipulate important information for solving a problem;
- processing speed is too slow for the task;
- attention wanders to irrelevant information;

- switching from one task to another is not efficient or causes a stoppage;
- information in WM is lost after the first few steps in problem solving or following directions;
- patterns, sequences, or algorithms that could serve as an efficient way to hold information are not recognized while working on a problem or multi-tasking; and
- information sent to LTM is disorganized, not connected to other information.

These are common processing challenges for students who struggle with mathematics, especially those with MLD. Additionally, teachers may observe that students have difficulty planning and initiating tasks, getting organized, completing tasks, understanding abstract concepts, expressing themselves fluently, and even demonstrating socially appropriate behaviors. Many studies have consistently confirmed a relation between one or more components of Baddeley and Hitch's (1974) model of working memory and mathematics achievement (Geary et al., 2019). The higher the capacity of the central executive, the better the performance on tasks across the entire mathematics domain. Central-executive deficits, especially working-memory components, are considered core underlying deficits of students with learning difficulties in mathematics (Geary, 2015).

In a meta-analysis of cognitive-profiling studies of thousands of individuals with mathematics difficulties, Peng et al. (2018) concluded that individuals with MD showed deficits in phonological processing, processing speed, working memory, attention, short-term memory, executive functions generally, and visuospatial skills. The severity of MD was related to processing-speed deficits. Deficits in processing speed and working memory were the most significant and stable cognitive markers of MD.

Variations in students' ages and the type and complexity of mathematics tasks will affect the roles and performances of various components of WM. These variations contribute to some of the inconsistencies within and across studies. For example, Meyer and colleagues (2010) found that the central executive and phonological loop predicted mathematics problem solving (reasoning) in second graders, but for third graders the visuospatial sketchpad scores were the predictors. Even when studies indicated issues within a component of WM, those components were difficult to disaggregate. Are problems within the phonological loop related to transforming symbols to verbal code, processing speed, attention allocation, or articulatory suppression?

While the evidence is compelling that deficits in executive functions are at the core of MLD, efforts to improve mathematics achievement though executive-function interventions would have little effect (Clements et al., 2016; Geary et al., 2019; Watts et al., 2015). The most promising interventions are those that build foundational mathematics knowledge and skills along with cognitive-strategy instruction, concept development, and practice features that support WM capacity and other executive-function deficits.

Components of Mathematics Learning Disabilities and Difficulties

Specific learning disabilities in mathematics (MLD) generally emerge in the early grades, although they tend to be identified later than those in reading, and continue throughout school and into adult life. Mathematical skill is a strong predictor, and may be even more important than literacy, for employment outcomes, wages, and productivity as an adult (Every Child a Chance Trust, 2009; Geary, 2015). Both environmental and genetic factors contribute to MLD, although specific contributions and student variability are not well understood. Some specific genetic disorders associated with difficulties in mathematics include Fragile X, traumatic brain injuries, spina bifida myelomeningocele, Williams syndrome, and Turner syndrome (Barnes & Raghubar, 2014; Mazzocco & Räsänen, 2013). Almost all disorders of childhood that affect the brain also affect mathematics because math is not a unitary skill and several neurocognitive systems are involved in learning.

Early work on mathematics difficulties by Russell and Ginsburg (1984) found that children with MD had generally adequate informal knowledge of mathematics, basic understanding of base-10 concepts, differing abilities related to number fact retrieval and use (that confounded the researchers), and skills for solving concrete word problems but difficulty with complex ones. These findings were initial glimpses into a complex and multidimensional array of characteristics. Students with MD typically demonstrate grade-level skills in some areas of mathematics but deficits in others (Geary et al., 2000; Jordan & Montani, 1997).

Geary, a cognitive psychologist working in the area of MLD since the 1980s, theorized three subtypes of mathematics disabilities (1993). The *procedural* subtype was characterized as use of immature procedures, frequent errors in carrying out procedures, poor concept understanding, and sequencing problems. These students' performances in mathematics were similar to those of younger students. The *semantic memory* subtype was described as problems retrieving math facts and filtering erroneous numbers. This subtype appeared to represent a cognitive difference and seemed to occur with phonetic-specific reading disabilities. The *visuospatial* subtype was theorized as difficulty representing mathematics concepts spatially and misinterpreting spatial information such as models, diagrams, graphs, or measurement estimates.

Since 1993 there has been substantial research into cognitive dimensions of mathematics disabilities, aided, in part, by technologies such as brain imaging and a closer connection between cognitive-psychological and educational-intervention research. In 2010 Geary reflected on the status of his theorized subtypes. He acknowledged that mathematics deficits were more complex and multidimensional than first thought. Geary's cognitive components of MLD that have been substantiated, or partly substantiated, by research include:

1. *Number sense*: fundamental deficit in early number sense that includes understanding the exact quantity of small collections (and the symbols that represent them), the approximate magnitude of larger quantities (mental number line), and implicit understanding of the effects of addition and subtraction on quantities.
2. *Semantic*: retrieval of arithmetic facts from long-term semantic memory.
3. *Procedural*: execution of procedures for solving arithmetic problems.
4. *Visuospatial*: representing and interpreting visual and spatial mathematical information.

In addition to these theorized subtypes, Geary (2015) also described deficits in mapping Arabic numerals, number words, and rational numbers onto associated quantities (a type of number sense) and poor understanding of some concepts such as fractions and other rational numbers.

Other researchers have proposed subtypes of MLD. Chan and Wong (2020) investigated the profiles and subtypes of students in the first two grades in school who fell below the 25th percentile on mathematics achievement measures. The researchers identified five clusters of students with moderate stability across the two grade levels: number-sense deficit, numerosity coding deficit, symbolic deficit, working-memory deficit, and mild difficulty. The authors concluded that children demonstrating persistent difficulties learning mathematics may show a range of cognitive deficits, especially in the preschool ages. Further, subtypes are not stable; students should be reassessed over time. Bartelet et al. (2014) identified six subtypes in students Grades 3 to 6: weak mental number line; weak ANS; spatial difficulties; access deficit (most severely impaired); flat performance (difficulties from other sources); and low nonverbal IQ. The researchers concluded that there is no one underlying core deficit for MLD.

In synthesizing subtype research, Karagiannakis et al. (2014) argued (for international audiences) against using the one-dimensional term *dyscalculia* and for the multidimensional term *mathematical learning difficulties*. They outlined characteristics and contributing cognitive systems for four subtypes of MD: core number deficit (arithmetic domain), memory deficit (retrieval and processing; all math domains), reasoning deficit (all domains), and visual-spatial deficit (arithmetic, geometry, algebra, analytical geometry, and calculus). Subtype research reinforces the view that MD present as an extremely complex, multifaceted, and heterogenous disorder. Most theorized subtypes have underlying cognitive deficits—either in general domains (working memory) or mathematics-specific processes (numerical understanding, spatial reasoning, execution of procedures). Individual students may have different profiles of strengths and weaknesses depending on their age, educational opportunities, and mathematics content. A diagnosis of specific learning disability in mathematics will not provide enough student-specific information for successful instruction. What is most critical for educators to realize is that frequent assessments of student strengths and weaknesses—in domain-general and domain-specific areas related to mathematics—should signal needed interventions.

Other research into mathematics disabilities and difficulties has compared the achievement of various subgroups. Increasingly, studies have identified distinct cutoff points on mathematics achievement tests for MLD and low-achieving groups (Lewis & Fisher, 2016). In exploring the differences between students with MLD and low achievers in mathematics (MLA), Geary (2013) found that MLD involved more persistent deficits across all working memory components while MLA involved problems with inhibitory control and task-switching components of the central executive. Desoete and colleagues (2012) found that MLA students caught up with their typically achieving (TA) peers between kindergarten and second grade on symbolic and nonsymbolic magnitude comparison tasks while students with MLD still had persistent deficits in word name and numeral-symbolic comparison tasks in second grade. Mazzocco et al. (2013) found differences between MLD and MLA groups of fourth to eighth graders on fraction magnitude-comparison tests. Both groups had difficulties, as compared with TA peers in fourth grade, but the MLA group had caught up with the TA group by fifth grade. Difficulties with fraction magnitude comparison persisted into eighth grade for the MLD group. This research on two groups of struggling learners indicates that the nature and trajectories of difficulties differ significantly on specific mathematics tasks.

Zhang et al. (2020) compared MLD and MLA groups (from kindergarten through fourth grade) and their typically achieving (TA) peers on a range of early cognitive skills. The measures of counting sequence knowledge (verbally counting numbers forward and backward), rapid automatized naming (RAN; naming symbols as quickly as possible), and spatial visualization were able to discriminate among the three groups, with MLD significantly lower than MLA and TA on these measures. These groups differed in specific numerical, language, and spatial skills and those skills each made unique contributions to students' mathematical difficulties. Many researchers have concluded that MLD "should be considered as a specific and definable impairment and not the lower end of a continuum of arithmetical ability" (Desoete et al., 2012, p. 75; see also Geary et al., 2012; Mazzocco et al., 2013; Stock et al., 2010). Students who are low achievers have different profiles than those with MLD and many students with MLD already had significant deficits in kindergarten. Students with MLD have more severe and persistent difficulties, slower growth rates, and difficulty with a wider range of mathematics tasks.

Another area of research attempting to discriminate among groups of students with mathematics disabilities is examining differences between MLD, learning disabilities in reading (RLD), and *comorbid* MLD/RLD. Researchers have estimated a prevalence of between 11% and 70% of students with MLD having a comorbid disability in reading (e.g., Moll et al., 2014; Willcutt et al., 2019), but prevalence results may depend on the mathematics subskills measured (Moll et al., 2019). Barnes et al. (2020) assessed almost 500 children before they began kindergarten for comorbid difficulties in math and reading and again at the point of school entry. Even at this young age, the researchers were able to discriminate risk groups. Children in the MLD/RLD-risk group performed lower than the MLD-risk group on mathematics-achievement measures. Both groups had lower levels of cognitive and reading skills than a not-at-risk group. Difficulties in ANS acuity discriminated children not at risk from those with severe math difficulties not at risk for reading difficulties. Phonological awareness was lower for MLD-risk and MLD/RLD-risk groups than the not-at-risk group. The only cognitive predictor that discriminated MLD risk from MLD/RLD risk was a direct measure of attention (vigilance over time). By the end of kindergarten, children with MLD/RLD risk had more severe deficits in mathematics skills than children with MLD risk only.

Raddatz et al. (2017) compared four groups of elementary students ages 6 to 11 (MLD, RLD, MLD/RLD, noLD) on basic number-processing and calculation skills. They concluded that students with MLD demonstrated impairments in a range of number-processing tasks, both verbal and nonverbal, while students with RLD showed math deficits only in tasks requiring verbal skills and Arabic number processing. The researchers confirmed the additive effect of comorbid MLD/RLD—that there is no shared cognitive deficit underlying a comorbid group—and recommended broader assessment for each disability area. Comorbidity of learning deficits in mathematics and reading is associated with greater severity of academic difficulties and lower response to some interventions. Research is still mixed on whether the relationship is related to underlying language skills or other executive-function factors such as attention or working memory. There is some agreement that students with comorbid deficits have more severe deficits in mathematics and that MLD and RLD have distinct characteristics; one disability is not the result of the other.

More research is needed on MLD characteristics across mathematics topics, such as rational numbers, algebra, and geometry, and the developmental trajectories of those domains. The fields of neuropsychology and cognitive neuroscience, and a new field of educational neuroscience, are contributing insights into cognitive deficits that impact mathematics learning with advances in functional neuroimaging technologies and molecular genetic investigation techniques (Räsänen et al., 2019). However, to have implications for instruction and intervention, neuroscience must collaborate with pedagogy to better understand the complexity of educational tasks and behaviors. Räsänen and colleagues (2019) warned about *neuromyths*, false simplifications that are well-marketed to eager educators as "brain-based interventions." Researchers publishing results should be explicit about limits to their research for application in the classroom.

Characteristics of Students who Struggle with Mathematics

The 14%–22% of students who struggle with mathematics, including those with specific mathematics disabilities, demonstrate a wide range of characteristics in the mathematics classroom. Some of the more common manifestations are discussed in this section, but teachers should be cognizant that not all students with difficulties learning mathematics will demonstrate all characteristics or have problems with all topics within mathematics and most will have strengths. Individual student profiles will change over time. Even with intensive interventions, some students with MLD will demonstrate persistent problems with achievement throughout the grade levels.

1. *Cognitive factors.* As discussed in the previous sections, cognitive-processing problems are central for students with MLD and other struggling students. Typical manifestations observable in the classroom include difficulty with: retrieving math facts, solving multistep word problems, recalling and applying procedures and algorithms, using strategies for estimation,

attending to the important parts of problems, applying previously learned information to new problems, visualizing information, and understanding concepts. Students with cognitive deficits may appear to perform like younger students, using less mature strategies, and requiring more time to reach automatic retrieval. However, they will show age-appropriate ability and even strengths in some areas of mathematics.

2. *Metacognitive factors.* Metacognition is an awareness of the skills, strategies, and resources that are needed to perform a task and the ability to use self-regulatory mechanisms, including adjustments, to complete the task (Borkowski & Burke, 1996). Sometimes called "thinking about one's own thinking," metacognition is the process involving being aware of and monitoring the use of executive and cognitive strategies. Students with metacognitive difficulties have trouble selecting and using effective learning strategies. They don't monitor their own use of strategies and have difficulty with generalization across time and setting. For example, a student may have trouble deciding how to solve a nonroutine word problem. Even if the student attempts the problem, he doesn't monitor the process or results. He doesn't ask himself, "Does this make sense? How can I change what I attempted?" Further, he can't draw on experiences with similar problems because they don't appear similar in his conceptualization of the problem.

Metacognitive issues impact any area of mathematics that involves planning, carrying out, and evaluating strategies for accomplishing tasks, including computation, problem solving, data analysis, measurement, and applying mathematics to real-world problems. It involves being aware of one's own cognitive resources and learning processes as well as skills in applying effective strategies for specific tasks. In a study of adolescents and adults with persistent MLD, Desoete (2009) found that metacognitive difficulties are often related to a lack of persistence and effort, little sustained attention, and problems with self-regulation. Classroom behaviors that may indicate issues with metacognition include disorganization, rushing through work, guessing answers, not being able to discuss strategies attempted, offering simplistic explanations when asked for reflection ("it just seemed like the best answer"), no evidence of planning before beginning a complex or multistep task, and poor or immature strategy selection.

3. *Language difficulties.* Problems understanding or using language underlie specific learning disabilities, including those in mathematics. Language difficulties in mathematics are also common in students with impoverished circumstances, poor prior instruction, and language differences. Language is our means for encoding information, storing semantic information, and communicating thought processes and information with others. Language allows us to talk to ourselves as we work through problems. In mathematics class, language difficulties are evident when students have trouble using symbols of math consistently and correctly, expressing math concepts clearly to others, and listening to explanations with understanding.

Mathematics has its own language of symbols and systems for recording information, with semantic (meaning or concept) and syntactic (structure or form) conventions, and involves many more symbols than reading. Each domain within mathematics has a specialized symbol system (e.g., arithmetic, algebra, set theory, geometry). There are directional, comparison, equality, and grouping symbols. Mathematics has its own systems of syntax, the relative positions of symbols within expressions. The position of numerals and other symbols is a critical indicator of the complete meaning.

Mathematics also has its own vocabulary. Some words are familiar and can have similar (equal, remainder) or completely different meanings (real, root). Some vocabulary will be novel (algorithm, integer). Language difficulties also appear with reading word problems and writing math expressions that reflect word problems. Understanding a word problem requires reading comprehension, connections with prior knowledge, focusing on important information while eliminating the extraneous, applying procedures for one or more steps while maintaining information in WM, and judging the reasonableness of the answer.

Verbal language can provide the bridge between the concrete representations of math and the more abstract and symbolic forms. As students advance in mathematics learning, they also use language to think—they manipulate concepts and ideas through language (oral or inner) without having to rely on concrete materials. Unfortunately, some students have few opportunities to talk about math. Teachers who limit lessons to lecture, demonstration, and worksheets are limiting their students' language development and related mathematics progress. Students should be responding frequently and discussing math problems and concepts with each other and the teacher. Students whose parents continue the dialog at home will have additional benefits.

4. *Motor factors.* Motor problems with written work (fine motor skills) are most evident in younger students but even adolescents with no physical disabilities can struggle with number and symbol formation. Motor skills involve more than one process. They may involve memory of the symbol along with its actual formation (visual and motor memories). They may involve visual perception and transfer (copying). Or they may involve integration of fine muscles with task demands, as with lining up digits within an algorithm. Indicators of motor issues are highly visible: poorly formed symbols, little control of spacing, excessive time for a task, and avoidance of written work.

Researchers have found a relationship between motor skills (fine and gross) and mathematics achievement (Kim et al., 2018; Pagani et al., 2010; Westendorp et al., 2011). While it is evident that motor skills are typically required for counting and developing number-line representations, the relationship between motor-skill development and mathematics achievement is not fully developed in research. One theory suggests an *embodied cognition*, that cognitive processes are grounded in the body's interaction with the environment (Pieters et al., 2015). Others suggest that motor and mathematics skills may both be fostered through common activities at young ages. In a study of young children kindergarten through Grade 2, Kim et al. (2018) found a reciprocal relationship between visuomotor integration and mathematics skills across kindergarten and Grade 1, but not into Grade 2, when mathematics skills shift more to fact retrieval. They found that fine-motor coordination contributed indirectly, through visuomotor integration. Once fine-motor coordination is mastered, requiring less attention, it is no longer strongly correlated. Rather than support narrow motor inventions to improve math skill (as was common in the 1970s), however, this research underscores the complex and multifaceted nature of cognitive development and the need to conduct comprehensive evaluations.

5. *Social and emotional factors.* Sometimes overlooked in the academic realm, social and emotional factors can cause as many learning difficulties as cognitive ones. The range of these factors is as diverse as the students served. Some students have trouble with peer or adult relationships, causing problems in cooperative-learning settings or seeking assistance. Others have self-concept and self-esteem issues that lower motivation, task persistence, and effort. Impulsive students make careless errors and do not take the time to understand deeper concepts and connections. Students who are inattentive allow irrelevant information to take over their attention and miss the important. Students with extreme anxiety—either toward mathematics or school in general—tend to avoid the source of their anxiety or perform at much lower levels than their abilities.

6. *Habits of learning.* A combination of environmental, cognitive, social, and emotional factors, habits of learning are formed from an early age but certainly can be modified throughout the lifespan. *Habits of learning* refers to how individuals view and participate in learning,

their self-discipline and self-motivation, goal setting, engagement in learning activities, and acceptance of challenges. Habits that could interfere with mathematical learning include avoidance, learned helplessness, impulsivity, little curiosity, poor assignment completion, disinterest, and working for the right answer rather than understanding. Even students with high abilities have habits, such as the drive for perfection, that can interfere with strong concept development and flexible problem solving.

The habits for good mathematics engagement promoted by national standards include perseverance in problem solving; attending to precision in communications as well as calculations; looking for patterns, connections, and structures within problems; reflecting and self-directing; seeking out alternative methods or solutions; and linking mathematics learning with real-world situations.

7. *Previous experiences.* A student's prior knowledge and previous experiences with mathematics are the best predictors of future success. Many of these experiences have been influenced by the factors described above. However, previous instructional experiences can also have a significant impact on achievement. If previous teachers did not explain concepts well, use effective teaching methods, or allow time for mastery and success, students' learning will be affected. If the curriculum and materials used weren't aligned with mathematics standards, learning might be superficial or limited. And if the student wasn't able to develop the deep concept understanding that comes from good teaching and sound curriculum, his or her mathematics achievement will suffer.

Students who have been served in separate special education settings for part or all of the school day may be affected by factors such as instruction from teachers without specific mathematics training, being "pulled out" of the regular classroom during critical instructional time, or having less than adequate time devoted to mathematics instruction. Students who changed schools frequently may have gaps in learning. Students with few opportunities for learning at home due to poverty or parent background may demonstrate significant delays. Those students whose primary language is not English may experience difficulties with the language demands of mathematics instruction.

TRY THIS

Observe students with mathematics learning difficulties in special, intervention, or general education settings and note the characteristics described in this section. Examine work samples and interact with the students if possible.

The beginning of this chapter introduced Zachary, Kristy, David, and Jessica, young children interested in the world around them. All of these children were involved in excellent preschool programs that offered rich environments with well-trained teachers. One of these children will be evaluated for a specific mathematics disability in second grade after spending a frustrating year unable to add and subtract, write numbers, or draw diagrams like the other first graders. This student's reading ability is progressing as expected but mathematics has become a source of failure, confusion, and anxiety. Although this child had a strong foundation in preschool, the formal pencil and paper tasks in first grade were not effective for continued progress in concept development. The student-evaluation team must plan assessments that will provide information on specific learning problems, as well as strengths, and plan for more effective interventions. The next chapter will explore assessment strategies for identification of problem areas, planning instruction, and monitoring student progress in mathematics.

References

American Psychiatric Association. (2013). *Diagnostic and statistical manual of mental disorders* (5th ed.). doi: 10.1176/appi.books.9780890425596

Annie E. Casey Foundation. (2019). *KIDS COUNT data book: State trends in child well-being.* www.aecf.org/m/resourcedoc/aecf-2019kidscountdatabook-2019.pdf

Atkinson, A., Waterman, A., & Allen, R. (2019). Can children prioritize more valuable information in working memory? An exploration into the effects of motivation and memory load. *Developmental Psychology, 55*(5), 967–980. doi: 10.1037/dev0000692

Baddeley, A. D. (1986). *Working memory.* Oxford University Press.

Baddeley, A. D. (2001). Is working memory still working? *American Psychologist, 56*(11), 851–864. doi: 10.1037/0003-066X.56.11.851

Baddeley, A. D. (2006). Working memory, an overview. In S. Pickering (Ed.), *Working memory and education* (pp. 3–26). Academic Press. doi: 10.1016/B978-0-12-554465-8.X5000-5

Baddeley, A. (2018). *Exploring working memory: Selected works of Alan Baddeley.* Routledge. doi: 10.4324/9781315111261

Baddeley, A. D., & Hitch, G. J. (1974). Working memory. In G. A. Bower (Ed.), *Recent advances in learning and motivation* (Vol. 8, pp. 47–90). Academic Press.

Bakker, M., Torbeyns, J., Wijns, N., Verschaffel, L., & De Smedt, B. (2019). Gender equality in 4- to 5-year-old preschoolers' early numerical competencies. *Developmental Science, 22*(2), 12718. doi: 10.1111/desc.12718

Banks, T. (2014). From deficit to divergence: Integrating theory to inform the selection of interventions in special education. *Creative Education, 5*(7), 510–518. doi: 10.4236/ce.2014.57060

Barnes, M. A., Clemens, N. H., Fall, A., Roberts, G., Klein, A., Starkey, P., McCandliss, B., Zucker, T., & Flynn, K. (2020). Cognitive predictors of difficulties in math and reading in pre-kindergarten children at high risk for learning disabilities. *Journal of Educational Psychology, 112*(4), 685–700. doi: 10.1037/edu0000404

Barnes, M. A., & Raghubar, K. P. (2014). Mathematics development and difficulties: The role of visual-spatial perception and other cognitive skills. *Pediatric Blood & Cancer, 61*(10), 1729–1733. doi: 10.1002/pbc.24909

Bartelet, D., Ansari, D., Vaessen, A., & Blomert, L. (2014). Cognitive subtypes of mathematics learning difficulties in primary education. *Research in Developmental Disabilities, 35*(3), 657–670. doi: 10.1016/j.ridd.2013.12.010

Baum, S. M., Schader, R. M., & Hébert, T. P. (2014). Through a different lens: Reflecting on a strengths-based, talent-focused approach for twice-exceptional learners. *Gifted Child Quarterly, 58*(4), 311–327. doi: 10.1177/0016986214547632

Berch, D. B., & Mazzocco, M. M. M. (Eds.). (2007). *Why is math so hard for some children? The nature and origins of mathematical learning difficulties and disabilities.* Paul H. Brookes.

Best, J. R., & Miller, P. H. (2010). A developmental perspective on executive function. *Child Development, 81*(6), 1641–1660. doi: 10.1111/j.1467-8624.2010.01499.x

Borkowski, J. G., & Burke, J. E. (1996). Theories, models, and measurements of executive functioning: An information processing perspective. In G. R. Lyon & N. A. Krasnegor (Eds.), *Attention, memory, and executive function* (pp. 235–262). Paul H. Brookes.

Bornemann, B., Foth, M., Horn, J., Ries, J., Warmuth, E., Wartenburger, I., & Van der Meer, E. (2010). Mathematical cognition: Individual differences in resource allocation. *ZDM, 42*(6), 555–566. doi: 10.1007/s11858-010-0253-x

Bredekamp, S. (Ed.). (1987). *Developmentally appropriate practice in early childhood programs serving children from birth through age 8.* National Association for the Education of Young Children.

Bruer, J. T. (1998). Education and the brain: A bridge too far. *Educational Researcher, 26*(8), 4–16. doi: 10.3102/0013189x026008004

Cattell, R. B. (1986). *Intelligence: Its structure, growth and action.* Elsevier. doi: 10.1016/s0166-4115(08)x6006-6

Center on the Developing Child at Harvard University (2016). *From best practices to breakthrough impacts: A science-based approach to building a more promising future for young children and families.* www.developingchild.harvard.edu

Chan, W. W. L., & Wong, T. T. (2020). Subtypes of mathematical difficulties and their stability. *Journal of Educational Psychology, 112*(3), 649–666. doi: 10.1037/edu0000383

Chu, F. W., vanMarle, K., & Geary, D. C. (2016). Predicting children's reading and mathematics achievement from early quantitative knowledge and domain-general cognitive abilities. *Frontiers in Psychology, 7*, 775. doi: 10.3389/fpsyg.2016.00775

Chu, F. W., vanMarle, K., Rouder, J., & Geary, D. C. (2018). Children's early understanding of number predicts their later problem-solving sophistication in addition. *Journal of Experimental Child Psychology, 169*, 73–92. doi: 10.1016/j.jecp.2017.12.010

Claessens, A., Engel, M., & Curran, F. C. (2014). Academic content, student learning, and the persistence of preschool effects. *American Educational Research Journal, 51*(2), 403–434. doi: 10.3102/0002831213513634

Clements, D. H., Sarama, J., & Germeroth, C. (2016). Learning executive function and early mathematics: Directions of causal relations. *Early Childhood Research Quarterly, 36*, 79–90. doi: 10.1016/j.ecresq.2015.12.009

Confrey, J., Maloney, A. P., & Nguyen, K. H. (2014). Introduction: Learning trajectories in mathematics. In A. P. Maloney, J. Confrey, & K. H. Nguyen (Eds.), *Learning over time: Learning trajectories in mathematics education* (pp. xi–xxii). Information Age Publishing.

Conway, C. M. (2020). How does the brain learn environmental structure? Ten core principles for understanding the neurocognitive mechanisms of statistical learning. *Neuroscience and Biobehavioral Reviews, 112*, 279–299. doi: 10.1016/j.neubiorev.2020.01.032

Coolidge, F., & Wynn, T. (2005). Working memory, its executive functions, and the emergence of modern thinking. *Cambridge Archaeological Journal, 15*(1), 5–26. doi: 10.1017/s0959774305000016

Cowan, N. (2001). The magical number four in short-term memory: A reconsideration of mental storage capacity. *Behavioral and Brain Sciences, 24*(1), 87–114. doi: 10.1017/s0140525x01003922

de Araujo, Z., Roberts, S., Willey, C., & Zahner, W. (2018). English learners in K-12 mathematics education: A review of the literature. *Review of Educational Research, 88*(6), 879–919. doi: 10.3102/0034654318798093

Desoete, A. (2009). Mathematics and metacognition in adolescents and adults with learning disabilities. *International Electronic Journal of Elementary Education, 2*(1), 82–100. www.iejee.com/index.php/IEJEE/article/view/259

Desoete, A., Ceulemans, A., De Werdt, F., & Pieters, S. (2012). Can we predict mathematical learning disabilities from symbolic and non-symbolic comparison tasks in kindergarten? Findings from a longitudinal study. *British Journal of Educational Psychology, 82*(1), 64–81. doi: 10.1348/2044-8279.002002

Every Child a Chance Trust (2009). *The long term costs of numeracy difficulties.* www.numicon.co.nz/uploads/66441/files/Numicon_research_ECC_paper.pdf

Fleer, M., & Hedegaard, M. (2010). A cultural–historical view of play, learning and development. In *Early Learning and Development: Cultural-historical Concepts in Play* (pp. 198–217). Cambridge University Press. doi: 10.1017/CBO9780511844836.015

Fletcher, J. M., Lyon, G. R., Fuchs, L. S., & Barnes, M. A. (2007). *Learning disabilities: From identification to intervention.* Guilford.

Frye, D., Baroody, A. J., Burchinal, M., Carver, S. M., Jordan, N. C., & McDowell, J. (2013). *Teaching math to young children: A practice guide* (NCEE 2014-4005). National Center for Education Evaluation and Regional Assistance (NCEE), Institute of Education Sciences, US Department of Education. https://ies.ed.gov/ncee/wwc/PracticeGuide/18

Fuchs, D., Fuchs, L. S., & Compton, D. L. (2012). Smart RTI: A next-generation approach to multi-level prevention. *Exceptional Children, 78*(3), 263–279. doi: 10.1177/001440291207800301

Fuchs, D., Mock, D., Morgan, P. L., & Young, C. L. (2003). Responsiveness-to-intervention: Definitions, evidence, and implications for the learning disabilities construct. *Learning Disabilities Research & Practice, 18*(3), 157–171. doi: 10.1111/1540-5826.00072

Garrett, R., & Hong, G. (2016). Impacts of grouping and time on the math learning of language minority kindergartners. *Educational Evaluation and Policy Analysis, 38*(2), 222–244. doi: 10.3102/0162373715611484

Geary, D. C. (1993). Mathematical disabilities: Cognitive, neuropsychological, and genetic components. *Psychological Bulletin, 114*(2), 345–362. doi: 10.1037/0033-2909.114.2.345

64 *Foundations of Mathematics Learning*

Geary, D. C. (2010). Mathematics disabilities: Reflections on cognitive, neuropsychological, and genetic components. *Learning and Individual Differences*, *20*(2), 130–133. doi: 10.1016/j.lindif.2009.10.008
Geary, D. C. (2011). Cognitive predictors of achievement growth in mathematics: A five-year longitudinal study. *Developmental Psychology*, *47*(6), 1539–1552. doi: 10.1037/a0025510
Geary, D. C. (2013). Learning disabilities in mathematics: Recent advances. In H. L. Swanson, K. Harris, & S. Graham (Eds.), *Handbook of learning disabilities* (2nd ed., pp. 239–255). Guilford Press.
Geary, D. C. (2015). The classification and cognitive characteristics of mathematical disabilities in children. In R. C. Kadosh & A. Dowker (Eds.), *The Oxford Handbook of Numerical Cognition* (pp. 751–770). Oxford University Press. doi: 10.1093/oxfordhb/9780199642342.013.017
Geary, D. C., Berch, D. B., & Mann Koepke, K. (2019). Cognitive foundations for improving mathematical learning. In D. C. Geary, D. B. Berch, & K. Mann Koepke (Eds.), *Mathematical cognition and learning*: Vol. 5, 1–36. Elsevier Academic Press. doi: 10.1016/B978-0-12-815952-1.00001-3
Geary, D. C., Hamson, C. O., & Hoard, M. K. (2000). Numerical and arithmetical cognition: A longitudinal study of process and concept deficits in children with learning disability. *Journal of Experimental Child Psychology*, *77*(3), 236–263. doi: 10.1006/jecp.2000.2561
Geary, D. C., Hoard, M. K., Nugent, L., & Bailey, D. H. (2012). Mathematical cognition deficits in children with learning disabilities and persistent low achievement: A five-year prospective study. *Journal of Educational Psychology*, *104*(1), 206–223. doi: 10.1037/a0025398
Geary, D. C., Nicholas, A., Li, Y., & Sun, J. (2017). Developmental change in the influence of domain-general abilities and domain-specific knowledge on mathematics achievement: An eight-year longitudinal study. *Journal of Educational Psychology*, *109*(5), 680–693. doi: 10.1037/edu0000159
Geary, D. C. & vanMarle, K. (2016). Young children's core symbolic and nonsymbolic quantitative knowledge in the prediction of later mathematics achievement. *Developmental Psychology*, *52*(12), 2130–2144. doi: 10.1037/dev0000214
Geary, D., vanMarle, K., Chu, F., Rouder, J., Hoard, M., & Nugent, L. (2018). Early conceptual understanding of cardinality predicts superior school-entry number-system knowledge. *Psychological Science*, *29*(2), 191–205. doi: 10.1177/0956797617729817
Ginsburg, H. P. (1977). *Children's arithmetic: The learning process.* Van Nostrand.
Goddu, M. K., & Gopnik, A. (2020). Learning what to change: Young children use "difference-making" to identify causally relevant variables. *Developmental Psychology*, *56*(2), 275–284. doi: 10.1037/dev0000872.supp
Gweon, H., & Schulz, L. (2011). 16-month-olds rationally infer causes of failed actions. *Science*, *332*(6037), 1524. doi: 10.1126/science.1204493
Hart, C. H., Burts, D. C., & Charlesworth, R. (Eds.). (1997). *Integrated curriculum and developmentally appropriate practice: Birth to age eight.* SUNY Press.
Hayes, J. (2000). A new framework for understanding cognition and affect in writing. In R. Indrisano & S. J. Squire (Eds.), *Perspectives on writing: Research, theory, and practice* (pp. 6–44). International Reading Association.
Hitch, G., Hu, Y., Allen, R., & Baddeley, A. (2018). Competition for the focus of attention in visual working memory: perceptual recency versus executive. *Annals of the New York Academy of Sciences*, *1424*(1), 64–75. doi: 10.1111/nyas.13631
Individuals with Disabilities Education Act, 20 USC § 1400 *et seq.* (1997). www2.ed.gov/policy/speced/leg/idea/idea.pdf
Johnson, E. S., Humphrey, M., Mellard, D. F., Woods, K., & Swanson, H. L. (2010). Cognitive processing deficits and students with specific learning disabilities: A selective meta-analysis of the literature. *Learning Disability Quarterly*, *33*(1), 3–18. doi: 10.1177/073194871003300101
Jordan, N. C., & Montani, T. O. (1997). Cognitive arithmetic and problem solving: A comparison of children with specific and general mathematics difficulties. *Journal of Learning Disabilities*, *30*(6), 624–634. doi: 10.1177/002221949703000606
Karagiannakis, G., Baccaglini-Frank, A., & Papadatos, Y. (2014). Mathematical learning difficulties subtypes classification. *Frontiers in Human Neuroscience*, *8*(57). doi: 10.3389/fnhum.2014.00057
Keats, E. J. (1962). *A snowy day.* The Viking Press.
Kim, H., Duran, C. A., K., Cameron, C. E., & Grissmer, D. (2018). Developmental relations among motor and cognitive processes and mathematics skills. *Child Development*, *89*(2), 476–494. doi: 10.1111/cdev.12752

Klibanoff, R. S., Levine, S. C., Huttenlocher, J.,Vasilyeva, M., & Hedges, L.V. (2006). Preschool children's mathematical knowledge: The effect of teacher "math talk." *Developmental Psychology, 42*(1), 59–69. doi: 10.1037/0012-1649.42.1.59

Krajewski, K. (2008).Vorschulische Förderung mathematischer Kompetenzen [Preschool furthering of mathematical competencies]. In F. Petermann & W. Schneider (Eds.), *Enzyklopädie der Psychologie, Reihe Entwicklungspsychologie, Bd. Angewandte Entwicklungspsychologie* (pp. 275–304). Hogrefe.

Krajewski, K., & Schneider, W. (2009). Early development of quantity to number-word linkage as a precursor of mathematical school achievement and mathematical difficulties: Findings from a four-year longitudinal study. *Learning and Instruction, 19*(6), 513–526. doi: 10.1016/j.learninstruc.2008.10.002

Lee, J. S., & Ginsburg, H. P. (2009). Early childhood teachers' misconceptions about mathematics education for young children in the United States. *Australasian Journal of Early Childhood, 34*(4), 37–45. doi: 10.1177/183693910903400406

Lewis, K. E. (2014). Difference not deficit: Reconceptualizing mathematical learning disabilities. *Journal for Research in Mathematics Education, 45*(3), 351–396. doi: 10.5951/jresematheduc.45.3.0351

Lewis, K. E., & Fisher, M. B. (2016). Taking stock of 40 years of research on mathematical learning disability: Methodological issues and future directions. *Journal for Research in Mathematics Education, 47*(4), 338–371. doi: 10.5951/jresematheduc.47.4.0338

Lewis, K. E., & Lynn, D. M. (2018). Against the odds: Insights from a statistician with dyscalculia. *Education Sciences, 8*(2), 63. doi: 10.3390/educsci8020063

Li, Y., & Geary, D. C. (2013). Developmental gains in visuospatial memory predict gains in mathematics achievement. *PLoS ONE, 8*(7): e70160.doi: 10.1371/journal.pone.0070160

Litowski, E. C., Duncan, R. J., Logan, J. A. R., & Purpura, D. J. (2020). When do preschoolers learn specific mathematics skills? Mapping the development of early numeracy knowledge. *Journal of Experimental Child Psychology, 195*, 104846–. doi: 10.1016/j.jecp.2020.104846

Martin, B. N., Fuchs, L. S., Crawford, L., & Smolkowski, K. (2019). The mathematical performance of at-risk first graders as a function of limited English proficiency. *Learning Disability Quarterly, 42*(4), 244–251. doi: 10.1177/0731948719827489

Mazzocco, M. M., Myers, G. F., Lewis, K. E., Hanich, L. B., & Murphy, M. (2013). Limited knowledge of fraction representations differentiates middle school students with mathematics learning disability (dyscalculia) versus low mathematic achievement. *Journal of Experimental Child Psychology, 115*(2), 371–387. doi: 10.1016/j.jecp.2013.01.005

Mazzocco, M., & Räsänen, P. (2013). Contributions of longitudinal studies to evolving definitions and knowledge of developmental dyscalculia. *Trends in Neuroscience and Education, 2*(2), 65–73. doi: 10.1016/j.tine.2013.05.001

Merz, E. C., & Noble, K. G. (2017). Neural development in context: Differences in neural structure and function associated with adverse childhood experiences. In E.Votruba-Drzal & E. Dearing (Eds.), *The Wiley handbook of early childhood development, programs, practices, and policies* (pp. 135–160). John Wiley & Sons. doi: 10.1002/9781118937334.ch7

Meyer, M. L., Salimpoor, V. N., Wu, S. S., Geary, D. C., & Menon, V. (2010). Differential contribution of specific working memory components to mathematics achievement in 2nd and 3rd graders. *Learning and Individual Differences, 20*(2), 101–109. doi: 10.1016/j.lindif.2009.08.004

Miyake, A., Friedman, N. P., Emerson, M. J., Witzki, A. H., Howerter, A., & Wager, T. D. (2000). The unity and diversity of executive functions and their contributions to complex "frontal lobes" tasks: A latent variable analysis. *Cognitive Psychology, 41*(1), 49–100. doi: 10.1006/cogp.1999.0734

Mix, K. S., & Cheng, Y. L. (2012). The relation between space and math: Developmental and educational implications. In J. B. Benson (Ed.), *Advances in child development and behavior* (Vol. 42, pp. 197–243). Elsevier. doi: 10.1016/b978-0-12-394388-0.00006-x

Mix, K. S., Levine, S. C., Cheng, Y. L., Young, C., Hambrick, D. Z., Ping, R., & Konstantopoulos, S. (2016). Separate but correlated: The latent structure of space and mathematics across development. *Journal of Experimental Psychology: General, 145*(9), 1206–1227. doi: 10.1037/xge0000182

Moll, K., Kunze, S., Neuhoff, N., Bruder, J., & Schulte-Körne, G. (2014). Specific learning disorder: Prevalence and gender differences. *PLoS ONE, 9*, e103537. doi: 10.1371/journal.pone.0103537

Moll, K., Landerl, K., Snowling, M. J., & Schulte-Körne, G. (2019). Understanding comorbidity of learning disorders: Task-dependent estimates of prevalence. *Journal of Child Psychology and Psychiatry, 60*(3), 286–294. doi: 10.1111/jcpp.12965

Morsanyi, K., Bers, B. M. C. W., McCormack, T., & McGourty, J. (2018). The prevalence of specific learning disorder in mathematics and comorbidity with other developmental disorders in primary school-age children. *British Journal of Psychology*, *109*(4), 917–940. doi: 10.1111/bjop.12322

Moschkovich, J. (2018). Talking to learn mathematics with understanding: Supporting academic literacy in mathematics for English learners. In A. L. Bailey, C. A. Maher, & L. C. Wilkinson (Eds.), *Language, literacy, and learning in the stem disciplines: How language counts for English learners* (pp. 13–34). Routledge. doi: 10.4324/9781315269610-2

Murray, E., & Harrison, L. J. (2011). The influence of being ready to learn on children's early school literacy and numeracy achievement. *Educational Psychology*, *31*(5), 529–545. doi: 10.1080/01443410.2011.573771

Murphy, M. M., Mazzocco, M. M. M., Hanich, L. B., & Early, M. C. (2007). Cognitive characteristics of children with mathematics learning disability (MLD) vary as a function of the cutoff criterion used to define MLD. *Journal of Learning Disabilities*, *40*, 458–478. doi: 10.1177/00222194070400050901

National Academies of Sciences, Engineering, and Medicine (2018). *How people learn II: Learners, contexts, and cultures.* The National Academies Press. doi: 10.17226/24783

National Association for the Education of Young Children (2002, 2010). *Early childhood mathematics: Promoting good beginnings. A joint position statement of the NAEYC and NCTM.* www.naeyc.org

National Association for the Education of Young Children (2020). *Developmentally appropriate practice: A position statement of the National Association for the Education of Young Children.* www.naeyc.org/resources/position-statements/dap/contents

National Center for Education Statistics (2019). *The nation's report card: 2019 mathematics and reading assessments.* www.nationsreportcard.gov/highlights/mathematics/2019/

National Council of Teachers of Mathematics (2000). *Principles and standards for school mathematics.* www.nctm.org/Standards-and-Positions/Principles-and-Standards/

National Governors Association Center for Best Practices & Council of Chief State School Officers. (2010). *Common core state standards for mathematics.* www.corestandards.org/Math/

National Joint Committee on Learning Disabilities (2010). *Comprehensive assessment and evaluation of students with learning disabilities.* www.ldonline.org/article/54711/

National Research Council (2000). *From neurons to neighborhoods: The science of early childhood development.* The National Academies Press. doi: 10.17226/9824

National Research Council (2012). *From neurons to neighborhoods: An update: Workshop summary.* The National Academies Press. doi: 10.17226/13119

National Research Council (2015). *Transforming the workforce for children birth through age 8: A unifying foundation.* The National Academies Press. doi: 10.17226/19401

Nelson, G. & Powell, S. R. (2018). A systematic review of longitudinal studies of mathematics difficulty. *Journal of Learning Disabilities*, *51*(6), 523–539. doi:10.1177/0022219417714773

Ng, S. S. N., & Rao, N. (2010). Chinese number words, culture, and mathematics learning. *Review of Educational Research*, *80*(2), 180–206. doi: 10.3102/0034654310364764

Nguyen, T., Watts, T. W., Duncan, G. J., Clements, D. H., Sarama, J. S., Wolfe, C., & Spitler, M. E. (2016). Which preschool mathematics competencies are most predictive of fifth grade achievement? *Early Childhood Research Quarterly*, *36*, 550–560. doi: 10.1016/j.ecresq.2016.02.003

Odic, D. (2018). Children's intuitive sense of number develops independently of their perception of area, density, length, and time. *Developmental Science*, *21*(2), p. e12533doi: 10.1111/desc.12533

O'Rear, C. D., & McNeil, N. M. (2019). Improved set-size labeling mediates the effect of a counting intervention on children's understanding of cardinality. *Developmental Science*, *22*(6), e12819. doi: 10.1111/desc.12819

Otero, T. M., & Barker, L. A. (2014). The frontal lobes and executive functioning. In S. Goldstein & J. A. Naglieri (Eds.), *Handbook of executive functioning* (pp. 29–44). Springer.

Pagani, L. S., Fitzpatrick, C., Archambault, I., & Janosz, M. (2010). School readiness and later achievement: A French-Canadian replication and extension. *Developmental Psychology*, *46*(5), 984–994. doi: 10.1037/a0018881

Peng, P., Lin, X., Ünal, Z. E., Lee, K., Namkung, J., Chow, J., & Sales, A. (2020). Examining the mutual relations between language and mathematics: A meta-analysis. *Psychological Bulletin*, *146*(7), 595–634. doi: 10.1037/bul0000231

Peng, P., Wang, C., & Namkung, J. (2018). Understanding the cognition related to mathematics difficulties: A meta-analysis on the cognitive deficit profiles and the bottleneck theory. *Review of Educational Research, 88*(3), 434–476. doi: 10.3102/0034654317753350

Piaget, J. (1952). *The origins of intelligence in children.* International Universities Press.

Pieters, S., Roeyers, H., Rosseel, Y., Van Waelvelde, H., & Desoete, A. (2015). Identifying subtypes among children with developmental coordination disorder and mathematical learning disabilities, using model-based clustering. *Journal of Learning Disabilities, 48*(1), 83–95. doi: 10.1177/0022219413491288

Pullen, P. C., van Dijk, W., Gonsalves, V. E., Lane, H. B., & Ashworth, K. E. (2019). Response to intervention and multi-tiered systems of support: How do they differ and how are they the same, if at all? In P. C. Pullen & M. J. Kennedy (Eds.), *Handbook of response to intervention and multi-tiered systems of support* (pp. 5–10). Routledge. doi: 10.4324/9780203102954-2

Purpura, D. J., Baroody, A. J., & Lonigan, C. J. (2013). The transition from informal to formal mathematics knowledge: Mediation by numeral knowledge. *Journal of Educational Psychology, 105*(2), 453–464. doi: 10.1037/a0031753

Purpura, D. J., Logan, J. A. R., Hassinger-Das, B., & Napoli, A. R. (2017). Why do early mathematics skills predict later reading? The role of mathematical language. *Developmental Psychology, 53*(9), 1633–1642. doi: 10.1037/dev0000375

Purpura, D. J., & Lonigan, C. J. (2013). Informal numeracy skills: The structure and relations among numbering, relations, and arithmetic operations in preschool. *American Educational Research Journal, 50*(1), 178–209. doi: 10.3102/0002831212465332

Purpura, D. J., Schmitt, S. A., & Ganley, C. M. (2017). Foundations of mathematics and literacy: The role of executive functioning components. *Journal of Experimental Child Psychology, 153*, 15–34. doi: 10.1016/j.jecp.2016.08.010

Raddatz, J., Kuhn, J., Holling, H., Moll, K., & Dobel, C. (2017). Comorbidity of arithmetic and reading disorder: Basic number processing and calculation in children with learning impairments. *Journal of Learning Disabilities, 50*(3), 298–308. doi: 10.1177/0022219415620899

Räsänen, P., Haase, V. G., & Fritz, A. (2019). Challenges and future perspectives. In A. Fritz, V. G. Haase, & P. Räsänen (Eds.). *International handbook of mathematical learning difficulties: From the laboratory to the classroom* (pp. 799–827). Springer. doi: 10.1007/978-3-319-97148-3

Rittle-Johnson, B., Fyfe, E. R., Hoffer, K. G., & Farran, D. C. (2017). Early math trajectories: Low-income children's mathematics knowledge from ages 4 to 11. *Child Development, 88*(5), 1727–1742. doi: 10.1111/cdev.12662

Rittle-Johnson, B., Zippert, E. L., & Boice, K. L. (2019). The roles of patterning and spatial skills in early mathematics development. *Early Childhood Research Quarterly, 46*, 166–178. doi: 10.1016/j.ecresq.2018.03.006

Russell, R. L., & Ginsburg, H. P. (1984). Cognitive analysis of children's mathematics difficulties. *Cognition and Instruction, 1*(2), 217–244. doi: 10.1207/s1532690xci0102_3

Sarama, J., & Clements, D. H. (2009). *Early childhood mathematics education research: Learning trajectories for young children.* Routledge. doi: 10.4324/9780203883785-10

Sarama, J., Clements, D. H., Barrett, J., Van Dine, D. W., & McDonel, J. S. (2011). Evaluation of a learning trajectory for length in the early years. *ZDM, 43*(5), 667–680. doi: 10.1007/s11858-011-0326-5

Seo, K., & Ginsburg, H. P. (2003). What is developmentally appropriate in early childhood mathematics education: Lessons from new research. In D. H. Clements & J. Sarama (Eds.), *Engaging young children in mathematics: Standards for early childhood mathematics education* (pp. 91–104). Erlbaum. doi: 10.4324/9781410609236

Shalev, R. S. (2007). Prevalence of developmental dyscalculia. In D. B. Berch & M. M. M. Mazzocco (Eds.), *Why is math so hard for some children? The nature and origins of mathematical learning difficulties and disabilities* (pp. 49–60). Paul H. Brookes.

Shanley, L., Clarke, B., Doabler, C., Kurtz-Nelson, E., & Fien, H. (2017). Early number skills gains and mathematics achievement: Intervening to establish successful early mathematics trajectories. *The Journal of Special Education, 51*(3), 177–188. doi: 10.1177/0022466917720455

Simanowski, S., & Krajewski, K. (2019). Specific preschool executive functions predict unique aspects of mathematics development: A 3-year longitudinal study. *Child Development, 90*(2), 544–561. doi: 10.1111/cdev.12909

Society for Neuroscience (2018). *Brain facts: A primer of the brain and nervous system* (8th ed.). www.brainfacts.org/the-brain-facts-book

Sophian, C. (2013). Mathematics for early childhood education. In O. N. Saracho & B. Spodek (Eds.), *Handbook of research on the education of young children* (3rd ed., pp. 169–178). Routledge. doi: 10.4324/9780203841198

Stock, P., Desoete, A., & Roeyers, H. (2010). Detecting children with arithmetic disabilities from kindergarten: Evidence from a 3-year longitudinal study on the role of preparatory arithmetic abilities. *Journal of Learning Disabilities, 43*(3), 250–268. doi: 10.1177/0022219409345011

Swanson, H. L. (2012). Cognitive profile of adolescents with math disabilities: Are the profiles different from those with reading disabilities? *Child Neuropsychology, 18*(2), 125–143. doi: 10.1080/09297049.2011.589377

Swanson, H., Olide, A., & Kong, J. (2018). Latent class analysis of children with math difficulties and/or math learning disabilities: Are there cognitive differences? *Journal of Educational Psychology, 110*(7), 931–951. doi: 10.1037/edu0000252

Toll, S. W. M., & Van Luit, J. E. H. (2014). The developmental relationship between language and low early numeracy skills throughout kindergarten. *Exceptional Children, 81*(1), 64–78. doi: 10.1177/0014402914532233

Träff, U., Olsson, L., Skagerlund, K., & Östergren, R. (2020). Kindergarten domain-specific and domain-general cognitive precursors of hierarchical mathematical development: A longitudinal study. *Journal of Educational Psychology, 112*(1), 93–109. doi: 10.1037/edu0000369

Watts, T., Duncan, G., Chen, M., Claessens, A., Davis-Kean, P., Duckworth, K., Engel, M., Siegler, R., & Susperreguy, M. (2015). The role of mediators in the development of longitudinal mathematics achievement associations. *Child Development, 86*(6), 1892–1907. doi: 10.1111/cdev.12416

Watts, T., Duncan, G., Siegler, R., & Davis-Kean, P. (2014). What's past is prologue: Relations between early mathematics knowledge and high school achievement. *Educational Researcher, 43*(7), 352–360. doi: 10.3102/0013189X14553660

Westendorp, M., Hartman, E., Houwen, S., Smith, J., & Visscher, C. (2011). The relationship between gross motor skills and academic achievement in children with learning disabilities. *Research in Developmental Disabilities, 32*(6), 2773–2779. doi: 10.1016/j.ridd.2011.05.032

Willcutt, E., McGrath, L., Pennington, B., Keenan, J., DeFries, J., Olson, R., & Wadsworth, S. (2019). Understanding comorbidity between specific learning disabilities. *New Directions for Child and Adolescent Development, 165*, 91–109. doi: 10.1002/cad.20291

Willoughby, M. T., Wylie, A. C., & Little, M. H. (2019). Testing longitudinal associations between executive function and academic achievement. *Developmental Psychology, 55*(4), 767–779. doi: 10.1037/dev0000664

Xenidou-Dervou, I., Van Luit, J. E. H., Kroesbergen, E. H., Friso-van den Bos, I., Jonkman, L. M., van der Schoot, M., & van Lieshout, E. C. D. M. (2018). Cognitive predictors of children's development in mathematics achievement: A latent growth modeling approach. *Developmental Science, 21*(6), e12671. doi: 10.1111/desc.12671

Xu, F., & Denison, S. (2009). Statistical inference and sensitivity to sampling in 11-month-old infants. *Cognition, 112*(1), 97. doi: 10.1016/j.cognition.2009.04.006

Zatorre, R. J., Fields, R. D., & Johnsen-Berg, H. (2012). Plasticity in gray and white: Neuroimaging changes in brain structure during learning. *Nature Neuroscience, 15*(4), 528–536. doi: 10.1038/nn.3045

Zhang, X., Räsänen, P., Koponen, T., Aunola, K., Lerkkanen, M. K., & Nurmi, J. E. (2020). Early cognitive precursors of children's mathematics learning disability and persistent low achievement: A 5-year longitudinal study. *Child Development, 91*(1), 7–27. doi: 10.1111/cdev.13123

3 Mathematics Assessment

Chapter Questions

1. What are the purposes of mathematics assessment?
2. How is the assessment cycle implemented in classrooms?
3. How can universal screening improve mathematics achievement for all students?
4. What are the forms and purposes for diagnostic assessment?
5. How should other assessment tools be used to inform instruction?
6. What factors should educators consider when making decisions about test accommodations?
7. What assessment practices are common within a multi-tiered systems of support framework?
8. How can educators collaborate to implement assessment systems that will improve instruction?

Jose Martinez, a 2nd grade teacher at Pinetops Elementary School, and Chris Johnson, the mathematics support teacher, are meeting before school begins for the year.

JOHNSON: Can you tell me a little about the assessment in your classroom for mathematics? How do you determine students' levels of instruction and progress in learning?

MARTINEZ: Well, during the first week of school each year I administer the textbook's accompanying assessment. Since there's no assessment in math for first grade, there's nothing else to go on. The textbook test includes items in place value, computation, measurement, and geometry, the first-grade standards. Then I plan instruction by units. Units for second grade start with number and place value and then go on to computation, data, measurement, geometry, and then some early algebra. Each unit has a pretest and posttest.

JOHNSON: What do you do for those students who have not mastered the first-grade standards?

MARTINEZ: I've been struggling with that and perhaps you can give me some ideas. Last year I tried to do a week of review for all students, but some students were still way behind.

JOHNSON: I could assist by providing some targeted instruction for students with gaps just prior to each unit of study if you can help me identify the gaps and outline the unit prerequisites. I also have some additional screening tools for those students you are concerned about.

MARTINEZ: Those are great ideas! I'll give you the list of students as soon as I've finished the pretest. Would you want to organize the targeted instruction during a special pull-out time or come into the classroom and work with students here?

JOHNSON: Let's start with a pull-out session about three times a week and see how the students respond to that.

DOI: 10.4324/9781003096733-3

MARTINEZ: OK, then my chapter tests will also indicate students who have not mastered material to the level necessary for the next concepts. Do I move ahead or hold everyone back while I reteach?

JOHNSON: That's a critical point. You could provide extension or application activities in the same content area for those who have met mastery while we can jointly plan some reteaching with other students. Or I could come in prior to the chapter tests and work with students in the class you've identified as needing extra assistance. Those who still don't meet mastery could spend extra time with my pull-out program.

MARTINEZ: Let's try the extension and application approach first because I believe those students need deeper connections made, then we can assess the results.

What assessment activities are described in this conversation? How are they related to mathematics standards and instructional planning? How might assessments in core mathematics classes and support settings differ?

Yogi Berra stated the essence of assessment in one profound sentence: "You can observe a lot by watching" (Berra, 1998, p. 123). Perhaps the most informative and useful assessment is that which is closest to the day-to-day instructional events in a classroom with a teacher watching and listening. However, with the current atmosphere of reform and accountability, the assessment demands on teachers and students are not so simple. This chapter examines the purposes of assessment in today's schools, mathematics assessment standards, specific methods for mathematics assessment of students with learning difficulties, and the assessment components of school-wide, multi-tiered systems of assessment and intervention.

Purposes of Mathematics Assessment

Students in elementary and secondary schools today are subjected to more assessments than at any time in the history of public education. High-stakes assessments are those that have serious implications for districts, schools, teachers, and students. The Elementary and Secondary Education Act of 2001 (No Child Left Behind Act) and its 2015 reauthorization (Every Student Succeeds Act) required annual statewide assessment of mathematics (and English/language arts) in Grades 3 through 8 and once in high school. These assessments may be summative (at the end of the school year) or formative, to inform instruction during the year; but they must be aligned with each State's academic standards and result in individual student summative reports (Council of Chief State School Officers, 2016). Many states and districts require additional assessments for various purposes, although there is growing concern about the time devoted to assessments reducing instructional time.

Purposes of high-stakes assessments range from grading individual schools to evaluating teachers and determining student promotions, so these measures weigh heavily on educators, students, and parents. However, since these assessments are more formal and the results might not be available for months, they may not be useful for planning instruction. High-stakes assessments have also promoted a "teach to the test" mentality in many schools, reducing quality instructional time for the broader curriculum. Parents and educators have pushed back on excessive formal testing in their schools in favor of more informative classroom assessments (Asp, 2018). There is evidence that formative assessment with learning supports is correlated with greater mathematics achievement (Wong et al., 2018), especially when teachers have training in using formative assessment for improving learning (Andersson & Palm, 2017). Classroom-based assessments that inform teachers, students, and parents about students' progress within the curriculum include various screening measures, informal assessments during instruction, and brief progress-monitoring measures. These assessments are more closely related to instruction with the purpose of improving learning.

The Joint Committee on Standards for Educational Evaluation developed the *Classroom Assessment Standards for PreK-12 Teachers* to provide teachers with research-based principles and guidelines for effective assessment of student learning (Klinger et al., 2015). The 16 standards are organized by domains: foundations, use, and quality. Under foundations, the first standard addresses the purposes of assessment: "Classroom assessment practices should have a clear purpose that supports teaching and learning" (F1). The second standard emphasizes the role of learning expectations (goals) as forming the foundation for aligning assessment practices with instruction.

The National Council of Teachers of Mathematics document *Assessment Standards for School Mathematics* outlined four purposes of assessment: monitor student progress, make instructional decisions, evaluate student achievement, and evaluate programs (NCTM, 1995). According to this document, assessment should:

- reflect the mathematics that all students need to know, be able to do;
- enhance mathematics learning;
- promote equity;
- be an open process;
- promote valid inferences about mathematics learning; and
- be a coherent process.

The Assessment Cycle

All mathematics assessments, according to the 1995 NCTM document, involve four interrelated phases: plan assessment, gather evidence, interpret evidence, and use results. This cycle is similar to many evidence-based assessment systems within schools, businesses, and other organizations. Consider a mathematics example while reviewing the phases depicted in Figure 3.1, a fifth-grade unit on solving problems involving addition and subtraction of fractions.

Planning assessment first requires teachers to translate curriculum standards into classroom and individual goals relevant for their students. Learning goals include aspects of

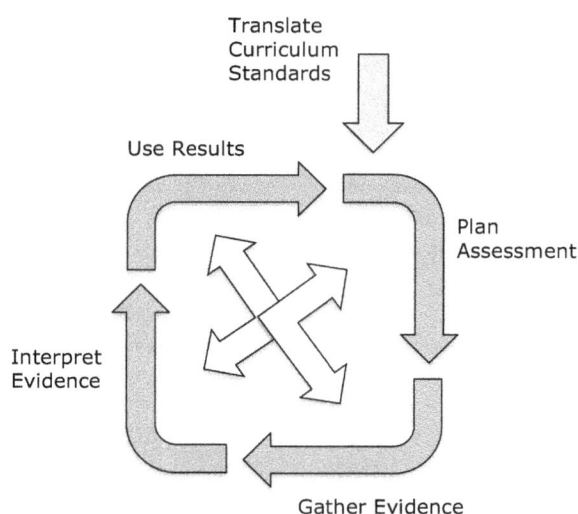

Figure 3.1 Assessment Cycle

factual information, conceptual understanding, skills, problem solving, applications, processes, and mathematical dispositions. These goals should be clear and specific for instructional and assessment purposes. For example, in the *Common Core State Standards for Mathematics* (CCSSM, NGA & CCSSO, 2010), two goals for fifth graders are: "Add and subtract fractions with unlike denominators by replacing given fractions with equivalent fractions" and "Solve word problems involving addition and subtraction of fractions referring to the same whole, including cases of unlike denominators" (p. 36). It is important that teachers consult district curriculum standards when setting learning goals, because grade-level goals may vary from those of textbooks and other materials.

For this example, the teacher knows that students should have already developed basic fraction concepts, equivalence understanding, and fluency with addition and subtraction of fractions with like denominators, but some students may have gaps. A pre-unit assessment would include items about these prerequisite concepts as well as items related to the goals of the unit. Other assessments for the unit should be planned as part of overall instructional planning for the unit. Teachers in most districts have a great deal of professional discretion about informal assessments that inform instruction. For these primarily teacher-created assessments, the teacher should consider: frequency (e.g., daily, once a week), form (e.g., interview, paper and pencil, performance task), weight (e.g., graded or ungraded), resources (e.g., time and materials), and previous experiences of students with assessments. In addition to the pre-unit assessment, the teacher in this example is planning brief checks of student understanding during each lesson (oral and written forms) and a post-unit assessment. An alternative assessment model for this unit would be to use different versions of the same outcome measure at specific intervals over the course of the unit. This progress-monitoring approach, called curriculum-based measurement, is discussed in a later section of this chapter.

Next, teachers should implement the assessment plan and *gather evidence* for individual students on a schedule that will be frequent enough to impact instruction but not take excessive amounts of instructional time. Assessment that is embedded within instruction best meets these criteria. The pre-unit assessment may indicate that some students require a review or reteaching of prerequisite skills and concepts. Other students may be very confident and fluent with equivalent fractions and adding and subtracting with like denominators and are already demonstrating concept understanding for operations with unlike denominators. These students will need more advanced goals for this unit of study, related to the same topics. Assessment assists the teacher in differentiating instruction to meet individual student needs. Assessment during the unit, embedded within lesson activities, should document individual student progress toward their goals. For example, a quick check of three fraction problems concludes one lesson. Two word problems for students to solve independently are embedded in another. The teacher asks students to explain their processes for the word problems to gain even more insight. The post-unit or summative assessment should confirm students' mastery of concepts and skills and readiness for the next unit of study.

The *interpretation* stage of the assessment cycle reminds teachers of validity, quality, and ethical issues. Can valid interpretations be made with the types of collection methods used? Does the evidence from a variety of sources converge to support inferences? Are assessment results useful for planning instruction? Do students and parents also understand results? The teacher in our example should consider whether the skill and concept domain of addition and subtraction of fractions was sampled adequately and whether any assessments were unclear or not reflective of student understanding because of language, format, item selection, or other factors. For example, were word problems within the post-unit assessment a sampling of whole number, fractions with like denominators, and fractions with unlike denominators so that students could demonstrate strong concept understanding on a range of applications? During this instructional unit, the teacher decided to add a word-problem checklist to use with the brief student interviews. The checklist served as a guide for checking concepts and skills as well as a data collection method. This decision illustrates the nonlinear nature of the assessment cycle.

Using assessment results to make instructional decisions also has multiple purposes. Teachers can examine the effects of instruction and the learning environment on student knowledge and skills. Individual student needs can be pinpointed and addressed. Students can view their own progress toward goals. *Formative* types of assessment, such as day-to-day monitoring of student performance to inform instructional decisions, and *summative* assessments, such as the unit posttest, should be linked to provide a coherent and integrated decision base. For the fractions with unlike denominators example, the pre-unit assessment informed the teacher of necessary remedial instruction or the need for individual student goals. The formative assessments within the unit informed the teacher whether a lesson's objectives were achieved or whether reteaching with an alternative approach should be in order. The post-unit assessment provided information for future instructional units and skill maintenance. Most importantly, students who could not solve these types of problems at the beginning of the unit could see that they now have new skills and concept understanding.

TRY THIS

Work through the assessment cycle with another example from mathematics curriculum standards.

Teachers should consider the real purpose of each assessment or evaluation activity in their classrooms. For those mandated assessment activities, how well is the curriculum aligned with assessment instruments? Can the results be useful in other ways? How much decision-making authority do teachers have for other assessments? For grading purposes, what measures will be fair to students? Do students understand which assessments are formative and which are summative? For instructional assessments, what methods will prove to be informative yet efficient? These questions prompt ongoing professional decisions.

Types of Mathematics Assessments

Mathematics assessments vary widely, depending on their purpose, and teachers should be adept in creating and selecting a range of assessments, administering and interpreting those within professional boundaries and expectations, and using the results for instructional planning. Mathematics differs from reading in that there is a broader range of concepts to be learned. The primary purpose of mathematics assessment should be to improve student achievement across domains. This section provides an examination of three major types of assessments for mathematics: 1) diagnostic assessment for the purpose of identifying student strengths and learning needs, also used within a comprehensive assessment for eligibility decisions for special education; 2) screening measures that are often used for early intervention and within multi-tiered systems of support (MTSS) frameworks; and 3) curriculum-based measurements (CBM), also used frequently within MTSS frameworks. The sections that follow examine other assessment strategies and issues for students with mathematics difficulties.

Diagnostic Assessment

Using terminology from the medical world, diagnostic assessment refers to a range of assessment tasks to determine students' strengths and weaknesses, learning abilities, and levels of knowledge, skills, and concept understanding prior to instruction. In a study of ten occupations where practitioners used diagnostic assessment in their work (from auto mechanic to physician and teacher), Alderson et al. (2015) concluded that diagnostic assessment should be situated within the range of other classroom assessment practices and the emphasis should be on strengths as well as weaknesses. The authors proposed five principles for the use of diagnostic assessment: sufficient teacher training; well-designed, valid instruments; integration of self-assessments; assessments embedded within classroom practices; and evaluation of recommended interventions. For mathematics instruction, diagnostic assessment can be used with individual students to discover specific strengths and weaknesses, learning gaps, misconceptions, and dispositions about mathematics. It can be used with groups of students prior to learning, as with a unit pretest. Diagnostic assessment is also used within the required comprehensive assessment for special education eligibility under the Individuals with Disabilities Education Improvement Act (IDEA, 2004):

> (2) In conducting the evaluation [for eligibility determination], the local educational agency shall—
>> (A) use a variety of assessment tools and strategies to gather relevant functional, developmental, and academic information, including information provided by the parent, that may assist in determining—
>>> (i) whether the child is a child with a disability; and
>>> (ii) the content of the child's individualized education program, including information related to enabling the child to be involved in and progress in the general education curriculum, or, for preschool children, to participate in appropriate activities; ….
> (3) … shall ensure that (C) assessment tools and strategies that provide relevant information that directly assists persons in determining the educational needs of the child are provided.
>
> (IDEA, 2004, 20 USC 14 § 1414 (b))

While the reauthorization of IDEA in 2004 allowed evaluation teams considering specific learning disabilities (SLD) to "use a process that determines if the child responds to scientific, research-based intervention as a part of the evaluation procedures" (20 USC § 1414 (b)(6)(B)), sometimes called responsiveness-to-intervention (RTI) within a MTSS model, the eligibility evaluation for SLD must still be comprehensive and include a range of assessments.

Formal Diagnostic Assessment

Diagnostic assessments for special education eligibility or classroom use may include formal and informal measures. Formal measures are those that are standardized (require a specific method of administration or protocol), have been field tested with a wide range of students, and offer technical evidence of quality, such as reliability and validity data. Example instruments for mathematics are depicted in Table 3.1, with notes about their features. The information in Table 3.1 represents only a sampling, not a comprehensive list of diagnostic instruments for mathematics.

Formal diagnostic measures offer many advantages for educators. They provide normative data that allow the comparison of individual students with their age- and grade-level peers. Many have been aligned with national curriculum standards, and therefore sample skills and concept understanding from the range of mathematics domains. These devices have been designed with floors and ceilings (places to begin and end the assessment session), in order to make broad assessment more efficient and less frustrating for students. Finally, they offer

Table 3.1 Selected Examples of Formal Diagnostic Measures

Instrument	Levels Time Forms	Content Areas	Response Mode	Purpose
Key Math™-3 Diagnostic Assessment, 3rd Ed.[1]	K-12 30–90 min two forms	10 subtests on concepts, operations, applications	Oral and written	Diagnosis, progress monitoring
Comprehensive Mathematics Abilities Test (CMAT)[2]	7–19 yrs 30–60 min (core) one form	Six core and six supplemental subtests, across curriculum	Oral and written	Screening, diagnosis
Test of Early Mathematics Ability, 3rd Ed. (TEMA-3)[3]	3 – 8 yrs 40 min two forms	Informal and formal number concepts & skills: comparison, literacy, facts, calculations, concepts	Oral, written, and modeling	Screening, diagnosis, progress monitoring
Woodcock-Johnson ® IV Tests of Achievement[4]	2 – 90 yrs 5–10 min per subtest three forms	Math subtests: applied problems, calculation, math facts fluency	Written	Screening, diagnosis, progress monitoring

Notes: [1]Connolly, 2007; [2]Hresko et al., 2003; [3]Ginsburg & Baroody, 2003; [4]Schrank et al., 2014.

good insights into general areas of strength and weakness, helping educators narrow down their focus of concern. However, since the instruments are broad, they may not be specific enough for planning instruction. Their protocols do not allow administrators to follow items with probes or additional questions. Most of these assessments have closed-end designs, not allowing for alternative responses by students. The assessment packages can be expensive for schools to acquire and keep current.

Formal diagnostic assessments can be followed with more specific informal probes or curriculum-based measures (formal or informal). These follow-up assessments may be needed to answer additional questions about specific student skills and concept understanding or to provide information about developing an individualized education program (IEP) and beginning interventions.

Informal Diagnostic Assessment

The assessment a teacher administers to a class at the beginning of the school year or for a newly enrolled student may be informal diagnostic assessments if those are used to identify strengths and gaps in learning. The pre-unit assessment that includes prerequisite concepts and skills is diagnostic. The assessment the teacher administers as part of the eligibility evaluation battery may be a teacher-developed, informal diagnostic assessment because it helps to answer questions about learning patterns. When informal assessments for mathematics are truly diagnostic, they typically have three phases: broad survey (or screening), specific probe, and diagnostic interview.

A broad survey includes a sampling of content within a domain, or, even broader, for grade-level goals in mathematics. The survey could include items along a trajectory for a content area within mathematics, across grade levels. In this case, the teacher is interested in pinpointing where concept understanding and skill development faltered. Figure 3.2 illustrates a broad survey for place-value and addition concepts following a trajectory from first- through second-grade concepts. Many curriculum-based measurement (CBM) instruments are broad enough to be considered surveys of curriculum domains, or broad skill and concept areas within domains.

Teacher directions for the survey (pause and wait after each):
Please write these numbers on your paper.
a. 128   b. 306
For (c) and (d), write the numbers in expanded form.
Put a greater than, equal to, or less than symbol between the pairs in (e) and (f).
For (g) and (h) write a number that will be ten more than the number on your paper.
For (i) and (j) find the sums.

Student work:

a.   128   b. 306

Write in expanded form:

c. 419   400 + 10 + 9

d. 205   200 + 0 + 5

Place <, =, or > between the numbers:

e. 349 < 352   f. 780 > 769

Write the number that is 10 more than:

g. 589   599   h. 392   412

Find the sums:

i.   29   j.   258
     + 15   + 317
      3 14   5615

Score: 7/10

Figure 3.2 Broad Survey of Place Value and Addition

A specific probe typically follows other assessments that raised additional questions about student learning. A probe includes a narrow range of skills and/or concepts and is very brief. Sometimes a probe assesses one instructional objective after a lesson or a range of skills within a specific unit. Probes can be administered to individuals, small groups, or the entire class, therefore there is great flexibility in conducting these brief, informative measures. A probe that might follow the broad survey in Figure 3.2 can be found in Figure 3.3. The teacher was concerned that he did not have enough information about the errors the student made on the survey, so the probe addressed applying place-value concepts within addition.

The diagnostic interview, also termed clinical or student interview, is a type of probe, but is usually conducted individually using student work, such as a performance task during a lesson or the results of a probe. The teacher asks the student to explain a math performance, such as work on an algorithm, completed task, or word problem. The interaction between teacher and student is guided by the teacher's open-ended questions. The goal of a diagnostic interview is to make cognitive processes of the student as external as possible so the teacher can observe

Teacher directions for the probe:

For problems (a) through (d), please write the number
that is 10 more than the number you see.
Now find the sums for the rest. Be sure to show your work.

Student work:

a. 39 + 10 = 49

b. 368 + 10 = 378

c. 390 + 10 = 400

d. 398 + 10 = 418

Find the sums:

e.
```
    2 6
  + 4 2
```
68

f.
```
    7 5
  + 3 4
```
109

g.
```
    1 4 8
  +   4 1
```
189

h.
```
    2 3 7
  +   4 5
```
2712

e.
```
    1 6 7
  + 2 1 3
```
3710

f.
```
    5 7 5
  + 6 1 8
```
8 1 4

1 1
___
8 2 5

Figure 3.3 Probe of Addition as Follow-Up to Survey

those through students' performances and explanations (Ginsburg & Dolan, 2011). Lewis and Fisher (2018) described setting up clinical interviews as deliberate diagnostic assessments with carefully selected tasks, materials, questions, and testing location; explaining the purpose to students; and treating the student as the "expert" during the interview. "Can you use a drawing to help me understand how to combine those values?" The interview should reveal strengths and misconceptions or gaps in learning as well as students' cognitive processes and strategies while engaged in mathematics tasks. The interview should lead to appropriate instructional interventions.

Sometimes a rubric is used for diagnostic interviews, as described by Poch et al. (2015) during problem-solving instruction using diagrams. The authors' rubric for interviews included the strands involved—conceptual understanding, procedural fluency, strategic competence, adaptive reasoning, and productive disposition—with places to indicate challenges and misconceptions as well as needed instructional supports. A rubric ensures data collection in an organized manner during the interview. Figure 3.4 illustrates a rubric that accompanies a diagnostic interview about the student's addition probe.

TRY THIS

Create a broad survey for one curriculum area and administer it to a student at the appropriate grade level. After reviewing the results, create and administer a probe for more information. Follow the probe with a diagnostic interview.

Screening Measures

Screening is an assessment approach that has been used for centuries and across disciplines, including for the selection of individuals for the military, as a preliminary step in candidate selection for employment, in college admissions, and in medicine to identify individuals who require additional evaluation. Educators responsible for mathematics instruction and interventions at all levels should be competent with screening tools and strategies. Screening activities are necessary in a wide range of situations within schools. Young children are screened for their readiness for school. Students are screened for their potential for accelerated or intervention programs, including *universal screening* (all students) within MTSS programs. Teachers often use broad screening tools at the beginning of the school year to better understand student instructional needs. Screening is used as an initial step in the diagnostic process, as with the broad surveys discussed in the previous section. Screening is used in secondary schools to determine course placement or to identify areas that require intervention. When individual students demonstrate unexpected learning or behavior issues, screening is often the first assessment approach in order to narrow the range of possible explanations.

Early Grades Screening

Screening for children at risk of later difficulties in mathematics is typically performed within preschool programs, as part of school-entry procedures, and in first grade. Schools with MTSS frameworks often screen all K-2 students three times a year. While it is common in Grades 3 and higher to use preceding-year state or district tests (Gersten et al., 2009; Klingbeil et al., 2019; VanDerHeyden et al., 2017), accurate screening tools for PreK though Grade 2 have been more elusive. One reason is the extremely wide range of environmental contexts and experiences of young children just beginning formal schooling, the quality of their informal learning. Another, as discussed in Chapter 2, is the increasing importance of domain-specific learning that takes place with formal schooling; mathematics-specific achievement becomes more predictive of future achievement (Geary et al., 2017; Lee & Bull, 2016). Young children who are delayed in acquiring their first mathematical concept—cardinal principle—may also have delays in executive functions such as attention and working memory that are important for informal and formal learning (Geary et al., 2018). Further, early screening is more likely to target literacy, language, and social-emotional benchmarks than those of mathematics (Kiss & Christ, 2019). Of the 35 states that required state-wide assessments in Grades K-2 in 2018, only 17 included a mathematics measure (CCSSO, 2019).

The importance of mathematics screening in early childhood settings has been emphasized for nearly two decades. Some early researchers recommended screening a single proficiency in mathematics such as a competency related to number sense (Clarke et al., 2008; Locuniak &

Teacher: Sam, can you tell me how you found the answer to a?
Sam: Is that wrong? I can change it.
Teacher: No, it's not wrong. I just want to find out more about your thinking. Most of your answers are right.
Sam: OK, I just looked at the 3 in the tens place and added one, to make a 4.
Teacher: Is that what you did for b, c, and d?
Sam: Yes, except in c I tried to add one to the nine and couldn't, so I just counted from 390 up to 400.
Teacher: What about d?
Sam: I knew I had to be in the 400s so I just changed the 9 to a 10 and added 8 and that made 18.
Teacher: What if you started with four hundred eighteen and subtracted 10 from that, what would you have?
Sam: Oh, there would be a zero in the tens place! It would go down to four hundred and the 8 would still be there in the ones.
Teacher: Good thinking, now tell me about how you worked e and f.
Sam: For e I first said two plus four is six and wrote that under, then I said six plus two is eight and wrote that under. I did the same for f.
Teacher: OK, tell me how you did h.
Sam: Well, I brought down the two because there's nothing under, then I plussed the three and the four and wrote down seven. Then I said seven plus five is twelve and wrote that down. I think I was supposed to write them down below the others, but I'm not sure.
Teacher: Is that what you did in f?
Sam: Yes, is that the right way?
Teacher: Well, it's almost the right answer. What method were you trying to use?
Sam: It's the one the teacher taught us last year, she called it left-to-right, but I forget where to put my numbers when I get these long problems.
Teacher: Now that I know what you're trying to do I can show you that method again. In fact, I'm going to show you two methods and let you pick which one makes the most sense for you. Then we'll practice that one to make sure you understand it.

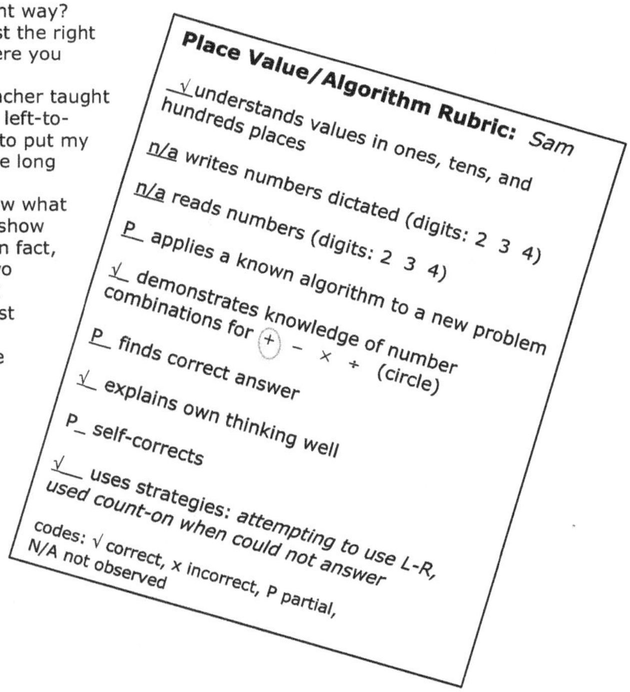

Place Value/Algorithm Rubric: Sam

√ understands values in ones, tens, and hundreds places

n/a writes numbers dictated (digits: 2 3 4)

n/a reads numbers (digits: 2 3 4)

P applies a known algorithm to a new problem

√ demonstrates knowledge of number combinations for (+) − × ÷ (circle)

P finds correct answer

√ explains own thinking well

P self-corrects

√ uses strategies: attempting to use L-R, used count-on when could not answer

codes: √ correct, x incorrect, P partial, N/A not observed

Figure 3.4 Diagnostic Interview with Rubric

Jordan, 2008), while others argued for multiple-proficiency approaches that included several aspects of number competence (Seethaler & Fuchs, 2010). Still others argued for a screening battery to include curriculum sampling beyond number concepts as well as cognitive or reasoning measures (Kroesbergen et al., 2012; Passolunghi & Lanfranchi, 2012). More recently, with attention on efficiency and efficacy, screening protocols have evolved to include gated assessments and brief evaluative interventions to more accurately identify students at risk.

Single-proficiency measures, as described by Clarke et al. (2014), are typically very quick measures (one to two minutes) of single number tasks, such as magnitude comparison, strategic counting, and basic-fact retrieval. Most of these measures have predictive validity ranging from 0.35 to 0.79 (very useful), with first-grade results stronger than those of kindergarten (Gersten et al., 2012). An example of a task with strong predictive validity for first graders on a later achievement test is "name the larger of two items, given number pairs with values between 0 and 99" (Clarke et al., 2011). The advantages of single-proficiency measures include efficiency and economy in identifying a subgroup of students who may be at risk for mathematics learning difficulties. But these measures alone will not yield enough information for planning interventions and yield high numbers of false positives (overidentification of children for intervention) and false negatives (failing to identify children who need intervention). Additionally, timed measures are essentially a measure of fluency, and fluency is highly correlated with nonmathematical tasks such as word naming (Polignano & Hojnoski, 2012).

Those instruments targeting several key proficiencies with young children typically focus on four primary areas as predictors of later mathematics achievement: magnitude comparison (comparing the relative magnitudes of numbers), strategic counting (using counting strategies such as *count on*), oral word problems involving simple operations, and retrieval of basic addition and subtraction number combinations (Gersten et al., 2012). Specific one-minute performance tasks that have demonstrated the greatest technical adequacy for early numeracy proficiency (K-1) are counting out loud, quantity discrimination, number identification, missing number, next number, and number facts. These components of many early grades screening tools were found to have few floor effects, no ceiling effects, and discrimination ability, confirming that a small number of items can provide information about which students are struggling the most (Lee et al., 2012). Developers of measures of multiple proficiencies argue that these number tasks are closer to mathematics content and learning trajectories than general cognitive-ability measures (Jordan et al., 2009). However, most of these multiskill measures are limited to number concepts.

Other researchers have pointed to additional research-based predictors of mathematics achievement that should be included in screening batteries, such as the domain-general, cognitive abilities of working memory, attention, processing speed, and language skills (Passolunghi & Lanfranchi, 2012; Powell et al., 2017). Working-memory measures tend to add precision to screening measures and teacher ratings of attentiveness can be strong predictors of achievement. In a study of children across two years of preschool (ages 3 to 5), Geary et al. (2018) found that children at risk for long-term mathematics difficulties could be identified before age four. They recommended an assessment battery that includes cardinal knowledge, count list, enumeration, letter (or numeral) recognition, and executive-function measures. The researchers also found that parental engagement with children in numeracy-related activities is important, especially for preschool children. Not having that involvement would be another risk factor for young children. There is disagreement, however, about the addition of cognitive measures to early screening protocols. Clarke et al. (2018) argued that these measures take more time and other resources to administer for little contribution to variance—as little as 1.1% for kindergarten children. Others suggested that measures of working memory, attention, and language abilities are important to include in batteries for PreK to Grade 2 children because deficits in these areas place children at increased risk for difficulties in acquiring early number skills and later mathematics achievement (Powell et al., 2017; Raghubar & Barnes, 2017).

Early grades screening measures for mathematics rarely address spatial abilities such as comparing sizes or reorienting objects. Number-line tasks have been added to some early screening measures as those assess both number knowledge and spatial aspects of number relations. For example, Clarke et al. (2020) added a 0 to 100 number-line task to the individually administered *Assessing Student Proficiency of Number Sense* (ASPENS, Clarke et al., 2011) and found an additional 7% variance for kindergarteners and 13% for first graders. Measures of number magnitude using a number line address certain aspects of spatial reasoning but do not consider others that are predictive of later mathematics achievement such as spatial visualization (Cheng & Mix, 2014; Uttal et al., 2013). There is some evidence that spatial skills of three-year-old children are predictive of their mathematics skills at age five (Verdine et al., 2017). Spatial abilities in kindergarten are predictive of mathematics performance in second grade (Frick, 2018). Specifically, mental rotation (visualizing an object in a different orientation) and spatial scaling (comparing spaces of different sizes) tasks accounted for 24%–43% variance in mathematics performance of second graders. Only a few early screening tools include robust spatial tasks, including the *Research-based Early Maths Assessments* (REMA-Brief, REMA-Short, REMA-Full; Clements et al., 2008) and *Tools for Early Assessment in Mathematics* (TEAM, Clements & Sarama, 2011/2016).

School-based assessment teams may decide to use the more efficient screening (5–15 min) of key proficiencies for universal screening in Grades K and 1 and follow up with more comprehensive screening batteries with those students in borderline areas (near cut score) or other risk factors. Sometimes called a multiple-gating procedure, a multiple-stage screening procedure could reduce time taken away from instruction and increase the numbers of students appropriately identified for more intensive intervention (Walker et al., 2014). A brief screening could be followed by broad content assessment, a comprehensive screening battery including cognitive measures, or even a period of intervention as a diagnostic tool. Geary et al. (2018) recommended assessing quantitative knowledge at the beginning of preschool and the *gains* in that knowledge throughout preschool as a better assessment for school readiness. Purpura et al. (2015) found that the *Preschool Early Numeracy Skills-Brief* (PENS-B) was highly correlated with a formal diagnostic test of numeracy skills (TEMA-3, Ginsburg & Baroody, 2003) and was sensitive to growth in young children, but still resulted in 4% false negatives and 14% false positives. The researchers recommended following the brief, five-minute screener with a period of instruction with frequent progress monitoring to truly identify the children most in need of intervention. Other researchers recommend using *dynamic assessment* as a second gate to improve screening accuracy (Seethaler et al., 2012). Dynamic assessment measures learning potential, a process of providing instruction for a *new* skill and measuring the student's response to that instruction and is discussed in more depth later in this chapter. VanDerHeyden et al. (2019) investigated the use of multiple subskill mastery measures twice a year (fall and midpoint) in Grades K, 1, 3, and 5 for screening followed by class-wide interventions. The subskill screening measures yielded a high number of false positives, as is typical. Adding the intervention not only increased the equity and accuracy of at-risk identification, but increased math proficiency as well. The percentage of at-risk students after the fall screening in kindergarten was 81% and only 28% after the class-wide intervention. Similar results were found for first grade (89% to 38%), third grade (57% to 33 %), and fifth grade (97% to 35 %).

In some districts, early grades screening activities include a broad array of performance tasks including mathematics, literacy, and language predictors as well as social and behavioral competencies. Purpura et al. (2015) pointed to the high comorbidity of mathematics and reading disabilities as well as the language demands of mathematics in recommending a broader risk-assessment battery that includes mathematics and literacy screening. It is likely that kindergarten students vary so widely in preschool experiences that screenings during kindergarten will not have the predictive utility of those administered during first grade, hence the need for more frequent screening and progress monitoring in these early grades.

Universal Screening in Schools

The screenings recommended for MTSS frameworks are not limited to the early grades because difficulties in mathematics and other areas can arise at any time across the grade levels. Students often encounter problems with multiplicative reasoning and fraction applications (about third to fourth grade) and with more abstract representations, such as algebraic expressions (about sixth grade) and linear equations (seventh grade). With regular screenings for patterns of strengths and weaknesses, teachers can implement interventions as quickly as needed and conduct progress monitoring to ensure students are achieving grade-level goals. These are not new concepts for general or special educators, but most schools have not had school-wide, comprehensive, multi-tiered systems until these recent initiatives.

A panel of experts in mathematics assessment and intervention recommended that elementary and middle schools screen all students at least twice each year to identify those at risk of potential mathematics difficulties (Gersten et al., 2009). These measures should be selected based on reliability, validity, efficiency, and other technical aspects (Gersten et al., 2012). The panel recommended that measures take no more than 20 minutes to administer and screenings should occur during the fall and mid-point of each academic year. Content for screening measures should reflect curriculum standards (content validity). For Grades 4 to 8, one source for screening can be the prior year's state or district assessment results. The panel addressed issues of time and other resources by recommending that trained school-based assessment teams conduct efficient individual screenings and that screening tools be selected carefully based on their technical specifications.

In selecting screening tools that are accurate and efficient, school-based MTSS or multi-disciplinary teams must consider the reliability and predictive validity of tools as well as the accuracy of cut scores for the local student population. After all, the purposes of screening are to validate core instruction and curricula and to identify which students are at risk of difficulties learning mathematics (Albers & Kettler, 2014). Test developers seek to maximize true positives (*sensitivity*, the proportion of students correctly identified as at risk relative to all students identified as at risk in the sample). Many developers set sensitivity at .90 (allowing for 10% error) in establishing cut scores. Proportion of true negatives or *specificity* of the instrument (the proportion of students correctly identified as not at risk) will decline as the sensitivity increases, thus are typically in the range of .35 to .75. However, if an instrument's sensitivity is only .62, its specificity may be above .90. The evaluation team must decide whether to accept too many false positives and provide interventions for too many students, to accept too many false negatives and miss students who need interventions, or to follow screenings with a second gate: diagnostic assessment, dynamic assessment, or brief evaluative intervention. Some researchers recommend adding additional statistics, such as area under the curve (AUC) or posttest probabilities (Klingbeil et al., 2019; Van Norman et al., 2017) to increase the accuracy of screening instruments. The MTSS school-level evaluation team should include a professional qualified to develop and interpret these statistics.

Cut scores are benchmarks for comparing student performance and delineating which students are likely to be at risk for later difficulties (Van Norman et al., 2017). Most screening-tool manuals recommend cut scores for at-risk students, as do state-wide tests when they indicate the score below which students are not proficient on the outcome measure. School-based evaluation teams should consider the test maker's population sample and compare that with the school's demographics as one means to evaluate the potential accuracy of an instrument's cut scores. Additionally, evaluating the results of using predetermined cut scores across the school year, using repeated measures, should inform the team about their accuracy. One district in Wisconsin partnered with a university to evaluate the use of the previous year state-wide test for Grades 6 to 8 as a screening tool (Klingbeil et al., 2019). The researchers found that the test scores were useful as an accurate and efficient screening measure, especially if local cut scores were established, due to the large proportion of high-achieving students in the district. Sensitivity was set at or above .90 for each grade and specificity values exceeded .70. Using an end-of-year test that is already required reduces assessment time and costs and may allow interventions to begin the first week of school, potentially discontinuing fall screening for most students. The researchers noted the drawback of using the previous year's population to establish each year's cut scores, emphasizing the need for ongoing review.

Universal screenings for Grades 3 through 12 can take several forms, including a two-stage screening process, use of district assessments as part of the screening system, and the use of CBMs for screening as well as progress monitoring. Fuchs and colleagues (2011) investigated a two-stage screening process for third graders. The first-stage universal screening comprised ten word problems of four types representing the third-grade curriculum, followed with a dynamic assessment session using algebra content (not familiar to third graders). The universal screening resulted in an accuracy rate of 56% (predicting students with and without difficulties). By adding the second-stage dynamic assessment, the accuracy rate increased to 74%, significantly reducing false positives. Even with this two-stage screening, the authors concluded that specificity was still unacceptably low and that a short period of CBM progress monitoring (5–8 weeks) or a more in-depth individually administered diagnostic assessment would be warranted for improved screening.

Lembke et al. (2012) recommended universal screening in mathematics three times a year (fall, winter, and spring), using a CBM as a universal screener, since those are available for most grade levels across mathematics content areas. These measures can be used for screening and the subsequent progress-monitoring and data-based intervention for a select group of students. Further diagnostic assessment, the authors suggested, should be used to uncover more information about students' strengths and weaknesses within areas targeted for instruction. Advantages for using CBM instruments for screening include the alignment with grade-level curriculum and moving smoothly into progress-monitoring and data-collection systems during interventions. Table 3.2 indicates which screening tools also include ongoing CBMs for progress monitoring. The next section provides more in-depth information about developing and implementing CBMs, one type of curriculum-based assessment.

TRY THIS

Review one of the screening devices in Table 3.2 or one used by a local school. What types of tasks and responses are required? What mathematics content is addressed?

Curriculum-Based Measurement

Curriculum-based measurement (CBM) is a specific assessment approach for monitoring student progress within the curriculum. Developed by Stanley Deno at the University of Minnesota, CBM was an outgrowth of Lindsley's and Lovitt's Precision Teaching of the 1970s that emphasized the direct and frequent measurement of academic skills being taught (Jenkins & Fuchs, 2012). After the passage of P. L. 94–142 in 1975, Deno and colleague Phyllis Mirkin developed the *Data-Based Program Modification Model* that was the precursor for CBM (1977). Deno's CBMs were simple, efficient, and inexpensive, yet demonstrated technical reliability and validity.

CBMs are brief and frequent measures of the *instructional outcome* for a unit or course of study. They do not assess lesson objectives as with Precision Teaching, or unit goals after instruction

Table 3.2 Selected Examples of Early Grades Screening Tools

Instrument	Levels	Scope/Forms	Response Mode	Purpose
Assessing Student Proficiency in Early Number Sense (ASPENS)[1]	K-1	Numeral identification, magnitude, missing number, number comparison, number combinations, base 10 (3 forms)	Constructed response (oral) 9 min	Screening, progress monitoring
Early Numeracy Skills Screener-Brief (PENS-B)[2]	PK-K	Numerals and number order 0–10, counting, number and set comparison, story problems, number combinations	Oral, motor <5 min	Screening
Tools for Early Assessment in Mathematics (TEAM)[3]	PK-2	Numerals, subitizing, counting, comparing number, simple arithmetic, fractions, shapes, patterning, measurement, data	Oral, motor 2 sessions 30–45 min each	Screening, progress monitoring
mCLASS®: Math[4]	K-3	Numerals, counting, comparing number, number combinations, numeral naming & writing (20 forms)	Oral, written, guided interview 2–10 min	Screening, progress monitoring
Number Sense Screener™[5]	K-1	Counting, number recognition, number comparison, non-verbal calculation, story problems, number combinations	Oral, modeling 15–20 min	Screening

Notes: [1]Clarke et al., 2011; [2]Purpura et al., 2015; [3]Clements et al., 2011/2016; [4]Lee et al., 2010; [5]Jordan & Glutting, 2012.

as with summative measures, or mastery of specific skills after instruction as with mastery measures. Rather, they are repeated measures of the same outcome concepts and skills that show how students are progressing toward proficiency. CBMs can be used within a system, such as MTSS, for screening, progress monitoring, and data collection. As CBMs are used to assess student progress toward specific learning goals, they can be used with grade-level curriculum standards or for special, individualized goals. Some CBMs repeatedly assess a task that is robust with multiple component skills, such as word-problem solving. In this approach, results of the performance assessed must have a high correlation with overall grade-level achievement. Alternatively, a CBM can systematically sample a grade-level's curriculum, with the full range of content sampled (Fuchs, 2004). For mathematics, with such a wide range of content domains, teachers could consider using CBMs for units of study or alternating CBMs for process skills (facts and procedures) and reasoning abilities (word problems and other applications).

Research-based CBMs differ from most teacher-developed measures in that they have demonstrated technical adequacy, reliability (multiple forms are equivalent and can be administered by classroom teachers) and validity (constructs within curriculum are measured with adequate scope and difficulty) "that equals or exceeds that of achievement tests" (Deno & Mirkin, 1977, p. 41). Additionally, they are sensitive to changes in student performance over time (Deno, 1985). Although the first CBMs with this technical quality were developed to measure progress in reading, mathematics has gained attention in recent years. Example CBMs for mathematics, with their target concept and skill areas, are noted in Table 3.3. All of these measures include multiple forms and data-collection systems for progress monitoring. Early CBMs in mathematics tended to focus on math facts and simple algorithms, but newer instruments sample the curriculum domains (e.g., the CCSSM critical areas for each grade level). While most CBMs focus on elementary topics, more are being developed for secondary concepts such as fractions (Rodrigues et al., 2019) and high-school algebra (Foegen &

Table 3.3 Selected Examples of Curriculum-Based Measurement Systems

Instrument	Levels	Scope	Response Mode/ Time
Monitoring Basic Skills Progress, 2nd Ed.: Basic Math[1]	1–6	Computation: grade-level sampling Concepts/Applications: grade-level sampling (30 per grade level)	Written, varies
STAR Math[2]	1–12	Four domains (32 subdomains): numbers & operations; algebra; geometry, measurement, and data; statistics and probability	Computer-adaptive[5], 20 min
System to Enhance Educational Performance- STEEP[3]	K–12	K: number naming, comparing, 1: advanced numeracy, computation 2–12: computation, advanced numeracy, concepts & applications	Computer- administered, varies
aimsweb®Plus Math[4]	K–8	Number naming fluency, number comparison fluency, math facts fluency, number sense fluency, quantity total fluency (3 benchmark and 20 progress monitoring forms per grade level)	Computer- administered, 1–7 min

Notes:
1 Fuchs et al., 1999;
2 Renaissance Learning, 2014;
3 Witt & VanDerHeyden, 2007 (iSTEEP);
4 Pearson, 2014;
5 *computer-adaptive* refers to automatic adjustment of item difficulty based on student responses (progress monitoring, not CBM).

Dougherty, 2010; Genareo et al., 2019). Some researchers have found that repeated measures of concepts and applications, but not simple computation, are strong CBM screening and progress-monitoring tools for middle school, predictive of outcome measures (Anselmo et al., 2017; Codding et al., 2016).

An example CBM for mathematics in a unit about proportions would include alternative forms of a 15-item test that takes 10 minutes to administer. The first version would be administered before instruction, to establish a baseline. Repeated administrations, every week during the six-week unit, should show a growth slope close to the goal line. If the growth slope is not adequate, a different instructional approach should be implemented. If growth exceeds the goal line, the goal may not be high enough (Lembke & Stecker, 2007). See Figures 3.5 (example CBM using CCSSM seventh-grade standards) and 3.6 (graph) for an illustration of the assessment approach for this unit on proportions. Some researchers recommend one- to two-minute measures, but for mathematics that may not be practical for content other than math facts and algorithms. Jitendra et al. (2014) successfully implemented word problem CBMs that included 8 items (in 10 minutes) every 2 weeks over a 12-week period. These unit-specific CBMs are often called probes, because they target a specific mathematics domain or skill area, not the entire grade-level curriculum.

Some teachers have difficulty using data from CBM progress graphs such as the one in Figure 3.6 for data-based decision making. Several research groups have studied preservice and inservice teachers' CBM graph comprehension, interpretation, and related factors using teacher think-aloud procedures (Fuchs, 2017). Although some teachers had experience with CBM graphs and their interpretation, they were not always proficient in graph interpretation and connecting the data to instructional decisions (Espin et al., 2017; van den Bosch et al., 2017). Teachers who had greater knowledge about CBM graphs were specific regarding instructional phases, integrated in describing student progress connected with instruction, and

CC A2   Name _____

Place a checkmark in front of the ratios that are equivalent:

___ a. 2:5 and 4:10   ___b. 1:5 and 3:15

___ c. 1:4 and 3:9   ___d. 3:5 and 4:6

___ e. 4:3 and 16:9   ___f. 2:3 and 8:12

Do these ratios form a proportion? If not, correct the second one.

___ g.   8 pets for every 5 students, 19 pets for every 13 students.

___ h.   3 volunteers for every 5 students, 9 volunteers for every 15 students.

i. This table shows how much money Ben earns babysitting. How much does he earn per hour? _____

Time spent in hours (x)	3	4
Amount earned (y)	$15	$20

j. Show the data about Ben on the coordinate plane on the paper provided.

k. What is the constant of variation for your graph?

Solve:

l. Mrs Adams drives 24 miles to school and back in two days. She drove 60 miles to school and back in five days. How far does she drive one way?

m. Henry's birthday party will cost $24 for 16 guests. How much will his party cost for 20 guests?

Write the linear equation that gives the rule for the tables below.

n.   o.

x	y
0	0
1	3
2	6
3	9

x	y
1	0
2	1
3	2
4	3

Figure 3.5 Curriculum-Based Measurement for Proportion Unit

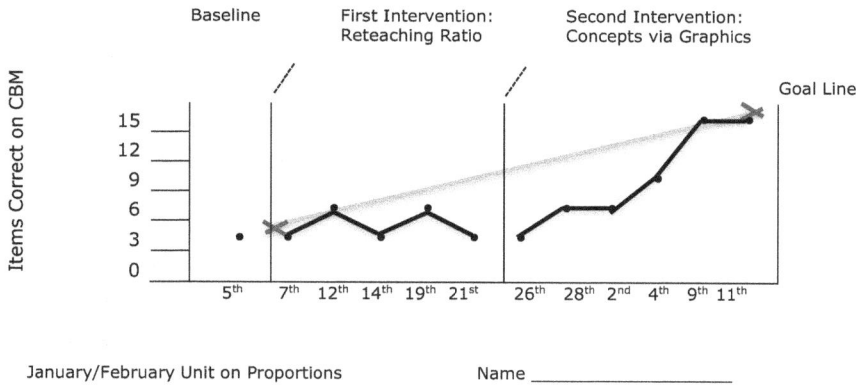

Figure 3.6 Progress–Monitoring Chart for Proportion Unit

coherent in making accurate reflections about the CBM process. Preservice teachers, although exposed to CBM graphing and interpretation in their course and field work, often do not develop proficiency by the end of their programs (Wagner et al., 2017). These researchers recommended more training in the use of CBM graphs for instructional decision making with more opportunities to practice with coaching.

If teachers are not able to locate research-validated CBMs for specific mathematics content, the following steps should be followed to develop curriculum-based measures. Note that without reliability and validity confirmation, teacher-developed measures may not assess content or growth adequately.

1. Identify the outcome goals for the unit of study.
2. Develop a table of specifications for the complete range of concepts and skills to be achieved by meeting the goals. Table 3.4 provides an example with a trajectory over Grades 2 to 4 for an intervention for a fourth grader with gaps going back to second-grade geometry.
3. Create test items for the range of concepts and skills, giving more weight to those with more emphasis in the curriculum. Test items might require paper and pencil responses, oral responses, demonstrations, or computer-based responses. For the example in Table 3.4, a CBM might include 20 items weighted by the percentages. A weight of 10% would yield two items, while items 5% or less might appear on alternate forms.
4. Seek the input of other educators in evaluating the quality and representative nature of the test items. Field-test items with a sample group of students if possible.
5. Develop multiple forms of the test that have the same number of items with the same weighting per the table of specifications.
6. Administer the CBMs on a predetermined schedule. Graph each student's results depicting baseline and instructional stages. Use the results to plan and modify instruction.

CBMs are used widely in today's classrooms, especially within multi-level intervention frameworks in general and special education settings. Increasingly CBMs are being developed and validated for mathematics applied to a wider range of topics and grade levels. Some teachers have found that the results of CBMs are easier to communicate with students, parents, and other teachers, because of the close connection to curriculum goals and data representations using graphs. It is easier to understand the results of progress toward a clearly

Table 3.4 Table of Specifications: Geometry Trajectory (2nd–4th)

Content	Types of Problems		
	Knowledge/ Skills/ Procedures	Applications/ Word Problems	Processes
(2) Identify 2- and 3-dimensional shapes.	2%		Reasoning, communicating,
(2) Identify sides, angles, edges, vertices, and faces of shapes.	3%		Reasoning, communicating
(2) Locate specific attributes of triangles, quadrilaterals, pentagons, hexagons, cubes.		6%	Reasoning and proving
(2) Draw shapes given specific attributes.		8%	Reasoning and modeling
(3) Categorize shapes by their attributes.	5%		Reasoning and proving
(3) Find the area and perimeter of quadrilaterals using unit squares.		10%	Making connections (to other domains), selecting strategies
(3) Solve word problems involving perimeter and area.		10%	Problem solving, making connections
(4) Identify acute, right, obtuse, and straight angles.	5%		Communicating, Reasoning
(4) Classify triangles by angles.	5%		Reasoning
(4) Draw lines, line segments, and rays.		8%	Modeling using tools
(4) Identify parallel, perpendicular, and intersecting lines.	5%		Reasoning, communication
(4) Construct angles of different types.		10%	Modeling using tools
(4) Measure angles in whole-number degrees using a protractor.		8%	Making connections (to measurement), modeling using tools
(4) Solve word problems involving angles.		10%	Problem solving
(4) Draw a line of symmetry through a figure.		5%	Reasoning
TOTALS	**25%**	**75%**	

identified curriculum goal using CBMs than standardized scores and other derived metrics from achievement tests (Hosp & Hosp, 2012). Schools using CBMs within school-wide frameworks should conduct regular program evaluations to ensure measures being used are appropriate for the school's student population and curriculum. Measures should indicate even small increments of student growth within the curriculum being taught and identify students who need more intensive intervention.

Other Assessment Strategies

A range of assessment strategies are used in conjunction with previously described diagnostic assessment, screening, and CBM tools as well as in other types of assessments in mathematics classrooms. This section describes some of the most common assessment tools—dynamic assessment, performance tasks, error analysis, and think-alouds—as well as other elements of classroom assessment systems.

Dynamic Assessment

Dynamic assessment (DA) shares many characteristics of other assessment-intervention models, such as Precision Teaching and RTI with CBM, but has a longer history (e.g.,

Thorndike, 1924; Vygotsky, 1934/1962). The roots of DA have been traced back to the individual differences assessment research of the 19th and early 20th centuries (Haywood, 2012). DA, as contrasted with static assessments conducted during single sessions, is an interactive approach to conducting assessment in a range of domains that focuses on the learner's ability to respond to instruction (Haywood & Lidz, 2007). Two defining characteristics of DA are: 1) intervention is embedded within the assessment; and 2) primary concern is the response of the individual being assessed (Lidz, 2014). There are variations of DA, but most involve assessment of the learner's processes during learning, information about the types of interventions that are successful for a learner, and assessment of learning potential (although potential for learning may be an elusive goal). Interventions are mediated in that students are provided feedback, guided prompts, and instructional scaffolds for moving to the next level. Lidz (2014) emphasized that the main challenge for the assessor in DA is to attend to the *processes* involved in the learning task—language, memory (all types), metacognition, reasoning, emotion, perception, attention, sensory, and motor—and determine the appropriate interventions that address those processes. It requires someone with skill in psychoeducational assessment and domain-specific intervention to be successful with this type of assessment.

DA offers a wide range of applications and iterations, ranging from standardized protocols to informal teacher-created episodes, from reading and math applications to speech/language and other fields (Gustafson et al., 2014; Lidz, 2014). Purposes for DA include screening, diagnostic classification, description of student performances, recommendations for interventions, and establishing baselines for follow-up instruction. DA is most commonly used with individuals but has been adapted for small-group and classroom-level interventions (e.g., Resing et al., 2012). Typically the student is presented with a task and accompanying instruction or instructional scaffolding. The amount of assistance needed to master the task is measured. In DA, the intervention is the means of assessment and the assessment results predict the intensity of instruction needed for future learning. As Fuchs and colleagues (2011) found, DA can be used within a multi-phase screening system to more accurately identify students' needs for intervention. Gustafson et al. (2014) suggested that DA should be used for screening as well in Tier 3 (in an MTSS/RTI framework) where intervention is more intensive and requires an almost clinical model of intervention and assessment.

Performance Tasks

Performance tasks require students to perform, create, construct, or produce something that the teacher can observe (McMillan, 2004). Mathematics performance tasks are embedded in formal and informal assessments and may include written, oral, and modeling responses. More formal tasks are often called performance assessments, with corresponding products or performances that are evaluated using predetermined criteria. Performance tasks can be designed for individual students, small groups, or the entire class. Examples of day-to-day performance tasks:

a. Please arrange the 24 blocks on your desk into rows and columns so that you have no blocks left over (individual or group response, modeling).
b. Solve the problem on the board for the value of x and show your work (individual or group, written).
c. Construct a ray that bisects the angle on your paper (individual or group, written).
d. Count to 50 by twos (individual or small group, oral).
e. Solve the word problem on your paper and be ready to explain your solution strategy (individual or group, written and oral).
f. Add the data we collected to your Excel spreadsheet. Discuss with your partner how to construct a formula to find the mean (pairs, written and oral).

Performance tasks can require specific responses, as with algorithms or multiple-choice items, or they may allow for a range of responses, as with open-ended or constructed-response tasks. These tasks can be embedded in lessons, used on formative assessments, or comprise a summative assessment, such as a unit posttest. Many mathematics textbooks include performance tasks in their teacher manuals or ancillary materials.

Curriculum goals are excellent sources for performance tasks. For example, one goal for fourth grade, within number and operation, is "Compare two fractions with different numerators and different denominators" (CCSSM, 4.NF.A.2.). A performance task could be: *There are five pairs of fractions on your page. Circle the larger fraction in each pair. You can use your reasoning, number lines, or computation, any method you choose, to determine which is larger.* For students with difficulty on this task, an individualized task could be: *Tell me which of the fractions,* $\frac{2}{3}$ *or* $\frac{3}{4}$*, is larger. Tell me in your own words how you figured that out.*

TRY THIS

Create a performance task for a specific curriculum standard.

Performance tasks are often the only assessment approach for measuring student abilities for the mathematics *processes*. For example, for the CCSSM practice standards listed below, we can create performance tasks for specific mathematics concepts and skills.

- *Problem solving, make sense of problems*: Paraphrase, put into your own words, the important propositions in this word problem.
- *Reasoning and proof, reason abstractly*: Put the information from the word problem into equation form, using the letter x for the unknown.
- *Using tools strategically*: Show me how you measured the largest angle of that triangle.
- *Model with mathematics*: Create an equation that shows the same results as we have on our coordinate plane.

Performance tasks that do not allow the teacher to evaluate much more than correct or incorrect responses can be modified to be more robust. For example, the following problem requires a single response and does not offer a window into student reasoning or other processes and possible misconceptions: $468 \times 5 =$ ____. A better task would ask the student to show or explain his work. An even better task, illuminating conceptual understanding, would be:

Patty was given three problems. She used a pattern to write three different problems so she could do the work mentally.
$250 \times 5 =$ ____. Since $125 \times 10 = 1250$, therefore $250 \times 5 = 1250$.
$688 \times 5 =$ ____. Since $344 \times 10 = 3440$, therefore $688 \times 5 = 3440$.
$240 \times 5 =$ ____. Since $120 \times 10 = 2400$, therefore $240 \times 5 = 2400$.
Explain Patty's pattern and why it works. Use her pattern to find $468 \times 5 =$ _____.
(Adapted from Hunsader et al., 2014, p. 212)

In core mathematics classrooms, more robust performance tasks are becoming the norm for a range of formative and summative assessment purposes. Teachers are encouraged to build in student self and peer assessments by providing rubrics and asking questions that require reflection and critique. Engaging performance activities help bridge instruction with assessment.

Teachers are challenged to select tasks aligned with grade-level curriculum and practice standards, and document student progress via rubrics or other systematic data-collection devices. Consider the previous task about patterns in multiplication problems. Which curriculum and practice standards are assessed with this task? How could the teacher code student responses to document understanding?

Teachers should find or create tasks that reflect the critical knowledge, skills, and concepts in the curriculum. Stenmark (2002) outlined eight criteria for good tasks: essential (a big idea of mathematics), authentic, rich, engaging, active, feasible, equitable, and open (more than one solution or approach). Sources for tasks include textbooks, journals, life situations, books of problems, and released state test items. Other sources include *Inside Mathematics* (www.insidemathematics.org), the *Mathematics Assessment Project* (www.map.mathshell.org), and NCTM.

Some sample tasks that meet these criteria:

a. *A performance task after a unit on measurement and geometry, to assess the connections between concepts*: We have been assigned the task of calculating the amount of paint needed to paint the hallway to our classroom. We would not paint doors or bulletin boards. Your group can solve this problem in a number of ways, but you should be able to explain your process.
b. *Open-ended task*: A baseball player has a batting average of .250 after playing in 20 games this year (average of 3 at-bats per game). Chart two different ways he can bring his average up to at least .300 over the next 10 games.
c. *Open-ended task from a state test*: Write out a real-life problem that would be answered by solving this equation: $20x + 4 = 150$.

Performance tasks should be closely integrated with classroom instruction and aligned with curriculum standards. Teachers should be able to interpret assessment information on the spot and provide constructive feedback to students (Heitink et al., 2016). Tasks should elicit students' prior knowledge and understanding so that teachers can scaffold their understanding to the next level by modifying tasks or developing additional tasks. Some tasks may be more complex or multistep and require a rubric for evaluation, but that should not preclude the teacher–student interaction that should take place during these assessment activities. Closely related to performance tasks for assessment is using these tasks for instruction (e.g., problem posing for instructional activity).

A performance task can also be a *mastery measure*, an assessment at the end of a segment or unit of study that is summative. This type of task should answer questions such as: Has the student mastered the skills and concepts for the unit of study? Can the student apply what has been learned to a problem situation or real-world context? For example, one mathematics goal for third graders is to represent data on a scaled bar graph and to solve "how many more" and "how many less" problems using the graph (CCSSM 3.MD.B.4). A performance task as a mastery measure could require students to collect data about after-school activities of classmates, represent the data on a scaled graph (e.g., each block representing 5 basketball hours or 5 dance hours), and create *compare* problems about their graphs.

Error Analysis

Error analysis is the process of examining student errors for clues to misconceptions or erroneous strategies. Identifying a pattern of errors can be especially enlightening. Error analysis should not be used to ensure students are solving math algorithms using a single method, but to uncover concept misunderstandings and faulty procedures. Figure 3.7 shows a student

a. $\frac{1}{2} + \frac{2}{3} =$ 1/2 + 3/5 = 4/5

b. $\frac{1}{4} + \frac{3}{5} =$ 1/4 + 4/9 = 5/9

c. $\frac{4}{5} + \frac{2}{5} =$ 4/5 + 6/10 = 10/10 = 1

d. $\frac{2}{3} + \frac{2}{7} =$ 2/3 + 4/10 = 6/10 = 3/5

Can you identify a pattern of errors?

Figure 3.7 Error Analysis

work sample with a discernable pattern of errors for addition of fractions. Error analysis can also be conducted with formal assessments as long as the test protocol is followed. If teachers are stumped by students' written work, a *diagnostic interview* or *think-aloud* can provide clues to difficulties and misunderstandings.

Think-Alouds

Ericsson and Simon (1993) traced the history of *think-aloud* techniques in psychology, cognitive science, and educational applications for assessment, instruction, and research methods in the text *Protocol Analysis: Verbal Reports as Data*, an elaboration of their theory on verbal thinking-aloud protocols as data, first published in 1980. The theorists credited Watson (1920) for the first documented analysis of thinking-aloud activity, published after a decade of work refuting introspection research methods. "A good deal more can be learned about the psychology of thinking by making subjects think aloud about definite problems, than by trusting to the unscientific method of introspection" (Watson, 1920, p. 91). Think-aloud strategies are used across a wide range of disciplines to better understand cognitive and metacognitive processes employed by problem solvers.

Research supports the use of think-alouds in mathematics to determine student concept understanding and metacognitive approaches of students with learning difficulties. For example, Lowenthal (1987) used interviewing to probe mathematics error patterns. Ginsburg (1997) demonstrated the use of clinical interviews to trace the mathematics concept understanding of individual students receiving the same instructional experiences over time. Think-alouds were used by Montague and Applegate (1993) to study the mathematics problem-solving approaches of students with specific mathematics learning disabilities (MLD) and by Rosenzweig et al. (2011) to study the metacognitive processes involved during problem solving by students with MLD. Gurganus and Shaw (2006) demonstrated the use of think-aloud protocols to probe big idea mathematics concept understanding of middle-school students with MLD.

These studies used audiotaped methods for coding think-alouds while students were engaged in math tasks, an extremely time-consuming method for analyzing student responses. If prepared carefully, think-aloud protocols can be used by teachers while listening to students thinking aloud concurrent to performing a mathematics task. Student comments could be coded as productive or nonproductive, types of metacognitive statements (self-statements, task understanding, strategy selection), and concept accuracy or misunderstanding. For thinking aloud during problem solving, the teacher should code for steps engaged, adequate paraphrasing, strategy used, and metacognitive strategies such as self-correction. Employing think-aloud strategies to check student understanding has the added benefit of enhancing student metacognition. Think-aloud strategies reinforce students' use of self-talk and their own progress monitoring while performing mathematics tasks.

For mathematics assessment, think-aloud strategies are similar to diagnostic interviews in that they both illuminate student thought processes. Diagnostic interviews are guided by the teacher through planned questions while think-alouds tend to be less teacher-directed during the mathematics task. The teacher prompts the student to say aloud what she is thinking while doing the task, "Please talk out loud while you are solving this problem. I want to listen to what you are thinking while you solve it." The teacher may not correct the student at all if the session is for assessment purposes. The teacher may decide to intervene, as with a *dynamic assessment* approach, during a think-aloud episode. Many teachers use their own think-alouds during instruction and modeling to demonstrate how thinking helps while performing mathematics tasks. A think-aloud approach has many applications for mathematics instruction and assessment, promoting more attention to cognitive and metacognitive strategies as well as concept understanding and misconceptions. Examine the think-aloud session illustrated in Figure 3.8 for concept understanding and metacognitive strategies.

Classroom Assessment Systems

Teachers responsible for monitoring the mathematics progress of groups of students, such as whole-class instruction or caseloads of students with mathematics goals on their IEPs, often develop group assessment systems to plan and document student progress. Systematizing assessment activities ensures their implementation and consistency and saves planning and instructional time. The features of a system will depend on caseload, amount of direct contact with students, the types of assessments conducted, access to technology, school-level requirements, and other setting-based variables. Potential elements of classroom assessment systems include:

* Charts of grade-level mathematics goals, especially the three or four focal points for each grade level, are displayed on bulletin boards or maintained in student notebooks.
* Schedules for administering CBMs, with corresponding data collection, are established and communicated with other teachers, students, and parents.
* A classroom assessment center is set up with graph paper (or dedicated computer), timers, calculators, probes, performance tasks, and other assessment materials that students can be taught to use. Individual student folders with IEP goals and assessment results should be kept confidential, for teacher, parent, and student use only.
* Portfolios for collecting student work samples can be organized at the beginning of the school year. Portfolios can include material that demonstrates student progress toward grade-level goals, student-selected work on key performance tasks, or structured portfolios for evaluation of student performances, and can add visual evidence for meetings with parents. Most portfolios include a structure and checklist or rubric by which to evaluate the artifacts. Teachers should be aware that portfolios have the advantage of providing

Mr Porter: DeVon, there is a problem on your paper, 493 divided by 8. I'd like you to work that while I watch. I'd also like for you to think out loud while you're working.

DeVon: What do you mean by think out loud?

Mr Porter: Just say out loud what you are thinking. Since I can't read your mind, it helps me to understand how you think while you're working.

DeVon: I'm not good with my 8s.

Mr Porter: That's OK, just work it the way you think you can find the quotient.

DeVon: I know you're supposed to think of eight into four first, you start that way.

Mr Porter: Go ahead.

DeVon: And it doesn't go so you think about eight into forty-nine. And I can't think of that table, and I know my 10s really well and eight times ten is eighty. So I just start subtracting 80s until I can't subtract any more [talks through each subtraction]. And I come to thirteen and I know I can take an eight from that. It leaves five for the remainder. Now I go back and count up the 80s, I mark those so I don't get lost. There are six and that goes first at the top. Then I count the eights; there is one. And a one goes after the six, then my remainder five. Or I could put 5/8 as the remainder and not write the little "r."

Mr Porter: Wow, DeVon. I'm really impressed. You figured out a strategy that works. You understand that division is like repeated subtraction. You know your multiples of ten. You subtracted without any mistakes. You also know your remainder concept and place value. You kept track of all this work as you were going. Do you think this is the most efficient method?

DeVon: No, I know you're supposed to think of eight into four, and then eight into forty-nine, but my eights aren't very good.

Mr Porter: Tell you what, why don't you make a list of the multiples of eights on the side of the paper and use that as you work. I can show you how to keep track while you're doing that procedure. I think it will save you some time and a lot of subtracting.

$$
\begin{array}{r}
61\ \text{r}\,5 \\
8\,\overline{)493} \\
-80 \\
\hline
413 \\
-80 \\
\hline
333 \\
-80 \\
\hline
253 \\
-80 \\
\hline
1\,73 \\
-80 \\
\hline
93 \\
-80 \\
\hline
1\,3 \\
-8 \\
\hline
5
\end{array}
$$

Figure 3.8 Think-Aloud Session

visual evidence of student performances and can be maintained by students, but they lack the technical standards of other types of assessments, thus can be biased and subjective.

- Maintenance checks for previously learned skills and concepts should be planned about a month after the initial instruction. These checks have the added benefit of reminding teachers and students of content connections.
- Assessment items that resemble the structure and response format of high-stakes assessments should be built into classroom assessments, along with any required accommodations.

Students need practice, with feedback, in making multiple-choice selections and writing out their reasoning for constructed-response items.

Teachers responsible for mathematics instruction need a full toolbox of assessment strategies and approaches. The *Classroom Assessment Standards for PreK-12 Teachers* is one validated source to assist teachers in examining their own classroom assessment practices (Klinger et al., 2015). Classroom and support teachers should be confident in administering formal instruments and designing and implementing informal measures. They should be skilled in data collection, charting student growth, designing performance tasks and rubrics, implementing think-alouds and diagnostic interviews, and analyzing student work. Most importantly, teachers should use mathematics assessment to improve instruction and student achievement.

Assessment Issues

Assessment is a controversial subject in today's context of accountability and high-stakes evaluations. Some issues, previously discussed, include the technical adequacy of instruments, curricular alignment, resources for assessment activities, and teacher skill. Additional issues that impact students with disabilities and those at risk for low achievement include accessibility and accommodations for mandated assessments and effectiveness of multi-tiered systems of support (MTSS) frameworks.

Accessibility and Accommodations

The 1997 reauthorization of IDEA and the 2001 No Child Left Behind Act (ESEA, NCLB) required students with disabilities to participate in state- and district-wide assessments. Schools were required to report student participation and performance by student groups, such as students with disabilities and English language learners (EL). In the 2016–2017 school year, states reported that an average of 95% of students with disabilities participated in statewide math assessments (Grades 3 through high school; Wu & Thurlow, 2019). Of these students with IEPs, about 57% were offered accommodations on statewide tests. The average proficiency rate for students with disabilities was approximately 14%, with third grade rates typically the highest (23%), decreasing over the grade levels.

The reported scores for up to 1% of students (about 10% of students with IEPs) are permitted from alternate assessments based on alternate achievement standards, for those students with the most severe cognitive disabilities. Until 2015, alternative assessments could be based on modified academic standards. The Every Student Succeeds Act (ESSA) of 2015 allowed states to continue to use alternate academic achievement standards for students with the most significant cognitive disabilities as long as those are aligned with the state's academic standards and promote access to the general curriculum.

The National Center on Educational Outcomes (Thurlow et al., 2019) found that all states offer alternate assessments and the three most common participation guidelines for alternate assessments based on alternate academic achievement standards (AA-AAAS) were "(a) significant cognitive disabilities or low intellectual and adaptive functioning; (b) extensive, intensive, individualized instruction and supports; and (c) use of an alternate or modified curriculum" (p. 1). States do not permit the use of disability label, SES factors, absences, or poor performance to qualify students for these assessments. An average of 8.9% of students with disabilities took AA-AAAS as reported by states for 2016–2017, an average of 1.28% of the total student populations in those states (Wu & Thurlow, 2019). The NCEO website includes links to state policies for alternate assessments as well as accommodations (http://nceo.info/).

The National Center on Educational Outcomes promotes six principles for inclusive assessment systems (Thurlow et al., 2016, p. v):

1. Every policy and practice reflects the belief that *all* students must be included in state, district, and classroom assessments.
2. Accessible assessments are used to allow all students to show their knowledge and skills on the same challenging content.
3. High-quality decision making determines how students participate in assessments.
4. Implementation fidelity ensures fair and valid assessment results.
5. Public reporting content and formats include the assessment results of all students.
6. Continuous improvement, monitoring, and training ensure the quality of the overall system.

Accessibility means the ability of an individual to reach or use something. Within disability education, accessibility refers to the design of products, devices, services, or environments to ensure ease-of-use and participation. Accessibility for assessment denotes that steps have been taken to ensure student participation in general assessments. Three major types of accessibility features are available: 1) universal features available for all students, such as zoom and highlight features on computers; 2) specially designed features such as embedded text; and 3) accommodations such as shortened testing periods for specific students. With the advent of technology-based assessments there are more opportunities for universal designs. More recent state assessment policies attempt to include all students in accessibility frameworks, noting that some general education students need accessibility features in order to demonstrate their knowledge and skills (Warren et al., 2015).

Accommodations provide students with disabilities and limited English access to regular assessments and are included in IEPs, *504 plans* \, and other individualized programs. Accommodations, changes in testing materials or procedures, differ from *modifications* in that accommodations do not change content standards. They allow students to show what they have learned without being handicapped by their disability. Accommodations should not change what is being measured, a validity issue. Teachers should implement the same accommodations for classroom-based assessments and instruction as will be used for high-stakes assessments. Each state must develop accommodation guidelines for assessment systems and all teachers must be trained in making appropriate accommodations (IDEA, 2004; ESSA, 2015). However, there remains wide variation in accommodation policies across states, from allowing all or most students to use accommodations, to restricting their use to students with IEPs, 504 plans, or limited English. Even for students with disabilities, there is a lack of consensus from state to state about accommodations to allow, restrict, or prohibit (Lai & Berkeley, 2012). Most assessment accommodations can be classified as presentation mode, timing or scheduling, response mode, setting, and equipment or materials (Lin & Lin, 2014).

A number of studies have examined the effects of assessment accommodations and the decision-making process for determining appropriate accommodations. Valid accommodations remove construct-irrelevant variance created by the student's individual characteristics and measure the same constructs as for students without accommodations. There are three methods for examining the validity of test accommodations: descriptive (logical analysis of the student's characteristics and disability with regard to testing requirements); comparative (comparing data bases of students with and without accommodations); and experimental (controlled studies of the effects of accommodations; Tindal, 1998). Teacher judgment, on the other hand, tends to over-identify accommodations for students (Fuchs et al., 2000). When IEP teams have difficulty making accommodation decisions, an assessment of students under various testing conditions can supplement teacher judgment.

One indicator for verifying an accommodation for items on a specific test is the *differential boost* (Phillips, 1994). Theoretically, an appropriate accommodation should increase the performance of students with disabilities but wouldn't make a difference in scores for students without disabilities. However, research has been limited and findings have been mixed in attempting to differentiate student performances using accommodations (e.g., Buzick & Stone, 2014; Fuchs et al., 2000). Some researchers have found benefits for students with disabilities on items with portions of visual models shaded (Zhang, 2012) and for changing response formats (Powell, 2012), but did not compare the responses of students without disabilities. Sometimes, if accommodations are not matched with student characteristics, performance can be negatively affected. Lin and colleagues (2016) found that sixth-grade students with attention or learning difficulties performed worse with a separate setting accommodation than in the classroom setting. In a meta-analysis of 11 studies, Cahan et al. (2016) found that extended time did not have a differential effect on math scores for students with MLD; the benefits were the same for students without disabilities (and all students should have extended time unless time is an important construct). Feldman et al. (2011) found that for eighth-grade students, accommodations improved the test performance of all students (no differential boost) and provided a boost for self-efficacy and motivation for students with MLD. Clearly a range of factors should be considered when planning accommodations.

Too often IEP teams use a checklist of accommodations and a general sense of a student's strengths and weaknesses to select accommodations, but they should make a more informed decision. Educators must consider the specific accommodations allowed by each test, the state's accommodation guidelines, and, most importantly, individual student characteristics. Accommodations should be tested with the student on similar items during classroom instruction and assessments to determine if those accommodations are needed and whether they have an impact on student outcomes. As more tests are provided via computer interface, students should be taught to select options that support the student's effort in test taking, not just because they are available.

Educators within schools need training in accommodation decision-making. The NCEO offers training modules and materials for IEP teams in data-based decision-making for accommodations (https://nceo.info). The first step is for everyone on IEP teams to expect students with disabilities to achieve grade-level standards. Then team members must learn about legal aspects of accommodation provision in federal and state educational law and case law (Peltier et al., 2018). For example, IDEA requires the IEP team to discuss possible accommodations during an IEP meeting that includes parental input. Team members should understand the range of instructional and assessment accommodations that are permitted in their state. They must make decisions about accommodations for individual students based on student, instructional, and assessment factors. Teachers implementing accommodations should evaluate their use, including feedback from the student, and make appropriate adjustments. Accommodations have the potential to allow students with disabilities to demonstrate their learning on regular assessments if accommodations are carefully selected and monitored for each student.

TRY THIS

Find your state's accommodation guidelines and review those related to mathematics. Find those here: https://nceo.info/state_policies

Assessment within Multi-Tiered Systems of Support

Multi-tiered systems of support (MTSS), as discussed in Chapter 1, are increasingly common in schools as comprehensive, school-wide approaches to identifying students at risk of poor outcomes in mathematics (and other academic and behavioral areas) and providing intensive interventions to improve students' achievement. In MTSS, assessment is fully integrated with instruction and interventions so that data-based decision making can guide all forms of assessment and instruction. Assessment provides information on which students require interventions and the success of those interventions. It informs professionals about needed instructional adjustments for students within various levels of the MTSS. Critical assessment components include:

- *Universal screening*, requiring brief assessments for all students in schools, using screening measures that are reliable and valid, as discussed in previous sections. Questions guiding mathematics screenings include: What knowledge, understandings, and skills are expected at this grade level? What instruments will best predict student outcomes? How often should we screen to determine which students are at risk of difficulties? How should we select cut scores? Which professionals and paraprofessionals will be involved in screening?
- *Diagnostic assessment*, as described in an earlier section, an in-depth assessment of students' strengths, weaknesses and learning needs. Questions guiding the use of diagnostic assessments include: Which students need further evaluation for identification or intervention purposes? What instruments are reliable and valid for our mathematics curriculum? Which of those instruments will provide enough information to plan instructional interventions? Which professionals will be involved in the administration and interpretation of diagnostic assessments?
- *Progress monitoring*, providing data on the effectiveness of instruction within any tier in a MTSS framework. A formative assessment approach, progress monitoring involves frequent and ongoing assessment to provide information on when effective instruction should be continued or when progress is below expectations, requiring instructional adjustments (Clemens et al., 2019). The most common tools for progress monitoring are curriculum-based measures (CBM) as those sample curriculum content, can be administered fairly quickly, and are sensitive to student growth. Questions guiding progress monitoring include: Which measures are aligned with our curriculum or the student's IEP goals? How often should those be administered? How are graphs prepared and interpreted with the measures we selected? How are these measures used within each level of our MTSS to improve student outcomes?
- *Systems of data collection*, important for timely decision making within MTSS. However, if data are collected from multiple sources and for differing purposes, teachers will struggle to manage the data collection, interpretation, and instructional decision making within the context of their regular teaching assignments, especially if they have little training in assessment administration and interpretation (Lopuch, 2018). Questions guiding establishing systems of data collection are: Which assessments are currently required by state and local regulations? Which additional assessments do we use in our MTSS? How can individual student data be accessed for these assessments while still protecting confidentiality? What decision rules will we use for individual students on each level of our system? How will we communicate student progress with parents?
- *Program assessment*, often overlooked, but critical for successful MTSS. This assessment requires a program-evaluation approach with multiple sources of data. At least 21 states

David eligible for SLD and special education services. The evaluation determined that David's working memory capacity is limited; he has trouble fluently using information we know he mastered. We think he has trouble holding new information in numerical form while operating on that information with previously learned concepts. He will need a longer period of intensive instruction using concrete and representational materials before he is able to work at the abstract level. His strong reading really supports word-problem solving, but he has trouble monitoring his work trying to solve the calculations and reasoning whether his answers make sense. If we can support his working memory by making sure David is fluent with his facts and by teaching him some strategies, I think he will have more success in math.

MARTINEZ: David is not hating math like he did at the beginning of the year. I'd like to keep him with the regular class if possible, but I know he needs extra assistance.

JOHNSON: He can still participate in the small-group sessions when we are targeting a skill he needs to work on.

BOSWELL: I'd definitely like to see David receive the supplemental instruction in the small groups, but I think he needs more intensive daily instruction in a special education setting to address the full range of his learning needs, hopefully in addition to his regular mathematics class. Let's set up a meeting with his parents and review the assessment results and develop an IEP that will provide the right combination of instruction in mathematics.

How did these teachers work together to plan and use assessment? What types of assessments were useful? How did collaboration contribute to understanding students' needs? Were David's parents involved at the appropriate point in the assessment process? Could a system of CBMs assist with decision-making?

As more students with learning difficulties are included in core mathematics classrooms with differing levels of support services, it is critical that teachers collaborate on an individual student level to plan, conduct, and interpret evaluations for the purposes of informing instruction and planning student interventions. Teachers can pool their expertise—in mathematics content understanding, student characteristics knowledge, and experience with assessment—to design and select tools that will reflect learning. Collaboration activities that will benefit assessments include the following:

- Discuss the types of assessments and the overall assessment systems within each setting—general, special, and mathematics support classes. A school-wide assessment system could form the overall framework.
- For similar assessment tasks, consider developing common rubrics. This approach would have the advantages of being consistent for students and the opportunity for inter-observer reliability measures.
- Share and discuss assessment results for individual students in terms of strengths and weaknesses and needed interventions. Strengths are just as important as weaknesses.
- Special education and other support teachers may be responsible for offering alternative settings for assessments with special accommodations such as extended time, read aloud, or small-group administration. It is important that accommodations used on district- and state-wide tests are also used for regular classroom assessments and that the tests be administered within the specified parameters of the accommodations.

Assessment has many purposes within mathematics education. It can assist teachers in understanding student-learning needs. It can inform communities of the mathematics performance of their children. The most important purpose is to inform teachers, students, and parents of individual student progress toward curriculum goals.

References

Albers, C. A., & Kettler, R. J. (2014). Best practices in universal screening. In P. L. Harrison & A. Thomas (Eds.), *Best practices in school psychology: Data-based and collaborative decision making* (pp. 121–132). National Association of School Psychologists. www.nasponline.org

Alderson, J. C., Brunfaut, T., & Harding, L. (2015). Towards a theory of diagnosis in second and foreign language assessment: Insights from professional practice across diverse fields. *Applied Linguistics, 36*(2), 236–260. doi: 10.1093/applin/amt046

Andersson, C., & Palm, T. (2017). Characteristics of improved formative assessment practice. *Education Inquiry, 8*(2), 104–122. doi: 10.1080/20004508.2016.1275185

Anselmo, G., Yarbrough, J., Kovaleski, J., & Tran, V. (2017). Criterion-related validity of two curriculum-based measures of mathematical skill in relation to reading comprehension in secondary students. *Psychology in the Schools, 54*(9), 1148–1159. doi: 10.1002/pits.22050

Asp, E. (2018). Back to the future: Assessment from 1990 to 2016 (pp. 551–575). In G. E. Hall, L. Quinn, & D. Gollnick (Eds.), *The Wiley handbook of teaching and learning*. Wiley Blackwell. doi: 10.1002/9781118955901.ch23

Berra, Y. (1998). *The Yogi book*. Workman Publishing.

Buzick, H., & Stone, E. (2014). A meta-analysis of research on the read aloud accommodation. *Educational Measurement Issues and Practice, 33*(3), 17–30. doi: 10.1111/emip.12040

Cahan, S., Nirel, R., & Alkoby, M. (2016). The extra-examination time granting policy: A reconceptualization. *Journal of Psychoeducational Assessment, 34*(5), 461–472. doi: 10.1177/0734282915616537

Cheng, Y.-L., & Mix, K. S. (2014). Spatial training improves children's mathematics ability. *Journal of Cognition and Development, 15*(1), 2–11. doi: 10.1080/15248372.2012.725186

Clarke, B., Baker, S., Smolkowski, K., & Chard, D. J. (2008). An analysis of early numeracy curriculum-based measurement: Examining the role of growth in students' outcomes. *Remedial and Special Education, 29*(1), 46–57. doi: 10.1177/0741932507309694

Clarke, B., Gersten, R., Dimino, J., & Rolfhus, E. (2011). *Assessing student proficiency in early number sense (ASPENS)* [Measurement instrument]. Instructional Research Group. www.inresg.org

Clarke, B., Haymond, K., & Gersten, R. (2014). Mathematics screening measures for the primary grades. In R. J. Kettler, T. A. Glover, C. A. Albers, & K. A. Feeney-Kettler (Eds.), *Universal screening in educational settings: Evidence-based decision making for schools* (pp. 199–221). American Psychological Association. doi: 10.1037/14316-008

Clarke, B., Shanley, L., Kosty, D., Baker, S., Cary, M., Fien, H., & Smolkowski, K. (2018). Investigating the incremental validity of cognitive variables in early mathematics screening. *School Psychology Quarterly, 33*(2), 264–271. doi: 10.1037/spq0000214

Clarke, B., Strand Cary, M., Shanley, L., & Sutherland, M. (2020). Exploring the promise of a number line assessment to help identify students at-risk in mathematics. *Assessment for Effective Intervention, 45*(2), 151–160. doi: 10.1177/1534508418791738

Clemens, N. H., Widales-Benitez, O., Kesitan, J., Peltier, C., D'Abreu, A., Myint, A., & Marbach, J. (2019). Progress monitoring in elementary grades (pp. 175–197). In P. C. Pullen & M. J. Kennedy (Eds.), *Handbook of response to intervention and multi-tiered systems of support*. Routledge. doi: 10.4324/9780203102954-13

Clements, D. H., Sarama, J. H., & Liu, X. H. (2008). Development of a measure of early mathematics achievement using the Rasch model: The research-based early maths assessment. *Educational Psychology, 28*(4), 457–482. doi: 10.1080/01443410701777272

Clements, D. H., Sarama, J., & Wolfe, C. B. (2011/2016). *TEAM—Tools for early assessment in mathematics* [Measurement instrument]. McGraw-Hill. www.mheducation.com

Codding, R., Mercer, S., Connell, J., Fiorello, C., & Kleinert, W. (2016). Mapping the relationships among basic facts, concepts and application, and common core curriculum-based mathematics measures. *School Psychology Review, 45*(1), 19–38. doi: 10.17105/SPR45-1.19-38

Connolly, A. J. (2007). *KeyMath™-3 diagnostic assessment* (3rd ed.) [Measurement instrument]. Pearson. www.pearsonassessments.com

Council of Chief State School Officers (2016, 16 December). *Final regulations on academic assessments under Title I, Part A of the Elementary and Secondary Education Act* [Memorandum]. https://ccsso.org/sites/default/files/2017-10/12-16-2016Memo-ESSAFinalRegulationsAcademicAssessments.pdf

Council of Chief State School Officers (2019). *K-2 assessments: An update on state adoption and implementa-tion.* https://ccsso.org/resource-library/k-2-assessments-update-state-adoption-and-implementation

Deno, S. L. (1985). Curriculum-based measurement: The emerging alternative. *Exceptional Children, 52*(3), 219–232. doi: 10.1177/001440298505200303

Deno, S. L., & Mirkin, P. (1977). *Data-based program modification: A manual.* Reston, VA: Council for Exceptional Children.

Elementary and Secondary Education Act (The No Child Left Behind Act) of 2001, Pub. L. No. 107–110, § 101 Stat. 1425 (2002). www.congress.gov/107/plaws/publ110/PLAW-107publ110.htm

Ericsson, K. A., & Simon, H. A. (1980). Verbal reports as data. *Psychological Review, 87*(3), 215–251. doi: 10.1037/0033-295x.87.3.215

Ericsson, K. A., & Simon, H. A. (1993). *Protocol analysis: Verbal reports as data* (Revised). Cambridge, MA: The MIT Press. doi: 10.7551/mitpress/5657.001.0001

Espin, C. A., Wayman, M. M., Deno, S. L., McMaster, K. L., & de Rooij, M. (2017). Data-based decision-making: Developing a method for capturing teachers' understanding of CBM graphs. *Learning Disabilities Research & Practice, 32*(1), 8–21. doi: 10.1111/ldrp.12123

Every Student Succeeds Act, 20 USC § 6301 (2015). www.congress.gov/114/plaws/publ95/PLAW-114publ95.pdf

Feldman, E., Kim, J.-S., & Elliott, S. N. (2011). The effects of accommodations on adolescents' self-efficacy and test performance. *Journal of Special Education, 45*(2), 77–88. doi: 10.1177/0022466909353791

Foegen, A. & Dougherty, B. (2010). *Algebra screening and progress monitoring.* Measurement (Goal 5) award from the Institute for Education Sciences, US Department of Education. Award Number: R324A110262. www.education.iastate.edu/aspm/

Frick, A. (2018). Spatial transformation abilities and their relation to later mathematics performance. *Psychological Research, 83*(7), 1465–1484. doi: 10.1007/s00426-018-1008-5

Fuchs, L. S. (2004). The past, present, and future of curriculum-based measurement research. *School Psychology Review, 33*(2), 188–192. doi: 10.1080/02796015.2004.12086241

Fuchs, L. S. (2017). Curriculum-based measurement as the emerging alternative: Three decades later [Special issue]. *Learning Disabilities Research & Practice, 32*(1). doi:10.1111/ldrp.12127

Fuchs, L. S., Compton, D. L., Fuchs, D., Hollenbeck, K. N., Hamlett, C. L., & Seethaler, P. M. (2011). Two-stage screening for math problem-solving difficulty using dynamic assessment of algebra learning. *Journal of Learning Disabilities, 44*(4), 372–380. doi: 10.1177/0022219411407867

Fuchs, L. S., Fuchs, D., Eaton, S. B., Hamlett, C. L., & Karns, K. M. (2000). Supplementing teacher judgments of mathematics test accommodations with objective data sources. *School Psychology Review, 29*, 65–85. doi: 10.1080/02796015.2000.12085998

Fuchs, L. S., Hamlett, C., & Fuchs, D. (1999). *Monitoring Basic Skills Progress (MBSP): Basic Math* (2nd ed.) [Measurement instrument]. Austin, TX: PRO-ED. www.proedinc.com

Geary, D. C., Nicholas, A., Li, Y., & Sun, J. (2017). Developmental change in the influence of domain-general abilities and domain-specific knowledge on mathematics achievement: An eight-year longi-tudinal study. *Journal of Educational Psychology, 109*(5), 680–693. doi: 10.1037/edu0000159

Geary, D., vanMarle, K., Chu, F., Rouder, J., Hoard, M., & Nugent, L. (2018). Early conceptual understanding of cardinality predicts superior school-entry number-system knowledge. *Psychological Science, 29*(2), 191–205. doi: 10.1177/0956797617729817

Genareo, V., Foegen, A., Dougherty, B., DeLeeuw, W., Olson, J., & Karaman Dundar, R. (2019). Technical adequacy of procedural and conceptual algebra screening measures in high school algebra. *Assessment for Effective Intervention, 153450841986202.* doi: 10.1177/1534508419862025

Gersten, R., Beckmann, S., Clarke, B., Foegen, A., Marsh, L., Star, J. R., & Witzel, B. (2009). *Assisting students struggling with mathematics: Response to Intervention (RTI) for elementary and middle schools* (NCEE 2009-4060). National Center for Education Evaluation and Regional Assistance, Institute of Education Sciences, US Department of Education. http://ies.ed.gov/ncee/wwc/

Gersten, R., Clarke, B., Jordan, N. C., Newman-Gonchar, R., Haymond, K., & Wilkins, C. (2012). Universal screening in mathematics for the primary grades: Beginnings of a research base. *Exceptional Children, 78*(4), 423–445. doi: 10.1177/001440291207800403

Ginsburg, H. P. (1997). *Entering the child's mind: The clinical interview in psychological research and practice.* Cambridge University Press. doi: 10.1017/cbo9780511527777

Ginsburg, H. P., & Baroody, A. J. (2003). *Test of early mathematics ability* (3rd ed.) [Measurement instrument]. PRO-ED. www.proedinc.com

Ginsburg, H. P., & Dolan, A. O. (2011). Assessment. In F. Fennell (Ed.), *Achieving fluency: Special education and mathematics* (pp. 85–104). NCTM.

Gustafson, S., Svensson, I., & Fälth, L. (2014). Response to intervention and dynamic assessment: Implementing systematic, dynamic, and individualised interventions in primary school. *International Journal of Disability, Development and Education, 61*(1), 27–43. doi: 10.1080/1034912x. 2014.878538

Gurganus, S. P., & Shaw, A. N. (2006, February). *Think-alouds for informal math assessment* [Paper presentation]. SC Council for Exceptional Children, Myrtle Beach, SC, United States.

Haywood, H. C. (2012). Dynamic assessment: A history of fundamental ideas. *Journal of Cognitive Education and Psychology, 11*(3), 217–229. doi: 10.1891/1945-8959.11.3.217

Haywood, H. C., & Lidz, C. S. (2007). *Dynamic assessment in practice: Clinical and educational applications.* Cambridge University Press. doi: 10.1017/cbo9780511607516.007

Heitink, M. C., Van der Kleij, F. M., Veldkamp, B. P., Schildkamp, K., & Kippers, W. B. (2016). A systematic review of prerequisites for implementing assessment for learning in classroom practice. *Educational Research Review, 17*, 50–62. doi: 10.1016/j.edurev.2015.12.002

Hosp, J. L., & Hosp, M. K. (2012). When the "emerging alternative" becomes the standard. In C. A. Espin, K. L. McMaster, S. Rose, & M. M. Wayman (Eds.), *A measure of success: The influence of curriculum-based measurement on education* (pp. 49–56). University of Minnesota Press. doi: 10.5749/minnesota/9780816679706.003.0005

Hresko, W. P., Schlieve, P. L., Herron, S. R., Swain, C., & Sherbenou, R. J. (2003). *Comprehensive mathematical abilities test* [Measurement instrument]. PRO-ED. www.proedinc.com

Hunsader, P. D., Thompson, D. R., & Zorin, B. (2014). Mathematical practices: Small changes in assessments = big benefits. In K. Karp & A. R. McDuffie (Eds.), *Annual perspectives in mathematics education 2014: Using research to improve instruction* (pp. 205–214). NCTM.

Individuals with Disabilities Education Act, 20 USC § 1400 *et seq.* (1997). www2.ed.gov/policy/speced/leg/idea/idea.pdf

Individuals with Disabilities Education Improvement Act, USC 20 §1400 *et seq.* (2004). www.congress.gov/108/plaws/publ446/PLAW-108publ446.pdf

Jenkins, J. R., & Fuchs, L. S. (2012). When the "emerging alternative" becomes the standard. In C. A. Espin, K. L. McMaster, S. Rose, & M. M. Wayman (Eds.), *A measure of success: The influence of curriculum-based measurement on education* (pp. 7–23*).* University of Minnesota Press. doi: 10.5749/minnesota/9780816679706.003.0002

Jitendra, A. K., Dupuis, D. N., & Zaslofsky, A. F. (2014). Curriculum-based measurement and standards-based mathematics: Monitoring the arithmetic word problem-solving performance of third-grade students at risk for mathematics difficulties. *Learning Disabilities Quarterly, 37*(4), 241–251. doi: 10.1177/0731948713516766

Jordan, N. C., & Glutting, J. J. (2012). *Number-sense screener (NSS), K-1* [Measurement instrument]. Brookes Publishing. https://brookespublishing.com

Jordan, N. C., Kaplan, D., Ramineni, C., & Locuniak, M. N. (2009). Early math matters: Kindergarten number concepts and later mathematics outcomes. *Developmental Psychology, 45*(3), 850–867. doi: 10.1037/a0014939

Kiss, A., & Christ, T. (2019). Screening for math in early grades: Is reading enough? *Assessment for Effective Intervention, 45*(1), 38–50. doi: 10.1177/1534508418766410

Klingbeil, D., Maurice, S., Van Norman, E., Nelson, P., Birr, C., Hanrahan, A., Schramm, A., Copek, R., Carse, S., Koppel, R., & Lopez, A. (2019). Improving mathematics screening in middle school. *School Psychology Review, 48*(4), 383–398. doi: 10.17105/SPR-2018-0084.V48-4

Klinger, D. A., McDivitt, P. R., Howard, B. B., Munoz, M. A., Rogers, W. T., & Wylie, E. C. (2015). *The classroom assessment standards for PreK-12 teachers.* Kindle Direct Press. https://evaluationstandards.org/classroom/

Kroesbergen, E. H., Van Luit, J. E. H., & Aunio, P. (2012). Mathematical and cognitive predictors of the development of mathematics. *British Journal of Educational Psychology, 82*(1), 24–27. doi: 10.1111/j.2044-8279.2012.02065.x

Lai, S. A., & Berkeley, S. (2012). High-stakes test accommodations: Research and practice. *Learning Disability Quarterly, 35*(3), 158–169. doi: 10.1177/0731948711433874

Lee, K., & Bull, R. (2016). Developmental changes in working memory, updating, and math achievement. *Journal of Educational Psychology, 108*(6), 869–882. doi: 10.1037/edu0000090

Lee, Y.-S., Lembke, E., Moore, D., Ginsburg, H. P., & Pappas, S. (2012). Item-level and construct evaluation of early numeracy curriculum-based measures. *Assessment for Effective Instruction, 37*(2), 107–117. doi: 10.1177/1534508411431255

Lee, Y.-S., Pappas, S., Chiong, C., & Ginsburg, H. (2010). *mCLASS®:MATH* [Measurement instrument]. Amplify Education. https://amplify.com/programs/mclass-math/

Lembke, E. S., Hampton, D., & Beyers, S. J. (2012). Response to intervention in mathematics: Critical elements. *Psychology in the Schools, 49*(3), 257–272. doi: 10.1002/pits.21596

Lembke, E. S., & Stecker, P. M. (2007). *Curriculum-based measurement in mathematics: An evidence-based formative assessment procedure.* RMC Research Corporation, Center on Instruction. www.academia.edu/29976252

Lewis, K., & Fisher, M. (2018). Clinical interviews: Assessing and designing mathematics instruction for students with disabilities. *Intervention in School and Clinic, 53*(5), 283–291. doi: 10.1177/1053451217736864

Lidz, C. S. (2014). Leaning toward a consensus about dynamic assessment: Can we? Do we want to? *Journal of Cognitive Education and Psychology, 13*(3), 293–307. doi: 10.1891/1945-8959.13.3.292

Lin, P.-Y., Childs, R. A., & Lin, Y.-C. (2016). Untangling complex effects of disabilities and accommodations within a multilevel RTI framework. *Quality & Quantity, 50*(6), 2767–2788. doi: 10.1007/s11135-015-0288-8

Lin, P.-Y., & Lin, Y.-C. (2014). Examining student factors in sources of setting accommodation DIF. *Educational Psychology and Measurement, 74*(5), 759–794. doi: 10.1177/0013164413514053

Locuniak, M. N., & Jordan, N. C. (2008). Using kindergarten number sense to predict calculation fluency in second grade. *Journal of Learning Disabilities, 41*(5), 451–459. doi: 10.1177/0022219408321126

Lopuch, J. (2018). Context matters: Insight on how school-based factors impact the implementation of response to intervention and achievement for students with learning disabilities. *Insights into Learning Disabilities, 15*(2), 207–221. https://files.eric.ed.gov/fulltext/EJ1203401.pdf

Lowenthal, B. (1987). Interviewing to diagnose math errors. *Academic Therapy, 23*(2), 213–217. doi: 10.1177/105345128702300221

McMillan, J. H. (2004). *Classroom assessment: Principles and practice for effective instruction.* Pearson.

Montague, M., & Applegate, B. (1993). Middle-school students' mathematical problem solving: An analysis of think-aloud protocols. *Learning Disability Quarterly, 16*(1), 19–32. doi: 10.2307/1511157

National Council of Teachers of Mathematics (1995). *Assessment standards for school mathematics.* www.nctm.org/Standards-and-Positions/More-NCTM-Standards/

National Governors Association Center for Best Practices & Council of Chief State School Officers (2010). *Common core state standards for mathematics.* www.corestandards.org/Math/

Passolunghi, M. C., & Lanfranchi, S. (2012). Domain-specific and domain-general precursors of mathematical achievement: A longitudinal study from kindergarten to first grade. *British Journal of Educational Psychology, 82*(1), 42–63. doi: 10.1111/j.2044-8279.2011.02039.x

Pearson (2014). *aimsweb®Plus Math.* [Measurement instrument]. www.pearsonassessments.com/professional-assessments/digital-solutions/aimsweb/about.html

Peltier, C., & Harrison, J. (2018). Selecting accommodations for mathematics assessments: Legal and practical considerations. *Preventing School Failure: Alternative Education for Children and Youth, 62*(4), 300–310. doi: 10.1080/1045988X.2018.1443425

Phillips, S. E. (1994). High-stakes testing accommodations: Validity versus disabled rights. *Applied Measurement in Education, 7*(2), 93–120. doi: 10.1207/s15324818ame0702_1

Poch, A. L., van Garderen, D., & Scheuermann, A. M. (2015). Students' understanding of diagrams for solving word problems: A framework for assessing diagram proficiency. *TEACHING Exceptional Children, 47*(3), 153–162. doi: 10.1177/0040059914558947

Polignano, J. C., & Hojnoski, R. L. (2012). Preliminary evidence of the technical adequacy of additional curriculum-based measures for preschool mathematics. *Assessment for Effective Intervention, 37*(2), 70–83. doi: 10.1177/1534508411430323

Powell, S. R. (2012). High-stakes testing for students with mathematics difficulty: Response format effects in mathematics problem solving. *Learning Disability Quarterly, 35*(1), 3–9. doi: 10.1177/0731948711428773

Powell, S. R., Cirino, P. T., & Malone, A. S. (2017). Child-level predictors of responsiveness to evidence-based mathematics intervention. *Exceptional Children, 83*(4), 359–377. doi: 10.1177/0014402917690728

Purpura, D. J., Reid, E. E., Eiland, M. D., & Baroody, A. J. (2015). Using a brief preschool early numeracy skills screener to identify young children with mathematics difficulties. *School Psychology Review, 44*(1), 41–59. doi: 10.17105/spr44-1.41-59

Raghubar, K. P., & Barnes, M. A. (2017). Early numeracy skills in preschool-aged children: a review of neurocognitive findings and implications for assessment and intervention. *The Clinical Neuropsychologist, 31*(2), 329–351. doi: 10.1080/13854046.2016.1259387

Renaissance Learning (2014). *STAR Math* [Measurement instrument]. www.renaissance.com/

Resing, W. C. M., Bosma, M., & Stevenson, C. E. (2012). Dynamic testing: Measuring inductive reasoning in children with developmental disabilities and mild cognitive impairments. *Journal of Cognitive Education and Psychology, 11*(2), 159–178. doi: 10.1891/1945-8959.11.2.159

Rodrigues, J., Jordan, N., & Hansen, N. (2019). Identifying fraction measures as screeners of mathematics risk status. *Journal of Learning Disabilities, 52*(6), 480–497. doi: 10.1177/0022219419879684

Rosenzweig, C., Krawec, J., & Montague, M. (2011). Metacognitive strategy use of eighth-grade students with and without learning disabilities during mathematical problem solving: A think-aloud analysis. *Journal of Learning Disabilities, 44*(6), 508–520. doi: 10.1177/0022219410378445

Schiller, E., Chow, K., Thayer, S., Nakamura, J., Wilkerson, S. B., & Puma, M. (2020). *What tools have states developed or adapted to assess schools' implementation of a multi-tiered system of supports/response to intervention framework?* (REL 2020–017). Institute of Education Sciences, National Center for Education Evaluation and Regional Assistance, Regional Educational Laboratory Appalachia. http://ies.ed.gov/ncee/edlabs

Schrank, F., Mather, N., McGrew, K., Wendling, B., & Woodcock, R. W. (2014). *Woodcock-Johnson® IV Tests of Achievement* [Measurement instrument]. Riverside Insights. www.riversideinsights.com/

Seethaler, P. M., & Fuchs, L. S. (2010). The predictive utility of kindergarten screening for math difficulty. *Exceptional Children, 77*(1), 37–59. doi: 10.1177/001440291007700102

Seethaler, P. M., Fuchs, L. S., Fuchs, D., & Compton, D. L. (2012). Predicting first graders' development of calculation versus word-problem performance: The role of dynamic assessment. *Journal of Educational Psychology, 104*(1), 224–234. doi: 10.1037/a0024968

Stenmark, J. K. (Ed.) (2002). *Mathematics assessment: Myths, models, good questions, and practical suggestions.* NCTM.

Thorndike, E. L. (1924). *An introduction to the theory of mental and social measurement.* John Wiley.

Thurlow, M. L., Albus, D. A., Larson, E. D., Liu, K. K., & Lazarus, S. S. (2019). *2018–19 participation guidelines and definitions for alternate assessments based on alternate academic achievement standards.* University of Minnesota, National Center on Educational Outcomes. https://nceo.umn.edu/docs/OnlinePubs/NCEOReport415.pdf

Thurlow, M. L., Lazarus, S. S., Christensen, L. L., & Shyyan, V. (2016). *Principles and characteristics of inclusive assessment systems in a changing assessment landscape* (NCEO Report 400). University of Minnesota, National Center on Educational Outcomes. https://nceo.umn.edu/docs/OnlinePubs/Report400/NCEOReport400.pdf

Tindal, G. (1998). *Models for understanding task comparability in accommodated testing.* Council of Chief State School Officers.

Uttal, D. H., Meadow, N. G., Tipton, E., Hand, L. L., Alden, A. R., Warren, C., & Newcombe, N. S. (2013). The malleability of spatial skills: A meta-analysis of training studies. *Psychological Bulletin, 139*(2), 352–402. doi: 10.1037/a0028446

van den Bosch, R. M., Espin, C. A., Chung, S., & Saab, N. (2017). Data-based decision-making: Teachers' comprehension of curriculum-based measurement progress-monitoring graphs. *Learning Disabilities Research & Practice, 32*(1), 46–60. doi: 10.1111/ldrp.12122

VanDerHeyden, A., Broussard, C., & Burns, M. (2019). Classification agreement for gated screening in mathematics: Subskill mastery measurement and classwide intervention. *Assessment for Effective Intervention*, 153450841988248. doi: 10.1177/1534508419882484

VanDerHeyden, A., Codding, R., & Martin, R. (2017). Relative value of common screening measures in mathematics. *School Psychology Review*, *46*(1), 65–87. doi: 10.17105/SPR46-1.65-87

Van Norman, E., Klingbeil, D., & Nelson, P. (2017). Posttest probabilities: An empirical demonstration of their use in evaluating the performance of universal screening measures across settings. *School Psychology Review*, *46*(4), 349–362. doi: 10.17105/SPR-2017-0046.V46-4

Verdine, B. N., Golinkoff, R. M., Hirsh-Pasek, K., & Newcombe, N. S. (2017). Links between spatial and mathematical skills across the preschool years. *Monographs of the Society for Research in Child Development*, *82*(1), 1–150. doi: 10.1111/mono.12280

Vygotsky, L. S. (1962). *Thought and language*. Cambridge, MA: MIT Press. (Original work published in 1934).

Wagner, D. L., Hammerschmidt-Snidarich, S. M., Espin, C. A., Seifert, K., & McMaster, K. L. (2017). Pre-service teachers' interpretation of CBM progress monitoring data. *Learning Disabilities Research & Practice*, *32*(1), 22–31. doi: 10.1111/ldrp.12125

Walker, H. M., Small, J. W., Severson, H. H., Seeley, J. R., & Feil, E. G. (2014). Multiple-gating approaches in universal screening within school and community settings. In R. J. Kettler, T. A. Gover, C. A. Albers, & K. A. Keeney-Kettler (Eds.), *Universal screening in educational settings: Evidence-based decision making for schools* (pp. 47–75). American Psychological Association. doi: 10.1037/14316-003

Warren, S., Christensen, L., Chartrand, A., Shyyan, V., Lazarus, S., & Thurlow, M. (2015). *Forum on implementing accessibility frameworks for ALL students*. National Center on Educational Outcomes. https://nceo.info/

Watson, J. B. (1920). Is thinking merely the action of language mechanisms? *British Journal of Psychology*, *11*(1), 87–104. doi: 10.1111/j.2044-8295.1920.tb00010.x

Witt, J. C., & VanDerHeyden, A. M. (2007). The system to enhance educational performance (STEEP): Using science to improve achievement (pp. 343–353). In S. R. Jimerson, M. K. Burns, & A. M. VanDerHeyden (Eds.), *Handbook of response to intervention*. Springer. doi: 10.1007/978-0-387-49053-3

Wong, T. K. Y., Tao, X., & Konishi, C. (2018). Teacher support in learning: Instrumental and appraisal support in relation to math achievement. *Issues in Educational Research*, *28*(1), 202–219. www.iier.org.au/iier28/wong.pdf

Wu, Y.-C. & Thurlow, M. (2019). *2016–2017 APR snapshot #19: State assessment participation and performance of students receiving special education services*. University of Minnesota, National Center on Educational Outcomes. https://nceo.info/Resources/publications/APRsnapshot/brief19

Zhang, D., Ding, Y., Stegall, J., & Mo, L. (2012). The effect of visual-chunking-representation accommodation on geometry testing for students with math disabilities. *Learning Disabilities Research & Practice*, *27*(4), 167–177. doi: 10.1111/j.1540-5826.2012.00364.x

4 Effective Mathematics Instruction

Chapter Questions

1. What overall principles should guide mathematics planning and instruction?
2. How do the principles guide unit planning?
3. What elements are most effective in mathematics lessons for students who struggle?
4. How should instruction be intensified for students with persistent learning difficulties?

As a new teacher at Balsam Middle School, Angela Smith feels a combination of excitement and anxiety. Angela walks down the hallway of her school, looking at student work displayed proudly on the walls, peering into classrooms and seeing students engaged in learning. "This is what I've always wanted to be a part of," thinks Angela, "making a difference in the lives of these students." Angela is excited about teaching, the relationships with students, and working with other professionals. During the internship semester, Angela's cooperating teacher and college mentor confirmed to Angela that her strengths include careful planning, keeping students engaged during instruction, and progress monitoring using student data. Angela's internship was in an elementary school, so Angela feels anxious about working with middle-school students and the mathematics curriculum at this level. How can she learn about so many students and get to know so many teachers in such a short time? Will she be able to keep up with the students in a middle-grades mathematics curriculum? How should she plan for her pull-out classes in mathematics special education? Will those students need a combination of grade-level material and remedial work to catch up with their peers?

Teachers such as Angela must plan and deliver mathematics instruction that will meet the needs of a range of students. Teacher–student interactions form the nexus of learning, with other factors contributing to outcomes secondarily. The teaching-learning model in Figure 4.1 illustrates the interaction of these factors. The overall context for instruction has an impact on learning and teachers have some influence over factors such as instructional time, resources, and classroom management systems. The teacher's characteristics have a major influence on student learning, including content knowledge, knowledge of general and content-specific pedagogy, knowledge of student characteristics, and other attributes such as experience, training, and dispositions (e.g., Connor et al., 2018; Hill et al., 2019; Hill & Chin, 2018). Finally, student characteristics impact learning, such as prior knowledge, attitudes, cognitive attributes, learning characteristics, language abilities, and external supports (e.g., Akyüz, 2014; Geary, 2015; Petty et al., 2013). Providing instruction that will result in student learning is a complex

DOI: 10.4324/9781003096733-4

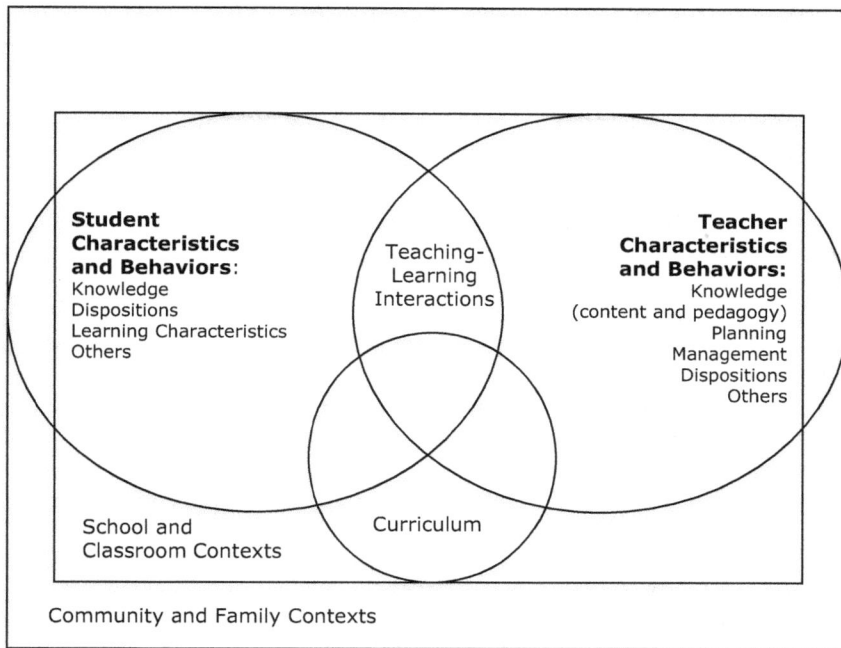

Figure 4.1 Teaching-Learning Model

enterprise. Teachers should endeavor to build and develop all those personal factors under their control while understanding and influencing student and contextual factors within the limits of professional practice.

This chapter's focus is on planning and delivering effective instruction, building on the previous chapters' discussions about mathematics dispositions, key curriculum concepts, student characteristics, and the role of assessment for improving learning. Teachers cannot provide successful instruction without understanding the characteristics of their students, knowing the curriculum, and having a toolbox of informative assessment strategies to integrate with instruction. This chapter begins with a broad view of effective instructional practices, synthesized from decades of research on mathematics instruction for students with learning difficulties. Next, instructional planning at the unit level is addressed, followed by lesson planning guidelines with examples. The final section addresses intensive intervention, highlighting research-based practices for multi-tiered systems of support (MTSS) frameworks.

Principles of Mathematics Instruction

Instructional principles serve as broad guidelines for planning and providing instruction (Box 4.1). They are reminders of the big picture that is frequently forgotten in the daily activities of teachers. They provide benchmarks for purposeful and reflective teaching. These principles describe the features of mathematics instruction that are required for success for students who struggle to learn, based on research. The principles are not discrete, they have areas of overlapping concepts, but those commonalities add coherence and reinforcement. The principles are articulated using classroom examples and can be used as tools for making decisions about instruction. They guide teachers' thinking during planning and actions during instruction.

Box 4.1   Principles of Mathematics Instruction

1. *Emphasize Concept Understanding*: Focus on a few key concepts, understand concepts deeply, teach for student understanding.
2. *Connect with Prior Knowledge*: Consider formal and informal, implicit and explicit prior knowledge. Use advance organizers, scaffolding, and other techniques.
3. *Promote Positive Attitudes*: Assist with goal setting, early success, and making meaningful applications. Model positive attitudes and establish a conducive environment.
4. *Be Mindful of Learner Characteristics*: Consider cognitive, meta-cognitive, affective, behavioral, developmental, and contextual factors.
5. *Design Systematic Sequences with Explicit Elements*: Provide for systematic development of new concepts and processes at unit, lesson, and task levels. Explicit elements in lessons promote success.
6. *Remember the Importance of Language*: Teach mathematics vocabulary and symbol systems in the context of concepts. Use language to promote math dialogue, think-alouds, and communication about concepts.
7. *Recognize the Power of Strategies*: Teach a range of strategies for processes and problem solving so students can use them independently.
8. *Keep Learners Engaged*: Promote active thinking, communicating, using tools, and modeling during mathematics instruction.
9. *Monitor Learning*: Employ methods for periodic progress monitoring as well as monitoring student responses during instruction. Teach students to self-monitor as they are engaged in mathematics.
10. *Ensure Transfer-of-Learning*: Plan for knowledge and skill maintenance, application, and generalization.

Emphasize Concept Understanding

The principle *emphasize concept understanding* incorporates three related aspects: focusing on a few key concepts for each grade level, emphasizing teachers' deep understanding of mathematics concepts, and teaching for student understanding. As discussed in Chapter 1, both the National Council of Teachers of Mathematics and the *Common Core State Standards for Mathematics* (CCSSM) called for fewer topics per grade level, in greater depth (NCTM, 2006; NGA & CCSSO, 2010). What are the most important concepts or *big ideas* of mathematics? Kame'enui and Carnine (1998) defined big ideas as "those concepts, principles, or heuristics that facilitate the most efficient and broadest acquisition of knowledge" (p. 8). They "represent major organizing principles, have rich explanatory and predictive power, help frame questions, and are applicable in many situations" (p. 95). The focus on big ideas is a correction to the overproliferation of narrow instructional objectives and massive numbers of concepts in textbooks, *a mile wide and an inch deep*. Examples of big ideas in mathematics include factor, operation, property, unit, and variable. For example, the concept of *factor* derives from multiplicative properties of number. A factor is any of the numbers or terms that form a product when multiplied together. A factor can divide a number or algebraic expression evenly without a remainder. For the number 24, the numbers 1, 2, 3, 4, 6, 8, 12, and 24 are factors. Knowledge of factors assists simple division, finding equivalent fractions, solving algebraic equations, and factoring expressions such as the difference of two squares $(a^2 - b^2) = (a + b)(a - b)$. The concept of factor is introduced in early elementary grades but is developed across grade levels.

The NCTM and CCSSM documents provide three or four major concepts per grade level. These key concepts are connected with other domains within the grade level and with similar concepts across grade levels. The NCTM's grade-level focal points were published in 2006 to provide target goals for each grade level. For example, the seventh-grade NCTM standards focus on three critical areas: 1) a unified understanding of number, including rational-number operations and negative integers extended to equations; 2) understanding of ratio concepts extended to proportions and percents; and 3) extending the concept of area to circles and surfaces of three-dimensional objects as well as the concept of volume. The CCSSM standards have similar areas of focus for seventh grade. Both sets of standards make connections with concepts in other domains (e.g., measurement, data analysis, algebra).

Teachers could convert these key concepts into areas of focus for each quarter and develop unit questions based on the standards. For example, driving questions for units about proportions could be:

- In what ways can we model proportional relationships?
- Can you identify a constant of proportionality (unit rate) in various depictions of proportional relationships?
- Can you create and solve word problems involving ratio and percent?

Teachers should share these broad questions with students and parents at the beginning of the school year with illustrations of activities in which the class will be engaged. A periodic review of these goals is also helpful for keeping the key concepts in mind.

TRY THIS

Select another grade level and investigate the three or four most important concepts for that grade level within the NCTM or CCSSM standards. Turn one concept into essential questions.

A second aspect of the concept-understanding principle is for teachers to develop deep understanding of the concepts they are teaching, across grade levels. Even special education teachers who may be supplementing core classroom instruction or co-teaching should prepare by examining the concepts within units of study, tracing the development of those concepts across grade levels, and making connections with related concepts. Relying on one's own schooling or a quick look at the textbook will not provide the depth of content knowledge needed for effective instruction. Consider the following situation:

Angela Smith, a special educator, is co-teaching in a seventh-grade mathematics class. Her co-teacher reminds Angela that the next unit of study is about the properties of circles, within the geometry domain and connected with measurement. Angela recalls terms such as radius and circumference. The class has just finished a challenging unit about triangles and angles so Angela knows she must put in some time studying circles. Some questions guide Angela's study: Why are circles special? How are they measured? How do circle concepts relate to those of triangles and angles? How should teachers explain π? Do concepts about expressions and equations from an earlier unit of study relate to circles? What real-world applications involve circles? What misconceptions might students have about circles and their properties?

These are great questions to guide Angela's study of the properties and applications of circles. Angela also needs to trace the development of related concepts across the grade levels. In kindergarten, students developed the concept of attributes of various shapes and could name basic shapes. First graders could compose and decompose shapes, such as half- and quarter-circles, and recognize three-dimensional shapes with circular faces (cones and cylinders). Second graders could draw shapes based on specified attributes and third graders began to develop the concepts of area and perimeter with straight-edged shapes. Fourth graders studied the properties of lines and angles and measured angles with references to the 360° circle in degrees as units. Fifth graders integrated geometry with measurement and data, measuring volume with various cubic units. Last year, in sixth grade, students extended their concepts of area and volume to other polygons and objects and plotted shapes on the coordinate plane. This year in seventh grade, students extend all this prior knowledge to concepts and problems related to area and circumference of circles. Additionally, what concepts and processes will pose difficulties for the students who struggle in this classroom? Angela knows she will not be ready to be an instructional partner for this unit of study until she has the content knowledge of these circle concepts.

The third aspect of concept understanding relates to student learning. Students should understand the concepts involved well enough to solve novel problems, make connections with other concepts and related procedures, and build on concepts to move to the next level. A focus on concept understanding does not reduce the importance of fluency and proficiency with procedures, however. Instruction should address both conceptual and procedural goals. For example, students with limited conceptual understanding often run into barriers when working with fractions. Reasoning about an *algorithm* for a problem such as $\frac{1}{2} \times \frac{1}{3}$ is difficult if you have a simplistic view of multiplication. If your view of multiplication is limited to objects in groups and you expect the product to be a larger number, you will not be able to apply fraction operations and move on to algebraic concepts. If you simply memorize the algorithm for multiplication of fractions, you will not understand when it applies.

The relationship between conceptual understanding and procedural knowledge appears to be bidirectional and iterative, with each supporting the other. In a review of research, Rittle-Johnson et al. (2015) concluded that mathematical competence depends on developing both conceptual and procedural knowledge, and that improvements in procedural knowledge often supports conceptual and vice versa. In a more recent review, Rittle-Johnson (2017) asserted that improvements in conceptual and procedural knowledge also support *procedural flexibility*, defined as "knowing more than one type of procedure for solving a particular type of problem and applying them adaptively to a range of situations" (p. 184). The three types of mathematics knowledge are all promoted by teaching-learning techniques such as comparing, explaining, and exploring.

It is not clear, however, whether instruction should be sequenced from conceptual to procedural or whether alternative orderings of conceptual and procedural segments would be most effective, as both approaches have been supported by research. Rittle-Johnson and Koedinger (2009) found that an iterative sequence for sixth graders, alternating concept lessons with procedural ones, produced more learning gains than beginning with concepts, particularly with applying procedures. L. S. Fuchs, Malone et al. (2017) found that fourth-grade students with lower working-memory capacity benefited from practicing fraction concepts using explanations with manipulatives over practice on procedural fluency, which benefitted most other students with mathematics learning difficulties (MD). However, fraction concept understanding was emphasized with all students in the researchers' interventions and yielded higher effect sizes than those in the control groups, which focused more on procedures. Opitz and colleagues (2017) implemented an intervention with low-achieving middle schoolers that addressed gaps in basic arithmetic concepts: the central ideas of the base-10 number system and the meaning of operations. The primary focus of conceptual understanding was complemented by procedural fluency practice with very compelling results for students in inclusive settings and small intervention groups. The appropriate emphasis and sequence may depend on the specific mathematics content and learning characteristics of students involved in the instruction. Clearly both conceptual and procedural knowledge about mathematics are essential for learning. Teachers should be aware that conceptual understanding is much more difficult to assess than procedural knowledge. Conceptual understanding forms the tripart basis for instructional planning: the intersection of teacher knowledge, curriculum focus, and teaching for student understanding.

Connect with Prior Knowledge

Prior knowledge that is related to new learning may be informal or formal, implicit or explicit. Explicit knowledge is knowledge one is aware of having and can verbalize, while implicit is unconscious (Ziori & Dienes, 2008). Prior knowledge may be acquired in formal settings, such as the classroom, or in everyday experiences. Biemans and Simons defined prior knowledge as "all knowledge learners have when entering a learning environment that is potentially relevant for acquiring new knowledge" (1996, p. 6). Prior knowledge for mathematics learning includes more than previous concepts in a learning trajectory. It includes related vocabulary, informal experiences, overall big ideas or generalizations, and conceptual frameworks. Prior knowledge can assist learning by providing connections for new learning, or it can impede learning through misconceptions.

Prior knowledge is one of the most important prerequisites for learning (Ausubel, 1968; Weinert & Helmke, 1998). A learner's prior knowledge explains between 30% and 60% of variance in learning outcomes (Dochy, 1992). David Ausubel (1968), a cognitive-learning theorist known for promoting *meaningful learning* of subject domains, rejected rote learning as not connected to prior learning and quickly forgotten because it is not connected with long-term memory. He also rejected pure discovery learning (radical constructivism) as not efficient, especially for students who have difficulty constructing connections with prior learning. Ausubel's learning theories held that we learn by constructing a network of concepts and adding to those. Learning must be meaningful—new information can be attached or anchored to what is already known—to be understood, remembered, and applied. Ausubel advocated the use of *advance organizers* for learners that help link prior knowledge to new information. Advance organizers are any introductory materials presented in advance of new information, such as an outline, questions, graphics (and even hypermedia), that present a framework for relating new information to prior knowledge (Ausubel, 1968). Other types of organizers, such as concept maps and other graphic organizers used during instruction, have clearly been influenced by Ausubel's theories.

Connecting prior knowledge is also related to the concept of *backward-reaching transfer*, defined by Salomon and Perkins as a context where "one faces a new situation and deliberately searches for relevant knowledge already acquired" (1989, p. 113). The individual forms an abstraction guiding the reaching back for relevant past learning and connections. The prior knowledge could be already abstracted or could be unformed and not connected within a framework yet. For example, a student is working with a new concept, *percent*, and the teacher prompts him to recall decimal fractions such as $\frac{3}{10}$ and $\frac{25}{100}$. At the time this student was introduced to decimal fractions, they didn't seem to be helpful and were related only to forming decimal numbers. He reorganizes his conceptions of fractions, decimal fractions, place value, and now adds percent concepts so those all fit into a conceptual framework. Some research on mathematics learning has found that new learning can sometimes interfere with prior knowledge in unproductive ways (MacGregor & Stacey, 1997; Van Dooren et al., 2004). For example, MacGregor and Stacey found that new learning about solving equations interfered with 11- to 15-year-olds' prior knowledge about simplifying expressions. One example was widespread misuse and overgeneralization of exponential notation (e.g., using x^3 instead of $3x$). In studying both children and adults and their learning using number-line estimating, Laski and Dulaney (2015) found that individuals with better inhibitory control—those who could suppress prior, naive knowledge that would interfere with new learning—demonstrated greater mathematics achievement. Teachers should be aware of potential unproductive effects of prior knowledge when students are involved in new learning.

Another related concept is *scaffolding*, building up new concepts and skills using prior knowledge and existing frameworks (first defined by Wood et al., 1976 with later work by Hogan & Pressley, 1997; van de Pol et al., 2010). Scaffolding is based on Vygotsky's concept of Proximal Zones of Development (PZD), the distance between the student's current knowledge and the potential knowledge if guided by adults or peers (Booth et al., 2017). Scaffolding involves three essential elements: 1) it is contingent on students' current understanding, 2) there is a gradual withdrawal of supports, and 3) responsibility transfers from the teacher to student. Intentional scaffolding is when the teacher deliberately guides the student's use of prior knowledge to develop new concepts. For example, a teacher is working with a student who cannot seem to understand the concept of a square root. She reaches back to explore what the student already understands and reminds him that they learned about squares, in both algebra (x^2) and geometry (area). If $5^2 = 25$, then $\sqrt{25} = 5$. She uses a plant's root analogy and diagram to assist the student's concept connections. Conscious scaffolding can be made by the student, self-direction to use previous learning to learn something new.

Research on mathematics teachers' awareness of the relationship between prior knowledge and new learning found that mathematics teachers tended to be explicitly aware of when students are or are not making connections with prior knowledge (Hohensee, 2016). However, teachers were not as aware of backward-transfer effects. For example, students are working on the new concept of quadratics, then begin demonstrating difficulties with the previously learned concepts of linear equations and linear data sets. Teachers often blame these issues on misconceptions of the original topic when, in fact, the new learning has negatively influenced prior understanding. Teachers should also be aware that optimal sequences of instruction may be influenced by prior knowledge. Fyfe et al. (2012) found that providing exploratory activities in problem solving prior to explicit instruction for second and third graders facilitated learning for students. However, students with little prior knowledge benefitted from feedback on strategies and outcomes during exploration while students with moderate prior knowledge benefited more from no feedback, an *expertise-reversal effect*. Student prior knowledge appears to have an impact on specific instructional elements, such as the use of exploration and feedback.

Teachers' understanding of student prior knowledge may differ, with some restricting it to prior applications (Lee et al., 2019). Teachers should view prior knowledge as knowledge to be developed further with new mathematics tasks. Teachers can assess students' prior knowledge through simple questions before a new concept is introduced. For example, before a question about linear measurement, ask, "What do you know about measuring? What tools do we use to measure? Have you measured something in centimeters before? How do those units compare with inches?" Other methods for assessing prior knowledge include known/unknown charts, pretests, think-alouds, interviews, and guided graphic organizers.

Finally, prior knowledge is related to the concept of learning trajectories. Within any mathematics domain there is a hypothesized trajectory of learning, beginning with the foundational skills and concepts related to a specific domain, and continuing through to target goals across months or years. Children move through "natural developmental progressions in learning and development" (Clements & Sarama, 2014, p. 3). Trajectories describe the goals of learning, the thinking and learning processes at each level, and learning activities that promote building concepts, skills, and processes (Clements & Sarama, 2014). This progression involves building competence over time, with new learning building upon the previous.

TRY THIS

Describe the prior knowledge needed for one of the concepts below. Use the curriculum standards or learning trajectory maps if needed.

a. coordinate-plane graphing   c. negative integers
b. percent   d. value of digits in hundredths' place

Promote Positive Attitudes

As discussed in Chapter 1, student and teacher attitudes about learning and mathematics have a direct effect on student learning. Teaching strategies that can build positive attitudes towards mathematics include: help students set goals, provide opportunities for early success, plan relevant applications of concepts, help students see the relationship between effort and result, model positive attitudes, and establish a classroom environment that promotes positive views and common goals. Having students set and monitor progress toward goals can have a significant effect on achievement in mathematics (e.g., Fadlelmula et al., 2015; Wang et al., 2019). However, goal-setting ability and follow-through varies significantly among students; it is a highly individual, dynamic, and context-dependent activity (Middleton et al., 2018). Students who develop more specific and learning-oriented goals are more likely to have positive outcomes in mathematics. Students who are able to envision distal goals (long-term) tend to manage time better, process information more efficiently, employ self-regulatory strategies more intensely, and achieve high grades in school.

Goal setting is particularly difficult for many students with disabilities. Some instructional programs such as Mercer and Miller's *Strategic Math Series* (1991–1994) include specific steps for goal setting. Students write personal achievement goals and chart their own progress toward those goals. This deliberate and visual type of goal setting has been linked to greater motivation toward learning and more success in reaching goals (Mercer & Miller, 1992). Self-monitoring and charting strategies also help students see the relationship between effort and success.

Motivation is the *why* behind behaviors and cognition. It is what focuses or energizes a student's attention, emotions, and activity (Mercer & Pullen, 2005). Motivators can be extrinsic or intrinsic, with a goal to promote more intrinsic, life-long, self-motivated habits of learning. Students with poor motivation towards school in general or specific subjects such as mathematics have usually experienced repeated failure, poor teaching, or low expectations. Additional techniques that are effective in promoting positive motivation include: give students choices and opportunities to help with planning, employ a variety of engaging instructional methods, make instruction personally relevant for students, challenge students at their instructional levels, provide clear and frequent feedback, focus on student effort and improvement, treat errors as a normal part of learning, communicate positive expectations, plan for student success, and teach with energy and enthusiasm (Anthony & Walshaw, 2009; Sharma, 2015; Walkington & Hyata, 2017). Motivated students tend to feel connected to school, are interested in mathematical tasks, and hold positive academic self-efficacy beliefs. In a study of middle-school mathematics students, Cleary and Kitsantas (2017) found that prior achievement in mathematics contributed the most variance in mathematics achievement, but self-efficacy and self-regulated learning contributed almost as much.

Teachers have an enormous influence over the affective aspects of the learning environment. By sending verbal messages of high expectations, providing opportunities to learn while allowing risk-taking and making mistakes, arranging productive student groupings, and planning meaningful learning activities, teachers create atmospheres that support learning.

TRY THIS

Visit a mathematics class and note the affective elements in the learning environment. What teacher and student behaviors promote or detract from a learning environment that encourages student risk-taking, opportunities for success, and positive attitudes about mathematics?

Be Mindful of Learner Characteristics

In the teaching-learning paradigm, learner characteristics have a significant impact on the success of teaching efforts. Characteristics include prior learning and attitudes about learning, as discussed for the previous principles. They also include linguistic and language abilities, cognitive factors such as working memory and attention, reading ability, the nature of a student's disability as it impacts learning, environmental factors such as home and community resources, developmental factors such as age and achieved milestones, and social and emotional traits that influence student participation. For students with specific disabilities in mathematics, teachers should attend closely to verbal and visual working memory, processing speed, attention, and other executive functions (Geary, 2015; Johnson et al., 2010). Some researchers suggest that instruction for students with mathematics learning disabilities (MLD) include tactics to reduce the demands on students' working memory, especially during practice sessions (e.g., L. S. Fuchs, Malone, et al., 2017). There is also evidence that those students with the lowest achievement measures prior to instruction and are nonresponders to instruction may have individual characteristics that impede learning and require even more intensive and individualized efforts (L. S. Fuchs et al., 2019).

For students who struggle with mathematics, a comprehensive assessment that includes the full range of factors that influence learning is critical for planning appropriate instruction and accommodations, regardless of the instructional setting. Once factors are identified, teachers should consider instructional designs that lessen or ease learning challenges (L. S. Fuchs, Fuchs, Powell, et al., 2008; L. S. Fuchs, Malone, et al., 2017). For example, for a unit about solving problems using multidigit multiplication, one student may lack basic number-combination fluency and require brief fluency instruction and drill before each lesson. Another student has verbal working-memory issues and cannot hold the product of 5×7 in her mind long enough to add the amount regrouped. A multiplication strategy for recording as she works would ease this challenge. A third student demonstrates attention problems during mathematics. An increased frequency of communication between teacher and student as well as carefully structured segments for instruction may assist this student. This student may also benefit from an accommodation such as designated seating. A fourth student has difficulty reading directions and word problems, so the teacher provides preinstruction in key vocabulary and checks for comprehension during instruction. See Chapter 6 for more discussion on instructional accommodations.

Students from different cultural or linguistic backgrounds may have different informal mathematics experiences and prior knowledge. Accurate and unbiased assessment of the mathematics background of these students is essential for planning instruction. Teachers should be aware of myths associated with cultural and linguistic differences:

- *Myth: Language differences are disabilities.* In fact, language differences may hide giftedness and true performance levels.
- *Myth: Mathematics is a culture-free area of study.* Language and culture are important factors in teaching and learning mathematics. For example, the history of geometry can shed light on terminology and tools used within geometry studies.
- *Myth: A student's ethnicity will indicate mathematics ability.* A teacher cannot assume that an Indian-American student will be gifted in mathematics or that a Native-American student will not be competitive. Each student's background and abilities should be assessed without the bias of these stereotypes. Those biased expectations can be as handicapping as a disability label.
- *Myth: A student with a background of poverty or a family with limited education will have little informal mathematics knowledge.* Again, individuals will have different experiences that influence their informal knowledge. A child in an impoverished environment may have some important mathematics experiences such as careful budgeting, playing counting games, or mental mapping of the neighborhood.

Teachers should take care to plan examples and real-life problems that link to individual student experiences yet extend a student's world knowledge. For students who are simultaneously learning English (EL), best practices have evolved from teaching math in the student's native language in pull-out classes to a more inclusive, but individualized approach based on the content, students' prior knowledge, and other student-level characteristics. Moschkovich (2013) advocated for equitable mathematics teaching practices: providing access to curricula; supporting mathematical reasoning, conceptual understanding, and discourse; and broadening participation for EL students. Teachers should focus on students' reasoning and practices, not on accuracy with language and vocabulary. Students' home language should be treated as a resource, not an obstacle. Mathematics tasks should not be watered down; they should be high-quality, grade-level tasks with multiple representations to support understanding and expression. Milyutin and Meyer (2018) recommended preteaching some vocabulary, providing multiple representations, building on students' current language and concepts, providing supports for group work, and expecting students to participate and succeed. These recommendations are similar to those for students with disabilities, to differentiate and individualize adaptations for access to grade-level standards.

Design Systematic Sequences with Explicit Elements

The sequence of instruction within a mathematics unit should provide for systematic development of new concepts and processes, based on research-supported approaches as well as individual student characteristics and responses to instruction. Research with students with mathematics difficulties over decades provides strong, compelling evidence for the need for *explicit* instruction that is systematic (Gersten, Beckmann, et al., 2009). There is also evidence that explicit instruction benefits typical students as well (Doabler et al., 2015; Marita

& Hord, 2017). Explicit instruction is overt and deliberate, not left to chance or discovery. Providing explicit instruction, however, is not equivalent to rote, unconnected, boring instruction without meaning. Explicit instruction does not require a teacher script; but must be carefully planned and sequenced. The call for explicit instruction must be understood in the context of its research support. Gersten, Chard, et al. (2009) noted that the term *explicit* may describe a wide array of instructional approaches. For a meta-analysis of mathematics instructional components, the researchers defined explicit instruction as including teacher step-by-step demonstration of problem solving, steps requisite for a specific type of problem, and students required to use the same steps. In their meta-analysis of studies from 1971 to 2007, Gersten, Chard, et al. (2009) identified 42 studies of interventions with students with MLD. The *effect sizes* \ for explicit instruction were large and significant, $g = 1.22$. The other effective interventions included systematic and explicit elements such as student verbalization of reasoning and using systematic and visual methods to solve problems. Dennis et al. (2016) conducted a meta-analysis of 25 studies identified from 2000 to 2014. The largest effect for students with MLD was peer-assisted learning, especially with younger students. Explicit, teacher-led instruction produced positive, moderate effects. Providing teachers and students with progress-monitoring data also had moderate effects on student learning. The authors concluded that specific systematic elements of interventions, such as sequencing tasks from easy to difficult and explicit explanations of concepts and procedures, were likely to benefit students with MLD.

The National Mathematics Advisory Panel defined explicit instruction: teachers provide clear models for solving problems using a range of examples, students receive extensive practice in strategies and skills, students have opportunities to think aloud, and students are provided with extensive feedback (NMAP, 2008). The NMAP noted that not all mathematics instruction needs to be explicit but that students who struggle with mathematics may need more regular explicit instruction. VanDerHeyden and Codding (2020) defined explicit instruction for school psychologists who may not be as familiar with the research as "… a systematic approach that incorporates previewing of previous skills and concepts, precise instructions, modeling, guided and independent practice, immediate feedback, and checks for maintenance of skills" (p. 23). They emphasized that explicit instruction allows for differentiation, scaffolding student understanding, and multiple formats for student responding. When planned carefully, it is highly engaging and provides students the opportunity to develop reasoning, problem solving, and other mathematical practices. In an overview of explicit instruction for students who struggle to learn content, Hughes et al. (2017) noted five essential elements: segmenting complex skills into smaller units, drawing student attention to important features of the content using modeling and think-alouds, promoting successful engagement by using systematically faded supports, providing opportunities for student response with feedback, and creating purposeful practice events.

Within explicit instruction, another sequence with instructional implications is the *Instructional Hierarchy* (IH), first proposed by Haring and Eaton (1978). According to this hierarchy, the stages of learning progress from *acquisition* of new knowledge to *fluency* with concepts and skills, then to *maintenance* over time and *application* in other contexts. New learning occurs within the acquisition stage and requires explicit modeling, guided practice, and frequent feedback. Students in the fluency phase, building speed and accuracy, need additional practice and possibly some special approaches such as incremental rehearsal and timed practice (Burns et al., 2010). In a meta-analysis of mathematics computation interventions, Burns et al. (2010) found that interventions should be adapted to the student's stage of skill development. Acquisition interventions demonstrated large effect sizes for students working with new concepts (frustration level) and yielded moderate effect sizes for students working at their instructional levels. Fluency interventions had only a small effect on students working

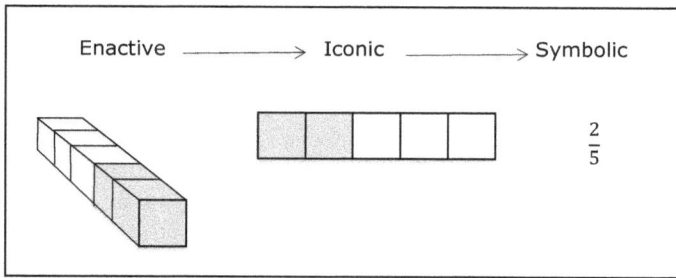

Figure 4.2 Modes of Representation

at a frustration level and were more effective at the instructional level. Curriculum-based measurement (CBM) probes, as discussed in Chapter 3, and many research-based mathematics programs offer recommendations for instructional levels for their content. Students' levels of skill should be taken into account when planning learning sequences.

Another sequencing strategy, also within explicit instruction, matches new concepts via representations to align with students' cognitive levels of understanding. This sequence for teaching new mathematics concepts and skills derives from Bruner's (1966) modes of representation or ways of knowing. Foundational to Bruner's work, Pestalozzi advocated the systematic use of objects to make number relationships clear, around 1800. Many methods texts of the early 20th century recommended the use of objects for mathematics instruction but the cognitive sequence that included a representational stage had not been developed until Bruner's work (Bidwell & Clason, 1970). Bruner's modes of representation are depicted in Figure 4.2 and represent movement from enactive to iconic and finally to symbolic.

The *enactive mode* refers to the child acting upon objects directly, through use of motor responses and not necessarily images or words. In mathematics this is translated into the *concrete* level where students manipulate materials to represent concepts. The *iconic mode* represents visuals, organized images, or icons that represent objects, but cannot be manipulated. This has been called the *representational* or *semi-concrete* level in mathematics. Examples include using drawings, graphs, tallies, diagrams, and other pictorial representations of math concepts. *Virtual manipulatives* on websites or apps, discussed in Chapter 7, are representational but have concrete elements. Finally, the *symbolic mode*, also termed *abstract*, uses rather arbitrary symbols to represent objects or concepts. In the symbolic mode, students are presented with verbal or written material to represent mathematical concepts, including numbers, symbols, and verbal explanations. Bruner's later work emphasized the importance of the teacher's communication and guidance in the learning process, not simply the student's interaction with objects and pictures (Bruner, 1982).

The concrete-representational-abstract (CRA) instructional sequence is supported by a large and robust body of research across grade levels for mathematics learning (e.g., Flores et al., 2014; Harris et al., 1995; Peterson et al., 1988; Sealander et al., 2012). Many students who have difficulty learning new math concepts and procedures have not had adequate experience with the concrete and representational levels of the sequence and bridging between levels that is provided by language: self-talk, dialog, and teacher-directed scaffolding. Concrete materials have the advantage of connecting with prior knowledge, being grounded in perceptual and motor experiences, and offering a correspondence between a form and its referent (Fyfe et al., 2014). Abstract materials have advantages in eliminating extraneous properties and representing structures more efficiently. Concrete materials, if used too long, can constrain transfer of knowledge (Sloutsky et al., 2005). Starting at the abstract level risks inefficient solution methods and the manipulation of symbols without conceptual understanding (Nathan, 2012). "By slowly decontextualizing the concrete materials, concreteness fading homes in on the pertinent structural features and results in a faded representation that can be a useful stand-in for a variety of specific contexts" (Fyfe et al., 2014, p. 13). The resulting knowledge is grounded and meaningful as well as abstract and transferable. There is evidence that direct instructional guidance (explicit instruction) is needed for students to move across CRA levels and transfer learning to more abstract applications (Kaminski & Sloutsky, 2009). The CRA sequence (concreteness fading) has been incorporated into special education mathematics programs (e.g., Mercer & Miller, 1991–1994) and at least three widely used curricula: *Singapore Math* (www.singaporemath.com), *Everyday Mathematics* (Carroll & Issacs, 2003), and *Building Blocks* for preschool-age children (Clements & Sarama, 2007).

Some teachers begin with concrete objects but move too quickly into representational and abstract forms. Others allow students to use concrete manipulatives far longer than needed for concept understanding. Many teachers omit the representational level altogether or limit student interaction with diagrams. Others, especially in upper grades, work exclusively at the abstract level. Textbooks begin at the representational level unless a manipulative set accompanies the text or the instructor's manual emphasizes the importance of the concrete level. Even with higher-level topics such as algebra and trigonometry, students need to work with models and representations of concepts in ways that are familiar as bridges to the demands of the abstract level. Students who have difficulties in mathematics may need supports at a previous level before continuing in the topic more abstractly.

An example of moving through Bruner's modes of representation is developed here with the concept of volume. At the concrete level students are manipulating interconnecting blocks, with each block representing a cubic unit. Combining blocks to form larger blocks results in *prisms* \ of differing dimensions from which students could compute volume in a number of ways. The next level, representation, employs diagrams depicting cubes and prisms labeled with dimensions. Students could use their concrete models to visualize the diagrams or apply rules developed using the models to the dimensions indicated on the diagrams. Finally, students working at the abstract level should be able to apply previous learning to a question such as, "What is the volume of a swimming pool with the dimensions 3 meters deep, 6 meters across, and 12 meters long?" They should use only the numbers and other mathematical symbols and formulas. Students who have trouble working entirely at the abstract level can fall back on drawing a diagram (representational) or making a model (concrete).

In a synthesis of research from 1975 to 2015, Bouck et al. (2018) identified 20 studies that examined the use of CRA for mathematics, addressing elementary and middle school students with MLD. Eight of these studies met the rigid criteria of methodologically sound research, therefore CRA is considered an evidence-based practice and should be considered for use in core mathematics classrooms and interventions for students with MLD. The CRA sequence has been effective when integrated with explicit strategy instruction for basic-fact combinations and procedures involving multidigit subtraction and multiplication (Flores et al., 2014; Mancl et al., 2012; Miller et al., 2011; Harris et al., 1995). CRA has also been effective for instruction in fractions (Butler et al., 2003) and introductory algebra topics (Maccini & Hughes, 2000; Maccini & Ruhl, 2000). Sometimes the CRA sequence can be implemented across one or two lessons, as with introducing the meaning for x^2 (using algebra tiles, then drawings) if students have previously worked with area concepts. In other situations, students may need two to three lessons at the concrete level and two to three at the representational level before working entirely at the abstract level with symbols. Even then, some students may need to move back to representational or concrete for additional work. Some concepts, such as fractions, may need iterative concrete (blocks or paper strips) and representational (number lines) work while students are working with symbols.

Carefully planned, systematic instruction with explicit elements has compelling research support over decades, especially for students who struggle in mathematics (Dennis et al., 2016; Gersten, Chard et al., 2009; Stevens et al., 2018). Systematic instruction involves carefully analyzing the content to be taught, breaking down topics into smaller pieces, and organizing content to connect to prior knowledge and develop sequentially. This instruction is enhanced through explicit elements, such as teacher modeling, think-alouds, scaffolding understanding, a sequence of visual supports, guided and differentiated practice with feedback, and engaging, success-oriented lessons.

TRY THIS

Select one of the following mathematics concepts and develop an example for each level of the concrete-representational-abstract sequence:

a. adding fractions with like denominators
b. subtraction with regrouping
c. computing the areas of quadrilaterals

Remember the Importance of Language

The language of mathematics allows us to manipulate difficult concepts and communicate with others. Language forms the bridge between the more visual representations of mathematics ideas to symbolic and abstract forms. In the teaching-learning activities of mathematics classrooms, language demands on students are formidable, even for students without language or learning disabilities. Mainstream mathematics educators and researchers have begun to focus on the essential role of language for mathematics teaching and learning (Radford & Barwell, 2016). They point to ways students' natural or informal language may interfere with or support mathematics understanding, how language advantages for some students may advance or hinder discussions within learning groups, and the challenges of diverse language classrooms for dialog and problem solving. Further, students with higher language proficiency outperform students with lower proficiency in mathematics classrooms (Erath et al., 2018; Prediger et al., 2019).

Receptive and expressive language demands, as well as their underlying cognitive facilities, should be understood and addressed by teachers. These include (Erath et al., 2018; Thomas et al., 2015; Thompson & Rubenstein, 2014):

- *Receptive language demands*: listening to the teacher during think-aloud, modeling, and directions; listening to peers during justifications, group work, and demonstrations; learning new vocabulary; understanding the meaning of the mathematics symbol system; reading from the textbook, test directions, or word problems; and general comprehension of new information presented via language.
- *Expressive language demands*: explaining work and justifying reasoning orally or in writing; participating in small group work with peers; serving as a tutor for a peer-assisted learning situation; using mathematics symbols and their syntax to express work on problems; using mathematics vocabulary accurately and precisely; describing tasks or models; and asking for assistance or clarification.

- *Cognitive demands related to language*: using aspects of memory for recalling prior semantic information (long-term); holding current information while using it (working); maintaining new knowledge for later use (sending to long-term); organizing new information for more effective use (e.g., number systems, word-problem types); and attending to critical aspects of new information while disregarding the extraneous.

Current mathematics process standards call for communication as a means for engaging in mathematics, for thinking and reflecting, and for understanding others. The CCSSM ask students to make conjectures, justify conclusions, communicate those to others, and respond to the arguments of others (2010). The NCTM (2000) standards described the role of communication across the grade levels. "Communication is an essential part of mathematics and mathematics education. It is a way of sharing ideas and clarifying understanding. Through communication, ideas become objects of reflection, refinement, discussion, and amendment" (p. 60).

Classroom discourse has been identified as a key element in developing mathematics process skills and concept understanding. But simply increasing the amount of talking in a classroom will not result in learning. Planned discourse that leads to learning is a complex activity (Walshaw & Anthony, 2008). In a review of mathematics classroom-discourse research with diverse students, Walshaw and Anthony organized findings around four purposeful and nested activities of teachers: 1) facilitating student participation by clarifying, establishing, and enforcing discourse participation rules; 2) careful listening and attentive noticing so that teachers can scaffold students' ideas forward; 3) fine-tuning student understanding through accurate mathematics language; and 4) shaping student debate and defense of ideas. The authors concluded that teachers who set up classroom conditions that facilitate these activities increase students' communication skills, mathematics understanding, and productive dispositions.

Research with students with difficulties in mathematics on increased student-to-student discourse has been mixed, while teacher-guided discourse, scaffolding, feedback, and the use of explicit language continues to receive support (Griffin et al., 2013; Haydon & Hunter, 2011). Doabler and colleagues argued that higher-quality practice opportunities through mathematically rich discourse can guide and scaffold students through explaining steps and reasoning about a solution, which lead to a higher-quality student verbalizations and subsequent achievement (Doabler, Clarke, Kosty, Kurtz-Nelson, et al., 2019). Other researchers recommended that teachers "tilt" the cognitive work (problem solving, reasoning) to students while providing scaffolds and asking questions, dynamically adapting their teaching responses (Xin et al., 2020). Scaffolding high-quality explanations, designing prompts that do not divert attention from concepts, and prompting students to explain why information is correct or incorrect also promote mathematics learning (Rittle-Johnson et al., 2017).

Drawing from cognitive science, Star and Verschaffel (2018) promoted the use of *explanatory questioning*, including elaborative interrogation and self-explanation, to improve students' mathematics learning and engagement with higher-order cognitive activities such as reasoning, hypothesizing, and justifying. Most of the research on elaborative interrogation has been conducted with reading, but this technique—asking a lot of "Why?" questions—shows promise for mathematics settings. Learners with low prior knowledge may struggle with "Why?" questions unless guided prompts and scaffolds are provided. Self-explanation helps learners identify gaps in their understanding and pushes them to modify or repair it.

Discourse within a mathematics classroom can be hampered by limited or imprecise mathematics vocabulary (Ricconni et al., 2015). Teachers should build student understanding of vocabulary (and related concepts) by connecting language use with models and diagrams, being consistent with terminology, and using math language redundantly as students gain understanding. Some students may need more explicit instruction on vocabulary, embedded

Figure 4.3 Confusing Mathematics Terminology

within concept or procedural instruction (Thomas, 2015). Teachers who lack a mathematics background should make an extra effort to research the most current and appropriate terminology before delivering instruction to students. For example, students in the 1960s and 1970s *borrowed* and *carried*. Today's teachers use the term *regrouping* for these place-value conversions. Teachers need to be aware that mathematics vocabulary often has confusing meanings for children (Figure 4.3) (Rubenstein & Thompson, 2002; Thompson & Rubenstein, 2000). For example, some words have different meanings in everyday English and mathematics: right, foot, and reflection. Other math words have more than one meaning within mathematics: round, square, side, second. Teachers can explore word origins (Rubenstein, 2000), use literature, and have children create their own definitions to enhance mathematics vocabulary learning.

To develop expressive language for mathematics, students should be asked to articulate math sentences, concepts, and processes as they are working. They should be asked to read mathematics sentences aloud, to explain the steps in a process, or share how they came up with a solution to a problem. Teachers should deliberately increase the verbal response rate of each student within learning contexts, whether they are in whole-group or small-group settings. Thompson and Rubenstein (2014) recommended tactics often employed by language-arts teachers to facilitate students' language: waiting for students to respond, revoicing a student's response, inviting student participation, probing student thinking, and creating opportunities for students to engage with another's reasoning.

Mathematics tasks that involve reading, such as textbook examples and word problems, may be particularly challenging for students who are learning English or those who have comorbid reading disabilities. The reading demands within mathematics assessments are also increasing (e.g., revised SAT, Murphy, 2015). Students should be taught specific reading strategies for mathematics applications, such as paraphrasing, explaining the material to a partner, building specialized vocabulary, and drawing schematic diagrams. Students who are learning English should be provided language supports (e.g., more visuals) in mathematics for language-heavy tasks such as teacher discussions and word-problem solving. It is critical that students' reading levels do not determine their mathematics levels and therefore block access to more advanced concepts.

TRY THIS

Explain to a colleague how to simplify one of the following expressions. Attend to the language demands.

$$\sqrt{9} + \frac{8x}{2} \qquad 3(x+6) + 7x$$

Recognize the Power of Strategies

A strategy is an individual's approach to a task, including how a student thinks and acts when planning, executing, and evaluating performance on the task (Deshler & Lenz, 1989). Strategies allow students to acquire, store, and use information in a variety of new settings, thereby gaining more control of their own learning. Strategies are most effective when they can be applied to a range of task demands. Strategies for mathematics include problem-solving approaches (Montague, 1992; Poylá, 1973), memory devices (e.g., simple mnemonic devices to remember sequences), strategies for finding number combinations (Mercer & Miller, 1992), and procedural strategies (e.g., a breaking-apart strategy for multidigit subtraction). While many students appear to develop strategies naturally, students with mathematics difficulties generally need explicit instruction in how to select and apply strategies.

Simple mathematics strategies can be demonstrated and practiced within one or two lessons. For example, an alternative strategy for solving multidigit multiplication problems is called lattice multiplication because the drawing resembles latticework (Broadbent, 1987). This strategy, thought to originate in 15th-century Europe, reduces the confusion of regrouping several times over the same column of numbers and limits the memory demands on the student (see Figure 4.4). A simple strategy such as lattice multiplication maintains fidelity to the underlying mathematics concepts but is limited to a narrow range of applications.

More complex strategies take longer to teach but can be applied to a wider range of situations. More complex strategies empower students to select and evaluate procedures for mathematics situations, to monitor their own performance, and to be aware of their executive processes during learning.

Teaching cognitive and metacognitive strategies has been cited as one of the most effective instructional approaches for students with disabilities (Lloyd et al., 1998; Swanson et al., 2014). Included in this instruction are effective instructional components such as modeling (with thinking aloud), providing guided practice with scaffolding and feedback, teaching for mastery, and teaching for generalization. Gradually the student assumes responsibility for practice, self-monitoring, and applying the strategy in actual learning settings. Students who have developed a repertoire of learning strategies are more successful in regular classroom placements (Montague, 1997). Strategic learners eventually personalize strategies or develop their own strategies for dealing with instructional setting demands. Losinski et al. (2019), in a meta-analysis of research with students with emotional disabilities and mathematics deficits, identified strategy instruction as a promising, evidence-based practice with strong evidence for teaching number sense, fractions, and problem solving. The authors emphasized the promise of the self-regulated strategy development (SRSD) approach with students with mathematics

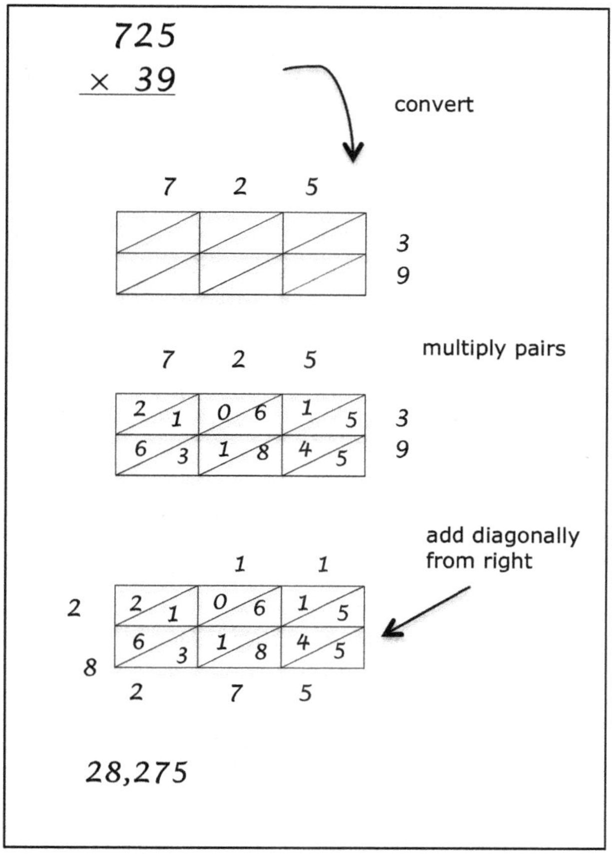

725
× 39

convert

7   2   5

3
9

7   2   5

multiply pairs

| 2/1 | 0/6 | 1/5 | 3 |
| 6/3 | 1/8 | 4/5 | 9 |

add diagonally
from right

1   1

2	2/1	0/6	1/5
	6/3	1/8	4/5
8			
	2	7	5

28,275

Figure 4.4 Steps for Lattice Multiplication

difficulties or at risk for poor outcomes (e.g., Ennis et al., 2014). The SRSD approach, also used for reading and writing interventions, is a systematic method for implementing strategy instruction using steps across teaching-learning episodes: develop background knowledge, discuss the strategy, model the strategy, ensure memorization of strategy, support students' learning (via scaffolding and feedback during practice toward goals), and establishing independent practice. These steps are similar to those of the Strategy Intervention Model (Deshler & Lenz, 1989).

Keep Learners Engaged

Students who struggle with mathematics often demonstrate problems maintaining attention and being active learners. They tend to be passive or fail to attend to the critical aspects of instruction. Mathematics curriculum standards are accompanied by process (NCTM) or practice (CCSSM) standards that have been called the verbs of mathematics. Those are what students should be doing while engaged in learning: problem-solving, reasoning, justifying, constructing, modeling, attending, communicating, making connections, representing, and expressing (Figure 1.1 in Chapter 1). Teachers should consider how to incorporate these processes when planning units and lessons, as is illustrated in Table 4.1 later in this chapter.

For daily mathematics lessons, teachers must plan learning activities that encourage a high rate of student responses and attention to learning tasks. Some effective techniques include group or choral responding (e.g., response cards, whiteboards), alternating individual responding with group responding, skillful questioning, and guided small group or peer interactions. For some types of content, such as factual or rule learning, responses can be quick and frequent. However, for concept development and processes, student responses tend to be less frequent so teachers should build in steps that require more interaction and engagement. Response rates will be higher for fluency building activities than for acquisition lessons, and even lower for written responses. Teachers who are skilled in asking questions use clear and precise language, pause before calling on students, and provide quick, encouraging feedback.

Teachers should also structure lessons in alternating segments that promote attention and engagement. For example, beginning with an active warm-up activity (e.g., leading choral mental math) could be followed by a teacher demonstration, then immediate student practice with feedback and a guided problem-solving episode in small groups. Some teachers establish routines so that students are active learners from the moment they enter the classroom to the time of departure. An example from Angela's classroom:

> Students are expected to pick up their individual folders, find their seat, and complete the review problem on the board while I am greeting students, checking parent notes, and taking attendance. Their folders indicate the class events for the day, either beginning with my group, practicing a skill using a game or the computer, or working in a peer-tutoring pair. The groups rotate so that I can monitor each student's learning and adjust the pair teaching and independent practice. At some point in the period I have a whole-group problem-posing session where students contribute at their own levels. At the end of class, while students return materials and folders, I ask each one to solve a problem, give an oral explanation of something we covered in class, or use a new vocabulary word in a sentence. I want them to leave feeling successful in learning something new each day.

Keeping learners engaged at this level requires extensive planning and knowledge of students. Keeping learners engaged is not the same as having a busy, talkative classroom of students. Engagement is not equivalent to active or on-task behavior. It is not simply offering some fun activities on Friday or using manipulatives. Students should be *intellectually* engaged, argued Peterson et al. (2013). Are the questions the teacher poses and the other classroom activities requiring students to think? Are activities and classroom dialog requiring students to make connections with prior learning, other concepts, and applications? Are lessons interesting but still addressing important concepts? For example, instead of stating, "Here we have one-half of a rectangle, can you show me one-half of this one?" try instead, "Let's think of several ways we use one-half every day. I want you to think for 30 seconds, until I give the signal, then explain your one-half to your partner." [Pause.] "Now let's see how many different examples for one-half we have" [adding specific examples through prompting].

Activities that engage learners promote positive dispositions, increase practice with language and concepts, and enhance overall mathematics achievement if those are planned carefully, geared to students' learning profiles, and monitored.

Monitor Learning

Teachers should monitor the effectiveness of instruction throughout units of study, described in the previous chapter as progress monitoring. Lesson-level progress monitoring may include warm-up activities that are monitored by the teacher, questioning and guided practice during a lesson, or independent work at some point during a lesson or afterward. Students receiving intensive instruction in a MTSS model or in special education settings require even closer, more frequent monitoring so that instruction can be adjusted. Often termed data-based instruction (DBI), or *data-based individualization*, systematic and frequent monitoring of students' progress within instruction and making adjustments or modifications to instruction based on individual student responses has a strong research base over four decades (Jung et al., 2018). Mathematics interventions using DBI have typically focused on computation and procedural skills (e.g., Allinder et al., 2000).

Teachers should also teach self-monitoring strategies so students can become more independent learners. Self-monitoring and other self-regulation strategies derive from cognitive and developmental theories of learning. Some research-based mathematics programs have self-monitoring components, such as *Solve It!* for word-problem solving (Montague, 2003). Common elements of these metacognitive-strategy interventions include assessment of current strategy use, explicit modeling of cognitive and metacognitive strategies with think-alouds by the teacher, student practice of strategies within target contexts, corrective feedback on student performances, and teaching for transfer. Students can be taught to self-monitor by using checklists, verbal rehearsals, and self-checking steps on assignments. During problem-solving instruction, a student could be taught to check off each step of the problem-solving process as she works through a word problem. Often the steps of a process are organized using a mnemonic device so that students can rehearse the steps and memorize them for future use. Students can also be prompted to check their own work for accuracy and identify where faulty procedures were used. Wang et al. (2019) investigated the effects of an embedded self-regulation (SR) component within a fraction magnitude intervention (third grade) that focused on explicit instruction using diagrams. The SR portions were four to seven minutes in tutoring pairs of students and addressed self-sufficiency, partner support, goal setting, taking responsibility, and tracking one's own progress. The SR group outperformed the control group on a distal measure of fraction items and word-problem solving. The SR group also performed better than an intervention group without SR, having the advantage of repeatedly assessing their own progress through graphs and charts and focusing on skills they needed to improve. The SR elements appeared to build "students' growth mindset, goal setting, planning, and perseverance" (p. 346). Students who learn and apply self-monitoring strategies will be more likely to persist with mathematics tasks, attempt back-up strategies when they are stuck, and successfully solve mathematics problems.

Ensure Transfer-of-Learning

A baseball player knows that without follow-through with the bat, the ball will drop short. Without follow-through activities for mathematics learning, new concepts and skills will be isolated learning or even forgotten. In the Instructional Hierarchy discussed earlier, after students have acquired and developed fluency with concepts and skills, the next stages are *maintenance* over time and *application* in other settings, both requiring the transfer of learning (Haring & Eaton, 1978).

The mathematics curriculum is cumulative and recursive, building on the same basic concepts over time to more advanced levels. However, students with learning difficulties may regress or forget concepts previously taught unless *maintenance* is an explicit goal. Frequently in intervention research a posttest (or proximal measure) will demonstrate significant learning has occurred, while a more distal measure (a few weeks later) will show an erosion of learning (Liu & Xin, 2017; Namkung & Bricko, 2020; Xin & Zhang, 2009). Techniques to promote skill and concept maintenance include building in periodic reviews and check points, teaching for understanding rather than isolated skills, and teaching memory and concept-organization strategies. When gaps or losses are noted, skills and concepts should be retaught.

Application of new skills and concepts in appropriate mathematics contexts is a goal of learning. When new concepts are taught, enough examples should be provided that students can view a range of possible applications. If concepts are taught in overly structured situations, their transfer power will be limited. For example, when teaching the concept of using a number line with whole numbers (0, 1, 2, 3, …), extend the line beyond the 0 on one end and the last whole number on the other, with arrows on each end. This depiction will assist students when they move to the full range of integers (negative and positive numbers and zero) as well as the coordinate plane.

Generalization is a type of application and refers to a broader range of situations and novel contexts. These situations may be in other subject areas in school, such as science and social studies, or in the world outside school. Teachers of different content areas should work together to identify and reinforce common concepts. Home, community, and work situations require mathematics skills ranging from the simplest computations to the more complex problem solving involving multiple concepts. Word problems and performance tasks can simulate these situational requirements. Problem-solving contexts have also been developed for computer-based applications for students with MLD (Bottge et al., 2015).

Teaching mathematics is not as simple as opening a textbook and following the teacher's manual. As will be discussed in Chapter 6, textbooks frequently lack critical elements of instruction and have too many topics. Teachers should plan units and lessons carefully and deliberately, keeping in mind these principles of instruction, student characteristics, assessment results, and content and process standards.

Planning and Delivering Instruction

The previous section provided global principles for providing mathematics instruction. Those guidelines can be used as planning tools for developing units of instruction or intensive interventions.

Planning with Principles in Mind

The ten principles of mathematics instruction are benchmarks for unit planning and offer elements for lesson planning. This section illustrates how to use the principles, turned into questions, for beginning unit planning. The principles described the importance of integrating conceptual and procedural learning, and scaffolding instruction by building on prior knowledge. How can teachers develop the mathematics processes at students' instructional levels while simultaneously building content knowledge? On-going assessments using a planning guide for the content focal areas as well as mathematics processes can assist integration of these processes within instruction. An integrated planning chart for two focal topics within 3rd-grade content is illustrated in Table 4.1. For example, students may have developed the concepts of specific two-dimensional shapes, such as squares and rectangles, and used rulers to measure units, but they have not applied measurement concepts to the areas and perimeters of these plane figures or used formulas to describe findings.

Box 4.2 illustrates how to turn the instructional principles into guiding questions for thinking about unit planning for a unit on equivalent fractions. Notice that process standards are embedded throughout this planning chart. This plan also reminds the teacher to consider

Table 4.1 Planning Chart with Focus on Mathematics Processes (Seventh Grade)

Content Focus	Units	How Processes* Will Be Addressed
Fractions and Fraction Equivalence	Understand unit fractions and what they represent: $\dfrac{1}{b}$.	PS—solve and pose fraction equivalent problems RP/AC—reason about equivalent fractions, justify conclusions, and critique work of others
	Represent fractions on a number line 0 to 1, recognizing whole and partitioning into equal parts.	MR—use fraction blocks, number lines, paper strips, and diagrams T— use rulers, number lines P—check equal partitions on number lines, use accurate symbols
	Understand equivalent fractions: recognize and generate simple equivalent fractions, including whole numbers as equivalent to fractions.	S— note structure and patterns of fractions represented on number lines C—connect with multiplicative reasoning, measurement, and whole number operations
	Compare fractions: same numerator, same denominator, and other fractions between 0 and 2 using <, +, and >.	
Geometry	Understand area as an attribute of plane figures.	PS—solve and pose problems related to area and perimeter RP—reason and demonstrate results of problems
	Measure area in various unit squares (cm^2, m^2, in^2, ft^2).	AC—justify solutions and critique the work of others MR—use concrete objects, graph paper, diagrams
	Understand perimeter as an attribute of plane figures.	P—apply accurate labels for attributes, use new vocabulary T—manipulate measurement and construction devices, virtual figures
	Measure perimeter with various units (cm, m, in, ft).	S—note the relationship between area and perimeter formulas
	Apply area and perimeter concepts to mathematical and real-world problems.	C—connect to fraction concepts (part of whole), prior knowledge of polygons, distributive property of multiplication

* Processes are a combination of NCTM processes and CCSSM practice standards.
Codes: PS – Problem Solving; RP – Reasoning & Proof; AC – Argument & Critique;
MR – Modeling & Representations; T – Using Tools; P – Applying Precision;
S – Using Structure; C – Making Connections

the students' prior knowledge and interests, to brush up on the concepts related to equivalent fractions, and to identify needed resources. Some textbooks introduce equivalent fraction concepts and skills by stating the definition of equivalent fractions, providing one or two rules for finding equivalents, offering drawings of number lines or partitioned shapes, and then providing practice finding equivalent fractions using specific procedures. That approach omits many elements that should be considered during unit planning, especially for students with difficulty in mathematics.

Box 4.2   Planning with Principles of Instruction

Equivalent Fractions (Grades 3 − 4)

Principles in Question Form	Unit Content
1. What concepts are emphasized in the unit of study?	Equivalency in broader terms, unit fractions, decimal fractions, comparison of fraction values, converting fractions to equivalent values for specific purposes (all processes as well)
2. What prior knowledge is essential for these concepts?	Addition, subtraction, multiplication, division concepts; concepts of factor and multiple; names and meanings of parts of fraction; simple fractions, how those can be modeled (prior development of processes)
3. How can we promote positive attitudes in this unit?	Create personal fraction bars; Fraction Track Game (NCTM Illuminations); student-created word problems
4. What learner characteristics should be planned for in this unit?	Linguistic: review and reteach vocabulary in context Working memory: recording methods to keep track of work Writing: use clear over/under fraction notation, fraction cards and diagrams
5. What sequence of instruction is needed within the unit?	Concrete to review concept: fraction bars and paper strips; Representations: number lines, fraction bar drawings to develop new concepts; Abstract symbols with clear language to describe fractions and connections; After equivalence concept is firm, teach algorithms explicitly using worked examples and prompts; Follow with applications with feedback
6. What mathematical language is critical for understanding?	Fraction, equivalent, numerator, denominator, terms of the fraction, proper and improper fractions, unit fraction, greatest common factor
7. What strategies would be helpful?	Develop strategies: 1. to compare two fractions, apply cross-multiplication 2. from one fraction, multiply or divide N and D by the same number 3. to find simplest form, divide N and D by GCF
8. How can learners be engaged?	Use materials: ribbon, blocks, paper; Begin each lesson with successful review; Plan guided small-group work
9. How will learning be monitored?	Pre- and post-unit tests, item checks for each lesson, oral think-alouds, charting progress for select students
10. When will this learning be applied and transferred to new learning?	Future units: fraction operations, geometric formulas, algebraic sentences; Review equivalence concept; Solve equations using concepts from fraction properties and operations

Unit Planning

Units of study for the year in core mathematics classrooms should be developed using grade-level curriculum standards and knowledge of students' learning needs. Often textbooks will have much more content than can be addressed at a deep level during an academic year. Further, most textbooks do not include material for intensive interventions or differentiated instruction, only general suggestions. Therefore, teachers should map out the year's curriculum goals, identify student-specific goals, and plan instructional units based on those maps. The textbook is only one resource for this planning. Grade-level curriculum mapping can be planned by groups of teachers and designed using a number of approaches. Units can be planned by content-focus domains, by integrated themes, or driven by projects or problems. For example, a year of thematic units for middle school could include units on cartography (measurement, ratios, graphing), cooking (measurement and fractions), architecture (measurement, geometry, proportions), and baseball (statistics, decimals, equations). All of these approaches will meet grade-level standards and allow for individual student interventions.

Teachers should consider a unit-planning approach that involves students and brings forward the big ideas, connections, and key processes within the unit. In addition to the planning matrix that accounts for instructional principles, a graphic unit organizer is helpful for teachers and students. *The Unit Organizer Routine* (Lenz et al., 1994) provides such a tool. By introducing students to unit concepts, relationships, sequences, and processes through a unit map and using the unit map as a guide and review tool throughout the unit, students are more aware of unit connections, learning goals, and expectations (see Figure 4.5). On field tests, students of teachers who used the *Unit Organizer Routine* consistently scored an average of 15 points higher on unit tests than students of teachers who used it only irregularly (Boudah et al., 2000). Professional development by a University of Kansas trained instructor is required for using this tool (https://sim.ku.edu).

As illustrated in the unit organizer, units should be planned with outcome goals in mind, often termed *backwards design* (Gagné et al., 1988; Wiggins & McTighe, 2005). You cannot know how you are going to teach until you know what you want students to learn. Identifying curriculum standards and key concepts assists in prioritizing goals for a unit of study. If a unit is for intensive intervention, the goals for the unit should be individualized and may cross grade levels within a content domain. Teachers should also identify prerequisite knowledge and skills to assist planning for preteaching or differentiation within the unit. For a unit on equivalent fractions for third graders, the outcome goals are:

> Compare fractions by reasoning about their size.
> Recognize and generate equivalent forms of fractions.

These goals can be turned into broad questions for the unit:

> Can you locate fractions on the number line (between 0 and 2)?
> Can you compare fractions as larger than, less than, or equal to?
> Can you write an equivalent number for a whole number, mixed number, or simple fraction?

Plan how these outcome goals and related questions will be assessed. Assessment can include frequent, formative assessments, an end-of-unit summative assessment, or a combination. For this unit on equivalent fractions we can plan a quick assessment within each lesson to ensure students understand concepts, use vocabulary and tools correctly, and apply new skills. The unit will conclude with a summative assessment, a sampling of items that represent the unit goals. There are many other options for assessment strategies, including think-alouds, performance tasks, quizzes, applied problems, and demonstrations. Two students who had gaps in learning for our example unit need extra assistance with basic fraction concepts. Their progress should be tracked on individual goals at least twice a week with those data informing intensive instruction until mastery.

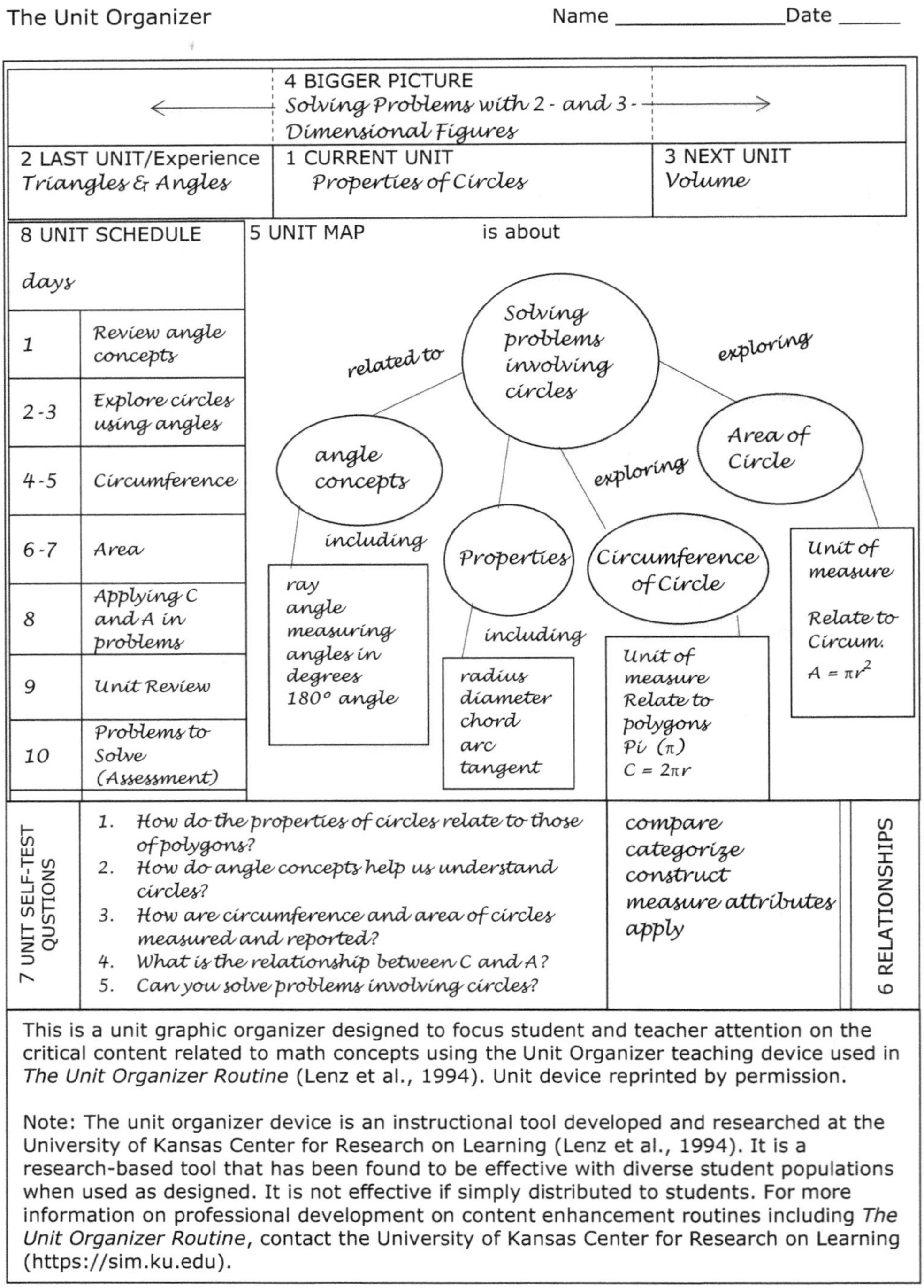

The Unit Organizer   Name _____ Date _____

	4 BIGGER PICTURE	
←	*Solving Problems with 2 - and 3 -*	→
	Dimensional Figures	
2 LAST UNIT/Experience	**1 CURRENT UNIT**	**3 NEXT UNIT**
Triangles & Angles	*Properties of Circles*	*Volume*

8 UNIT SCHEDULE

days

1	*Review angle concepts*
2-3	*Explore circles using angles*
4-5	*Circumference*
6-7	*Area*
8	*Applying C and A in problems*
9	*Unit Review*
10	*Problems to Solve (Assessment)*

5 UNIT MAP   is about

Solving problems involving circles

related to   exploring

angle concepts   exploring   *Area of Circle*

including   *Properties*   *Circumference of Circle*   *Unit of measure*

ray
angle
measuring
angles in
degrees
180° angle

including   *Relate to Circum.*

$A = \pi r^2$

radius
diameter
chord
arc
tangent

Unit of measure
Relate to polygons
Pi (π)
$C = 2\pi r$

7 UNIT SELF-TEST QUSTIONS

1. *How do the properties of circles relate to those of polygons?*
2. *How do angle concepts help us understand circles?*
3. *How are circumference and area of circles measured and reported?*
4. *What is the relationship between C and A?*
5. *Can you solve problems involving circles?*

compare
categorize
construct
measure attributes
apply

6 RELATIONSHIPS

This is a unit graphic organizer designed to focus student and teacher attention on the critical content related to math concepts using the Unit Organizer teaching device used in *The Unit Organizer Routine* (Lenz et al., 1994). Unit device reprinted by permission.

Note: The unit organizer device is an instructional tool developed and researched at the University of Kansas Center for Research on Learning (Lenz et al., 1994). It is a research-based tool that has been found to be effective with diverse student populations when used as designed. It is not effective if simply distributed to students. For more information on professional development on content enhancement routines including *The Unit Organizer Routine*, contact the University of Kansas Center for Research on Learning (https://sim.ku.edu).

Figure 4.5 Unit Organizer Routine

TRY THIS

Select one of the following unit goals, turn it into one to three questions, and explain how the goal would be assessed during a unit of study.

a.   Understand that the two digits of a two-digit number represent amounts of tens and ones (first grade).
b.   Graph points on the coordinate plane to solve real-world problems (fifth grade).
c.   Understand that there are numbers that are not rational and approximate those using rational numbers (eighth grade).

Next, the learning activities for the unit should be outlined, including a strategy for beginning the unit of study, the use of various student groupings for activities, the activities for the CRA levels applied within this unit, needed learning supports, and a unit-closure approach. There is support for beginning units in mathematics in a number of ways: problem posing, exploration, advance organizers, real-world examples, review activities, teacher-led discussions, or other anticipatory sets (Ma & Papanastasiou, 2006). Approaches that are not as effective for introducing a new topic include looking at the textbook while the teacher talks and having the teacher explain rules and definitions. Units of study within textbooks often begin with a problem or situation where the concepts and skills in the unit will be applied, as an introduction to a meaningful context. Sometimes introductory materials include unit goals and vocabulary for teacher guidance. However, textbooks often lack information on pedagogical approaches for introducing units and planning other portions of instruction (Doabler et al., 2012).

Whether planning using an adopted textbook, a range of resources, or simply grade-level curriculum standards, teachers should begin units with introductory methods that pique student interest, stimulate thinking, trigger prior knowledge, and establish explicit learning goals. Some examples of approaches with these merits:

*   *Problem posing*: Pose a problem or problem situation that introduces the content, is accessible for students, and creates interest and curiosity. For example, for a unit on decimal operations: *You have 4.6 pounds of frog food. If you feed your frog 0.1 pound of frog food a day, how many days will your frog food last?* (Kent, 2014, pp. 92–93). As the teacher listens to group work, he can assess concept understanding and strategy use. The problem creates a context for similar problems to be posed by students. In a meta-analysis of problem-posing research, Rosli et al. (2014) found strong mean effects on mathematics achievement, problem-solving skills, and attitudes towards mathematics across Grades 3 to 12.
*   *Exploration*: Similar to problem posing, exploration requires the teacher to set up a situation with resources and provide parameters for student work. The teacher should pose questions to guide student investigations. Students can investigate data sets, an erroneous problem, a real-world situation, a model, an artifact, or a diagram. For example, the teacher could distribute baseball statistical charts and ask students, with guidance, to uncover the formulas behind statistics such as *bases on balls*. This exploration could lead into a unit on algebraic equations.

- *Advance organizer:* Advance organizers for units of study can include outlines, questions, learning goals, and graphic organizers. These organizers should be referenced throughout the unit and used for review.
- *Teacher-led discussion*: Before a unit commences, a teacher-led discussion helps to uncover prior knowledge as well as to preview unit goals and activities. For a unit about statistics, "We are beginning a new unit today. Has anyone heard these terms: survey, random sampling, error of measurement? Tell me what you understand about those." [Solicit several responses for each.] "In our unit on statistics you will be involved in collecting data using survey techniques. You will enter data into a computer spreadsheet and analyze the data using some new metrics. We will also critique surveys from the popular press."

Units of study do not have to be presented in real-world situations to be interesting for students. "Interest can be created and stimulated" and intellectual activities themselves can be rewarding if materials are well organized and presented (Bruner, 1960, p. 117). "The best introduction to a subject is the subject itself" (Bruner, 1966, p. 155). A seemingly uninteresting topic, algebraic equations, could be introduced by showing a simple equation with variables:

> This is an algebraic equation because we have small letters, called *variables*, representing a range of numbers, and the equal sign to show equality of the values of the sides: $3 + x = 2 - y$. There are many possible numbers for x and y; the values for x and y can vary. Can your group come up with several? Try thinking "if x is a specific number, what would y be?" Don't forget to try fraction and decimal numbers. You can use any method as long as you end up with values for x and y that fit this rule.

When planning activities for the unit about equivalent fractions, the teacher decided to use the concrete level for review of basic fraction concepts such as part of a whole, unit of the whole, and describing mixed fractions. Concrete materials included interlocking blocks and fraction strips. Then he moved to a combination of concrete and representational forms for locating fractions on the number line and magnitude comparisons. At the representational stage, number lines and diagrams of fraction bars supported learning as symbols were introduced. These images were connected with verbal descriptions using precise mathematics language, and with records of student work using fraction notation. Other activities for this unit included: using virtual manipulatives, working in small groups to create word problems, converting fractions, and developing mnemonic devices to support strategies for generating equivalent fractions.

This unit concluded with a quick review of major concepts and procedures before the unit test. Other units could conclude with projects, group problems, and even products, such as a class book or posters. It is important that students realize they have new concepts, skills, vocabulary, and processes as a result of their efforts in this unit. Some teachers pull out the advance organizer as a tool for review and to remind students of their new learning. Other teachers have a brief discussion about what students learned and are able to do.

For planning this unit, the principles chart (Box 4.2) was used to clarify outcome goals, unit questions, assessment strategies, and activities for developing the concepts and skills within the unit. The principles served as a reminder of essential considerations for unit and lesson planning. An example of the outline for this unit on equivalent fractions is depicted in Box 4.3. Next we turn to lesson planning, developing the day-to-day lessons within this unit of study. We will refer to the principles chart and unit outline as we develop these lessons.

Box 4.3   Unit Outline: Equivalent Fractions

Unit Goals	1. Compare fractions as <, >, or = by reasoning about their size. 2. Recognize and generate equivalent forms of fractions.
Unit Questions	1. Can you write an equivalent fraction for a whole number? 2. Can you write an equivalent fraction for a simple fraction? 3. Can you locate fractions on a 0 to 1 number line? On a 0 to 2 number line? 4. How can you compare two fractions?
Assessment Strategies	Pre-test on prerequisite concepts and skills. Check understanding within lessons: think-alouds, independent items. Post-test sampling of unit questions. Progress monitoring with charting for two students (on unit goals).
Unit Introduction	Problem: Show students rulers on their desks, marked to 16ths. Discuss what each mark means and introduce unit focus. Advance organizer: Present unit questions using a graphic unit organizer and review prior knowledge.
Lesson Sequence (1–3 days each)	1. Exploring Simple Fractions with Blocks, Fraction Strips 2. Identifying other Equivalent Fractions using Strips 3. Locating Equivalent Fractions on Number Line Pairs 4. Strategy for Finding Simplest Form of Fraction 5. Strategy for Creating an Equivalent Fraction 6. Continue with Strategies, Applications in Problems 7. Review comparing fractions, equivalent fractions, and strategies.
Unit Closure	Revisit unit organizer for review of vocabulary, concepts, and strategies. Unit test (sampling of 15 items based on unit questions).

Lesson Planning

Planning at the lesson level requires teachers to develop a sequence of lessons using the unit outline and principles for mathematics instruction for guidance. A single lesson may be designed for one mathematics period or cover two or three days for the same lesson objective. If a lesson spans more than one class, elements should be added so that each day's portion begins with review and ends with closure. A lesson plan is also developed with a backward-planning approach: identify lesson objectives or outcomes, decide how the objectives will be assessed during the lesson, then plan the sequence of activities within the lesson. Other elements include listing resources needed for the lesson, providing for individual student differences and accommodations, and anticipating misconceptions. Teachers should also plan vocabulary and carefully sequenced examples to use during the lesson.

Teachers must decide whether a lesson should be very explicit, a teacher-directed lesson using direct instruction, or less explicit in the beginning with teacher-guided exploration followed by explicit instruction. The explicitness of a lesson's design falls along a continuum from exogenous to endogenous construction of knowledge as illustrated in Table 4.2, an adaptation of Moshman's (1982) constructivist variations. Endogenous constructivism, where learners discover their own knowledge, is impractical and idealistic, based on theoretical views that true knowledge is *only* that which the learner constructs for him or herself (Mercer & Pullen, 2005). The other end of the continuum, reductionism, is theoretically where all knowledge is reduced to isolated skills and procedures taught by rote, without meaning and connections. That extreme would result in isolated knowledge that could not be applied or

Table 4.2 Continuum for Construction of Knowledge

	Reductionism	Exogenous Constructivism	Dialectical Constructivism	Endogenous Constructivism
	Rote Instruction	Explicit Instruction	Guided Exploration	Child-Determined Exploration
Teacher	Presenter	Directive	Supportive	Peripheral
Student	Passive	Responsive and Engaged	Interactive and Engaged	Self-Regulated
Content	Isolated skills, procedures, definitions	Skills, concepts, procedures, relationships	Skills, concepts, procedures, relationships	Concepts, relationships

Teaching–Learning Zone

connected with other knowledge. Most teachers with diverse classrooms will plan lessons in the exogenous to dialectical ranges, planned and guided by the teacher, while ensuring high levels of student engagement. Curriculum developers and mathematics educators, with some radical exceptions, now support this balanced view for teaching and learning (Cole & Washburn-Moses, 2010; Hennessey et al., 2012; NMAP, 2008; Rittle-Johnson, 2017; Sweller, 2016).

Two lessons for the unit on equivalent fractions are provided in Boxes 4.4 and 4.5. The lesson in Box 4.4 falls into the dialectical range, with guided exploration followed by more explicit instruction. This lesson includes effective instructional elements of frequent, high-quality instructional interactions, high levels of student verbalizations (productive discourse), and clear and consistent demonstrations of new concepts (Doabler et al., 2014). The lesson in Box 4.5 illustrates a direct-instruction lesson from the beginning. The same lesson elements appear in both lessons; the first lesson begins with more discourse and exploration while the second begins with teacher modeling. The direct-instruction lesson includes modeling, guided practice with feedback, and independent practice with instruction at the representational level. Note that the section titled *special provisions* would be developed based on student characteristics from a given classroom.

Box 4.4   Exploration Then Explicit Instruction Lesson

Lesson: Exploring Simple Fractions

Lesson Objective(s)

1. Students will make statements about fraction comparisons (fractions between 0 and 1).
2. Students will identify equivalent fractions to fractions provided (halves, fourths, eights, and 16ths).

Curriculum Standard/IEP Goal

3a. Understand two fractions as equivalent if they represent the same point on the number line.
3b. Recognize and generate simple equivalent fractions.

Assessment Method(s)

1. Listen to student discussions in small groups for concept understanding, vocabulary use, and misconceptions.
2. Given three simple fractions on cards, students should show three equivalent fractions.

Sequence of Instruction

1. Review fraction concepts and terms from previous unit: a fraction is a number expressed with a numerator and denominator. For the fraction $\frac{2}{3}$ the denominator 3 represents the total number of portions or units and the numerator 2 represents the part of the whole. Review the meaning of several simple fractions using a 0 to 1 number line: $\frac{1}{4}, \frac{4}{5}, \frac{1}{3}, \frac{1}{2}$, and $\frac{3}{4}$. Review that you can only combine fractions if they have the same denominator.

2. *You are in your groups of three or four today with some strips of paper for each student. Please fold a strip in half* (demonstrate). *Please draw a line in your fold and label each part $\frac{1}{2}$. Now take another strip and fold that into four parts. Draw lines in your folds and label each part $\frac{1}{4}$.* [Continue with $\frac{1}{8}$ and $\frac{1}{16}$.] *Check your work with your group as you go. If you are satisfied, tape down your set of strips in order onto the chart with the 1 at the top.*

3. Circulate among groups prompting and scaffolding as needed. Prompt students to use their fraction language with each other. Listen for misconceptions.

4. *Now, in your group, I would like you to look for the patterns of your fraction strips. What do you notice? Can your group find one rule or discovery to share with us?*

5. Ask each group to share a finding. Have someone in the group justify their finding. Ask students in the other groups whether the finding is valid. Listen to and scaffold language used during this dialog.

6. *Now we are going to write some equivalent statements using what you have found. What does equivalent mean? What does the = sign represent?* (review)

7. Model the statement: $\frac{1}{2} = \frac{1}{4} + \frac{1}{4}$. Think aloud: *On my fraction strip chart $\frac{1}{2}$ is the same as two of the $\frac{1}{4}$ parts. So I write my equation like this. How can I simplify this side? Yes, so my equation is now $\frac{1}{2} = \frac{2}{4}$. We can say that one-half is equivalent to two-fourths because they represent the same number. Any time I use one-half, I could use two-fourths instead.* Repeat model step with another example.

8. Guided practice: *At your table, please write an equation with another fraction.* Show $\frac{2}{8}$ =? *Let's see what we have.* (Report out, provide corrective feedback.) *Let's do two more together before we wrap up. Try these: $\frac{4}{16}$ = ? and? = $\frac{3}{4}$ and $\frac{8}{8}$ =?*

9. *OK, good working today! Let's check what you learned. Who can put into their own words what an equivalent fraction is?* Call on two or three students, with corrective feedback. *Now I'm going to hold up a fraction card. Please take a blank card and write an equivalent fraction for mine. Hold up your fraction card so I can check.* Show: $\frac{1}{4}, \frac{2}{16}, \frac{4}{4}$ *and* $\frac{1}{2}$. Hold up more if needed.

Materials

Strips of white paper 11" long and 2" wide (standard paper).
Plain white, standard paper lined off every 2 inches horizontally. Write the numeral 1 in center of the top portion:

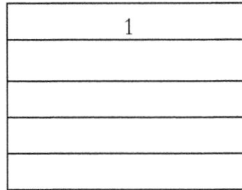

Fraction 3 by 5 cards (and blank cards): $\frac{1}{4}, \frac{2}{16}, \frac{1}{2}, \frac{4}{4}, \frac{3}{4}, \frac{8}{16}$.

Special Provisions

Use colored pencils, different colors for halves, fourths, eights, and 16ths.
Have students shade in sections of the fraction parts.
Assign roles to student groups. More review of basic fraction concepts if needed.
Some students may need extra time with adding fractions: $\frac{1}{8} + \frac{1}{8} + \frac{1}{8} + \frac{1}{8} = ?$
If some students are firm with these equivalents, offer thirds or fifths (and additional paper strips).

Anticipated Misconceptions

Confusing parts of a fraction with individual numbers. The entire fraction is a number.
Attempting to add two fractions with unlike denominators.
Difficulty expressing a whole (1, 2) as a fraction equivalent.
Looking at the folded lines between parts on the fraction strips rather than the unit as a portion.

Lesson Vocabulary (most review)

equivalent, value, numerator, denominator, portion, halves, fourths, eighths, 16ths

Planned Examples

Most examples are within lesson steps. Additional equivalent examples if needed, ordered from simplest to most difficult:

$\frac{1}{2} = \frac{?}{8}$ and $\frac{2}{8} = \frac{?}{4}$ and $\frac{4}{16} = ?$ and $\frac{2}{?} = \frac{1}{8}$

Box 4.5   Direct Instruction Lesson

Lesson: Comparing Fractions Using Number Lines

Lesson Objective(s)

1. Students will locate fractions on the number line (0 to 2 number line).
2. Students will compare two fractions on the number line and indicate greater, less than, or equivalent to using the symbols >, <, and =.

Curriculum Standard/IEP Goal

3a. Understand two fractions as equivalent if they represent the same point on the number line.
3c. Compare two fractions and record the results using < = > symbols.

Assessment Method(s)

1. Listen to student reasoning and justifications during practice.
2. Independently compare five fractions using < = > symbols.

Sequence of Instruction

1. Review concepts from previous lessons: equivalence, parts of a fraction, fractions equivalent to halves, fourths, eighths, thirds, fifths, sixths, etc. on paper strip models.
2. *Today we are going to use number lines to locate the position of fractions. This will be similar to your paper strips but we will label points on number lines. You have number lines on your desk. Everyone point to the 0, 2, 1 that are already labeled on your number line.* [Check responses.]
3. *Your number lines have dots between each number. One number line will be halves, one for quarters, and one for thirds. What is the distance that divides the number line between 0 and 1 exactly into two parts?* [Point to the dot on the chart/projected image.] *Yes, there are two halves, so we label this dot* $\frac{1}{2}$.

   We will label the 0 point $\frac{0}{2}$ *and the 1 is how many halves?*

   Yes, label that $\frac{2}{2}$. *The next point, between the 1 and 2 we will label* $\frac{3}{2}$, *and the 2 is* $\frac{4}{2}$. Call on students to name points on a blank number line (referring to their personal one). Point out that the denominators stay the same because the line is divided into equal portions within each whole.

4. *Repeat with fourths, comparing back to halves.*
5. *Repeat with thirds, again comparing.*
6. *Now we're going to compare fractions with different denominators using our number lines. Place a marker on your fraction for* $\frac{1}{2}$. *Look across number lines and see if there is a fraction with the denominator of 4 in the same position. Yes* $\frac{2}{4}$. [Write $\frac{1}{2}=\frac{2}{4}$ on the board.] *These fractions are equivalent, they represent the same point on the number line, the same value. Let's find more equivalent fractions.*

7. Now we're going to find fractions larger and smaller than $\frac{1}{2}$. [Continue to locate fractions, record pairs using <, >, and =.]
8. *Let's review some of the fractions we compared.* [Call on students to compare and justify. Encourage accurate fraction language.]
9. *There are five pairs of fractions on the board. Independently write those on your paper and use your number lines to compare those. Record the <, >, or = between each pair.*

Materials

Paper with five parallel number lines for each student, with dots and whole numbers indicated (0 to 2 with arrows on each end).
Chart of number lines, or projection on white board. Interlocking blocks. Discs.

Special Provisions

Some students are limited to halves and fourths until those are firm. Other students may want to extend to fifths and sixths. Use a concrete model (connecting blocks) to show equal portions if the connection is not made to the equal portions on the number line. Some students may benefit from a linear measurement connection, such as a ruler. Provide pairs on paper (step 9) if students have difficulty copying accurately from the board.

Anticipated Misconceptions

Students often fail to consider the 0 as a point on the number line, mark any location between 0 and 1 for simple fractions (misunderstand that the fractional portion of the line from 0 is the fraction). They may think of the 1 and 2 in the fraction ½ as numbers, but the entire fraction is a number, with terms (numerator and denominator). Teach students to consider whether a fraction is equivalent to a whole number, between 0 and 1, or between 1 and 2.

Vocabulary

whole number, halves, half, simple fraction, quarter(s), improper fraction, third(s), numerator, denominator, partition, portion, greater than, less than, equal, equivalent, linear, compare, comparison, terms (of fraction)

Planned Examples

Step 9:

A.	B.	C.	D.	E.
$\frac{1}{4}$ $\frac{1}{2}$	$\frac{3}{4}$ $\frac{5}{4}$	$\frac{2}{3}$ $\frac{3}{4}$	$\frac{5}{3}$ $\frac{6}{4}$	$\frac{3}{2}$ $\frac{6}{4}$

Additional pairs if needed:

$\frac{4}{3}$ $\frac{3}{3}$	$\frac{3}{4}$ $\frac{2}{4}$	$\frac{2}{3}$ $\frac{4}{6}$	$\frac{4}{6}$ $\frac{6}{6}$	$\frac{2}{5}$ $\frac{3}{4}$

Other elements must be considered when planning lessons within a unit framework. Teachers should attend to the nature of content and specific instructional strategies while planning and implementing lessons.

Nature of Content

The teacher must consider whether lesson objectives address discrimination, factual learning, rules, procedures, concept learning or problem solving (Mastropieri & Scruggs, 1994; also termed *content form*). These cognitive skills, derived from Robert Gagné's (1956) hierarchy of learning, emphasize the levels needed for specific content, and require different instructional considerations.

Discrimination learning is simply discriminating one stimulus from another. Students must discriminate letters, numbers, and other visual symbols, colors, shapes, objects, names, and so forth. Discrimination learning is an important foundation for other types of learning. To teach students to discriminate, teachers use examples and non-examples, modeling, and prompts with feedback. For example, students may be discriminating rectangles from other two-dimensional shapes. The teacher would show several examples of rectangles as well as shapes that are not rectangles. Then students would be asked to identify examples as rectangles or not. This example of discrimination involves concept formation, so it is at a higher level than discriminating a 2 from a 5.

Factual learning is establishing a basic association that follows a stimulus–response pattern. For example, given $7 + 4$, the response is 11. The perimeter of a polygon is the sum of its sides. One yard is the same as 36 inches. Factual learning should be supported by concept learning, but facts are learned to fluency so they can be recalled without burdening working memory. Factual learning is usually accomplished through repetition, distributive practice, and mnemonic strategies.

Rule learning incorporates discrimination and factual learning but allows students to generalize learning to new examples. For example, a rule for the order of operations when completing a problem such as $2 \times 3 + 7 \times 8$ is that multiplication and division must be completed before addition and subtraction ($6 + 56 = 62$). Math rules are generally consistent, but they evolve over the grade levels to incorporate new concepts. When students first learn adding negative integers, with rules such as "adding two negatives gives a negative," mathematics notation uses addition and subtraction symbols, in addition to the negative symbol: $-7 + (-2) = -9$, but as they move into expressions and equations the notation changes, confusing some students: $-7 - 2 = -9$ and $5 - (-2) = 7$. Rules are taught through examples, modeling, strategies, and distributive practice.

Procedural learning is important for the execution of a sequence of tasks such as applying an algorithm or solving a multistep problem. It often involves discrimination, factual, and rule components, but depends on strong concept understanding. Students discriminate among problem types, apply algorithms or other solution strategies, and keep track of their steps. Procedures are taught by modeling, providing application activities, investigating worked examples, thinking aloud, and giving corrective feedback. In mathematics there is usually more than one effective procedure for achieving a solution.

Concepts can range from very simple (shape) to complex (fractals). Concept understanding is difficult to measure, even at the simplest levels. Students may demonstrate understanding by identifying the concept in a novel situation, making connections between concepts, or using the concept in an analogy or broader concept. Some concepts are quite concrete: linear measurement, percent, frequency distribution, and area. The most difficult concepts are abstract: dimension, probability, proof. Concept understanding is developed through the use of examples and non-examples, rules, guided experiences, models, analogies, and concept maps.

A final content type is *problem solving*. In the mathematics standards, problem solving is a primary goal of instruction. Problem-solving situations in today's curriculum are realistic, complex, and frequently require interdisciplinary approaches. They require reading comprehension and metacognitive strategies. Problems require the application of rules, facts, procedures, and concepts. Problem solving is usually taught through strategic approaches: analysis of problem types, the use of schemas or diagrams, and instruction in estimation and self-evaluation (Peltier et al., 2018; Poylá, 1973).

Effective Instructional Strategies

Research over the past 40 years has consistently supported specific instructional elements of lessons that are effective with students with disabilities and other learning difficulties. The body of research from which these findings are drawn has been called the effective instruction literature but is now more often termed *evidence-* or *research-based practices* or *high-leverage instructional practices*. Two instructional approaches with significant research support for students with disabilities—direct instruction and cognitive-strategy instruction—incorporate these elements. Many of these practices originate from the instructional design and learning theories of B. F. Skinner, S. Engelmann, R. Gagné, D. Meichenbaum, and L. S. Vygotsky are essential elements of explicit instruction. The research-supported elements of lessons and interventions were embedded within the principles of mathematics instruction discussed at the beginning of this chapter. They include:

- *Modeling and demonstration.* Teacher modeling and demonstration has been correlated with significant learning gains for students with MLD, in combination with careful sequencing and encouraging students to follow the same steps of the model (Gersten, Beckmann, et al., 2009; Hughes et al., 2017).
- *Carefully selected and sequenced examples.* A meta-analysis of studies with students with MLD found a large effect size for the use of carefully selected and sequenced examples (Gersten, Chard, et al., 2009). Most researchers recommend a sequence of examples that will gradually scaffold student understanding of new concepts and procedures while interspersing easier, previously learned material (Gersten, Chard, et al., 2009). Teachers should show students carefully selected *worked examples* in combination with self-explanation prompts to guide students through the process for solving a problem (Booth et al., 2017; Clark et al., 2011). Figure 4.6 shows a worked example with self-explanation prompts and both correct and incorrect worked examples with questions (e.g., McGinn et al., 2015; Star et al., 2014).
- *Student verbalizations.* Encouraging students to verbalize—whether during diagnostic assessments, while justifying responses, or when using self-talk to monitor work—is an effective strategy that will help students anchor their skills, decrease impulsivity, and facilitate self-regulation (Gersten, Chard et al., 2009).
- *Guided practice that is rich and varied*: Classrooms with higher rates of practice opportunities are more beneficial for students with mathematics difficulties (Clarke et al., 2011). There is strong research support for *distributed practice*, practice that is spread out across a learning session or across multiple sessions (Booth et al., 2017). Mathematics practice should be varied to promote attention, accuracy, and fluency, and can include written work, verbalizations, computer practice, and simple games. Homework should be assigned only after guided practice reaches an independent level. Box 4.6 lists examples for providing varied practice and homework.

Worked Example with Prompts

$2(5x + 6) = 20$

$2(5x + 6) = 20$
$\div 2 \qquad\quad \div 2$
$5x + 6 = 10$
$\quad - 6 \qquad -6$
$\quad\;\; 5x = 4$
$\quad\;\; \div 5 \qquad \div 5$

$x = \dfrac{4}{5}$

Why did this student divide both sides of the equation by 2?

Why did the student subtract 6 from both sides?

Why did the student divide both sides by 5?

Incorrect Worked Example with Question

$2(5x + 6) = 20$

$2(5x + 6) = 20$
$\div 2 \qquad\quad \div 2$
$5x + 6 = 10$
$\qquad - 6$
$\qquad\; 5x = 10$
$\qquad \div 5 \;\div 5$
$\qquad\quad x = 2$

This student did not find the correct solution. Why not?

Worked Example with Question (Comparing Solution Methods)

$2(5x + 6) = 20$

$\dfrac{2(5x + 6)}{2} = \dfrac{20}{2}$

$5x + 6 = 10$
$5x + 6 - 6 = 10 - 6$
$\dfrac{5x}{5} = \dfrac{4}{5}$

$x = \dfrac{4}{5}$

This student has the correct solution but used different notation. Why is it accurate?

Your turn: $3(2x + 9) = 15$

Figure 4.6 Worked Examples with Prompts

- *Instructional feedback*: Feedback is one of the most effective instructional methods, more powerful than prior ability or socioeconomic status (Center on Instruction, 2008), and consistently associated with large effect sizes in instructional research. Teachers' feedback during instruction should be as specific and immediate as possible (Kehrer et al., 2013) and be provided for successful as well as unsuccessful responses, calling attention to the concepts and processes involved.

Box 4.6   Varied Practice

Practice Alternatives

- Assign task cards
- Drill in pairs
- Have students create problems
- Use simple dice practice
- Plan computer-based practice
- Create models or diagrams
- Develop tasks with number grids
- Practice with flash cards
- Allow for problem choices
- Adapt board games
- Use grid paper
- Create foldable graphic organizers

Independent practice only after successful guided practice.

Homework Ideas

- Take a survey
- Use a newspaper to create problems
- Use kitchen items
- Make a chart
- Make up a problem
- Take measurements
- Do a puzzle
- Finish a booklet
- Teach someone else
- Practice only 2 or 3 items

Homework should not be a new skill or concept!

- *Visual representations:* Visual representations of mathematics concepts and processes support new learning and are most effective when connected with other instructional components such as teacher modeling and verbalizations (Gersten, Chard, et al., 2009). Teachers should be careful with using multiple visual representations of a concept, however, especially if students fail to understand what is depicted in a representation or fail to make connections among representations (Rau & Matthews, 2017).
- *Mindful review:* Review of previously learned material plays a critical role in developing new concepts as well as maintaining and generalizing new concepts and procedures. Review should be *cumulative*, built into lessons over time, and include prior learning as well as new knowledge. Kame'enui et al. (2002) warned about mindless drill-and-practice review, however, and called for *judicious review*, that which is selective and used purposefully.

These effective elements of lessons can be found within the high-leverage practices developed by the Council for Exceptional Children (CEC) and the CEEDAR Center (McLeskey et al., 2017, 2019). High-leverage practices (HLP) are a set of instructional practices that are crucial for beginning and early career teachers to be able to do well (TeachingWorks, n.d.). These are the practices that are used frequently in classrooms and have been shown to improve student outcomes if implemented successfully. The CEC/CEEDAR document lists 12 instructional practices that are high leverage (see Box 4.7). Compare these practices to the principles of mathematics instruction and essential elements of lessons discussed in this chapter. Other organizations have adapted HLP to their content areas or age levels.

Box 4.7   High-Leverage Practices

Collaboration

1. Collaborate with professionals to increase student success.
2. Organize and facilitate effective meetings with professionals and families.
3. Collaborate with families to support student learning and secure needed services.

Assessment

4. Use multiple sources of information to develop a comprehensive understanding of a student's strengths and needs.
5. Interpret and communicate assessment information with stakeholders to collaboratively design and implement educational programs.
6. Use student assessment data, analyze instructional practices, and make necessary adjustments that improve student outcomes.

Social/Emotional/Behavioral

7. Establish a consistent, organized, and respectful learning environment.
8. Provide positive and constructive feedback to guide students' learning and behavior.
9. Teach social behaviors.
10. Conduct functional behavioral assessments to develop individual student behavior support plans.

Instruction

11. Identify and prioritize long- and short-term learning goals.
12. Systematically design instruction toward specific learning goals.
13. Adapt curriculum tasks and materials for specific learning goals.
14. Teach cognitive and metacognitive strategies to support learning and independence.
15. Provided scaffolded supports.
16. Use explicit instruction.
17. Use flexible grouping.
18. Use strategies to promote active student engagement.
19. Use assistive and instructional technologies.
20. Provide intensive instruction.
21. Teach students to maintain and generalize new learning across time and settings.
22. Provide positive and constructive feedback to guide students' learning and behavior.

Source: McLeskey, J., Barringer, M-D., Billingsley, B., Brownell, M., Jackson, D., Kennedy, M., Lewis, T., Maheady, L., Rodriguez, J., Scheeler, M. C., Winn, J., & Ziegler, D. (2017, January). *High-leverage practices in special education*. Arlington, VA: Council for Exceptional Children & CEEDAR Center.

Teachers and administrators often seek out the "magic" textbook or program that will result in student learning. Regardless of its research base, a textbook will not ensure learning; it is only a resource. Further, many topics within the mathematics curriculum do not have research-supported programs across grade levels for students who struggle to learn. If a selected program has been supported with high-quality research, it must be implemented with fidelity by well-trained teachers and with similar students in order to expect the same results. It is not the textbook or program, rather the instructional elements that make a difference in learning. This point is especially important for students with difficulties learning mathematics, who need intensive, individualized instruction. After meta-analyses of elementary-, middle-, and high-school mathematics programs with research support, Slavin and colleagues concluded, "educators as well as researchers might do well to focus more on how the classroom is organized to maximize student engagement and motivation rather than expecting that choosing one or another textbook will move students forward by itself" (Slavin & Lake, 2008; Slavin et al., 2009, p. 887).

Student interns and novice teachers often feel overwhelmed by the planning demands of instruction. Indeed, there are many aspects of the teaching-learning process that must be considered when planning effective instruction, especially for students who have difficulties in mathematics. Taking the long view first, studying curriculum standards and student needs will help teachers prioritize their efforts. The demands of planning units, lessons, and activities within those lessons can all be mitigated by collaborating with other professionals and taking advantage of sound resources. Novice teachers often make up for a lack of experience with meticulous planning. Even experienced teachers, with planned units and lessons and strong instructional techniques, should revisit their plans with curriculum standards, student characteristics, and effective instructional principles in mind.

Intensive Intervention

Unit planning and lesson-planning strategies also apply to short-term intensive interventions for small groups of students. *But intensifying involves bumping it up a few notches!*

> Several students in David's second-grade classroom are struggling with place-value concepts applied to two-digit addition and subtraction, even with differentiated examples and practice guided by the teacher, Mr Martinez. Because of the importance of these concepts for mathematics achievement, Ms Boswell, the special educator for lower grades, sets up a small group of five students in the corner of Mr Martinez' classroom for 25 minutes, three times a week during the class flex time. She structures this instruction very carefully, with a warm-up drill, daily goal setting for individuals and the group, explicit place-value instruction using concrete materials and visual representations, and a high level of verbalizations with practice. She emphasizes place-value concepts but also explicitly teaches additive-fluency strategies. Because of the small group size, Ms Boswell is able to individualize practice questions, scaffolding, reteaching, and attention redirection. The daily independent practice is checked immediately with feedback for understanding and self-correction. Each session ends with a quick, targeted review and goal checking. Students look forward to these lessons because they enjoy the activities and follow their own progress on graphs. This intervention lasted for six weeks for four of the five students, when they were able to achieve their intervention goal of 80% on the CBM of place-value concepts and related two-digit addition and subtraction. The fifth student needed additional time and support with these concepts and skills.

In the multi-tiered systems of support (MTSS) framework, as described in earlier chapters, students who are not making progress within the core mathematics curriculum may need more individualized and intensive interventions. Tier 1 is the research-based core curriculum within general education for all students. The majority of students should be making expected grade-level progress at this level. If not, there may be an issue with the curriculum and/or instruction for a classroom. Tier 1 should include universal screening, differentiated learning, appropriate accommodations for individual students, and ongoing progress monitoring. Tier 1 approaches for students with mathematics difficulties are discussed in Chapter 6, as are student-grouping approaches and co-teaching. Approximately 15%–20% of students may require more intensive intervention in mathematics than can be provided in Tier 1 programs, even with accommodations and other adaptations (Bouck et al., 2019). That could be as many as five students in a class of 25.

Students requiring special interventions at Tiers 2 and 3 may have major gaps in learning, need more individualized instruction, or need more intense interventions in order to make progress within the curriculum. Interventions can take place within the general education setting, in small-group settings for supplemental instruction, in special education settings, in specially designed alternative classes, or even one-on-one settings, depending on the school's staffing and resources (Bouck et al., 2019). Increasing *intensity* refers to increasing levels of support for student learning, not the setting. Greater intensity can refer to more dosage, an increase in the number of instructional sessions for a longer period of time in a smaller group to allow for more guided responses. It also refers to elements of instruction that are increasingly individualized and targeted: focusing on specific subskill and concept deficits; employing explicit instructional techniques; attending to attention and motivation; and embedding explicit language, routines, and opportunities for transfer (L. S. Fuchs, Fuchs et al., 2017; National Center on Intensive Intervention, NCII, 2013).

MTSS implementation will necessarily differ across grade levels (Jitendra & Krawec, 2019). Early elementary interventions (K to 2) tend toward prevention, for students at risk of difficulties, while those in Grades 3 to 5 may address very targeted remediation of earlier concepts and skills. Interventions in middle and high schools typically involve students who have larger deficits, lack basic conceptual understanding from earlier topics, have a history of difficulties with mathematics, and may even have identified mathematics disabilities. Secondary settings also involve highly specialized teachers, established schedules, and specific criteria for earning credit for coursework. Further, some secondary mathematics topics have not received the research attention as those for elementary students so there are fewer research-based standardized or packaged interventions. The following sections describe Tier 2 and 3 interventions with research support and examples.

Tier 2 Interventions

Tier 2 interventions *supplement* Tier 1 instruction and can be provided within the core mathematics classroom (e.g., co-teacher, station teaching) or in small pull-out groups based on student profiles. These interventions are time limited and offer moderate intensity (L. S. Fuchs, Fuchs, & Malone, 2017). For example, in a third-grade classroom, five students were identified with weakness in multiplicative reasoning after the mid-year screening. A mathematics support teacher pulled the students aside while the other students were engaged in flex-time activities. The teacher worked with these students three times a week for 25-minute sessions over nine weeks using a research-based intervention program. Some students mastered the conceptual and procedural objectives after six weeks, while others required the entire program.

Powell and Fuchs (2015) argued that Tier 2 interventions should be research-validated, standardized intervention programs or packages, but noted that validated programs are not available for all grade levels and specific mathematics topics. They recommended a systematic format for each lesson at this tier:

- warm-up activity (brief);
- explicit instruction (modeling, guided practice, immediate and corrective feedback, and concept-procedure connections);
- review; and
- a motivational component (embedded somewhere in the lesson to maintain attention and interest).

The authors described intensification at this level as adding elements such as smaller steps, more precise or repeated language, student think-alouds, modeling, manipulatives, worked examples, repeated practice, error correction, fading support, and fluency building (Powell & Fuchs, 2015).

Most early research on Tier 2 interventions focused on number and place-value concepts, number-combination fluency, procedural knowledge and skill, and problem solving at the elementary level (Bryant, Bryant, Gersten, Scammacca, & Chavez, 2008; L. S. Fuchs et al., 2007; L. S. Fuchs, Fuchs, Powell et al., 2008). Further, much of the early mathematics intervention research was single-subject, multiple-baseline design, resembling the intensity and individualization of Tier 3 interventions (e.g., Maccini & Hughes, 2000; van Garderen, 2007). More recently, intervention research has expanded considerably to include more focus on concept understanding as well as a wider range of mathematics topics. There are also more standardized, research-based intervention programs available for implementation with small groups, especially at the elementary level. Examples of research-based Tier 2 interventions are listed in Table 4.3.

These studies into the efficacy and efficiency of Tier 2 interventions with elementary and middle-school students with MLD share some features. Most effective interventions targeted specific mathematics topics and emphasized concept understanding integrated with procedural fluency. Most interventions employed a systematic, explicit instructional approach with frequent practice opportunities, scaffolding techniques, multiple representations, high levels of verbalizations, and frequent progress monitoring with academic feedback. Some interventions included motivational or attention elements, problem solving, speeded practice, or CRA sequences. These temporary interventions were typically 25–30-minute sessions, three or four times a week, for 10–20 weeks, depending on mathematics content and student responses. Unfortunately, many intervention studies involved researcher-implemented instruction measured by researcher-designed outcome assessments, but a few have included professional development for classroom teachers who implemented the intervention and nationally normed assessments. Most of the studies identified a cut-point for inclusion, typically below the 25th percentile (DeFouw et al., 2019). Studies also identified moderators of student outcomes, such as preintervention math skills, attentive behaviors, and language abilities.

D. Fuchs et al. (2014) summarized MTSS (RTI) research by pointing out that many, but not all students respond to Tier 1 and Tier 2 interventions. Even when these interventions were implemented with fidelity, the overall rate of inadequate responders in mathematics has not decreased. The success rate of older students, grade four and higher, participating in Tier 2 interventions was even lower than that of primary students. The authors recommended intensifying Tier 2 interventions by decreasing group sizes and increasing the duration of interventions. Additionally, well-trained teachers applying explicit, systematic interventions with frequent progress monitoring are needed to bring Tier 2 interventions up to a level of producing few nonresponders.

Tier 3 Interventions

Students who do not respond to Tier 2 interventions, or those who demonstrate very low levels of mathematics achievement even within Tier 1, may need interventions designed for

Table 4.3 Examples of Research-Based Tier 2 Interventions

Gr	Tier 2 Interventions	Group
K	*ROOTS:* Procedural fluency, concept understanding of numbers & quantities, early addition and subtraction operations, place value concepts (11 to 19). (50 lessons, 4–5 times a week, 16–20 weeks) Explicit instruction, overt demonstrations, visual representations, frequent practice, student verbalizations, specific academic feedback.	Clarke[1]
1	*Fusion:* Whole number concepts of numbers to 100, number combination fluency, addition and subtraction, problem solving. (60 lessons, 30 min, 3 times a week, 20 weeks) Small group explicit instruction, systematic practice, math models, scaffolded examples, academic feedback.	Clarke[2]
1	*Precision Mathematics Level 1:* Conceptual understanding and problem-solving skills for measurement and data analysis. (8 units over 8 weeks, 30-min lessons, 4 days a week) Integrated science, print and technology-based activities, explicitly addressed mathematics language, vocabulary, and word-problem subtypes.	Doabler[3]
1–2	*Early Numeracy Intervention:* Early mathematics skills and concepts—numbers, operations, basic facts, place value. (15- to 25-min booster lessons, 3 times a week, 18–23 weeks; 11 units) Small group tutoring with highly structured CRA approach, effective instruction elements, daily progress monitoring, group contingencies.	Bryant[4]
2	*Individualizing Student Instruction in Mathematics* (ISI-Math): Numeracy, measurement, geometry, algebra, depending on student needs. (regular class, 20 min supplementary over school year) Station approach for small groups, grouped by learning needs, systematic explicit instruction, scaffolding understanding, high levels of practice opportunities, progress monitoring.	Connor[5]
3–4	*Fraction Face-Off!* Measurement interpretation of fractions, comparing, ordering, placing on number line, and equivalencies. A schema-based word-problem-solving element. (3 times a week, 35 min, 12 weeks, 36 lessons) Tutors for pairs of students, systematic and explicit instruction, CRA, supported explanations, systematic review.	Fuchs[6]
5–7	*Concept Understanding:* Basic arithmetic concepts (place value, operations, word problems) focused on conceptual understanding, procedural fluency. (90 min a week plus 45 min independent practice; 14 weeks) Small group instruction, developing meaning then fluency, manipulatives and carefully selected representations, scaffolded abstraction, systematic and explicit instruction. Additional independent practice targeted individual needs within general classroom.	Opitz[7]
6–8	*Schema-Based Instruction:* Word problem solving involving proportion and multiplicative compare. (60-min sessions, 3–4 times a week, 12 sessions) Small group tutoring, systematic domain-specific focus, visual representations for problem types, self-regulation strategies, explicit instructional elements.	Xin[8]
6–8	*Explicit Inquiry Routine:* to understand and solve one-variable equations. (daily, 55-min sessions, 10 lessons) Small group instruction, explicit content sequencing, scaffolded inquiry, CRA sequence.	Scheuermann[9]

1 Clarke et al., 2016; Clarke et al., 2017; Doabler, Clarke, Kosty, Kurtz-Nelson, et al., 2019.
2 Clarke et al., 2014; Doabler et al., 2020.
3 Doabler, Clarke, Kosty, Turtura, et al., 2019.
4 Bryant, Bryant, Gersten, Scammacca, & Chavez, 2008; Bryant, Bryant, Gersten, Scammacca, Funk, et al., 2008; Bryant et al., 2011.
5 Connor et al., 2018.
6 L. S. Fuchs, Malone, et al., 2017; L. S. Fuchs, Malone, et al., 2016; L. S. Fuchs, Schumacher, Malone, & Fuchs, 2015.
7 Opitz et al., 2017.
8 Xin et al., 2005.
9 Scheuermann et al., 2009.

delivery at the Tier 3 level. Up to approximately 5% of students may need Tier 3 academic intervention in mathematics and/or reading (NCII, 2013; Zumeta, 2015). In an early study, L. S. Fuchs, Fucks, Craddock et al. (2008) found that a tutoring intervention (Tier 2) reduced the proportion of students with mathematics difficulties from 6.8% of the general third-grade population to 3.9% who would require more intensive instruction. School intervention teams should decide how to define nonresponsiveness to Tier 2 interventions and consider moving the lowest performers within Tier 1 directly to Tier 3 supports.

> Jean and Jack, students in Ms Smith's resource class at Balsam Middle School, have been struggling with fraction computations, even with extra short-term interventions on multiplying and dividing fractions. Jean's progress is consistently below 40% on weekly measures over eight weeks, while Jack's progress is uneven, ranging from 10% to 70%. Ms Smith recommends that these students receive a Tier 3 intervention for multiplicative reasoning and early fraction concepts such as magnitude comparison and equivalence. Ms Smith will instruct the two students together in daily 30-minute sessions during the school's flex period (math lab) and take advantage of peer interactions for explicit language skills, self-monitoring goals, and motivation. She will conduct a more specific concept and skills assessment, going back to elementary topics, and review their prior assessments. She plans to build students' knowledge and confidence by using accurate visual representations (e.g., paper strips and number lines), scaffolding understanding, and monitoring progress twice a week. Additional strategies include explicit language and distributed practice with immediate feedback. She will adjust the instructional methods when either student is not making progress. Ms Smith will also check with mathematics teachers to ensure she is framing concepts accurately and using language and skill sequences that are appropriate for fractions and other connected domains.

Harlacher (2016) described the major differences between Tier 2 and Tier 3 interventions as the amount of instructional time, group size, nature and frequency of progress monitoring, the use of individualized assessment to plan interventions, and degree of manipulation of key instructional variables such as scaffolding and feedback on practice attempts. Others make the distinction by noting that intervention at this level is within special education and involves the implementation of data-based individualization (DBI; Dennis & Gratton-Fisher, 2020; Lemons et al., 2018; NCII, 2013). While most special education research using single-subject, multiple-baseline designs (typically 1–5 students) offer evidence of effective intensive interventions in mathematics, they may not be "packaged" as interventions for Tier 3. Intervention programs for small groups of students (Tier 2) can be adapted to be more intensive (Powell & Stecker, 2014). Examples of Tier 3 interventions are displayed in Table 4.4.

Interventions at Tiers 2 and 3 require increasing intensity with additional resources, personnel, and instructional time. The benefits of focusing effective instruction with learning supports within core mathematics classrooms (Tier 1) are evident. For the approximately 15%–25% of students who require Tier 2 instruction and 3%–5% who require Tier 3, it is important that interventions not only employ effective instructional strategies but focus on critical mathematics topics with the most effective approaches (e.g., using number lines for fraction concepts rather than part–whole representations). The *Content Strands* in this text include example intensive interventions for content related to each strand.

This chapter initiated a three-chapter focus on effective instruction in mathematics for students with learning difficulties. The next chapter discusses research-based approaches for problem-solving instruction across the grade levels, followed by a chapter on instruction, accommodations, and collaboration models for core mathematics classes (Tier 1). For students who struggle to learn mathematics concepts and skills, it clearly isn't adequate to deliver

Table 4.4 Examples of Research-Based Tier 3 Interventions

Gr	Tier 3 Interventions	Group
1	*Number Rockets (Galaxy Math):* Whole number concepts of number combinations, sets, double-digit addition and subtraction, word problems. (30- to 40-min lessons, 3 times a week, 16 weeks, 5 units) One-to-one (or small group) tutoring, developing interconnected knowledge of numbers, CRA, speeded or nonspeeded practice.	Fuchs[1]
2	*Number and Operations Intervention:* Number magnitude and comparison, addition and subtraction combinations, place value, two-digit addition and subtraction. (daily 35-min lessons, 12 sessions) One-to-one, high level of interaction, prompts, student verbalizations.	Dennis[2]
3	*Hot Math:* Schema-broadening instruction for math problem solving, 4 word-problem types. (30-min, daily, 16 weeks in 3-week units) Small group tutoring, most difficult concepts targeted, manipulatives, scaffolding, self-regulated strategies.	Fuchs[3]
6–7	*Strategy Instruction:* Ratio equivalency and proportional reasoning focused on additive and multiplicative relation, build-up, and unit rate strategies. (15 lessons, 25 min each) One-to-one, explicit instruction with meaningful practice, CRA, conceptual instruction, think-alouds with scaffolding.	Hunt[4]
9	*DBI Intervention:* Basic arithmetic skills—multidigit operations with whole numbers, fractions, decimals, some geometry and initial algebra. (50-min lessons, 2 times a week, 16–20 lessons) One-to-one, individual scope and sequences, explicit elements, focus on data-based individualization (DBI).	Dennis[5]

Note: Some Tier 2 interventions can be intensified for Tier 3.
1 L. S. Fuchs et al., 2013; L. S. Fuchs et al., 2019.
2 Dennis, 2015.
3 L. S. Fuchs, Fuchs, Craddock, et al., 2008.
4 Hunt & Vasquez, 2014.
5 Dennis & Gratton-Fisher, 2020.

one-size-fits-all instruction. Nor is it adequate to provide interesting activities without other effective elements. Instruction that will result in learning must be carefully planned, with each student's characteristics in mind, and delivered with attention to student growth and understanding. Applying the principles of data-based individualization for interventions will promote the most achievement in the shortest amount of time.

References

Akyüz, G. (2014). The effects of students and school factors on mathematics achievement in TIMSS 2011. *Education and Science, 39*, 150–162.

Allinder, R. M., Bolling, R. M., Oats, R. G., & Gagnon, W. A. (2000). Effects of teacher self-monitoring on implementation of curriculum-based measurement and mathematics computation achievement of students with disabilities. *Remedial and Special Education, 21*(4), 219–226. doi: 10.1177/074193250002100403

Anthony, G., & Walshaw, M. (2009). Characteristics of effective teaching of mathematics: A view from the West. *Journal of Mathematics Education, 2*, 147–164.

Ausubel, D. P. (1968). *Educational psychology: A cognitive view.* Holt, Rinehart and Winston.

Bidwell, J. K., & Clason, R. G. (1970). *Readings in the history of mathematics education.* NCTM.

Biemans, H. J. A., & Simons, P. R. J. (1996). A computer-assisted instructional strategy for promoting conceptual change. *Instructional Science, 24*(2), 157–176. doi: 10.1007/bf00120487

Booth, J. L., McGinn, K. M., Barbieri, C., Begolli, K. N., Chang, B., Miller-Cotto, D., Young, L. K., & Davenport, J. L. (2017). Evidence for cognitive science principles that impact learning in mathematics. In D. C. Geary, D. B. Berch, R. Ochsendorf, & K. M. Koepke (Eds.), *Acquisition of complex arithmetic skills and higher order mathematics concepts: Volume 3*. Elsevier. doi: 10.1016/B978-0-12-805086-6.00013-8

Bottge, B. A., Toland, M. D., Gassaway, L., Butler, M., Choo, S., Griffen, A. K., & Ma, X. (2015). Impact of enhanced anchored instruction in inclusive math classrooms. *Exceptional Children*, *81*(2), 158–175. doi:10.1177/0014402914551742

Bouck, E. C., Park, J., Bouck, M., Alspaugh, J., Spitzley, S., & Buckland, A. (2019). Supporting middle school students in tier 2 math labs: instructional strategies. *Current Issues in Middle Level Education*, *24*(2). doi: 10.20429/cimle.2019.240203

Bouck, E. C., Satsangi, R., & Park, J. (2018). The concrete-representational-abstract approach for students with learning disabilities: An evidence-based practice synthesis. *Remedial and Special Education*, *39*(4), 211–228. doi: 10.1177/0741932517721712

Boudah, D. J., Lenz, B. K., Bulgren, J. A., Schumaker, J. B., & Deshler, D. D. (2000). Don't water down! Enhance content learning through the unit organizer routine. *TEACHING Exceptional Children*, *32*(3), 48–56. doi: 10.1177/004005990003200308

Broadbent, F. W. (1987). Lattice multiplication and division. *Arithmetic Teacher*, *34*(5), 28–31.

Bruner, J. S. (1960). *The process of education*. Harvard University Press.

Bruner, J. S. (1966). *Toward a theory of instruction*. W. W. Norton & Company.

Bruner, J. S. (1982). The language of education. *Social Research*, *49*, 835–853.

Bryant, D. P., Bryant, B. R., Gersten, R., Scammacca, N., & Chavez, M. M. (2008). Mathematics intervention for first- and second-grade students with mathematics difficulties. *Remedial and Special Education*, *29*(1), 20–32. doi: 10.1177/0741932507309712

Bryant, D. P., Bryant, B., Gersten, R., Scammacca, N., Funk, C., Winter, A., Shih, M., & Pool, C. (2008). The effects of tier 2 intervention on the mathematics performance of first-grade students who are at risk for mathematics difficulties. *Learning Disability Quarterly*, *31*(2), 47–63. doi: 10.2307/20528817

Bryant, D. P., Bryant, B., Roberts, G., Vaughn, S., Pfannenstiel, K., Porterfield, J., & Gersten, R. (2011). Early numeracy intervention program for first-grade students with mathematics difficulties. *Exceptional Children*, *78*(1), 7–23. doi: 10.1177/001440291107800101

Burns, M. K., Codding, R. S., Boice, C. H., & Lukito, G. (2010). Meta-analysis of acquisition and fluency math interventions with instructional and frustration level skills: Evidence for a skill-by-treatment interaction. *School Psychology Review*, *39*, 69–83. doi: 10.1080/02796015.2010.12087791

Butler, F. M., Miller, S. P., Crehan, K., Babbitt, B., & Pierce, T. (2003). Fraction instruction for students with mathematics disabilities: Comparing two teaching sequences. *Learning Disabilities Research & Practice*, *18*(2), 99–111. doi: 10.1111/1540-5826.00066

Carroll, W. M., & Issacs, A. C. (2003). Achievement of students using the university of Chicago school mathematics project's everyday mathematics. In S. Senk & D. Thompson (Eds.), *Standards-based school mathematics curricula* (pp. 9–22). Routledge. doi: 10.4324/9781003064275-4

Center on Instruction (2008). *Synopsis of "The Power of Feedback."* RMC Research Corporation. www.centeroninstruction.org

Clark, R. C., Nguyen, F., & Sweller, J. (2011). *Efficiency in learning: Evidence-based guidelines to manage cognitive load*. John Wiley & Sons.

Clarke, B., Doabler, C. T., Kosty, D., Kurtz Nelson, E., Smolkowski, K., Fien, H., & Baker, S. K. (2017). Testing the efficacy of a kindergarten mathematics intervention by small group size. *AERA Open*, *3*(2), 1–16. doi:10.1177/2332858417706899

Clarke, B., Doabler, C. T., Smolkowski, K., Baker, S. K., Fien, H., & Strand Cary, M. (2016). Examining the efficacy of a tier 2 kindergarten intervention. *Journal of Learning Disabilities*, *49*, 152–165. doi:10.1177/0022219414538514

Clarke, B., Doabler, C. T., Strand Cary, M., Kosty, D., Baker, S. K., Fien, H., & Smolkowski, K. (2014). Preliminary evaluation of a tier-2 mathematics intervention for first-grade students: Using a theory of change to guide formative evaluation activities. *School Psychology Review*, *43*, 160–177. doi: 10.1080/02796015.2014.12087442

Clarke, B., Smolkowski, K., Baker, S., Fein, H., Doabler, C., & Chard, D. (2011). The impact of a comprehensive tier 1 core kindergarten program on the achievement of students at-risk in mathematics. *Elementary School Journal, 111*(4), 561–584. doi: 10.1086/659033

Cleary, T., & Kitsantas, A. (2017). Motivation and self-regulated learning influences on middle school mathematics achievement. *School Psychology Review, 46*(1), 88–107. doi: 10.1080/02796015. 2017.12087607

Clements, D. H., & Sarama, J. (2007). Effects of a preschool mathematics curriculum: Summative research on the building blocks project. *Journal for Research in Mathematics Education, 38*(2), 136–163.

Clements, D. H., & Sarama, J. (2014). *Learning and teaching early math: The learning trajectories approach* (2nd ed.). Routledge. doi: 10.4324/9780203520574

Cole, J. E., & Washburn-Moses, L. H. (2010). Going beyond "the math wars." *TEACHING Exceptional Children, 42*(4), 14–20. *doi: 10.1177/004005991004200402*

Connor, C. M., Mazzocco, M. M. M., Kurz, T., Crowe, E. C., Tighe, E. L., Wood, T. S., & Morrison, F. J. (2018). Using assessment to individualize early mathematics instruction. *Journal of School Psychology, 66*, 97–113. doi: 10.1016/j.jsp.2017.04.005

DeFouw, E. R., Codding, R. S., Collier-Meek, M. A., & Gould, K. M. (2019). Examining dimensions of treatment intensity and treatment fidelity in mathematics intervention research for students at risk. *Remedial and Special Education, 40*(5), 298–312. doi: 10.1177/0741932518774801

Dennis, M. S. (2015). Effects of tier 2 and tier 3 mathematics interventions for students with mathematics difficulties. *Learning Disability Research & Practice, 30*(1), 29–42. doi: 10.1111/ldrp.12051

Dennis, M., & Gratton-Fisher, E. (2020). Use data-based individualization to improve high school students' mathematics computation and mathematics concept, and application performance. *Learning Disabilities Research and Practice.* doi: 10.1111/ldrp.12227

Dennis, M. S., Sharp, E., Chovanes, J., Thomas, A., Burns, R. M., Custer, B., & Park, J. (2016). A meta-analysis of empirical research on teaching students with mathematics learning disabilities. *Learning Disabilities Research and Practice, 31*(3), 156–168. doi: 10.1111/ldrp.12107

Deshler, D. D. & Lenz, B. K. (1989). The strategies instructional approach. *International Journal of Disability, Development, and Education, 36*(3), 203–224. doi: 10.1080/0156655893603004

Doabler, C., Baker, S., Kosty, D., Smolkowski, K., Clarke, B., Miller, S., & Fien, H. (2015). Examining the association between explicit mathematics instruction and student mathematics achievement. *The Elementary School Journal, 115*(3), 303–333. doi: 10.1086/679969

Doabler, C., Clarke, B., Kosty, D., Kurtz-Nelson, E., Fien, H., Smolkowski, K., & Baker, S. (2019). Examining the impact of group size on the treatment intensity of a tier 2 mathematics intervention within a systematic framework of replication. *Journal of Learning Disabilities, 52*(2), 168–180. doi: 10.1177/0022219418789376

Doabler, C., Clarke, B., Kosty, D., Turtura, J., Firestone, A., Smolkowski, K., Jungjohann, K., Brafford, T., Nelson, N., Sutherland, M., Fien, H., & Maddox, S. (2019). Efficacy of a first-grade mathematics intervention on measurement and data analysis. *Exceptional Children, 86*(1), 77–94. doi: 10.1177/0014402919857993

Doabler, C., Clarke, B., Kosty, D., Turtura, J., Sutherland, M., Maddox, S., & Smolkowski, K. (2020). Using direct observation to document "practice-based evidence" of evidence-based mathematics instruction. *Journal of Learning Disabilities, 2221942091137–.* doi: 10.1177/0022219420911375

Doabler, C., Fien, H., Nelson-Walker, N., & Baker, S. (2012). Evaluating three elementary mathematics programs for presence of eight research-based instructional design principles. *Learning Disability Quarterly, 35*(4), 200–211. doi: 10.1177/0731948712438557

Doabler, C. T., Nelson, N. J., Kosty, D. B., Fien, H., Baker, S. K., Smolkowski, K., & Clarke, B. (2014). Examining teachers' use of evidence-based practices during core mathematics instruction. *Assessment for Effective Intervention, 39*(2), 99–111. doi: 10.1177/1534508413511848

Dochy, F. J. R. C. (1992). *Assessment of prior knowledge as a determinant for future learning: The use of prior knowledge state tests and knowledge profiles.* Utrecht: Lemma B.V.

Ennis, R. P., Harris, K. R., Lane, K. L., & Mason, L. H. (2014). Lessons learned implementing self-regulated strategy development with students with emotional and behavioral disorders in alternative educational settings. *Behavioral Disorders, 40*(1), 68-77. doi: 10.17988/0198-7429-40.1.68

Erath, K., Prediger, S., Quasthoff, U., & Heller, V. (2018). Discourse competence as important part of academic language proficiency in mathematics classrooms: the case of explaining to learn and learning to explain. *Educational Studies in Mathematics, 99*(2), 161–179. doi: 10.1007/s10649-018-9830-7

Fadlelmula, F. K., Cakiroglu, E., & Sungur, S. (2015). Developing a structural model on the relationship among motivational beliefs, self-regulated learning strategies, and achievement in mathematics. *International Journal of Science and Mathematics Education, 13*(6), 1355–1375. doi: 10.1007/s10763-013-9499-4

Flores, M. M., Hinton, V., & Strozier, S. D. (2014). Teaching subtraction and multiplication with regrouping using the concrete-representational-abstract sequence and strategic instruction model. *Learning Disabilities Research & Practice, 29*(2), 75–88. doi:

Fuchs, D., Fuchs, L. S., & Vaughn, L. S. (2014). What is intensive instruction and why is it important? *TEACHING Exceptional Children, 46*(4), 13–18. doi: 10.1177/0040059914522966

Fuchs, L. S., Fuchs, D., Craddock, C., Hollenbeck, K., Hamlett, C., & Schatschneider, C. (2008). Effects of small-group tutoring with and without validated classroom instruction on at-risk students' math problem solving: Are two tiers of prevention better than one? *Journal of Educational Psychology, 100*(3), 491–509. doi: 10.1037/0022-0663.100.3.491

Fuchs, L. S., Fuchs, D., & Gilbert, J. K. (2019). Does the severity of students' pre-intervention math deficits affect responsiveness to generally effective first-grade intervention? *Exceptional Children, 85,* 147–162. doi:10.1177/0014402918782628

Fuchs, L. S., Fuchs, D., & Hollenbeck, K. (2007). Extending responsiveness to intervention to mathematics at first and third grades. *Learning Disabilities Research & Practice, 22*(1), 13–24. doi: 10.1111/j.1540-5826.2007.00227.x

Fuchs, L. S., Fuchs, D., & Malone, A. S. (2017). The taxonomy of intervention intensity. *TEACHING Exceptional Children, 50*(1), 35–43. doi: 10.1177/0040059917703962

Fuchs, L. S., Fuchs, D., Powell, S. R., Seethaler, P. M., Cirino, P. T., & Fletcher, J. M. (2008). Intensive intervention for students with mathematics disabilities: Seven principles of effective practice. *Learning Disability Quarterly, 31*(2), 79–92. doi: 10.2307/20528819

Fuchs, L. S., Geary, D. C., Compton, D. L., Fuchs, O, Schatschneider, C, Hamlett, C. L., De Selms, J., Seethaler, P. M., Wilson, J., Craddock, C. F., Bryant, J. D., Luther, K., & Changas, P. (2013). Effects of first-grade number knowledge tutoring with contrasting forms of practice. *Journal of Educational Psychology, 105*(1), 58–77. doi:10.1037/a0030127

Fuchs, L. S., Malone, A. S., Schumacher, R. F., Namkung, J., Hamlett, C. L., Jordan, N. C., Seigler, R. S., Gersten, R., & Changas, P. (2016). Supported self-explaining during fraction intervention. *Journal of Educational Psychology, 108*(4), 493–508. doi: 10.1037/edu0000073

Fuchs, L. S., Malone, A. S., Schumacher, R. F., Namking, J., & Wang, A. (2017). Fraction intervention for students with mathematics difficulties: Lessons learned from five randomized controlled trials. *Journal of Learning Disabilities, 50*(6), 631–639. doi: 10.1177/0022219416677249

Fuchs, L. S., Schumacher, R. F., Malone, A., & Fuchs, D. (2015). *Fraction face-off!* https://frg.vkcsites.org/what-are-interventions/math_intervention_manuals/

Fyfe, E. R., McNeil, N. M., Son, J. Y., & Goldstone, R. L. (2014). Concreteness fading in mathematics and science instruction: A systematic review. *Educational Psychology Review, 26*(1), 9–25. doi: 10.1007/s10648-014-9249-3

Fyfe, E. R., Rittle-Johnson, B., & DeCaro, M. S. (2012). The effects of feedback during exploratory mathematics problem solving: Prior knowledge matters. *Journal of Educational Psychology, 104*(4), 1094–1108. doi: 10.1037/a0028389

Gagné, R. M. (1956). *The conditions of learning* (2nd ed.). Holt, Rinehart, and Winston.

Gagné, R., Briggs, L., & Wager, W. (1988). *Principles of instructional design* (3rd ed.). Holt, Rinehart, and Winston.

Geary, D. C. (2015). The classification and cognitive characteristics of mathematical disabilities in children. In R. C. Kadosh & A. Dowker (Eds.), *The Oxford Handbook of Numerical Cognition* (pp. 751–770). Oxford University Press. doi: 10.1093/oxfordhb/9780199642342.013.017

Gersten, R., Beckmann, S., Clarke, B., Foegen, A., Marsh, L., Star, J. R., & Witzel, B. (2009). *Assisting students struggling with mathematics: Response to intervention (RtI) for elementary and middle schools* (NCEE 2009–4060). National Center for Education Evaluation and Regional Assistance, Institute of Education Sciences, US Department of Education. https://ies.ed.gov/ncee/wwc/PracticeGuide/2

Gersten, R., Chard, D. J., Jayanthi, M., Baker, S. K., Morphy, P., & Flojo, J. (2009). Mathematics instruction for students with learning disabilities: A meta-analysis of instructional components. *Review of Educational Research, 79*(3), 1202–1242. doi: 10.3102/0034654309334431

Griffin, C. C., League, M. B., Griffin, V. L., & Bae, J. (2013). Discourse practices in inclusive elementary mathematics classrooms. *Learning Disability Quarterly, 36*(1), 9–20. doi: 10.1177/0731948712465188

Haring, N. G., & Eaton, M. D. (1978). Systematic instructional technology: An instructional hierarchy (pp. 23-40). In N. G. Haring, T. C. Lovitt, M. D. Eaton, & C. L. Hansen (Eds.), *The fourth R: Research in the classroom*. Merrill.

Harlacher, J. E. (2016). *Distinguishing between Tier 2 and Tier 3 instruction in order to support implementation of RTI*. National Center for Learning Disabilities, RTI Action Network. www.rtinetwork.org/

Harris, C. A., Miller, S. P., & Mercer, C. D. (1995). Teaching initial multiplication skills to students with disabilities in general education classroom. *Learning Disabilities Research & Practice, 10*, 180–195.

Haydon, T., & Hunter, W. (2011). The effects of two types of teacher questioning on teacher behavior and student performance: A case study. *Education and Treatment of Children, 34*(2), 229–245. doi: 10.1353/etc.2011.0010

Hennessey, M. N., Higley, K., & Chesnut, S. R. (2012). Persuasive pedagogy: A new paradigm for mathematics education. *Education Psychological Review, 24*(2), 187–204. doi: 10.1007/s10648-011-9190-7

Hill, H. C., Charalambous, C. Y., & Chin, M. J. (2019). Teacher characteristics and student learning in mathematics: A comprehensive assessment. *Educational Policy, 33*(7), 1103–1134. doi: 10.1177/0895904818755468

Hill, H. C., & Chin, M. J. (2018). Connections between teachers' knowledge of students, instruction, and achievement outcomes. *American Educational Research Journal, 55*(5), 1076–1112. doi: 10.3102/0002831218769614

Hogan, K., & Pressley, M. (Eds.). (1997). *Scaffolding student learning: Instructional approaches and issues*. Brookline Books.

Hohensee, C. (2016). Teachers' awareness of the relationship between prior knowledge and new learning. *Journal for Research in Mathematics Education, 47*, 17–27.

Hughes, C. A., Morris, J. A., Therrien, W. J., & Benson, S. K. (2017). Explicit instruction: Historical and contemporary contexts. *Learning Disabilities Research & Practice, 32*, 140–148. doi:10.1111/ldrp.12142

Hunt, J. H., & Vasquez, E. (2014). Effects of ratio strategies intervention on knowledge of ratio equivalence for students with learning disability. *The Journal of Special Education, 48*(3), 180–190. doi: 10.1177/0022466912474102

Jitendra, A. K., & Krawec, J. (2019). Effects from secondary interventions and approaches for the prevention and remediation of mathematics difficulties (pp. 271–290). In P. C. Pullen & M. J. Kennedy (Eds.), *Handbook of Response to Intervention and Multi-Tiered Systems of Support*. Routledge. doi: 10.4324/9780203102954-19

Johnson, E. S., Humphrey, M., Mellard, D. F., Woods, K., & Swanson, H. L. (2010). Cognitive processing deficits and students with specific learning disabilities: A selective meta-analysis of the literature. *Learning Disability Quarterly, 33*(1), 3–18. doi: 10.1177/073194871003300101

Jung, P., McMaster, K., Kunkel, A., Shin, J., & Stecker, P. (2018). Effects of data-based individualization for students with intensive learning needs: A meta-analysis. *Learning Disabilities Research and Practice, 33*(3), 144–155. doi: 10.1111/ldrp.12172

Kame'enui, E. J., & Carnine, D. W. (1998). *Effective teaching strategies that accommodate diverse learners*. Pearson.

Kame'enui, E. J., Carnine, D. W., Dixon, R. C., Simmons, D. C., & Coyne, M. D. (2002). *Effective teaching strategies that accommodate diverse learners* (2nd ed.). Pearson.

Kaminski, J. A., & Sloutsky, V. M. (2009). The effect of concreteness on children's ability to detect common proportion. In N. Taatgen & H. van Rijn (Eds.), *Proceedings of the conference of the cognitive science society* (pp. 335–340). Erlbaum.

Kehrer, P., Kelly, K., & Heffernan, N. (2013). Does immediate feedback while doing homework improve learning. In Y. Boonthum-Denecke (Ed.), *Proceedings of the Twenty-Sixth International Florida Artificial Intelligence Research Society Conference* (pp. 542–545). AAAI Press.

Kent, L. B. (2014). Students' thinking and the depth of the mathematics curriculum. *Journal of Education and Learning, 3*(4), 90–95. doi: 10.5539/jel.v3n4p90

Laski, E., & Dulaney, A. (2015). When prior knowledge interferes, inhibitory control matters for learning: The case of numerical magnitude representations. *Journal of Educational Psychology, 107*(4), 1035–1050. doi: 10.1037/edu0000034

Lee, H., Coomes, J., & Yim, J. (2019). Teachers' conceptions of prior knowledge and the potential of a task in teaching practice. *Journal of Mathematics Teacher Education, 22*(2), 129–151. doi: 10.1007/s10857-017-9378-y

Lemons, C., Vaughn, S., Wexler, J., Kearns, D., & Sinclair, A. (2018). Envisioning an improved continuum of special education services for students with learning disabilities: Considering intervention intensity. *Learning Disabilities Research and Practice, 33*(3), 131–143. doi: 10.1111/ldrp.12173

Lenz, B. K., Bulgren, J. A., Schumaker, J. B., Deshler, D. D., & Boudah, D. A. (1994). *The unit organizer routine.* Edge Enterprises.

Liu, J., & Xin, Y. P. (2017). The effect of eliciting repair of mathematics explanations of students with learning disabilities. *Learning Disability Quarterly, 40*, 132–145. doi:10.1177/0731948716657496

Lloyd, J. W., Forness, S. R., & Kavale, K. A. (1998). Some methods are more effective than others. *Intervention in School and Clinic, 33*(4), 195–200. doi: 10.1177/105345129803300401

Losinski, M., Ennis, R. P., Sanders, S. A., & Wiseman, N. (2019). An investigation of SRSD to teach fractions to students with disabilities. *Exceptional Children, 85*(3), 291–308. doi: 10.1177/0014402918813980

Ma, X., & Papanastasiou, C. (2006). How to begin a new topic in mathematics: Does it matter to students' performance in mathematics? *Evaluation Review, 30*, 451–480. doi: 10.1177/0193841X05284090

Maccini, P., & Hughes, C. A. (2000). Effects of a problem-solving strategy on the introductory algebra performance of secondary students with learning disabilities. *Learning Disabilities Research & Practice, 15*(1), 10–21. doi: 10.1207/sldrp1501_2

Maccini, P., & Ruhl, K. L. (2000). Effects of a graduated sequence on the algebraic subtraction of integers by secondary students with learning disabilities. *Education and Treatment of Children, 23*, 465–489.

MacGregor, M., & Stacey, K. (1997). Students' understanding of algebraic notation: 11–15. *Educational Studies in Mathematics. 33*(1), 1–19. doi: 10.1023/A:1002970913563

Mancl, D. B., Miller, S. P., & Kennedy, M. (2012). Using the concrete-representational-abstract sequence with integrated strategy instruction to teach subtraction with regrouping to students with learning disabilities. *Learning Disabilities Research & Practice, 27*(4), 152–166. doi: 10.1111/j.1540-5826.2012.00363.x

Marita, S., & Hord, C. (2017). Review of mathematics interventions for secondary students with learning disabilities. *Learning Disability Quarterly, 40*(1), 29–40. doi: 10.1177/0731948716657495

Mastropieri, M. A., & Scruggs, T. E. (1994). *Effective instruction for special education.* (2nd ed.). PRO-ED.

McGinn, K., M., Lange, K. E., & Booth, J. L. (2015). A worked example for creating worked examples. *Mathematics Teaching in the Middle School, 21*(1), 26–33. doi: 10.5951/mathteacmiddscho.21.1.0026

McLeskey, J., Barringer, M-D., Billingsley, B., Brownell, M., Jackson, D., Kennedy, M., Lewis, T., Maheady, L., Rodriguez, J., Scheeler, M. C., Winn, J., & Ziegler, D. (2017, January). *High-leverage practices in special education.* Council for Exceptional Children & CEEDAR Center. https://ceedar.education.ufl.edu/wp-content/uploads/2017/07/CEC-HLP-Web.pdf

McLeskey, J., Billingsley, B., Brownell, M. T., Maheady, L., & Lewis, T. J. (2019). What are high-leverage practices for special education teachers and why are they important? *Remedial and Special Education, 40*(6), 331–337. doi: 10.1177/0741932518773477

Mercer, C. D., & Miller, S. P. (1991–1994). *Strategic math series (A series of seven manuals).* Edge Enterprises.

Mercer, C. D., & Miller, S. P. (1992). Teaching students with learning problems in math to acquire, understand, and apply basic math facts. *Remedial and Special Education, 13*(3), 19–35, 61. doi: 10.1177/074193259201300303

Mercer, C. D., & Pullen, P. C. (2005). *Students with learning disabilities* (6th ed.). Pearson.

Middleton, J., Jansen, A., & Goldin, G. A. (2018). The complexities of mathematical engagement: Motivation, affect, and social interactions. In J. Cai (Ed.). *Compendium for research in mathematics education.* NCTM. www.nctm.org

Miller, S. P., Kaffar, B. J., & Mercer, C. D. (2011). *Addition with regrouping.* Edge Enterprises.

Milyutin, E., & Meyer, L. (2018, 27 February). *How do we support English language learners' success in mathematics?* NCTM. https://my.nctm.org/blogs/evgeny-milyutin/2018/02/27/how-do-we-support-english-learners-success-in-math

Montague, M. (1992). The effects of cognitive and metacognitive strategy instruction on the mathematical problem solving of middle-school students with learning disabilities. *Journal of Learning Disabilities, 25*(4), 230–248. doi: 10.1177/002221949202500404

Montague, M. (1997). Cognitive strategy instruction in mathematics for students with learning disabilities. *Journal of Learning Disabilities, 30*(2), 164–177. doi: 10.1177/002221949703000204

Montague, M. (2003). *Solve It! A practical approach to teaching mathematical problem-solving skills.* Exceptional Innovations.

Moschkovich, J. (2013). Principles and guidelines for equitable mathematics teaching practices and materials for English language learners. *Journal of Urban Mathematics Education, 6*(1), 45–57. doi: 10.21423/jume-v6i1a204

Moshman, D. (1982). Exogenous, endogenous, and dialectical constructivism. *Developmental Review, 2*(4), 371–384. doi: 10.1016/0273-2297(82)90019-3

Murphy, J. S. (2015, 20 January). New SAT, new problems. *The Atlantic.* www.theatlantic.com/

Namkung, J. M., & Bricko, N. (2020). The effects of algebraic equation solving intervention for students with mathematics learning difficulties. *Journal of Learning Disabilities, 54*(2), 111–123. doi: 10.1177/0022219420930814

Nathan, M. J. (2012). Rethinking formalisms in formal education. *Educational Psychologist, 47*(2), 125–148. doi: 10.1080/00461520.2012.667063

National Center on Intensive Intervention (2013). *Data-based individualization: A framework for intensive intervention.* Office of Special Education Programs, US Department of Education. https://intensiveintervention.org

National Council of Teachers of Mathematics (2000). *Principles and standards for school mathematics.* www.nctm.org/Standards-and-Positions/Principles-and-Standards/

National Council of Teachers of Mathematics (2006). *Curriculum focal points for prekindergarten through grade 8.* www.nctm.org/curriculumfocalpoints/

National Governors Association Center for Best Practices & Council of Chief State School Officers (2010). *Common core state standards for mathematics.* www.corestandards.org/Math/

National Mathematics Advisory Panel (2008). *Foundations for success: The final report of the National Mathematics Advisory Panel.* US Department of Education. www2.ed.gov/about/bdscomm/list/mathpanel/report/final-report.pdf

Opitz, E. M., Freesemann, O., Prediger, S., Grob, U., Matull, I., & Hußmann, S. (2017). Remediation for students with mathematics difficulties: An intervention study in middle schools. *Journal of Learning Disabilities, 50*(6), 724–736. doi: 10.1177/0022219416668323

Peltier, C. J., Vannest, K. J., & Marbach, J. J. (2018). A meta-analysis of schema instruction implemented in single-case experimental designs. *The Journal of Special Education, 52*(2), 89–100. doi: 10.1177/0022466918763173

Peterson, B. E., Corey, D. L., Lewis, B. M., & Bukarau, J. (2013). Intellectual engagement and other principles of mathematics instruction. *The Mathematics Teacher, 106*(6), 446–450. doi: 10.5951/mathteacher.106.6.0446

Peterson, S. K., Mercer, C. D., & O'Shea, L. (1988). Teaching learning disabled students place value using the concrete to abstract sequence. *Learning Disabilities Research & Practice, 4*, 52–56.

Petty, T., Wang, C., & Harbaugh, A. P. (2013). Relationships between student, teacher, and school characteristics and mathematics achievement. *School Science and Mathematics, 113*(7), 333–344. doi: 10.1111/ssm.12034

Powell, S. R., & Fuchs, L. S. (2015). Intensive intervention in mathematics. *Learning Disabilities Research & Practice, 30*(4), 182–192. doi: 10.1111/ldrp.12087

Powell, S. R., & Stecker, P. M. (2014). Using data-based individualization to intensify mathematics intervention for students with disabilities. *TEACHING Exceptional Children, 46*(4), 31–37. doi: 10.1177/0040059914523735

Poylá, G. (1973). *How to solve it* (39th ed.). Princeton University Press.

Prediger, S., Erath, K., & Opitz, E. M. (2019). The language dimension of mathematical difficulties (pp. 437–455). In A. Fritz, V. G. Haase, & P. Räsänen (Eds.), *International handbook of mathematical learning difficulties*. Springer. doi: 10.1007/978-3-319-97148-3

Radford, L., & Barwell, R. (2016). Language in mathematics education research (pp. 275–313). In A. Gutiérrez, G. C. Leder, & P. Boero (Eds.), *The second handbook of research on the psychology of mathematics education*. Sense Publishers. doi: 10.1007/978-94-6300-561-6_8

Rau, M. A., & Matthews, P. G. (2017). How to make 'more' better? Principles for effective use of multiple representations to enhance students' learning about fractions. *ZDM Mathematics Education*, *49*, 531–544. doi: 10.1007/s11858-017-0846-8

Riccomini, P., Smith, G., Hughes, E., & Fries, K. (2015). The language of mathematics: the importance of teaching and learning mathematical vocabulary. *Reading & Writing Quarterly*, *31*(3), 235–252. doi: 10.1080/10573569.2015.1030995

Rittle-Johnson, B. (2017). Developing mathematics knowledge. *Child Development Perspectives*, *11*(3), 184–190. doi: 10.1111/cdep.12229

Rittle-Johnson, B., Loehr, A. M., & Durkin, K. (2017). Promoting self-explanation to improve mathematics learning: A meta-analysis and instructional design principles. *ZDM Mathematics Education*, *49*, 599–611. doi: 10.1007/s11858-017-0834-z

Rittle-Johnson, B., & Koedinger, K. R. (2009). Iterating between lessons on concepts and procedures can improve mathematics knowledge. *British Journal of Educational Psychology*, *79*(3), 483–500. doi: 10.1348/000709908x398106

Rittle-Johnson, B., Schneider, M., & Star, J. R. (2015). Not a one-way street: Bidirectional relations between procedural and conceptual knowledge of mathematics. *Educational Psychology Review*, *27*(4), 587–597. doi: 10.1007/s10648-015-9302-x

Rosli, R., Capraro, M. M., & Capraro, R. M. (2014). The effects of problem posing on student mathematical learning: A meta-analysis. *International Education Studies*, *7*(13), 227–241. doi: 10.5539/ies.v7n13p227

Rubenstein, R. N. (2000). Word origins: Building communication connections. *Mathematics Teaching in the Middle School*, *5*, 493–498.

Rubenstein, R. N., & Thompson, D. R. (2002). Understanding and supporting children's mathematical vocabulary development. *Teaching Children Mathematics*, *9*(2), 107–112.

Salomon, G., & Perkins, D. N. (1989). Rocky roads to transfer: Rethinking mechanisms of a neglected phenomenon. *Educational Psychologist*, *24*(2), 113–142. doi: 10.1207/s15326985ep2402_1

Scheuermann, A. M., Deshler, D. D., and Schumaker, J. B. (2009). The effects of the explicit inquiry routine on the performance of students with learning disabilities on one-variable equations. *Learning Disability Quarterly*, *32*(2), 103–120. doi: 10.2307/27740360

Sealander, K. A., Johnson, G. R., Lockwood, A. D., & Medina, C. M. (2012). Concrete-semiconcrete-abstract (CSA) instruction: A decision rule for improving instructional efficacy. *Assessment for Effective Intervention*, *38*, 53–65. doi: 10.1177/1534508412453164

Sharma, S. (2015). Promoting risk taking in mathematics classrooms: The importance of creating a safe learning environment. *The Mathematics Enthusiast*, *12*, 290–306.

Slavin, R. E., & Lake, C. (2008). Effective programs in elementary mathematics: A best-evidence synthesis. *Review of Educational Research*, *78*(3), 427–515. doi: 10.3102/0034654308317473

Slavin, R. E., Lake, C., & Groff, C. (2009). Effective programs in middle and high-school mathematics: A best-evidence synthesis. *Review of Educational Research*, *79*(2), 839–911. doi: 10.3102/0034654308330968

Sloutsky, V. M., Kaminski, J. A., & Heckler, A. F. (2005). The advantage of simple symbols for learning and transfer. *Psychological Bulletin and Review*, *12*(3), 508–513. doi: 10.3758/bf03193796

Star, J. R., Rollack, C., Durkin, K., Rittle-Johnson, B., Lynch, K., Newton, K., & Gogolen, C. (2014). Learning from comparison in algebra. *Contemporary Educational Psychology*, *40*, 41–54. doi: 10.1016/j.cedpsych.2014.05.005

Star, J. R., & Verschaffel, L. (2018). Providing support for student learning: Recommendations from cognitive science for the teaching of mathematics (pp. 292–307). In J. Cai (Ed.), *Compendium for research in mathematics education*. NCTM.

Stevens, E. A., Rodgers, M. A., & Powell, S. R. (2018). Mathematics interventions for upper elementary and secondary students: A meta-analysis of research. *Remedial and Special Education, 39*(6), 327–340. doi: 10.1177/0741932517731887

Swanson, H. L., Orosco, M. J., & Lussier, C. M. (2014). The effects of mathematics strategy instruction for children with serious problem-solving difficulties. *Exceptional Children, 80*(2), 149–168. doi: 10.1177/001440291408000202

Sweller, J. (2016). Cognitive load theory, evolutionary educational psychology, and instructional design (pp. 291–306). In D. C. Geary & D. B. Berch (Eds.), *Evolutionary perspectives on child development and education*. Springer International Publishing. doi: 10.1007/978-3-319-29986-0_12

TeachingWorks. (n.d.). *High-leverage practices*. University of Michigan. www.teachingworks.org/work-of-teaching/high-leverage-practices

Thomas, C. N., Van Garderen, D., Scheuermann, A., & Lee, E. J. (2015). Applying a universal design for learning framework to mediate the language demands of mathematics. *Reading & Writing Quarterly, 31*(3), 207–234. doi: 10.1080/10573569.2015.1030988

Thompson, D. R., & Rubenstein, R. N. (2000). Learning mathematics vocabulary: Potential pitfalls and instructional strategies. *Mathematics Teacher, 93*, 568–574.

Thompson, D. R., & Rubenstein, R. N. (2014). Literacy in language and mathematics: More in common than you think. *Journal of Adolescent & Adult Literacy, 58*(2), 105–108. doi: 10.1002/jaal.338

van de Pol, J., Volman, M., & Beishuizen, J. (2010). Scaffolding in teacher–student interaction: A decade of research. *Educational Psychology Review, 22*(3), 271–296. doi: 10.1007/s10648-010-9127-6

Van Dooren, W., De Bock, D. Hessels, A., Janssens, D., & Verschaffel, L. (2004). Remedying secondary school students' illusion of linearity: A teaching experiment aiming at conceptual change. *Learning and Instruction, 14*(5), 485–501. doi: 10.1016/j.learninstruc.2004.06.019

van Garderen, D. (2007). Teaching students with learning disabilities to use diagrams to solve mathematical word problems. *Journal of Learning Disabilities, 40*(6), 540–553. doi: 10.1177/00222194070400060501

VanDerHeyden, A. M., & Codding, R. S. (2020). Belief-based versus evidence-based math assessment and instruction. *Communiqué (National Association of School Psychologists), 48*(5), 1, 20–25.

Walkington, C., & Hayata, C. (2017). Designing learning personalized to students' interests: balancing rich experiences with mathematical goals. *ZDM, 49*(4), 519–530. doi: 10.1007/s11858-017-0842-z

Walshaw, M., & Anthony, G. (2008). The teacher's role in classroom discourse: A review of recent research into mathematics classrooms. *Review of Educational Research, 78*(3), 516–551. doi: 10.3102/0034654308320292

Wang, A., Fuchs, L., Fuchs, D., Gilbert, J., Krowka, S., & Abramson, R. (2019). Embedding self-regulation instruction within fractions intervention for third graders with mathematics difficulties. *Journal of Learning Disabilities, 52*(4), 337–348. doi: 10.1177/0022219419851750

Weinert, F. E., & Helmke, A. (1998). The neglected role of individual differences in theoretical models of cognitive development. *Learning and Instruction, 8*(4), 309–323. doi: 10.1016/s0959-4752(97)00024-8

Wiggins, G., & McTighe, J. (2005). *Understanding by design* (2nd ed.). Association for Supervision and Curriculum Development. www.ascd.org

Wood, D., Bruner, J. S., & Ross, G. (1976). The role of tutoring in problem solving. *Journal of Child Psychology & Psychiatry and Allied Disciplines, 17*(2), 89–100. doi: 10.1111/j.1469-7610.1976.tb00381.x

Xin, Y. P., Chiu, M. M., Tzur, R., Ma, X., Park, J. Y., & Yang, X. (2020). Linking teacher-learner discourse with mathematical reasoning of students with learning disabilities: An exploratory study. *Learning Disability Quarterly, 43*(1), 43–56. doi: 10.1177/0731948719858707

Xin, Y. P., Jitendra, A., & Deatline-Buchman, A. (2005). The effects of mathematical word problem solving instruction on middle school students with learning problems. *The Journal of Special Education, 39*(3), 181–192. doi: 10.1177/00224669050390030501

Xin, Y. P., & Zhang, D. (2009). Exploring a conceptual model-based approach to teaching situated word problems. *The Journal of Educational Research, 102*, 427–441. doi: 10.3200/JOER.102.6.427-442

Ziori, E., & Dienes, Z. (2008). How does prior knowledge affect implicit and explicit concept learning? *The Quarterly Journal of Experimental Psychology, 61*(4), 601–624. doi: 10.1080/17470210701255374

Zumeta, R. (2015). Implementing intensive intervention: How do we get there from here? *Remedial and Special Education, 36*(2), 83–88. doi: 10.1177/0741932514558935

5 Problem-Solving Instruction

Chapter Questions

1. Why are problem-solving skills emphasized in mathematics standards and assessments?
2. What traits do good problem solvers share?
3. How do students with mathematics difficulties perform on problem-solving tasks?
4. How should problem solving be taught to students with learning difficulties?
5. What sequences and strategies are most effective for solving routine and nonroutine problems?
6. How can problem solving be addressed within MTSS frameworks?

One train departs from Miami at the same time as another train departs from New York.

Many adults break out in a sweat just hearing the first line of this memorable word problem. Of course, "word" or "story" problems aren't representative of most problem-solving demands in real life, but they are the way problem situations are presented to students in mathematics textbooks and on standardized tests. This chapter explores strategies for teaching students how to interpret and solve word problems. But first, read the following problem and think about how it could be solved:

> *You have been contracted to paint the town's water tower—a huge metal-covered cylinder with a height of 15.5 meters and the radius at the circular base of 7.1 meters. Because the cylinder is on a platform, its base will not be painted. What is the surface area of the portion to be painted? If it takes 1 gallon to paint 20 square meters, how many gallons of paint should you order?*

> Think: What are the reading demands for understanding this problem? Is personal experience useful? In how many different ways can this problem be solved?

The water-tower problem is pragmatic, from the real world. It is taken from an actual problem situation a painter might encounter. Another problem type is the abstract problem—one that challenges thinking for the sake of the mental exercise. Try this example:

> *The letter "E" represents any even digit (2, 4, 6, 8) and the letter "D" represents any odd digit (1, 3, 5, 7, 9). If E + D = 15 and 2E + D = 21, what are the values of E and D?*

> Think: What makes this problem different than the previous one? What are the reading and experiential demands? In how many ways can this problem be solved?

DOI: 10.4324/9781003096733-5

Problem solving is one of the most complex, ubiquitous, and critical skills developed in mathematics curricula. The next section elaborates on this importance.

The Importance of Problem-Solving Skills

Problem solving, broadly defined, is "using what you know to learn what you don't" (A. E. Gurganus, personal communication, 7 September 2020). Problem solving in mathematics refers to tackling problem situations or "tasks that have the potential to provide intellectual challenges that can enhance students' mathematical development…. [Problems] can promote students' conceptual understanding, foster their ability to reason and communicate mathematically, and capture their interests and curiosity" (Cai & Lester, 2010, p. 1). Problem solving "is not only a goal of learning mathematics but also a major means of doing so" (National Council of Teachers of Mathematics, NCTM, 2000, p. 52). Both the NCTM standards and the *Common Core State Standards for Mathematics* (CCSSM; NGA & CCSSO, 2010) included problem solving as the first process standard. Problem solving—that is making sense of problems, considering various solution strategies, understanding the problem-solving approaches of others, representing and modeling problems, and finding solutions to mathematical and everyday problems—is central to every grade level, every domain, and every topic in the mathematics curriculum.

One compelling argument for the centrality of problem solving in the mathematics curriculum is found in the NCTM standards document. "We live in a time of extraordinary and accelerating change … The need to understand and be able to use mathematics in everyday life and in the workplace has never been greater and will continue to increase" (2000, p. 4). Problem-solving abilities developed in mathematics can be applied to other subject areas and learning situations. Problem solving is an essential requirement of the workplace, from hourly positions to the professions. Adults in the workforce must solve challenging problems, typically in collaboration with others. They are required to define problems, gather information, and generate and evaluate solutions.

Problem solving should be the basis for mathematics understanding, not an afterthought or later application of skills learned in isolation. Consider the format of mathematics textbooks more than 20 years ago. Lessons typically began with an algorithm and a few examples for working through the procedure. The algorithm was followed by items for in-school and homework practice. Two or three word problems, obviously related to the same algorithm, were positioned at the end of the lesson. This approach is sometimes called "teaching by modeling procedures." The focus of this lesson is on the procedure and not concept development, connection with prior knowledge, or application.

Consider a lesson that begins with problem solving—or, more specifically, a problem situation that needs solving.

> *The weather station workers need to prepare a monthly report on the weather data they have been collecting. The precipitation amounts over the past month were: 1.2 cm, 0.8 cm, 1.075 cm, and 0.25 cm in four precipitation events. This month's precipitation was 0.7 cm greater than the previous month's. How can we find this month's total precipitation?*

A skillful teacher would turn it over to the students, with guidance, for small working groups to discuss *how* to go about finding the total amount. The teacher would prompt students to discuss what they know about the problem, drawing from experiences. The questions could be, "What do we know? What are we looking for? What information is extra, not needed to answer the question? What do you already know about how precipitation is measured and reported?" Then she would ask for some ideas, without judgment, on possible strategies for solving this problem. As students listen to others, they tend to re-form ideas or variations on ideas. The teacher has a view of her students' thought processes by listening to their reasoning. This lesson begins with exploration and inquiry, with structure, to encourage student engagement, connections to prior learning, and deeper concept understanding. It allows all students to be active learners, engaged in problem-solving processes.

As each group of students settles on a way to solve the problem, the teacher encourages each to apply its strategy. One group may draw the rain collector on graph paper and draw in each precipitation event. Another group may use one centimeter as a reference point and use a number line to see how close to one centimeter each week's measurement falls. Another group finds a calculator and simply plugs in the numbers and quickly arrives at a solution, then spends time examining why the solution makes sense by working backwards through algorithms. The fourth group attempts to line up the numbers and add them up—arriving at 12.775 centimeters. Then they exclaim, "That can't be right!" and begin discussing what went wrong and where to put the decimal points.

Eventually groups begin to settle on the same solution from a variety of directions. As they share their processes with each other, the meanings of the concepts involved deepen. At this point the teacher could demonstrate the most common algorithm and reinforce the place-value concepts involved. Guided practice with two or three similar problems will help students feel confident with the algorithm and problem structure. At this point in the lesson, particularly for students who struggle in mathematics, it is important for the teacher to explicitly describe the problem type or pattern and strategies used for solving the problem. It is also critical that the teacher not allow a faulty strategy, such as using the keyword *total*, that might work in this problem but not another. And, of course, a master teacher would intentionally build on this problem type (several like quantities *combine* into one amount) later in the curriculum.

TRY THIS

Select one mathematics concept or algorithm and develop a problem-solving context in which it could be introduced. Use one from a student's IEP or select one from the following:

1. adding simple fractions,
2. graphing a set of binary coordinates, or
3. computing the area of a rectangle.

Problem solving is a critical part of today's high-stakes testing. High-stakes testing is a term that refers to those district- or state-wide tests with significant implications for students, teachers, schools, and/or districts. The outcomes of these tests may determine a student's promotion, a teacher's paycheck or contract, and a school's leadership and funding. Most states' required tests are heavily weighted with mathematics problem solving. In some cases, students must show their work and explain their answers in writing. This emphasis on problem solving over calculation is a result of mathematics standards that emphasize problem solving and related reasoning, modeling, and communication skills. Problem solving is often how students are required to demonstrate concept and procedural knowledge. See Box 5.1 for examples of test items that involve problem solving.

Box 5.1   Fifth-Grade End-of-Grade Test Items

1. Joseph and Marie are putting together the ribbons and medals for field day awards. They need to lace 200 medals with 30 centimeters of ribbon each. They have seven spools of ribbon, each with 10 meters. How much ribbon will they have left over?
2. The Parent-Teacher Organization sold raffle tickets during the field day for $3.25 for a chance at a new bicycle. Belen's parents have $20.00 to purchase raffle tickets. How many tickets can they buy?
3. In the 100-meter dash, the top fifth grader had a record race of 11.5 seconds. This result was 35% faster than the top second-grader's time. Write an equation for finding the second-grader's time, using x (fifth-grader's time) and y (second-grader's time).

Source: created by author.

TRY THIS

Find examples of specific state or district mathematics test items for a given grade level. Analyze the sample for problem-solving requirements. What type of problem solving is required? What is the student asked to do (processes) and know (content) when solving the problems?

Problem solving is an integral part of life, not just within the study of mathematics. Problem-solving abilities are some of the most important outcomes of education. For example, the National Association of Colleges and Employers' *Job Outlook* report for 2020 surveyed 150 employers about the attributes they seek on a candidate's resume (NACE, 2020). Problem-solving skills were rated most important for employment (91.2% of employers), followed by the ability to work in a team (86.3%), strong work ethic (80.4%), and analytical/quantitative skills (79.4%), with a range of other job-related skills following those. The Association of American Colleges and Universities contracted for a similar survey of 400 employers with at least 25 employees targeting two- and four-year college graduates' preparation (Hart Research Associates, 2018). The employers nearly universally agreed that to achieve success, a candidate's demonstrated capacity to think critically, communicate clearly, and solve complex problems was more important than the undergraduate major (91% agreed).

The Organisation for Economic Co-operation and Development's *Programme for International Student Assessment* (PISA) measures 15-year-olds' abilities to use their reading, mathematics, and science knowledge and skills to meet real-life challenges (OECD, 2018). Mathematics assessments focus on measuring students' capacity to formulate, use, and interpret mathematics in a variety of real-life contexts (shopping, sports, occupational, societal, and scientific). Students in the US performed significantly below the international average in mathematics, ranking between 32nd and 39th of 79 participating nations (with measurement error), and boys in the US outperformed girls by a statistically significant nine points (OECD, 2018). Seventy-three percent of students in the US attained Level 2 or higher, able to interpret and recognize how a simple situation can be represented mathematically. Only 8 percent achieved a Level 5 or higher, able to model complex situations mathematically and select, compare, and evaluate appropriate problem-solving strategies for dealing with them.

If the outcomes of our educational programs are adults who are productive and successful in their careers, members of functional families, and contributing and lawful citizens within communities, then problem-solving abilities must be developed throughout the curriculum in schools and reinforced at home. Regardless of the subject area or social context, active problem solving requires the solver to move through a sequence of thinking and decision-making steps.

Generally speaking, these are:

1. Identify what the problem is and really attempt to understand it. Consider the type of problem and its context.
2. Develop a plan for solving the problem. Consider relevant information, previous experience, and ways to represent the problem. Is more information required?
3. Implement a problem-solving solution method. This could involve an equation, algorithm, chart, model, or even gathering more data.
4. Examine what resulted for its effectiveness and validity.
5. Reflect back on the whole process. If the process wasn't successful, what went wrong? If it was successful, why did it work and will it be useful in the future?

Although these actions are listed sequentially, a problem solver typically loops back to previous steps during the process. Some of the processes involved in problem solving are cognitive, such as reading with understanding, analyzing data, visualizing, and computing. Others are meta-cognitive: self-instructing, self-questioning, and self-monitoring (Montague, 2003).

TRY THIS

Apply the first two steps of the generic problem-solving steps listed above to at least one of the following problems. Or if a real problem is available, try all five steps.

a. Two students in your class frequently bicker and call each other names. The situation is disruptive to instruction and distracts the other students.
b. A social studies class has been challenged to find the most cost-effective and viable way to provide humanitarian aid to a draught-stricken African country.
c. In science class, groups of students must identify as many characteristics as possible of several unknown liquids. The only information they have is that none of the substances is hazardous for touch, smell, heating, or cooling situations.

Characteristics of Problem Solvers

In developing recommendations for problem-solving instruction in mathematics, many researchers looked to the characteristics of excellent problem solvers (e.g., Erbas & Okur, 2012; Jitendra, Petersen-Brown, et al., 2015; Kikas et al., 2019; Lucangeli et al., 1997; Scheid, 1993; Schoenfeld, 1992, 2013; van Garderen, 2016). Good problem solvers in mathematics:

- have knowledge that is well connected and structured (not isolated);
- tend to focus on structural features of problems and perceive those structures rapidly and accurately;
- recognize patterns in making sense of problems;
- are successful in monitoring and regulating their efforts;

- exhibit flexibility during problem solving;
- have good estimation (prediction) skills;
- have strong reading comprehension;
- are able to create mental images, visualize situations;
- tend to use powerful content-related processes (rather than general ones);
- display beneficial attitudes such as persistence and curiosity;
- demonstrate higher self-concept;
- use a range of strategies effectively and are able to switch strategies as needed;
- use metacognitive verification to ensure they answer the question(s);
- are able to generate full descriptions of their work on problems; and
- learn from each problem-solving experience.

Students who struggle with mathematics word-problem solving (WPS) demonstrate some common characteristics. They often misunderstand the language within word problems (WP), are not able to distinguish important from irrelevant information, have difficulty selecting appropriate algorithms, and are not able to generalize strategies across problem types (L. S. Fuchs et al., 2010; Gersten et al., 2009; Shin & Bryant, 2013). Poor problem solvers have trouble representing the information from problems in diagrams or other models and often rely on the surface story of the problem or less-sophisticated solution strategies such as trial-and-error (van Garderen et al., 2012). These students may have deficits in working memory and attention, impacting their focus on the important aspects of a problem, following through with selecting operations, and performing multistep computations (Andersson, 2008; Swanson et al., 2008). Poor WP solvers frequently have gaps in mathematics concept understanding, a deficit that prevents them from making connections and recognizing patterns. They use fewer metacognitive strategies to plan, execute, and monitor their work (Montague & Applegate, 1993). Poor problem solvers spend less time with understanding the problem and translating the problem into useful representations. They tend to "grab the numbers" and perform familiar operations without making sense of the problem and their results (Cook & Riser, 2005).

Characteristics of Students with Mathematics Disabilities and Difficulties

Students with learning disabilities in mathematics (MLD) share many of the characteristics of poor and novice problem solvers. Some studies have compared the characteristics of students with identified MLD with their typically achieving (TA) peers. Other researchers have attempted to distinguish those students with more persistent difficulties in mathematics problem solving ($\leq$ 11th percentile) from low achievers with difficulties in mathematics (mathematics difficulties (MD), between 11th and 25th percentiles). Swanson et al. (2018) found, within a heterogenous sample of third graders, identifiable groups of learners at risk for difficulties in mathematics. One group—those with combined calculation, problem solving, and reading comprehension difficulties ($\leq$ 25th percentile)—demonstrated average intelligence and word identification skills. Another discrete group had mathematics problem-solving deficits ($\leq$ 25th percentile)—predicted by knowledge of problem components, magnitude estimation tasks (long-term memory retrieval), and working memory (WM) measures—but showed average performance in calculation and reading. The researchers confirmed the usefulness of the 25th percentile for risk status for third graders and the impact of reading comprehension difficulties on problem solving for some students. Setting the cut-off score at the 11th percentile would have overlooked half the students at risk for a range of mathematics difficulties.

In a meta-analysis of 110 studies on the relation between mathematics and working memory, Peng et al. (2016) found that word problem-solving ability was significantly correlated with WM capacity ($r = .37$). For individuals with mathematics difficulties (MD), the correlation was also significant ($r = .25$). The researchers concluded that word-problem solving draws on domain-general WM resources that are not specifically language-based. The central executive of WM for directing attention to relevant information, suppressing irrelevant, and for coordinating multiple cognitive processes simultaneously (e.g., fact retrieval, text comprehension), may be the most important role for word-problem solving.

Other researchers have studied the effects of reading comprehension and language deficits on WPS abilities. Fuchs et al. (2019) described higher-order comorbidity (MLD and RLD) as difficulty across reading comprehension and word-problem solving, as opposed to the lower-order comorbidity of word reading and calculation skills. The researchers linked reading comprehension with WPS via the language processes required to understand complex problem text to generate accurate and helpful problem representations. The researchers found that language comprehension strongly and uniquely predicted later WPS for second graders (Fuchs et al., 2018). In a study of sixth-grade students, Boonen et al. (2016) compared the reading comprehension abilities of students classified as successful and less successful word-problem solvers. By varying the semantic complexity of word problems, the researchers found that less successful students had trouble with inconsistent WP where the phrasing required more inference. Both lower performing and more successful students had difficulties with complex semantics, such as the use of *more than* and *less than* within WP statements. For some students, especially those with concurrent learning disabilities in reading comprehension, the linguistic challenges of WPS are substantial. Difficulties may include parsing relevant from irrelevant information, interpreting the question or problem, updating information held in WM, and decoding unfamiliar vocabulary (Cornoldi et al., 2012; Fuchs & Fuchs, 2002; Swanson et al., 2015).

Translating WP into diagrams or other visual representations is challenging for some students. In comparing students in Grades 4 to 7, van Garderen et al. (2012) found that students with MLD drew about the same number of diagrams as their typically achieving peers but drew more pictorial diagrams (surface features) and fewer schematic ones. The use of schematic diagrams, those that show the deep structure of problems, is correlated with successful problem solving (van Garderen & Montague, 2003).

Students with MD also frequently have deficits in metacognitive strategies required for successful WPS. These include self-monitoring during problem-solving stages, self-checking solutions, and evaluating the success of various solution strategies. Rosenzweig et al. (2011) used think-alouds to compare the metacognitive verbalizations of students with MLD, MD, and TA students. Each group of students made approximately the same number of metacognitive verbalizations while problem solving, however, students with MLD made more nonproductive verbalizations, such as "I'm confused" or "I need a calculator." As problems became more difficult, TA students increased their productive metacognitive verbalizations while students with MLD and the low achievers increased their nonproductive verbalizations.

To summarize, the problem-solving characteristics of students with MLD and MD are multifaceted and these students may exhibit varying profiles. Some difficulties arise from the nature of the word problem: number of steps, irrelevant information, linguistic challenges, and required response format. Other difficulties are related to reading comprehension ability and the development of well-connected mathematics concepts. Many students with MLD lack the WM resources to focus attention, hold important information in WM while processing it, and discard irrelevant information. Some students lack strategies—both cognitive and metacognitive—at a level required for problem-solving situations. Difficulty translating a problem into a workable representation such as a schematic diagram or number line will inhibit problem-solving success. Difficulty self-monitoring and self-regulating during the problem-solving process will likely result in incorrect solutions. These characteristics tend to be persistent, especially in those students with the lowest problem-solving performances. This body of literature about student characteristics has guided significant research into successful interventions.

Problem-Solving Approaches

Problem-solving approaches, or how the problem solver goes about the problem-solving process, have been virtually the same, give or take a few intermediate phases, since the work of Pólya in the 1940s (1945). The 1980 NCTM yearbook included the Pólya problem-solving approach on the flyleaf (Krulik, 1980). Problem-solving literature in such diverse fields as political science, psychology, chemistry, physics, computer science, and educational leadership from the 1970s cited his model. However, criticisms at the time noted that Pólya's strategies were descriptive rather than prescriptive; they had only face validity and lacked rules for implementation (Schoenfeld, 1987). That is likely what Pólya intended—for students to discover, with teacher guidance, which approaches work for various problems.

Pólya's approach for problem solving was offered for teachers as a way to engage students in a carefully selected sequence of problems so that students would build confidence in approaching new problems and develop a range of problem-solving strategies or heuristics. The term *heuristic* comes from the Greek *heuriskein* meaning to find or discover (Romanycia & Pelletier, 1985). In its most basic form, as applied to learning, it means to discover the solution to problems, to self-inform. Pólya, a student of Greek mathematics and philosophy, among other fields, saw himself as reviving a more modern heuristic, the study of methods and rules of discovery, with emphasis on the mental operations useful for solving problems (Romanycia & Pelletier, 1985). In his presentations and writings, Pólya addressed the important roles of attention, memory, and motivation during the problem-solving process. "Heuristic is a kind of tactics of problem solving" (Pólya, 1971, p. 624). He asserted that the first task of heuristics is to "collect and classify typical problem-solving procedures," resulting in a tool chest of tools and procedures for problem solving (p. 628). Pólya would argue that developing heuristics includes the struggle, the mental operations, the prior knowledge brought to a problem, and the personal experience of working through a problem using a variety of tools. In that sense, the teacher is the guide by showing examples and possible procedures, not the instructor of a step-by-step, linear process. The teacher assists students in building a toolbox of familiar strategies that can be applied across a range of novel problems. The heuristic is the approach, the mind-set to take on a new problem.

Therefore, Pólya's four principles are not a step-by-step process, rather guidance through four phases of problem solving that can be iterative with questions to stimulate thinking, prior knowledge, and cognitive and metacognitive strategies. Pólya's approach can be invoked when teachers and students are faced with new and unfamiliar problem types, not standard algorithms (Schoenfeld, 1992). This approach to problem solving was intended to teach students to think, to demystify the process. The general principles with sample guiding questions are:

1. *Understand the problem.* The problem solver should ask: What is the unknown? What are the data? What is the condition?
2. *Devise a plan.* The problem solver should ask and do: Have I seen it before? Do I know a related problem? Look at the unknown and find a connection. Can I solve a related problem or part of the problem?
3. *Carry out the plan.* The problem solver should ask: Is the plan working with each step? Have I checked each step? Do I need to retrace steps?

4. *Look back.* The problem solver should ask: Can I check the result? Does the result make sense in the problem's context? Can I use the result, or the method, for some other problem?

For word-problem solving (WPS), most mathematics textbooks introduce a problem-solving sequence, typically patterned on Pólya's principles, in an early chapter (or separate reference book) and provide general heuristics (tactics) throughout the book and series. Some general heuristics might include making lists or tables, preparing diagrams, writing number sentences, reasoning, solving a simpler problem, working backward, guess and check, and find a pattern. These suggestions seem vague, especially without examples, and many core mathematics teachers do not teach more systematic or explicit strategies. For students who struggle with WPS, general prompts will not be sufficient. These students require systematic, explicit problem-solving instruction that is integrated with mathematics content at each grade level and for each domain. The four principles can effectively guide student approaches to problem solving if taught and reinforced systematically with explicit cognitive and metacognitive strategies. This chapter will not use the term *heuristics,* because of its general nature, to refer to systematic and explicit procedures for solving a problem, or for specific, explicitly taught strategies within the process, such as creating a diagram or making a list.

Word problem-solving research with low-achieving students and those with MLD has yielded significant recommendations about the essential elements for intervention. The most effective programs and interventions include explicit instruction in:

- specific strategies for understanding the problem, and all parts of the problem;
- methods for translating the problem information onto a schematic diagram or other representation of the problem's quantitative relationships or structure;
- recognizing problem types and applying related solution strategies; and
- applying cognitive and metacognitive strategies during the problem-solving process.

In addition, the most successful interventions are teacher directed and systematic, allowing enough time (over weeks) to develop concepts, practice strategies, and make applications. They employ scaffolding, thinking aloud, progress monitoring, and corrective feedback. Some programs use other elements, such as peer work and brief review of basic skills and concepts, within problem-solving lessons.

Priming the Problem-Structure Approaches

In psychology, the process of *priming* (from Lashley, 1951) refers to the preparation of mental representations for their efficient use in needed situations, and has been applied within linguistics, behavioral sciences, and other cognitive fields. For mathematics problem solving, Jitendra coined the phrase *priming the problem structure* to describe those types of instructional approaches that use strategy instruction to explicitly teach students to use their prior knowledge and the information in word problems to understand the underlying structure of new word problems (Jitendra, Petersen-Brown, et al., 2015). Priming activates certain schemas in long-term memory, already organized and easier to access, such as a previous problem type that would offer assistance for a new problem. WPS approaches that prime problem structures through strategy instruction are considered evidence-based practices and have moderate to strong effects on WPS achievement for students with MD or MLD across Grades 1 through 8 (Fuchs, Seethaler, et al., 2020; Lein et al., 2020; Peltier & Vannest, 2017).

Schema derives via Latin from the Greek *skhêma*, meaning form or shape. Broadly speaking, schema refers to a structured framework or plan. Schema can refer to a diagrammatic presentation or the mental codification of experience in an organized way (Merriam-Webster, n.d.). A schema can be a mental representation, a diagram, an outline, a framework, or other ways to understand and manipulate novel information using prior knowledge and experience. It is not a static problem type. Creating a schema involves the mental representation of a problem using information from the problem and prior knowledge, thus can be highly individual. Schemas are dynamic, changing with new information and experience. When complex information is organized into a schema, mentally or in diagram form, there is a reduction of working memory load allowing for more efficient problem solving.

Significant and ongoing research supports the use of *strategy instruction priming the problem structure* interventions with students with mathematics difficulties and disabilities in problem solving for Tiers 1, 2, and 3 within multi-tiered systems of support (MTSS) models. Specific intervention packages include schema-based instruction (e.g., Jitendra et al., 1998; Jitendra et al., 2019) and schema-based word-problem intervention (e.g., Fuchs et al., 2002; Wang et al., 2019). These approaches involve strategies for reading and understanding problem text, priming the underlying problem structure, using visual representations of mathematical relationships, providing explicit and systematic instruction, scaffolding student reasoning, and employing cognitive and metacognitive strategies for the phases of problem solving.

Jitendra and colleagues, building on the schema work of Marshall (1995), developed the schema-based instruction (SBI) approach for students with mathematics disabilities or difficulties, applied across Grades 2 to 8 (e.g., Jitendra & Hoff, 1996; Jitendra et al., 2007). This approach is based on research from cognitive psychology, studies about expert problem solvers, and the effective instruction literature for students at risk for MLD. The unique features of SBI include four essential components established in problem-solving research (Jitendra, 2019). Students are taught to recognize the underlying mathematical structure of a word problem as one of several problem categories (e.g., change, group, compare for addition and subtraction problems) by reading and paraphrasing to understand the problem situation. Next, students are taught to recognize or draw a visual schematic representation of the problem that shows quantitative relationships within the problem (illustrated in a later section of this chapter), supported by linguistic scaffolding. Students then plan how to solve the problem by translating the visual schema into quantitative representations (equations) and solve for the unknown. Throughout the process teachers use questioning to encourage students' use of metacognitive strategies during problem solving, such as monitoring their strategy use, reflecting on the process, and evaluating the outcome. SBI also incorporates effective instructional practices of systematic, explicit instruction; scaffolding to support learning; think-alouds; frequent response with feedback; and progress monitoring.

More recent work by Jitendra and colleagues has taken SBI into inclusive settings, taught by general education mathematics teachers (Jitendra & Star, 2011; Jitendra et al., 2019), and studied the use of tutors for small-group interventions (Jitendra et al., 2013). The investigators also researched SBI protocols for ratio, proportion, and percent types for middle grades (Jitendra et al., 2016; Jitendra, Harwell, et al., 2015; Jitendra et al., 2017). These SBI interventions added estimation components and contrasting multiple solution methods for these challenging topics. Further research examined the multiplication and division problem types of equal groups, unit rate, and array using SBI with visual diagrams and supported reasoning (Alghamdi et al., 2020). The SBI program can be accessed through Jitendra's books *Solving Math Word Problems* (2007) and, with J. Star (2017), *Solving Ratio, Proportion, and Percent Problems Using Schema-Based Instruction*.

The schema-based WP interventions developed by Fuchs and colleagues began with an examination of a problem-solving tutoring treatment that explicitly taught problem structures and transfer to a broader range of schemas (Fuchs et al., 2002). The schema-broadening approach was an outgrowth of the work of Cooper and Sweller (1987) on problem-solving transfer and that of Salomon and Perkins (1989) on broadening schemas to explicit problem-solving instruction. In schema-broadening, students were taught to recognize problems of the same type even if they are presented in a different format, contain irrelevant information, include unfamiliar vocabulary, or display information differently, as in charts or graphs. Students were taught to organize information from problems in different ways (e.g., equations or charts). In studies incorporating small-group tutoring, students receiving schema-broadening instruction (12–15 weeks, three times a week) significantly outperformed students with no tutoring and those with number-combination tutoring (Fuchs et al., 2009; Powell, 2011). The schema-broadening approach included a general WP attack strategy, instruction on recognizing problem structures, visuals such as manipulatives and diagrams, presentation of meta-equations to represent the schema (e.g., $B - s = D$; bigger amount minus smaller amount equals the difference), explicit instruction on essential language components, the use of worked examples during instruction, and broadening problem types to promote transfer. Some of the meta-equations for addition and subtraction (e.g., $P1 + P2 = T$; part one plus part two equals the total) may be confusing as students transition to multiplicative reasoning and related equations.

More recently, the Fuchs research group has examined the effects of other elements within schema-based interventions such as self-regulated instruction (Wang et al., 2019), supported self-explanation (Fuchs et al., 2017), and language comprehension strategies (Fuchs, Seethaler, et al., 2020), expanding their schema-based intervention model across Grades 1 to 5 with word problems involving basic operations as well as fractions. The whole- and small-group programs *Pirate Math* and *Hot Math* for Grades 2 and 3 are available through the Fuchs Research Group at Vanderbilt University's Peabody College. *Super Solvers*, an extension of *Fraction Face-Off!*, addresses fraction concepts, operations, and problem solving from third through fifth grades (Malone et al., 2019; Wang et al., 2019).

Other Research-Supported Approaches

Montague et al. (2000) created an instructional program for the secondary grades titled *Solve It!* that incorporates problem-solving steps with prompts for self-questioning during the process. The steps begin with verbs that prompt cognitive and metacognitive actions: read (the problem), paraphrase (information within the problem), visualize, hypothesize, estimate (the answer), compute, and check (Montague, 2003). The program was developed through three studies involving middle- and high-school students with MLD (Montague, 1992; Montague, Applegate, & Marquard, 1993; Montague & Bos, 1986). More recent research with the *Solve It!* program compared middle-school students with and without MLD over a seven-month period in mathematics classes taught by general education teachers (Montague et al., 2011). Intervention included a three-day intensive instruction period followed by weekly problem-solving practice sessions. Elements of the program included explicit, teacher-directed instruction, think-alouds, individualized feedback, practice using self-regulation strategies, and monthly curriculum-based measures (CBMs). All students in the treatment groups showed greater growth in problem solving over the school year than comparison groups. Although students with MLD performed lower than TA and MD students in the treatment groups, they outperformed all ability groups in the comparison classrooms by the end of the year. Other researchers have continued to study the *Solve It!* approach with other elements. Gonsalves and Krawec (2014) described the use of number lines to enhance the visualize step, as a more flexible representation (with some limitations). Krawec and Huang (2017) investigated increasing the explicitness of instruction, expanding the role of meta-cognition, reviewing concepts, and adding visual cues, resulting in the intervention group significantly outperforming the control group after six days of initial instruction and 12–16 weekly practice sessions. The *Solve It!* program is available through Exceptional Innovations (www.exinn.net).

Some researchers studying mathematics WPS with students with mathematics difficulties have examined the use of the "Singapore" bar model as a bridge between the word problem text and subsequent mathematics equations for solving (Morin et al., 2017; Xin, 2019). Singapore math, a term coined in the US for the approach used in Singapore's textbooks since 1992, emphasizes concept understanding by using the concrete-representational-abstract (CRA) sequence (Ho & Lowrie, 2014). The bar model, at the representation stage, uses rectangles of different lengths (strip diagrams) to model mathematics concepts as well as the quantitative relationships within word problems, but has not had a research base. For students who struggle with WPS, Morin et al. (2017) and Xin (2019) independently investigated the use of the bar model within schema-based instruction for third-grade students employing single-subject, multiple-baseline research designs. Morin et al. (2017) attempted instruction using bar models for addition, subtraction, and multiplication word problems across seven lessons, showing an increase in problem-solving skills for each of the six students but inconsistent application of the cognitive strategies involved in WPS. Xin (2019) focused on five types of addition word problems to bridge students' text understanding to translation into a mathematics equation with an unknown driving the solution strategy. All three students reached 100% mastery by the end of the instruction phase and maintained at least 91% at posttest.

Bar models have also been used with middle-grades students for percent (described by Parker, 2004) and algebraic reasoning (described by Beckmann, 2004). In some cases, such as with multiplication and division of fractions, other representations may be more effective and efficient than bar models. Students with mathematics difficulties will need more explicit instruction in modeling quantitative relations, a more systematic introduction to various problem types with scaffolding, more opportunities for practice, and assistance drawing an accurate model of a specific mathematical relationship.

TRY THIS

Explain how the bar model assists the learner in understanding each word problem.

a)   Two students have 15 raffle tickets. If one has 9 tickets, how many tickets does the other student have?

b)   What is $\frac{3}{5}$ of 30?

The research base on problem-solving instruction for students with MLD and at risk for MD has offered compelling evidence for educators on the most effective approaches for teaching these complex tasks. Programs and interventions for mathematics problem solving should provide systematic, explicit instruction in cognitive and metacognitive strategies involved in problem solving as well as instruction in understanding problems, identifying problem types, and making structural representations of critical problem information. Students should be active learners in this instruction, identifying structures and components, thinking aloud about the process, discussing problem-solving approaches with each other, and generating new problems.

Teaching Problem-Solving Strategies

For solving real-world problems, the most challenging step for most people is defining the problem and understanding its components. The homeowner who notices water collecting around the foundation of her house must seek out the problem. The scientist who identifies a decline in a fish species in a river must first clarify the problem. The automobile mechanic whose client is complaining of a strange engine noise must diagnose the problem. The teacher who is working with a student with significant learning problems in mathematics must diagnose specific gaps in learning or difficulty areas. A clear understanding of the problem is essential before solutions can be considered. Often in real-world problem situations additional information must be collected.

In mathematics classrooms, word problems are common ways to present problem situations to students. Word problems are sometimes introduced after specific mathematics concepts and procedures are taught, as an application of new skills. That could be beneficial if these problems were connected with prior addition and subtraction problem types with mixed practice. More frequently, word problems introduce a unit of study as a method of garnering interest and focusing on concepts and alternative solution strategies. In this case word-problem solving would be followed by concept and procedural development, then more WPS. For example, a unit about linear equations could begin with a word problem about the water levels in a river decreasing at a rate of 0.5 feet a day from a level of 34 feet over a week. This and similar problems connect with prior concepts of rate, ratio, and coordinate plane graphing, leading to setting up and solving linear equations in abstract forms. The problem types are connected with solution strategies.

When faced with novel word problems, students should have experience with a basic approach, such as Pólya's four principles of the problem-solving process, Montague and Bos' (1986) elaborations on those four phases, or other similar methods for initiation and activation. Teachers should employ explicit and systematic instruction of the process and support cognitive and meta-cognitive strategies that will promote independence. The following sections provide specific tactics for each phase of this problem-solving process. These are the cognitive and metacognitive strategies incorporated in the research-based programs discussed in the previous section, with examples.

Strategies for "Understand the Problem"

To understand mathematics word problems, the solver must first read the problem and understand its parts. This is a reading comprehension task with linguistic and vocabulary challenges for some students. While reading any text, students use their prior knowledge to form mental images of action or situations. For word problems, student attention should be directed to problem parts by carefully posed teacher questions: What does the problem ask? What information is provided that is important? What information is not useful? Can you put into your own words what the problem is asking? Can you tell me what your mental image of this problem shows? Does this problem remind you of a problem type you have seen? Students who have difficulty with this phase may need more explicit instruction in paraphrasing, locating important information, and determining problem type.

Reading Comprehension Strategies

A paraphrasing strategy, such as the one described by Swanson et al. (2015), can promote problem understanding. Students with MLD require explicit and graduated instruction in making paraphrases of the problem's propositions—those important elements of any problem. Propositions include the information required for solving the problem, the irrelevant information, and the question or challenge. For example, a word problem reads: *Bobby's sister had six pairs of white socks. She lost one pair at the gym. She goes to the gym four days a week. How many pairs does Bobby's sister have now?* The paraphrase might be: "There's a girl who has six pairs of socks. That's important. The girl lost one pair and that's also important. It's not important where she lost the socks or how many days she goes to the gym or even that they are white socks. I know that a pair is two socks, but that's not important here either. I want to know how many pairs of socks she still has." A good paraphrase is in the student's own words and specifies the types of propositions within the problem.

Some teachers ask students to underline, highlight, or code parts of problems to indicate whether they include important information, irrelevant information, or the question phrase. By drawing attention to critical aspects of the problem and interacting with those components, students have a better understanding of the problem as a whole. They are not as likely to "number grab" or guess. Teachers should avoid ineffective methods, such as finding "key" or "clue" words. This approach is superficial, does not lead to understanding the underlying structure of a problem, and often misrepresents the necessary operation. For example: *When Leslie and Sarah put their baseball cards together there were 24 cards. If 10 belonged to Leslie, how many were Sarah's?* Notice that the word *together* does not mean to add. Again, keywords can be deceptive and are not reliable problem-solving tools. Students should be encouraged to consider the structure of the problem (Gurganus & Del Mastro, 1998). Some researchers have asked students to write out their paraphrase of the problem propositions (Kong & Swanson, 2019; Morin, 2017). Not only does this activity place stress on students' attention and interest in word problems, it is unlikely to be an effective strategy for transfer.

Some word problems have more challenging linguistic constructions. Problems may be heavily loaded with mathematics vocabulary rather than presented in everyday language. The vocabulary and language phrasing may need to be pretaught, with many examples for practice. For example: *Let A and B be two finite sets such that n(A) = 20, n(B) = 28 and n(A ∪ B) = 36, find n(A ∩ B).* Virtually every word or phrase has a specific mathematics meaning. Some word problems lack visible numbers, for example: *Kirby has a dozen pairs of shoes in her closet. Her dog Eve Claire decided to chew on a shoe. How many pairs can Kirby wear now?* Other word problems include indirect language or inferences, for example: *At the grocery store, a bottle of olive oil costs 7 dollars. That is 2 dollars less than at the corner market. If you need to buy five bottles of olive oil, how much will it cost at the corner market?* Boonen et al. (2016) found that indirect phrases (e.g., that is) and *less than* comparisons were difficult for many students and reading comprehension was correlated with WPS. The researchers recommended earlier reading comprehension instruction specifically for semantically complex word problems and incorporating that within schema-based instruction.

Recognizing Problem Types

Mathematics problems can be categorized as *routine*, those for which basic mathematics operations can be directly applied, and *nonroutine*, those problems that require alternative strategies. Problems also can be classified as one-step, two-step, or multistep. Consider the characteristics of the problems depicted previously in Box 5.1. Which problems are routine and nonroutine? Which require more than one step?

Routine problems typically require one or two steps of the application of basic operations. Mathematicians in the mid-1970s to early 1980s studied children's solution processes for addition and subtraction problems and related those to the semantic structure of problems (see discussion in Carpenter et al., 1983). This research resulted in three basic problem types: *combine* (also called part-part-whole and static); *change* (also called join/separate and dynamic); and *compare* (also called static relationship and measure).

Combine problems involve two or more quantities that are combined. The unknown is either the total or the amount in one of the subsets. For example: *Janet placed two cups, two plates, and four utensils into her picnic basket. How many objects are in her basket?* The structure of combine problems is part + part (+ part) = total. If the total is the unknown, we add because we expect a larger number. If one of the parts is unknown, we subtract the other parts from the total.

Change problems involve some quantity being increased or decreased. The unknown may be the starting amount, the change amount, or the ending amount. For example, *Erica had a dozen blocks before giving half to Amy. How many did Erica have left?* The change involved giving away blocks. The unknown is the ending amount. This problem is also interesting because the numbers are not obvious. *If Mr. Brown filled up his tank with 10 gallons and it holds a total of 25 gallons, how many gallons of gas were in the tank originally?* What is the change amount? What is the unknown?

Compare problems involve a comparison between two quantities using the relations more than or less/fewer than. The unknown may be the referent, the compared quantity, or the difference between the two quantities. For example, *Whitney had four more books to carry than her sister. If her sister had three books, how many did Whitney have?* The referent is three books, the difference is four (designated with "more books than"), and the compared quantity is unknown (Whitney's number of books). We don't know the compared quantity so we add. Our reasoning tells us that Whitney will have the larger amount.

Figure 5.1 illustrates how the addition-subtraction problem structure for change problems, dynamic situations, would be depicted by diagrams from three research perspectives. The example word problem includes relevant information, an unknown, and one irrelevant piece (one at a time). Explicit instruction in connecting problem types with diagrams is warranted for students with MD because without this instruction, students tend to draw pictures of the surface story, fail to show connections among problem components, and use diagrams in a less strategic manner than their peers when simply asked to "draw a picture" (van Garderen et al., 2012).

In studying multiplication and division problem types applied to integers, Greer (1992) described four main classes of situations: *equal groups* (also called vary, equivalent sets, and unit rate); *multiplicative comparison* (restate, part-whole); *Cartesian product* (product of measures, combinations); and *rectangular area* (or array). An example for *equal groups*, where two clearly different factors are identified: *The book sale committee has collected 10 boxes of books with 20 books in each box.* "Boxes" is the multiplier that operates on the multiplicand or "books in each." If the product were known first, this example would generate two types of division problems—partitive division (if the number in each set were unknown) and quotative division (if the number of sets were unknown). The equal groups problem type includes the expression of a per unit value or unit rate. The relationship can be centered on one property of an object: a) *You want to purchase 9 yards of cloth that costs $4 per yard. What will you pay?* b) *A boat moves at a steady speed of 4 meters per second. How far does it move in 30 seconds?* Or the relationship can be between two different entities: c) *For every 5 students we have 2 balls. How many balls do we have for 25 students?* Both equal groups and unit rate types of problems have the same underlying structure: factor × factor = product.

The *multiplicative comparison* (restate, part-whole) problem type involves a relationship that can be expressed between two distinct but similar things such as two children, two bicycles, or two buildings. Each entity has some common property that can be counted (age, cost, length, weight, height). One is described in two different ways (restated). For example, *a new bicycle costs three times as much as a used one. If a new bike costs $150, how much is the used one?* The price of a new bike is stated in two ways in this problem: it costs $150 and it is three times the cost of a used bike. Another example: *Crystal has four times as many quarters as her sister. If her sister has five quarters, how many does Crystal have?* This is the division version—if the problem were

Change Problem Schemas

Example Problem:

Neal carried eight chairs, one at a time, into the dining room. He saw that there wasn't enough room so he carried two back. How many chairs were left in the dining room?

Schema-Based Instruction Diagram

Conceptually Based Model of Problem Solving

Number Line Representation

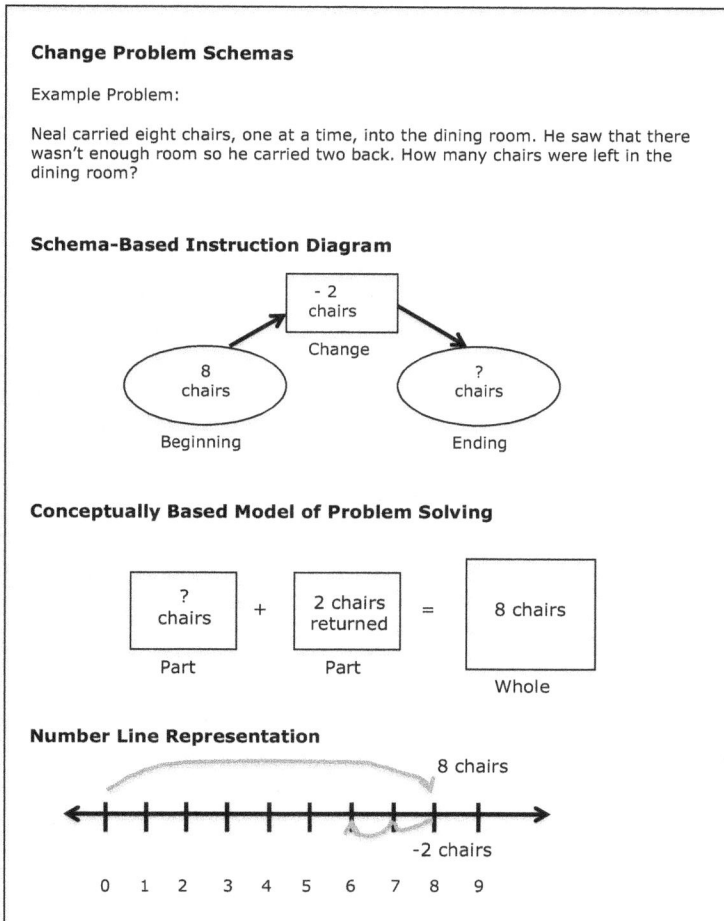

Figure 5.1 Change Problem Schemas

SBI Diagram for Change adapted from *Schemas in Problem Solving* (p. 135), by S. P. Marshall, 1995, New York: Cambridge University Press. Copyright 1995 by Cambridge University Press. Adapted by permission. COMPS Diagram adapted from Y. P. Xin, B. Wiles, & Y.-Y. Lin, *The Journal of Special Education*, *42*(3), p. 171, copyright 2008 by SAGE Publications. Adapted by permission of SAGE Publications.

worded the other way around, it would require multiplication. English (1997) termed this problem type "asymmetrical" because one entity has x times as many things as the other entity.

As seen with the multiplicative comparison and equal groups problem types above, multiplication and division problems are more difficult because they involve more than one kind of object or unit (Reed, 1999). Multiplication and division problems involve a transformation of referents and are based on the concept of intensive quantity (Schwartz, 1988). An intensive quantity is not measurable alone, rather it is the ratio of two measurable quantities. For example, 50 miles per hour is actually the ratio of a measurable distance (in miles) divided by a measurable amount of time (in hours).

A third multiplication-division type is the symmetrical *Cartesian product*, where two sets are matched by all possible ordered pairs. *If you are buying a new bicycle and have the choices of 10 colors and 3 models, how many options do you have in all?* Each color is matched with each model resulting in 30 options. The division version would provide the total options and one set of choices would be unknown. A word problem could introduce more sets, ordered triples and so forth. For example: *The blue-plate special offers 3 choices of protein, 5 choices of vegetables, and 2 carbohydrate choices. How many unique combinations could be ordered?*

Finally, the *rectangular area* or *array* model, of length by width can require multiplication or division, depending on the unknown. This type of problem is considered symmetrical because the factors have interchangeable roles in equations (Carpenter et al., 2015). Consider these examples: *The area of a rectangular floor is 60 feet. If one side measures 12 feet, what is the length of the connecting side?* The equation could be $12 \times s = 60$ or $60/12 = s$. *The bakery sheet pan holds 8 rows and 4 columns of cookies. How many cookies does the sheet pan hold?*

Middle-school mathematics is also dependent on multiplicative reasoning, with topics such as ratio, proportion, and percent. Greer (1994) proposed a larger classification system, to include the previous integer problem types, as well as problem types for other rational numbers (fractions and decimals). These include ratio, rate, and proportion problem types. Of course, the full domain of multiplicative types extends well beyond these concepts (e.g., exponentiation, linear function), but these four basic multiplicative structures are critical foundations. Figure 5.2 illustrates multiplication-division problem types.

Underlying problem structures for ratio, proportion, and percent problem types have been investigated by Jitendra and colleagues (e.g., 2014) and Xin and colleagues (e.g., 2011). Students need practice with a wide range of contexts (e.g., measurements, prices, rates) in order to develop conceptual understanding of proportional reasoning. Proportion problems provide one set of data by which to compare one or more other sets of data. For example, *Amy drew a picture of a building on a scale of 1 cm to 2 meters. If a window is 1.5 cm tall in Amy's drawing, how tall is the actual window?* A more complex proportion problem: *We are planning a party for 20 people and will need 5 gallons of punch and 3-dozen cookies. How much food would we need for 30 people?*

Word problems in middle and high school are also connected with a more formal study of algebra. Algebra is another way of representing the elements within a problem. Algebra deals with general statements of relationships, using letters and other symbols to represent parts and relationships of a problem. Consider the word problem in Figure 5.3 and examine how that is expressed using a diagram, number line, and algebraic notation. For students with difficulty translating a word problem directly into algebraic form, a diagram can serve as an intermediate step. For example, the proportion diagram in Figure 5.3 resembles the algebraic structure of the problem.

Many algebraic word problems are relational. For example, *One swimmer swims 100 meters farther than another swimmer. If the total distance by both swimmers is 1,500 meters, how far did each swimmer swim?* The following diagram could assist the learner in converting problem information into an algebraic equation: $(s + 100) + s = 1500$.

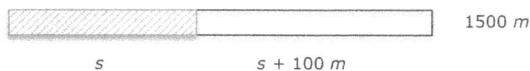

Two-variable, two-equation problems are more complex. For example: *You're going shopping with $200 you received for birthday money. A store is advertising all slacks for $25 and all tops for $50. You want to purchase six items so you can mix and match. How many of each can you buy to spend all your money (before taxes)?* These problems are best represented with two equations, coordinate plane graphing, or tables. Representing the problem statements as equations would result in s + t = 6; and 25s + 50t = 200. Then we express either t or s in terms of the other. Graphically, this is the intersection of the two linear equations at on a coordinate plane.

Equal Groups (Vary)

The third-grade class is going on a field trip to the art museum. There are a total of 160 students on five buses. How many students are on each bus if each has the same number of students?

SBI Representation

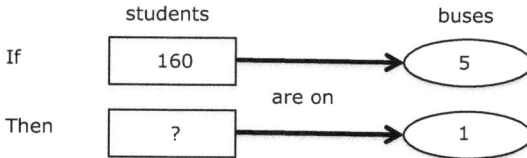

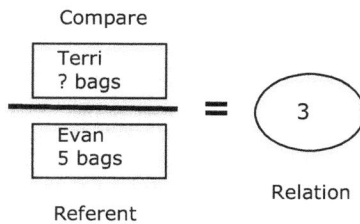

students   buses

If   160 ———————▶ 5

are on

Then   ? ———————▶ 1

COMPS Representation

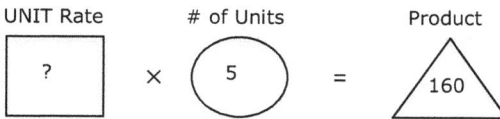

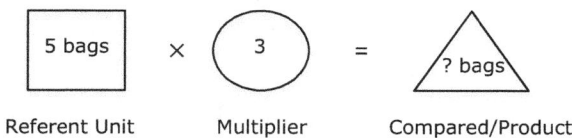

UNIT Rate   # of Units   Product

?   ×   5   =   160

Multiplicative Comparison

Evan's family donated five bags of clothes to the winter clothing drive. Terri's family donated three times as many bags as Evan. How many bags did Terri's family donate?

SBI Representation

Compare

Terri
? bags
——————   =   3

Evan
5 bags

Referent   Relation

COMPS Representation

5 bags   ×   3   =   ? bags

Referent Unit   Multiplier   Compared/Product

Figure 5.2 Common Multiplication/Division Problem Types

SBI Diagram for Vary adapted from *Schemas in Problem Solving* (p. 135), by S. P. Marshall, 1995, New York: Cambridge University Press. Copyright 1995 by Cambridge University Press. Adapted by permission. SBI Diagram for Multiplicative Comparison adapted from *Solving Math Word Problems* (p. 173), by A. K. Jitendra, 2007, Austin, TX: PRO-ED. Copyright 2007 by PRO-ED, Inc. Adapted with permission. COMPS Diagrams Equal Group and Multiplicative Comparison adapted from Y. P. Xin, B. Wiles, & Y.-Y. Lin, *The Journal of Special Education, 42*(3), p. 171, copyright 2008 by SAGE Publications. Adapted by permission of SAGE Publications.

A recipe for oatmeal cookies calls for 2 cups of flour for every 3 cups of oatmeal. How much oatmeal is needed if 7 cups of flour are used?

Diagram

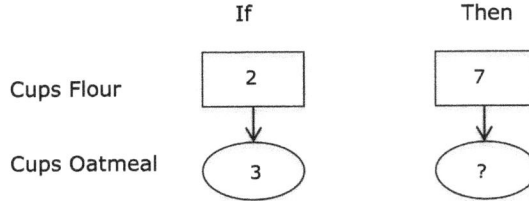

Number Lines

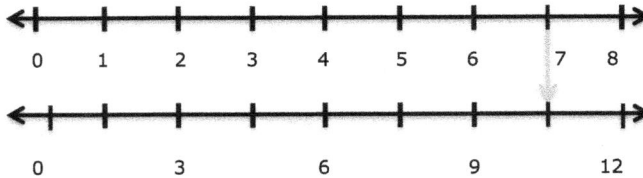

Equation

$$\frac{2\ c\ flour}{3\ c\ oatmeal} = \frac{7\ c\ flour}{?\ c\ oatmeal}$$

Figure 5.3 Ways to Represent a Proportion

Diagram adapted from A. K. Jitendra, J. R. Star, M Rodrigues, M. Lindell, & F. Someki (2011). Improving students' proportional thinking using schema-based instruction. *Learning and Instruction*, 21, 737. Proportion diagram used with permission of Elsevier and adapted by permission of first two authors.

Apply "understand the problem" strategies to the following problem:

> *The parent-teacher organization of the school has decided to replace a concrete sidewalk in front of the school that has been in place 20 years, but is now cracking, with stone blocks. The sidewalk is 36 feet long and 9 feet wide. Each stone block is 15 inches square and there will be a one-half inch gap between blocks. About how many blocks should the school order?*

Is this a routine or nonroutine problem? What is its structural type? Can you visualize the project described? Can you paraphrase the question? What information is important for solving the problem and what is not needed? Can you represent the problem with a diagram or algebraic equation?

Students need practice with understanding the problem, without actually solving the problem, using effective strategies. Discussing problems within small groups will assist students in the cognitive processes involved. Students find it interesting to discuss a problem's features without having to actually solve the problem (all numbers provided, or no numbers provided exercises). Further, students who begin grasping a specific problem type can solidify that understanding by creating their own problems for others to identify.

Strategies for "Develop a Plan"

The next stage in problem solving is to develop a plan for finding a solution or solutions. By the time students are in high school, they should have a repertoire of strategies. Some strategies (using objects) can be introduced as early as first grade, while others (applying a formula) are more abstract and are introduced in later grades. Additionally, some strategies introduced early can become more sophisticated with higher-level applications.

Many of these strategies, such as draw a picture, working backward, setting up equations, and using a related problem, were proposed by Pólya (1945) in his "Short Dictionary of Heuristic," a chapter in *How to Solve It*. The Pólya heuristics, however, are too broad to be offered as strategies and should be considered as "families of approaches," as one can identify specific strategies within each general approach (Schoenfeld, 2013). His heuristics were a list of the general approaches most problem solvers attempt informally when faced with unfamiliar problems, the mental operations that can be useful but might be fallible, leading the problem solver to try another approach.

The following section offers more specific strategies, with examples, that can enhance a student's toolbox for solving mathematics problems. Teachers and students should realize, however, that the act of problem solving is attempting to make sense of a situation, bringing all prior knowledge and mathematics tools that the learner has developed to that point, with additional scaffolding and guidance by the teacher. There is usually more than one way to solve a problem. One student may view a problem differently than other students. The teacher's role is to understand students' prior conceptions and to guide new learning, while allowing for alternative methods as long as those are mathematically sound and do not promote misconceptions.

Students should first consider the problem type, as developed in "understand the problem." If the problem is routine and a previous problem structure can be recalled, then that solution strategy can be implemented—typically applying one or more of the four operations to the selected information provided in the problem. If the problem type is clearly specialized content—such as interpreting or creating a graph, finding a ratio, measuring a figure, determining perimeter or volume, constructing lines or angles, or solving an equation—then the problem directions and format should prompt the strategy. Finally, if the problem is nonroutine or a previous structure cannot be recalled, students should apply one of the following strategies.

Create Visual Representations

Using problem information and relationships among quantitative elements, students should be encouraged to create an accurate visual representation of the problem situation. This representation could be a schematic diagram, number line, table, or bar diagram, whichever best represents the problem in a way the student understands. The visual representation should not be a simple picture of the story of the problem but show the problem structure. For example, which of the following three problems has a similar structure? How would diagrams show these problems?

a. A couple set out to hike the Hadrian's Wall Path of 84 miles. They walked an average of 12 miles a day with sightseeing about 2 hours each day. How many days did the hike take to complete?

b. Two people hiked the coastal pathway of 120 miles. If they walked 10.2 miles the first day, 8.4 miles the second day, and 12 miles the third day, how many miles have they hiked by the end of the third day?

c. If a factory worker can package 45 boxes in one hour, how many boxes can he package in a 40-hour week?

Problems (a) and (c) are equal group (unit rate) problem types. Problem (b) has the same "cover story" as problem (a) but is structurally different. It is a combine (addition) problem.

Other nonroutine problems call for using a diagram in problem solving. Teachers should use think-alouds to model interpreting these problems and creating diagrams for solving them. What language in the following problems prompts specific constructions?

a. Study the pattern of the figures shown below. Draw the next three figures in the pattern.

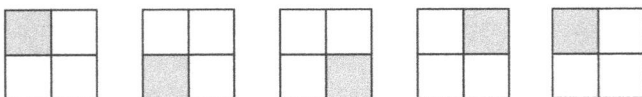

b. Draw a ray that bisects the angle below:

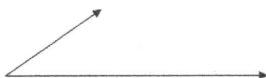

c. The following are coordinates of points in a coordinate plane: $(-1, 3)$, $(3, 3)$, and $(3, -1)$. What would the coordinates of a fourth point have to be so that a square will be formed by connecting the points with line segments?

Relate to Familiar Problem

When faced with a new problem, students should search their long-term memory for a familiar structure. One benefit of teaching problem types is their use in future problem-solving activities. Saloman and Perkins (1989) identified two types of problem transfer. Forward-reaching involves teaching structures and abstractions when the problem type is initially learned. Backward-reaching transfer requires students to think back to previous problem types for common structures. Students with difficulties in WPS may need explicit instruction in making both types of transfers. For example, fourth-grade students had previously learned the combine, change, and compare problem types with integers. Now they are faced with this problem: *A recipe calls for $\frac{3}{4}$ cup of milk, $\frac{1}{2}$ cup of heavy cream, and $\frac{1}{8}$ cup of melted butter. Would a 2-cup measure hold all these liquids? Explain why.* In a small group, one student recognizes that they need to combine the fractional amounts and another student offers that after they get that amount, they can compare it with two cups. A third student points out that this problem is harder because they must put all the fractions into the same terms, in order to combine and compare them. Another student rules out using addition and subtraction approaches for a more direct fraction-bar approach, similar to an activity using fraction bars the previous week. The first student points out that whether you convert the fractions first or simply line them up using fraction bars, it is still a combine and compare problem.

Estimate

Sometimes called *guess and check*, estimation is a critical skill for step four in the problem-solving process (look back) but also can be part of an effective problem-solving strategy. This strategy is particularly helpful for mathematics tests with multiple-choice options. Consider the following problem from Grade 4 (no calculators allowed):

a. Nora wants to buy 3 granola bars. Each bar costs $1.89. If she uses a coupon for $1.00 off
    the price of one bar, how much will Nora owe for the 3 bars?
    A. $2.67   B. $5.67   C. $4.67   D. $7.07

We can estimate by using $2 a bar for 3 bars would be $6 and a dollar off would leave $5. This
estimate would quickly eliminate A and D. We know our estimate of $5 is a bit high because
we rounded up, so that would eliminate B. The answer must be C. The student with time for
an exact calculation could check this estimate by multiplying 3 by $1.89 and subtracting $1.
   Some problems ask for estimation rather than an exact answer. For example:

b. According to the 1970 United States Census, Florida had an average of 126 people per
    square mile. The state of Florida is listed as 53,927 square miles. Which of the following is
    the best *estimate* of the total population?
    A. 5,000,000   B. 6,800,000   C. 8,100,000   D. 10,800,000

The previous problem forced students to select from four estimates. When problems do not
provide a choice for answers and require an estimate, students with learning difficulties often
try to find the exact answer, regardless of the directions. The concept and process of estimating
can be more difficult than conducting a routine computation for these students. Consider the
following problem:

c. *Mr. Sterner is shopping for a few items in the grocery store. If the items cost $4.95, $1.75, $2.20,
    $8.10, and $6.89, provide an* estimate, *to the nearest dollar, of the total cost.*

Teachers could scaffold understanding by reviewing rounding skills. Do we want to estimate
to the nearest dime or dollar for this problem? What are the rules for rounding? What does
"to the nearest dollar" mean?

Solve a Simpler Problem

The *solving simpler strategy* is effective when large numbers or complex situations are presented,
and simplification can be combined with other strategies. Once the student can set out a solution for and solve a simpler version of the problem, the same solution strategy can be applied
to the actual problem. Consider the following problems and how they can be simplified.

a. A rectangular park measures 587 feet wide and 1,398 feet long. How long would a diag-
    onal pathway (from one corner to the far corner) be in feet?
b. The surface area of the United States (50 states and District of Columbia) is 9,631,418
    square kilometers. Of this area, 9,161,923 km² is land and 469,495 km² is water. What
    percentage of the total area is water?

For problem a), if the student is unsure of the algorithm, he could attempt the problem by
first drawing the rectangle and diagonal and labeling the rectangle with 6 units and 14 units.
The diagonal would be a bit longer than the 14-unit side. The drawing reminds the student
of the Pythagorean Theorem. Finding the hypotenuse of a triangle within the rectangle ($a^2 +$
$b^2 = c^2$) yields about 15 units. The solution with larger numbers (1516.2 feet) makes sense. For
problem b), dividing 0.5 by 9.6 would provide an approximation, but students would need to
be firm with place value and estimation skills, as well as decimal notation equivalent to percent ($0.05 = 5\%$).

Use Objects

Some problems lend themselves to the use of objects because they are about objects (example a). Abstract objects, such as cubes or straws, can be used to represent real-life objects in some problems (example b), especially if a CRA sequence is being used for WPS.

a. If Ramon had 10 crayons after giving 6 to his brother, how many crayons did he start with?
b. Gail, Hannah, Isabelle, Joyce, and Kay are standing in line for baseball tickets. Gail and Isabelle are next to each other. Joyce is last. Kay is not next to either Gail or Hannah, nor is she first. List these baseball fans in the order they are in line.

Act It Out

Word problems that involve action or change can be acted out with students to model what happened in the problem. Younger students may need to act out problems involving simple addition and subtraction to see the difference (example a). Even older students need convincing sometimes with problems such as example (b). Acting these out can assist students with keeping up with the steps of multistep problems.

a. Angela had 10 books on her desk. She loaned 3 to her sister. How many books did she have left? Her sister returned one book. How many does she have now?
b. Ms. Shaw saw a porcelain vase at a flea market and bought it for $30. Her friend, Ms. Leggett, begged her to sell it so she sold it for $40. Ms. Shaw changed her mind but had to pay $50 to get the vase back. A dealer spotted the vase on the way out of the market and offered Ms. Shaw $60. If Ms. Shaw takes this offer, will she come out ahead, even, or behind? By how much?

Create a List, Table, or Chart

Problems that present a lot of data or ask about combinations, a number of possible answers, a sequence of events, or even proportions lend themselves to making lists, tables, or charts. Even if another strategy is eventually needed for solving the problem, this graphic representation of data can assist the problem solver in understanding critical aspects of the problem.

a. A town's population in 2015 was 6,200. If the population grows by 400 people a year, approximately when will it reach 12,000 people?
b. A local shopping mall is being renovated so that each store's owner will have a choice for colors for the front, door, and shutters. The possible choices are front (gray, cream, or teal); shutters (maroon or yellow); and doors (black or red). Show all possible outcomes of the color choices.
c. A photo shop offers frames for photos that measure 5 inches by 7 inches and up. If the frames must maintain the same proportions for photographs, find the sizes of frames with the shorter side of 6, 7, 8, 9, and 10 inches.

Solve Part of the Problem

Problems that are multistep or have complex structures lend themselves to decomposition. However, Pólya cautioned about "a very foolish and bad habit with some students to start working at details before having understood the problem as a whole" (1945, p. 72). Once students consider the complete problem by asking, "What is not known? What is the information?" they should ask, "Would it be helpful to separate parts for better understanding or partial solutions?" The examples that follow are problems that can be decomposed and recombined.

a. On Monday Marty's mother gave him school money for the week. He spent $2.25 for lunch every day for five school days. He paid a $0.35 book fine at the library and bought school supplies for $1.50. If Marty had $1.90 left at the end of the week, how much did his mother give him on Monday?

b. Two classrooms of students each have 24 children. If $\frac{3}{4}$ of the first class and $\frac{2}{3}$ of the second class board a bus to travel to a museum, how many more children in the first class board the bus than in the second class?

Use a Formula

The most common problems requiring the application of formulas are those in geometry and measurement. Formulas describe the relationships among points, lines, and planes for two- and three-dimensional objects. They describe other relationships such as conversions between measurement systems, trigonometric functions, and even compounded interest, calorie intake calculations, and batting averages. Many students with MD have trouble remembering specific formulas because concepts underlying formulas are not understood or they view each formula as a separate, abstract memorization task. Consider the two examples below, both addressing surface area:

a. What is the area of a room 12 feet long by 8 feet wide?
b. What is the area of a parallelogram whose two long sides measure 12 inches and the height measures 8 inches?

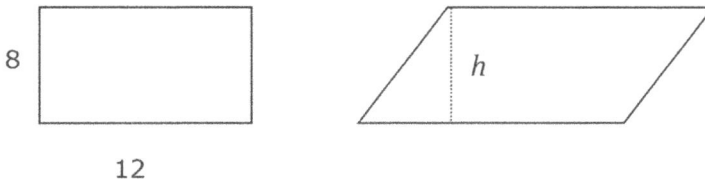

If students understand the concept that area is a square unit measure of length by width in very rough terms, they will readily determine that the floor is 96 square feet. If they also understand that this rough length by width can be applied to the parallelogram by visually chopping off one side with a perpendicular line (shown dotted above) and placing it on the other side, the parallelogram's area is 96 square inches. Hands-on tools such as geoboards or graph paper can assist this spatial understanding.

Other formulas are called for when a standard reference is provided in the problem. What is the temperature, interest payment, batting average, length in meters, volume, or sum of squares, given specific data that can be plugged into a formula?

c. At 3 pm, the temperature outside was 75° Fahrenheit. What was the temperature in degrees Celsius? (Apply the formula $C = (F - 32)/1.8$)
d. Nolan Ryan pitched 149 innings in the 1981 regular season and gave up 28 earned runs. What was his earned run average (ERA) that season? Apply the formula: (number of earned runs × 9) divided by (number of innings pitched).

Some statewide tests provide formulas for students to apply (example e), others expect students to recall formulas, then apply them (example f). A few states provide a page at the front of the test booklet listing common formulas and conversions. Both the American College Test (ACT) and the Scholastic Aptitude Test (SAT) require students to apply formulas they have memorized.

Students need practice in using formulas and connecting those concepts with the mathematics constructs involved.

e. The variables x and y are related by the following formula:

$$\frac{(x+10)}{6} = y \qquad \text{If } x = 8, \text{ what is the value of } y?$$

f. The slope of a ramp is 30° as pictured below. If the ramp is 40 feet long, how high is the ramp at the tallest point?

Write an Equation

Similar to applying a formula, an equation uses symbols to represent the known and unknown information in a problem. Some problems require two or more simultaneous equations because there is more than one unknown or more than one data statement (see example b). For students who struggle to make connections, it is often helpful to use special letters to represent unknowns (e. g., s for student, t for teacher), as long as those symbols aren't confused with other symbols or letters (5 and +). One letter should represent one value. For example, it wouldn't follow standard notation practice to represent yellow birds with yb, because that means y multiplied by b.

a. Barb has 10 times as many songs downloaded as Connie. Write an equation that shows this relationship.
b. The third-grade class is planning a party for parents. If 30 sodas and four-dozen cookies are ordered, the cost would be $40. If 40 sodas and five-dozen cookies are ordered, the cost would be $51.25. How much do sodas and cookies cost?

Many state-level tests include items that require students to create equations (example c) or select from a list of equations (example d).

c. Rosie bought 12 yards of fabric for a total of $59.40. Write an equation that could be used to compute the cost per yard.
d. Mora downloaded 15 songs to her personal player. Sara downloaded 3 times as many as Mora. Allison has 4 songs more than Sara. Which equation shows how many songs Allison has (z)? (In this problem, students may confuse the multiplication symbol with an x.)
    A) $15 + 3 + 4 = z$ B) $4 - (3 \times 15) = x$ C) $4 \times (15 - 3) = x$
    D) $(15 \times 3) - 4 = z$ E) $(15 \times 3) + 4 = z$

As is discussed in Strand F, many students hold a restricted view of variables and tend to consider them concrete objects or labels rather than values in equivalence statements that demonstrate functional relationships (Küchemann, 1978).

Work Backward

The working-backward strategy has been used for years by students who discovered the answers to some problems at the end of their math textbooks. It is also an excellent strategy for solving problems when a reasonable answer can be estimated or may even be known but a solution strategy is not apparent. Some problems begin with the "final" information and ask for the pieces along the way that led to the answer (examples a and b).

a. The book sale yielded $925 for the school. If 637 hardback books were sold at $1 and paperbacks cost 50 cents, how many paperbacks were sold?
b. The diver needs at total of 287 points to beat the other divers at competition. The first two dives yielded a total of 190 points from the scores of 10 judges on a 10-point scale. What should the average score by judges be for the final dive to win the competition?
c. How can you bring the amount of liquid in a container up to six quarts when you have only two containers, one four quarts and one nine quarts. (From Pólya, 1945, pp. 198–201.)

Look for Patterns

Patterns are systematic repetitions. They may be in numerical, visual, or event (action) forms. Patterns are used in problems that call for predictions and generalizations. The following examples demonstrate problems calling for numerical (example a), visual (example b), and event (example c) forms.

a. The first five terms in a sequence are 1, 2, 4, 7, and 11. What will be the eighth term in this sequence?
b. Draw the design that will be fifth in this pattern.

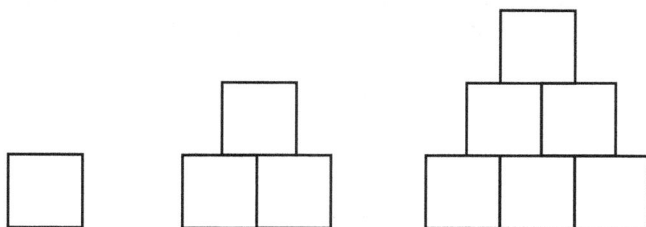

c. Mr. Smith has decided to sell his XYZ stock when it reaches $100 per share. The chart below shows the growth of his stock by weeks. When should he sell his stock if it continues to grow at the same rate?

Week	Price per Share ($)
1	5
2	10
3	20
4	40

Students with learning difficulties should be taught these alternative strategies explicitly, with many examples connecting the solution strategy to problem types. Some teachers use code words to prompt strategy use, for example: Patterns, Simpler Case, Draw, Table, or Model.

Teachers should *avoid* surface descriptions for problem types: the racetrack type, the boxes type. Students should also be taught critical vocabulary in word problem directions, such as solve, compute, simplify, estimate, graph, construct, measure, select, find, which expression, or which statement.

TRY THIS

Identify at least two useful problem-solving strategies for the following problem:

> *Brenda wanted to cut a square piece of cardboard into a rectangle. She cut two inches from one side and three inches from the other. If the area of the new rectangle decreased by 50 percent, what was the approximate length of the original square?*

Strategies for "Carry Out the Plan"

Selecting a solution strategy will not result automatically in the correct solution. Students who are poor problem solvers generally apply a strategy without monitoring its effectiveness. They should be taught the essential metacognitive processes that accompany problem solving. Montague (1992) defined metacognitive processes for mathematics as awareness and regulation of cognitive processes. Metacognition is the executive control part of the information-processing system and includes selecting appropriate strategies, monitoring progress during work, evaluating whether a solution strategy was successful, and reflecting back on what was learned in the process.

First, teachers should model think-aloud strategies for students and encourage them to use this mental talk during the problem-solving process. Students with MD may need question starters to prompt this self-talk: Am I applying the strategy in a logical manner? Do I have all the necessary information? Have I calculated accurately? Am I keeping track of each step of my work in case I have to back track? Do I need to change my approach or try a different strategy?

Keeping track of work is especially important in multistep problems. Students should begin recognizing problems that have several steps. Some may actually be short in terms of wording, but it is evident that information has not been presented in a direct form without a first or second step of information manipulation. For example, *Joe bought 5 gallons of gas for his lawn-mower for $9.75. If he burns 1 gallon in 1.5 hours, how much should he compute for the cost of gas in a 3-hour job?*

Even when a solution strategy isn't successful, students should understand they can sometimes use part of the process or at least learn from the process. Again, prompting questions help guide this type of thinking. Did any part of the solution make sense? Can I use some of the information? At what point did the solution strategy break down? Approaches for rescuing a partially successful strategy include going back to ask if all relevant information was used, estimating the solution to gauge the work ahead, considering whether there may be more than one solution, and applying a new strategy (such as charting or drawing a diagram) at the mid-point. Approaches for starting over after a failed process include starting with a different perspective on the problem, simplifying the problem, and using a different strategy.

Strategies for "Look Back"

Students with difficulties in mathematics are often satisfied with any answer. They write it down and are finished. Good problem solvers, however, engage in a final step that not only increases the probability of a correct solution but also informs them more deeply on problem type and solution strategy for future reference.

Once a solution (or solutions) is determined, the problem solver should compare it with expectations. Does it make sense in the context of the problem? Is it close to the original estimate? Has the problem question or implied question been answered? Can I check the solution by inserting it back into the problem or working backward? Generally, the more pragmatic and real-world the problem, the more sense making can be made. However, students can fail to complete this step in even the most obvious problems. A physics teacher was teaching about electric current (newton, joule, watt, volt, ohm). She used an example of a 100 W light bulb (100 Watt-hours). She asked the class to compute the cost of burning a light bulb all night (6 hours) if a kWh costs $.0577. The results should have been energy times rate or .6 kWh × .0577 = $ .03462 (just over 3 cents). Some students returned with results such as $34.62 and even $346,200!

Students should be encouraged to study the problem type for future reference (forward-reaching transfer). Was this a compare or combine problem? Why did a chart strategy work with this information? Could I have used a simpler strategy and still achieved the solution?

Some state-wide tests require students to explain their answers in writing. For example: *What can be the greatest number of days in two months? Explain your answer.* This task is especially challenging for students with a combination of mathematics and language disabilities. For success, students must engage in practice throughout the year in providing oral explanations of problem solutions, listening to other students provide reasoning, and occasionally writing out these explanations with corrective feedback. Students with MD often provide faulty or superficial reasoning such as "it was the best answer" or "because that's what it came out to be" or "because one number was bigger than the other."

TRY THIS

Solve and write out an explanation for the following problem.

Mr Jones grows four different crops on his 200-acre farm. Each acre has the same number of plants. Of the total acres, 25 percent is spring wheat, 35 percent is corn, 20 percent is barley, and the rest are soybeans. Mr Jones thinks that if he plants wheat in all of the acres that are now soybeans, more than half of his farm would be wheat. Is he correct? Explain your answer.

Curriculum Concerns

Teaching problem solving requires the integration of assessment, content standards, instruction, and practice across mathematics topics and grade levels. Some content areas lend themselves to specific solution strategies such as formulas in geometry and equations in algebra, as long as the formulas and equations are connected with accurate concepts. Younger students need more concrete strategies, while older students should have developed a large repertoire of problem-solving approaches.

Assessing Word Problem-Solving Abilities

Assessing word problem-solving ability can be challenging because the process of problem solving is complex and time-consuming. Poch et al. (2015) recommended the use of diagnostic assessments that can help inform instruction. Through the use of flexible interviews, the researchers collected student work samples and asked students what they understood about diagrams and how they used them in problem solving. A checklist targeted five key areas: conceptual understanding (of diagrams); procedural fluency (for translating material to the diagram); strategic competence (understanding various uses of diagrams); adaptive reasoning (explaining their reasoning); and productive disposition (confidence in using diagrams as a problem-solving tool). Through these informal interviews, teachers can identify misconceptions and challenges students demonstrate. The checklist provides a format for data collection for informing instruction and documenting student progress.

Other researchers have developed curriculum-based measures (CBMs) to monitor student progress during problem-solving instruction biweekly over several weeks or monthly for the school year. For example, Jitendra et al. (2014) administered problem-solving probes of eight items (6 one-step and 2 two-step problems) every two weeks during a 12-week problem-solving intervention. The probes were a direct measure of the problem types across the instructional unit. The CBMs were able to measure student growth over the instructional period with third graders, even for a task as complex as problem solving using diagrams.

Many researchers assessed mathematics vocabulary, reading comprehension skills, and working memory capacity as part of a broad assessment of the factors that contribute to difficulties (Fuchs, Powell, et al., 2020; Yip et al., 2020). For example, Fuchs, Seethaler, and colleagues (2020) took a sampling of 50 vocabulary words from preprimer to first grade reading lists to assess first-graders' word-reading skill and assessed specific word-problem language through the oral reading of word problems with oral questions about phrases that assist solutions. The researchers found that students selected as having mathematics difficulties (<25th percentile) were a full standard deviation below TA peers on language scores and benefitted from WP language instruction embedded within a schema-based WP intervention.

Yip et al. (2020) assessed second-grade students' WP reasoning using a brief word-problem typology measure where students matched simple addition and subtraction word problems with a diagram of their type. This measure was part of a screening battery that also included first and second grade-level mathematics items, a reading achievement measure, basic arithmetic, magnitude representation, verbal and spatial WM, nonverbal intelligence, and attention ratings. The six-item WP reasoning measure significantly differentiated students with and without MLD (sensitivity at 71%). When combined with the 28-item arithmetic operations task, sensitivity increased to 89%. If teachers have access to comprehensive assessments, multiple measures may help fill out individual student profiles related to WPS. If not, brief screening measures of WP knowledge and procedural skills will help identify students requiring interventions, even in the early grades.

Intensive Interventions

Most students with mathematics difficulties or disabilities are served in core mathematics classrooms with general curricula and textbooks, termed *Tier 1* in MTSS models. If some students are struggling with problem solving within that setting, the first assessment should determine whether the instructional approaches and curricular materials used are evidence-based, taught by highly qualified teachers. Are most students (> 75%) in the classroom succeeding with the mathematics instruction, including instruction on problem solving? Are mathematics problem-solving abilities included in class-wide screenings? Are students developing conceptual knowledge and procedural fluency connected with WPS? Are they developing a range of problem-solving strategies? If the instruction in the core mathematics classroom is providing explicit WPS instruction consistently across mathematics domains, but a small percentage of students are not succeeding, then more intensive intervention may be needed for those students.

Tiers 2 and 3 Interventions

Tier 2 interventions for problem solving can be provided within the core mathematics class-room or in pull-out supplementary sessions. As each instructional unit changes, the basic WPS approach and prior problem types would be reinforced with new concepts, and new problem types and strategies added to reflect the content. For example, a second-grade class has completed a unit on multidigit addition and subtraction with related problem types. The next unit on linear measurement involves using rulers to measure lengths in centimeters and inches. Students can apply addition and subtraction problem types in this unit (e.g., *The edge of this desk is 30 inches long and that desk is 36 inches. If I put them together how long will they be?*). The measurement unit is also preparing students for multiplicative reasoning problem types, with the introduction of the iteration of units. Within the core classroom, a teacher could set up stations or flex time to work with a small group of students on WPS elements. Researchers have also validated systematic peer tutoring for WPS interventions. For example, Powell et al. (2015), in a whole-class intervention on double-digit addition/subtraction WP types with second-grade students, supplemented that instruction with structured pair practice for Tier 2 (30 min, 3 times a week). In paired practice, the higher-performing student begins as the coach modeling three problems, then the other student thinks aloud solving three more problems with feedback from the coach, and then the peer tutors switch roles, for a total of 12 WP. Not only did students improve in WPS, but also in computation skills.

Practice sessions for WPS should be scheduled across the school year, as mathematics concepts and problem types change. Krawec and Huang (2017) implemented an intensive week-long intervention on a problem-solving approach, followed by weekly practice sessions and new problem types in fifth- and sixth-grade classrooms taught by core mathematics teachers. The intervention was equally effective for all student groups (MLD, MD, and TA) and, although the intervention group was lower than the control group at the beginning, they performed 20% higher by the final CBM. Even after a year of practice sessions, students with MLD did not reach mastery (70%), but improved their WPS performance, indicating the need for more intensive interventions across mathematics domains throughout the school year.

Some students (3%–5%) may need even more intensive instruction at Tier 3, in smaller groups or with one-to-one interventions (Zumeta, 2015). These students will likely require more than WPS instruction on grade-level mathematics concepts. They may need remedial work for gaps in concept understanding or procedural fluency. They may need explicit instruction in problem types and WPS strategies that should have been mastered earlier. For example, problems with proportional reasoning depend on experience with multiplicative problem types. Multiplicative reasoning requires a shift in thinking from additive problem types. Problems involving algebraic solutions may first require instruction in using visual representations before the abstract solution strategies. It is important that Tier 3 interventions for mathematics integrate problem solving with concept and procedural development, not address WPS as a separate topic. Special educators providing Tier 2 or 3 interventions may consider using one of the research-based programs described earlier in this chapter, or develop their own interventions using the essential elements from these programs.

Selecting Word Problems

Teachers providing interventions supplemental to or supplanting core mathematics instruction may need to select and sequence word problems so students develop skill with problem types and their representations incrementally (Miller & Mercer, 1993; Swanson, 2016). Problem types need to be developed gradually, with simple and direct WP first, followed by those that increase in complexity (indirect, irrelevant information, multistep). Problem structure

Table 5.1 Word-Problem Characteristics

Word Problems:

1. Walter brought 7 books back to the library on Monday. That was 4 more than he brought on Friday. How many books did he return on Friday?
2. Karen wants to buy five apples. The sign says six apples for $3.25 and bananas for $.99 a bunch. How much would she pay for five apples?
3. A recipe calls for three times as much flour as milk. Make a list of the amounts of flour needed for 1, 1 1/2, 2, and 2 1/2 cups of milk.
4. Jane and Bob are driving from Raleigh to Orlando. If they average 56.5 miles per hour for the trip, how long will the drive take?

Source: teacher made

	Problem 1	*Problem 2*	*Problem 3*	*Problem 4*
Math Content	Whole number operations	Rational number operations (decimals)	Rational number operations (fractions), measurement concepts	Rational number operations (decimals), measurement concepts, rate
Problem Type	Compare	Proportion	Multiplicative comparison	Equal groups or proportion
Vocabulary	No specialized vocabulary	No specialized vocabulary	Cups	Miles per hour
Question	Direct	Direct	Indirect	Direct
Steps	One	Two	Multiple	Two
Information	All provided and needed	Contains extraneous	All provided and needed	Distance needed
Possible Strategies	Diagram, act it out	Diagram, proportion	Diagram, equation, pattern, table	Diagram, equation

knowledge should be extended to a range of examples to promote transfer (Fuchs et al., 2010). In addition to textbooks, word problems are readily available on websites for problem solving and in professional journals. When examining a word problem, teachers should consider: What content and problem type is reflected? Is the problem stated in a question or implied? How many steps are involved? Does the problem contain extraneous material or omit necessary information? Are students prompted to use a specific strategy? Have students been exposed to the mathematics terms and concepts in the problem? Table 5.1 illustrates a checklist for evaluating problem characteristics. Teachers should work through each problem they plan to teach, considering possible strategies and anticipating difficulties, before posing problems for students. For teachers who do not have a strong mathematics background, it is especially important to study problems ahead of time, anticipate difficulties, and plan additional scaffolds.

Word problems for a specific type should match the learning trajectory of the concept (Huang et al., 2019). For addition/subtraction comparison problems a research-based trajectory is: 1) finding the difference given two quantities, 2) finding the bigger number given smaller number and how many more, and 3) finding the smaller number given the bigger number and how many less. A CRA approach is often implemented, with language supports and fading at each level. Some students may need to practice comparing bigger and smaller quantities on number lines. Others may need practice with equivalence statements (e.g., 8 is n more than 5), the structure of comparison problems (biggest = smaller + difference), and varying positions for the unknown (e.g., $8 + u = 15$; $u - 4 = 10$). Attention to language comprehension, cognitive and metacognitive strategies, and supports for WM must also be addressed when sequencing WP practice.

Problem Solving across the Curriculum

The problem-solving sequence employed in mathematics is useful for other applications, including science, social studies, writing, social skills, and everyday problems. The wide applicability of this process has the potential to mold a problem-solving approach in students that will serve them in many situations, both in school and in other settings.

In science, problems related to physical, chemical, and/or biological aspects of the world and beyond also involve practices for science and engineering such as asking questions, using models, planning and carrying out investigations, and constructing explanations (NRC, 2012). These practices require students (and scientists) to define and model problems, plan and carry out a systematic investigation, revise and retest explanations, interpret data, construct explanations based on evidence, and communicating information. These practices are more interconnected (less linear) than the former scientific method but retain many elements of a complex problem-solving process.

A reading and study-skills strategy related to the problem-solving process is the KWL procedure (Ogle, 1986). Before reading, either as an individual or with a group, examine what you already *know* about the subject. Determine what you *want* to learn (and how you will find out). Read the material and reflect on what you *learned*. Many writing processes include problem-solving stages: planning, organizing, writing, revising/editing, sharing or publishing (e.g., Harris & Graham, 1996). The *process* of writing is emphasized in self-regulated, strategic approaches.

For social skills, decision-making, or interpersonal dilemmas, a common strategy involves recognizing and defining the problem, listing possible solutions with probable outcomes, selecting and implementing a solution, and evaluating the results. For example, a student may have thought other students didn't want to chat with her because of her ethnicity when the problem was actually her social cues (e.g., nonverbal facial expressions). Defining the problem led to solution strategies. Some schools have implemented school-wide problem-solving procedures that employ similar problem-solving steps for behavioral issues as well as curricular applications (e.g., Kramer et al., 2014). In these schools everyone—teachers, students, administrators, and staff—is using the same language and holding students responsible for applying these strategies in a wide range of settings.

Research on effective strategies for solving mathematics word problems and their use by students with learning difficulties has yielded important results for educators. Teachers should connect problem solving with related mathematics content and introduce a problem-solving sequence that can be applied to both routine and nonroutine problems. There is clear evidence that students at risk for difficulties in word-problem solving require explicit, systematic instruction for each problem domain with regular practice across the school year. To build a toolbox of strategies across the grade levels, students should employ valid and consistent approaches to problem solving, including schema-based and cognitive strategies. Problem solving cannot be taught as a discrete unit or over a week or two. It requires on-going instruction and practice, provided within core mathematics instruction or in intervention sessions.

Problem solving can be the most challenging and rewarding aspect of the mathematics curriculum. Teachers should model the critical dispositions of curiosity, persistence, interest, and enthusiasm they want their students to develop.

The trains passed each other in Charleston, South Carolina 12 hours after their departures. What was the average speed of each train?

References

Alghamdi, A., Jitendra, A. K., & Lein, A. E. (2020). Teaching students with mathematics disabilities to solve multiplication and division word problems: The role of schema-based instruction. *ZDM, 52*(1), 125–137. doi: 10.1007/s11858-019-01078-0

Andersson, U. (2008). Mathematical competencies in children with different types of learning difficulties. *Journal of Educational Psychology, 100*, 48–66.

Beckmann, S. (2004). Solving algebra and other story problems with simple diagrams: A method described in grade 4–6 texts used in Singapore. *The Mathematics Educator, 14*(1), 42–46.

Boonen, A., de Koning, B., Jolles, J., & van der Schoot, M. (2016). Word problem solving in contemporary math education: A plea for reading comprehension skills training. *Frontiers in Psychology, 7*, 1–10. doi: 10.3389/fpsyg.2016.00191

Cai, J., & Lester, F. (2010). *Why is teaching with problem solving important to student learning?* NCTM. www.nctm.org/Research-and-Advocacy/Research-Brief-and-Clips/Problem-Solving/

Carpenter, T. P., Fennema, E., Franke, M. L., Levi, L., & Empson, S. B. (2015). *Children's mathematics: Cognitively guided instruction* (2nd ed.). Heinemann.

Carpenter, T. P., Hiebert, J., & Moser, J. M. (1983). The effect of instruction on children's solutions of addition and subtraction word problems. *Educational Studies in Mathematics, 14*, 55–72.

Cook, J. L., & Rieser, J. J. (2005). Finding the critical facts: Children's visual scan patterns when solving story problems that contain irrelevant information. *Journal of Educational Psychology, 97*, 224–234.

Cooper, G., & Sweller, J. (1987). Effects of schema acquisition and rule automation mathematical problem-solving transfer. *Journal of Educational Psychology, 79*(4), 347–362. doi: 10.1037/0022-0663.79.4.347

Cornoldi, C., Drusi, S., Tencati, C., Giofre, D., & Mirandola, C. (2012). Problem solving and working memory updating difficulties in a group of poor comprehenders. *Journal of Cognitive Education and Psychology, 11*(1), 39–44. doi: 10.1891/1945-8959.11.1.39

English, L. D. (1997). Children's reasoning processes in classifying and solving computational word problems. In L. D. English (Ed.), *Mathematical reasoning: Analogies, metaphors, and images* (pp. 191–220). Erlbaum.

Erbas, A. K., & Okur, S. (2012). Researching students' strategies, episodes, and metacognitions in mathematics problem solving. *Quality & Quantity, 46*(1), 89–102. doi: 10.1007/s11135-010-9329-5

Fuchs, L. S., & Fuchs, D. (2002). Mathematics problem-solving profiles of students with mathematics disabilities with and without comorbid reading disabilities. *Journal of Learning Disabilities, 35*(6), 563–573. doi: 10.1177/00222194020350060701

Fuchs, L. S., Fuchs, D., Hamlett, C. L., & Appleton, A. C. (2002). Explicitly teaching for transfer: Effects on the mathematical problem-solving performance of students with mathematics disabilities. *Learning Disabilities Research & Practice, 17*(2), 90–106. doi: 10.1111/1540-5826.00036

Fuchs, L. S., Fuchs, D., Seethaler, P. M., Cutting, L. E., & Mancilla-Martinez, J. (2019). Connections between reading comprehension and word-problem solving via oral language comprehension: Implications for comorbid learning disabilities. In L. S. Fuchs & D. L. Compton (Eds.), *Models for innovation: Advancing approaches to higher-risk and higher-impact learning disabilities science. New directions for child and adolescent development, 165*, 73–90. doi: 10.1002/cad.20288

Fuchs, L. S., Gilbert, J. K., Fuchs, D., Seethaler, P. M., & Martin, B. (2018). Text comprehension and oral language as predictors of word-problem solving: Insights into word-problem solving as a form of text comprehension. *Scientific Studies of Reading, 22*, 152–166. doi: 10.1080/10888438.2017.1398259

Fuchs, L. S., Malone, A. S., Schumacher, R. F., Namkung, J., & Wang, A. (2017). Fraction intervention for students with mathematics difficulties: Lessons learned from five randomized controlled trials. *Journal of Learning Disabilities, 50*(6), 631–639. doi: 10.1177/0022219416677249

Fuchs, L., Powell, S., Fall, A., Roberts, G., Cirino, P., Fuchs, D., & Gilbert, J. (2020). Do the processes engaged during mathematical word-problem solving differ along the distribution of word-problem competence? *Contemporary Educational Psychology, 60*, 101811–. doi: 10.1016/j.cedpsych.2019.101811

Fuchs, L. S., Powell, S. R., Seethaler, P. M., Cirino, P. T., Fletcher, J. M., Fuchs, D., & Zumeta, R. O. (2009). Remediating number combination and word problem deficits among students with mathematics difficulties: A randomized control trial. *Journal of Educational Psychology, 101*(3), 561–576. doi: 10.1037/a0014701

Fuchs, L., Seethaler, P., Sterba, S., Craddock, C., Fuchs, D., Compton, D., Geary, D., & Changas, P. (2020). Closing the word-problem achievement gap in first grade: Schema-based word-problem intervention with embedded language comprehension instruction. *Journal of Educational Psychology*. doi: 10.1037/edu0000467

Fuchs, L., Zumeta, R., Finelli, R., Schumacher, R., Powell, S., Seethaler, P., Hamlett, C., & Fuchs, D. (2010). The effects of schema-broadening instruction on second graders' word problem performance and their ability to represent word problems with algebraic equations: A randomized control study. *Elementary School Journal*, *110*(4), 440–463. doi: 10.1086/651191

Gersten, R., Chard, D., Jayanthis, M., Baker, S., Murphy, P., & Flojo, J. (2009). Mathematics instruction for students with learning disabilities: A meta-analysis of instructional components. *Review of Educational Research*, *79*(3), 1202–1242. doi: 10.3102/0034654309334431

Gonsalves, N., & Krawec, J. (2014). Using number lines to solve math word problems: A strategy for students with learning disabilities. *Learning Disabilities Research & Practice*, *29*(4), 160–170. doi: 10.1111/ldrp.12042

Greer, B. (1992). Multiplication and division as models of situations. In D. A. Grouws (Ed.), *Handbook of research on mathematics teaching and learning* (pp. 276–295). Macmillan.

Greer, B. (1994). Extending the meaning of multiplication and division. In G. Harel & J. Confrey (Eds.), *The development of multiplicative reasoning in the learning of* mathematics (pp. 61–85). State University of New York Press.

Gurganus, S. & Del Mastro, M. (1998). Mainstreaming kids with reading and writing problems: Special problems of the mathematics classroom. *Reading and Writing Quarterly*, *14*(1), 117–125. doi: 10.1080/1057356980140107

Harris, K. R., & Graham, S. (1996). *Making the writing process work: Strategies for composition and self-regulation*. Brookline Books.

Hart Research Associates (2018, July). *Fulfilling the American dream: Liberal education and the future of work*. The Association of American Colleges and Universities. www.aacu.org/research/2018-future-of-work

Ho, S., & Lowrie, T. (2014). The model method: Students' performance and its effectiveness. *The Journal of Mathematical Behavior*, *35*, 87–100. doi: 10.1016/j.jmathb.2014.06.002

Huang, R., Zhang, Q., Chang, Y., & Kimmins, D. (2019). Developing students' ability to solve word problems through learning trajectory-based and variation task-informed instruction. *ZDM*, *51*(1), 169–181. doi: 10.1007/s11858-018-0983-8

Jitendra, A. K. (2007). *Solving math word problems: Teaching students with learning disabilities using schema-based instruction*. Pro-Ed.

Jitendra, A. K. (2019). Using schema-based instruction to improve students' mathematical word problem solving performance (pp. 595–609). In A. Fritz, P. Räsänen, & V. G. Hasse (Eds.). *International handbook of mathematical learning difficulties*. Springer. doi: 10.1007/978-3-319-97148-3_35

Jitendra, A. K., Dupuis, D. N., Star, J. R., & Rodriguez, M. C. (2016). The effects of schema-based instruction on the proportional thinking of students with mathematics difficulties and with and without reading difficulties. *Journal of Learning Disabilities*, *49*(4), 354–367. doi: 10.1177/0022219414554228

Jitendra, A. K., Dupuis, D. N., & Zaslofsky, A. F. (2014). Curriculum-based measurement and standards-based mathematics: Monitoring the arithmetic word problem-solving performance of third-grade students at risk for mathematics difficulties. *Learning Disability Quarterly*, *37*(4), 241–251. doi: 10.1177/0731948713516766

Jitendra, A. K., Griffin, C., Haria, P., Leh, J., Adams, A., & Kaduvetoor, A. (2007). A comparison of single and multiple strategy instruction on third-grade students' mathematical problem solving. *Journal of Educational Psychology*, *99*(1), 115–127. doi: 10.1037/0022-0663.99.1.115

Jitendra, A. K., Griffin, C. C., McGoey, K., Gardill, M. C., Bhat, P., & Riley, T. (1998). Effects of mathematical word problem solving by students at risk or with mild disabilities. *The Journal of Educational Research*, *91*(6), 345–355. doi: 10.1080/00220679809597564

Jitendra, A., Harwell, M., Dupuis, D., & Karl, S. (2017). A randomized trial of the effects of schema-based instruction on proportional problem-solving for students with mathematics problem-solving difficulties. *Journal of Learning Disabilities*, *50*(3), 322–336. doi: 10.1177/0022219416629646

Jitendra, A. K., Harwell, M. R., Dupuis, D. N., Karl, S. R., Lein, A. E., Simonson, G., & Slater, S. C. (2015). Effects of a research-based mathematics intervention to improve seventh-grade students' proportional problem solving: A cluster randomized trial. *Journal of Educational Psychology, 107,* 1019–1034. doi: 10.1037/edu0000039

Jitendra, A. K., Harwell, M. R., Im, S., Karl, S. R., & Slater, S. C. (2019). Improving student learning of ratio, proportion, and percent: A replication study of schema-based instruction. *Journal of Educational Psychology, 111*(6), 1045–1062. doi: 10.1037/edu0000335

Jitendra, A. K., & Hoff, K. (1996). The effects of schema-based instruction on the mathematical word-problem-solving performance of students with learning disabilities. *Journal of Learning Disabilities, 29*(4), 422–431. doi: 10.1177/002221949602900410

Jitendra, A. K., Petersen-Brown, S., Lein, A. E., Zaslofsky, A. F., Kunkel, A. K., Jung, P.-G., & Egan, A. M. (2015). Teaching mathematical word problem solving: The quality of evidence for strategy instruction priming the problem structure. *Journal of Learning Disabilities, 48*(1), 51–72. doi: 10.1177/0022219413487408

Jitendra, A. K., Rodriguez, M., Kanive, R. G., Huang, J.-P., Church, C., Corroy, K. C., & Zaslofsky, A. F. (2013). The impact of small-group tutoring interventions on the mathematical problem solving and achievement of third grade students with mathematics difficulties. *Learning Disability Quarterly, 36*(1), 21–35. doi: 10.1177/0731948712457561

Jitendra, A. K., & Star, J. R. (2011). Meeting the needs of students with learning disabilities in inclusive mathematics classrooms: The role of schema-based instruction. *Theory into Practice, 50*(1), 12–19. doi: 10.1080/00405841.2011.534912

Jitendra, A. K. & Star, J. R. (2017). *Solving ratio, proportion, and percent problems using schema-based instruction*™. Center on Teaching and Learning, University of Oregon. https://dibels.uoregon.edu/market/sbi

Kikas, E., Mädamürk, K., & Palu, A. (2019). What role do comprehension-oriented learning strategies have in solving math calculation and word problems at the end of middle school? *British Journal of Educational Psychology, 90*(S1), 105–123. doi: 10.1111/bjep.12308

Kong, J., & Swanson, H. (2019). The effects of a paraphrasing intervention on word problem-solving accuracy of English learners at risk of mathematic disabilities. *Learning Disability Quarterly, 42*(2), 92–104. doi: 10.1177/0731948718806659

Kramer, T. J., Caldarella, P., Young, K. R., Fischer, L., & Warren, J. S. (2014). Implementing *Strong Kids* school-wide to reduce internalizing behaviors and increase prosocial behaviors. *Education and Treatment of Children, 37*(4), 659–680. doi: 10.1353/etc.2014.0031

Krawec, J., & Huang, J. (2017). Modifying a research-based problem-solving intervention to improve the problem-solving performance of fifth and sixth graders with and without learning disabilities. *Journal of Learning Disabilities, 50*(4), 468–480. doi: 10.1177/0022219416645565

Krulik, S. (Ed.). (1980). *Problem solving in school mathematics: Yearbook of the National Council of Teachers of Mathematics.* NCTM.

Küchemann, D. (1978). Children's understanding of numerical variables. *Mathematics in School, 7*(4), 23–26.

Lashley, K. S. (1951). The problem of serial order in behavior (pp. 112–131). In L. A. Jeffress (Ed.), *Cerebral mechanisms in behavior.* Wiley.

Lein, A., Jitendra, A., & Harwell, M. (2020). Effectiveness of mathematical word problem solving interventions for students with learning disabilities and/or mathematics difficulties: A meta-analysis. *Journal of Educational Psychology, 112*(7), 1388–1408. doi: 10.1037/edu0000453

Lucangeli, D., Coi, G., & Bosco, P. (1997). Metacognitive awareness in good and poor math problem solvers. *Learning Disabilities Research & Practice, 12,* 209–212.

Malone, A., Fuchs, L., Sterba, S., Fuchs, D., & Foreman-Murray, L. (2019). Does an integrated focus on fractions and decimals improve at-risk students' rational number magnitude performance? *Contemporary Educational Psychology, 59,* 101782–. doi: 10.1016/j.cedpsych.2019.101782

Marshall, S. P. (1995). *Schemas in problem solving.* Cambridge University Press.

Merriam-Webster (n.d.). *Merriam-Webster.com dictionary.* www.merriam-webster.com

Miller, S. P., & Mercer, C. D. (1993). Using a graduated word problem sequence to promote problem-solving skills. *Learning Disabilities Research & Practice, 8,* 169–174.

Montague, M. (1992). The effects of cognitive and metacognitive strategy instruction on mathematical problem solving of middle-school students with learning disabilities. *Journal of Learning Disabilities, 25,* 230–248. doi: 10.1177/002221949202500404

Montague, M. (2003). *Solve It! A practical approach to teaching mathematical problem-solving skills.* Exceptional Innovations. www.exinn.net/

Montague, M., & Applegate, B. (1993). Mathematical problem-solving characteristics of middle-school students with learning disabilities. *The Journal of Special Education, 27*(2), 175–201. doi: 10.1177/002246699302700203

Montague, M., Applegate, B., & Marquard, K. (1993). Cognitive strategy instruction and mathematical problem-solving performance of students with learning disabilities. *Learning Disabilities Research & Practice, 8,* 223–232.

Montague, M., & Bos, C. S. (1986). The effect of cognitive strategy training on verbal math problem solving performance of learning-disabled adolescents. *Journal of Learning Disabilities, 19*(1), 26–33. doi: 10.1177/002221948601900107

Montague, M., Enders, C., & Dietz, S. (2011). Effects of cognitive strategy instruction on math problem solving of middle-school students with learning disabilities. *Learning Disability Quarterly, 34*(4), 262–272. doi: 10.1177/0731948711421762

Montague, M., Warger, C., & Morgan, T. H. (2000). Solve it! Strategy instruction to improve mathematical problem solving. *Learning Disabilities Research & Practice, 15*(2), 110–116. doi: 10.1207/sldrp1502_7

Morin, L., Watson, S., Hester, P., & Raver, S. (2017). The use of a bar model drawing to teach word problem solving to students with mathematics difficulties. *Learning Disability Quarterly, 40*(2), 91–104. doi: 10.1177/0731948717690116

National Association of Colleges and Employers (January 2020). *Job outlook 2020.* www.naceweb.org

National Council of Teachers of Mathematics (2000). *Principles and standards for school mathematics.* www.nctm.org/Standards-and-Positions/Principles-and-Standards

National Governors Association Center for Best Practices & Council of Chief State School Officers (2010). *Common core state standards for mathematics.* www.corestandards.org/Math/

National Research Council (2012). *A Framework for K-12 Science Education: Practices, Crosscutting Concepts, and Core Ideas.* The National Academies Press. doi: 10.17226/13165.

Ogle, D. S. (1986). K-W-L group instructional strategy. In A. S. Palincsar, D. S. Ogle, B. F. Jones, & E. G. Carr (Eds.), *Teaching reading as thinking.* (Teleconference Resource Guide, pp. 11–17. Association for Supervision and Curriculum Development).

Organisation for Economic Co-operation and Development (2018). *Programme for international student assessment (PISA): Results from PISA 2018.* www.oecd.org/pisa/publications/PISA2018_CN_USA.pdf

Parker, M. (2004). Reasoning and working proportionally with percent. *Mathematics Teaching in the Middle School, 9*(6), 326–330. www.jstor.org/stable/41181930

Peltier, C., & Vannest, K. (2017). A meta-analysis of schema instruction on the problem-solving performance of elementary school students. *Review of Educational Research, 87*(5), 899–920. doi: 10.3102/0034654317720163

Peng, P., Namkung, J., Barnes, M., & Sun, C. (2016). A meta-analysis of mathematics and working memory: Moderating effects of working memory domain, type of mathematics skill, and sample characteristics. *Journal of Educational Psychology, 108*(4), 455–473. doi: 10.1037/edu0000079

Poch, A. L., van Garderen, D., & Scheuermann, A. M. (2015). Students' understanding of diagrams for solving word problems: A framework for assessing diagram proficiency. *TEACHING Exceptional Children, 47,* 153–162. doi: 10.1177/0040059914558947

Pólya, G. (1945). *How to solve it: A new aspect of mathematical method.* Princeton University Press.

Pólya, G. (1971). Methodology or heuristics, strategy or tactics? *Archives de philosophie, 34*(4), 623–629.

Powell, S. R. (2011). Solving word problems using schemas: A review of the literature. *Learning Disabilities Research & Practice, 26*(2), 94–108. doi: 10.1111/j.1540-5826.2011.00329.x

Powell, S. R., Fuchs, L. S., Cirino, P. T., Fuchs, D., Compton, D. L., & Changas, P. C. (2015). Effects of a multitier support system on calculation, word problem, and prealgebraic performance among at-risk learners. *Exceptional Children, 81*(4), 443–470. doi: 10/1177/0014402914563702

Reed, S. K. (1999). *Word problems: Research and curriculum reform.* Erlbaum.

Romanycia, M. H. J., & Pelletier, F. J. (1985). What is a heuristic? *Computational Intelligence, 1*(1), 47–58. doi: 10.1111/j.1467-8640.1985.tb00058.x

Rosenzweig, C., Krawec, J., & Montague, M. (2011). Metacognitive strategy use of eighth-grade students with and without learning disabilities during mathematical problem solving: A think-aloud analysis. *Journal of Learning Disabilities, 44*(6), 508–520. doi: 10.1177/0022219410378445

Salomon, G., & Perkins, D. N. (1989). Rocky roads to transfer: Rethinking mechanisms of a neglected phenomenon. *Educational Psychologist, 24*(2), 113–142. doi: 10.1207/s15326985ep2402_1

Scheid, K. (1993). *Helping students become strategic learners: Guidelines for teaching.* Brookline Books.

Schoenfeld, A. H. (1987). Pólya, problem solving, and education. *Mathematics Magazine, 60*(5), 283–291. doi: 10.2307/2690409

Schoenfeld, A. H. (1992). Learning to think mathematically: Problem solving, metacognition, and sense making in mathematics. In D. A. Grouws (Ed.), *Handbook of research on mathematics teaching and learning* (pp. 334–370). Macmillan.

Schoenfeld, A. H. (2013). Reflections on problem solving theory and practice. The *Mathematics Enthusiast, 10,* 9–34.

Schwartz, J. L. (1988). Intensive quantity and referent transforming arithmetic operations. In J. Hiebert & M. Behr (Eds.), *Number concepts and operations in the middle grades* (Vol. 2, pp. 41–52). NCTM.

Shin, M., & Bryant, D. P. (2013). A synthesis of mathematical and cognitive performance of students with mathematics learning disabilities. *Journal of Learning Disabilities, 48*(1), 96–112. doi: 10.1177/0022219413508324

Swanson, H. (2016). Word problem solving, working memory and serious math difficulties: Do cognitive strategies really make a difference? *Journal of Applied Research in Memory and Cognition, 5*(4), 368–383. doi: 10.1016/j.jarmac.2016.04.012

Swanson, H. L., Jerman, O., & Zheng, X. (2008). Growth in working memory and mathematical problem solving in children at risk and not at risk for serious math difficulties. *Journal of Educational Psychology, 100*(2), 343–379. doi: 10.1037/0022-0663.100.2.343

Swanson, H. L., Lussier, C. M., & Orosco, M. J. (2015). Cognitive strategies, working memory, and growth in word problem solving in children with math difficulties. *Journal of Learning Disabilities, 48*(4), 339–358. doi: 10.1177/0022219413498771

Swanson, H., Olide, A., & Kong, J. (2018). Latent class analysis of children with math difficulties and/or math learning disabilities: Are there cognitive differences? *Journal of Educational Psychology, 110*(7), 931–951. doi: 10.1037/edu0000252

van Garderen, D. (2016). Spatial visualization, visual imagery, and mathematical problem solving of students with varying abilities. *Journal of Learning Disabilities, 39*(6), 496–506. doi: 10.1177/00222194060390060201

van Garderen, D., & Montague, M. (2003). Visual-spatial representations and mathematical problem solving. *Learning Disabilities Research & Practice, 18,* 246–254.

van Garderen, D., Scheuermann, A., & Jackson, C. (2012). Examining how students with diverse abilities use diagrams to solve mathematics word problems. *Learning Disability Quarterly, 36*(3), 145–160. doi: 10.1177/0731948712438558

Wang, A., Fuchs, L., Fuchs, D., Gilbert, J., Krowka, S., & Abramson, R. (2019). Embedding self-regulation instruction within fractions intervention for third graders with mathematics difficulties. *Journal of Learning Disabilities, 52*(4), 337–348. doi: 10.1177/0022219419851750

Xin, Y. (2019). The effect of a conceptual model-based approach on "additive" word problem solving of elementary students struggling in mathematics. *ZDM, 51*(1), 139–150. doi: 10.1007/s11858-018-1002-9

Xin, Y. P., Wiles, B., & Lin, Y.-Y. (2008). Teaching conceptual model-based word problem story grammar to enhance mathematics problem solving. *The Journal of Special Education, 42*(3), 163–178. doi: 10.1177/0022466907312895

Xin, Y. P., Zhang, D., Park, J. Y., Tom, K., Whipple, A., & Si, L. (2011). A comparison of two mathematics problem-solving strategies: Facilitate algebra-readiness. *Journal of Educational Research, 104*(6), 381–395. doi: 10.1080/00220671.2010.487080

Yip, E., Wong, T., Cheung, S., & Chan, K. (2020). Do children with mathematics learning disability in Hong Kong perceive word problems differently? *Learning and Instruction, 68,* 101352–. doi: 10.1016/j.learninstruc.2020.101352

Zumeta, R. (2015). Implementing intensive intervention: How do we get there from here? *Remedial and Special Education, 36*(2), 83–88. doi: 10.1177/0741932514558935

6 Instruction and Collaboration for General Education Settings

Chapter Questions

1. In planning universally designed instruction, what aspects of the mathematics classroom should be considered?
2. How are instructional accommodations selected and documented?
3. How can teachers plan lessons using core mathematics textbooks more effectively?
4. How do various collaboration models for instruction support students with learning difficulties?
5. How can teachers promote partnerships with parents for improved mathematics learning?

Joseph Lopez, Pine Road High School's newest mathematics teacher, has a challenging first-year assignment. His schedule includes two Geometry classes, two Algebra I classes, and one integrated Mathematics I course. The Mathematics I course integrates algebra, geometry, and statistics in a new curriculum option for entering ninth graders, and Joseph is excited to be part of this curriculum revision at Pine Road. The Geometry and Algebra I courses are a more traditional approach to high-school topics, but Joseph knows those courses will necessarily include connections to the other domains and is confident planning lessons across domains. Joseph is more concerned about his ninth- and tenth-grade students. With a caseload of 140 students in the five courses, it is difficult to get to know each student's background, abilities, and needs in mathematics. He knows that 20 of these students have identified disabilities, with 12 of those attending a learning support class three times a week. Another 10 students are new to the country and began learning English in seventh or eighth grade. Checking last spring's test scores, Joseph notes that 25% scored below basic levels in mathematics and only 15% scored above proficient, able to reason and problem solve with grade-level content. How can he plan lessons that will challenge all the students in his classes, meet the needs of students with special learning needs, and still cover the material to meet grade-level standards? Joseph decides he needs to seek out the special educators and other support staff at Pine Road High School for some assistance with planning.

Instruction for students who struggle in mathematics takes place primarily in general education classrooms by mathematics teachers with support and collaboration from other professionals, such as special education teachers and mathematics-support specialists. The percentage of students with disabilities served in regular classrooms most of the school day increased from 31.7% in 1989 to 63.4% in 2018 (Institute of Education Sciences, 2018). Seventy-one percent

DOI: 10.4324/9781003096733-6

of students with learning disabilities (SLD) and 67% of students with other health impairments (OHI, including ADHD) were served in regular classrooms for 80% or more of the school day by 2017. Further, the percentage of public-school students who were English language learners (EL) increased from 9.7% in 2008 to 10.1% in 2016, with expectations to meet the same academic content and achievement standards that all students are expected to meet (IES, 2018).

A typical mathematics class of 25 students, using very rough estimates, may include three students with identified disabilities, two students who are EL, and one or two students identified as gifted. Of the entire class, only ten students are performing at expected levels in mathematics. Classrooms in high-poverty or diverse minority areas may see an even greater percentage of students with learning difficulties. Planning instruction that will be effective for all students in this classroom is challenging. This core mathematics classroom is also considered Tier 1 in districts with multi-tiered systems of support (MTSS) programs, required to provide high-quality instruction using an evidence-based curriculum with supports to meet the learning needs of all students in the classroom. The teacher in this classroom should also implement frequent measures of student progress to inform instructional decisions including screening for students who may need more intensive interventions.

This chapter addresses some of the challenges for teachers such as Joseph Lopez planning core mathematics instruction for diverse groups of students and the other professionals supporting that instruction. Topics include universal design for learning, differentiated instruction, instructional accommodations, textbook suitability, and teacher collaboration models.

Instructional Adaptations

Teachers providing mathematics instruction for diverse student groups should first design instruction that will be accessible for *all* students. Universal design elements benefit more students and are less obvious and time-consuming than individually designed adaptations. A few students may still require individualized accommodations and other supports within the mathematics classroom for instruction to be effective.

Universal Design for Learning

The concept of *universal design*, first developed for architectural applications, is also applied to concepts of curriculum interface where diverse learners seek access to the same curriculum standards as their peers. Universal design for learning (UDL) "means the design of instructional materials and activities that allow the learning goals to be achievable by individuals with wide differences in their abilities to see, hear, move, read, write, understand English, attend, organize, engage, and remember" (Orkwis & McLane, 1998). The Higher Education Opportunity Act of 2008 defined UDL as:

> a scientifically valid framework for guiding educational practice that: (A) provides flexibility in the ways information is presented, in the ways students respond or demonstrate knowledge and skills, and in the ways students are engaged; and (B) reduces barriers in instruction, provides appropriate accommodations, supports, and challenges, and maintains high achievement expectations for all students, including students with disabilities and students who are limited English proficient.
>
> (HEOA, 2008)

This definition was referenced in the Every Student Succeeds Act of 2015, requiring the application of principles of UDL for the development of student academic assessments, comprehensive literacy instruction, and the use of technology to support learning needs of all students.

Each aspect of the classroom environment should be considered when employing UDL, including the physical environment, instructional materials, methods for interactions, and approaches to assessment. The Center for Applied Special Technology (CAST), founded in 1984 to explore ways technology can enhance access for people with disabilities, has been a leader in promoting accessibility within education. CAST's framework is based on the brain's neural systems with three guidelines: 1) provide multiple means of engagement; 2) provide multiple means of representation; and 3) provide multiple means of action and expression. Each of these guidelines is discussed here with examples from mathematics classrooms.

1. *Provide Multiple Means of Engagement.* Student engagement in learning can be affected by factors such as motivation, prior knowledge, the student's background, and other student and classroom characteristics. The CAST guidelines recommend providing learning options that optimize relevance, choice, value, and authenticity, and promote self-regulation and persistence. Teachers designing instruction for mathematics should consider student interests and background when planning choices for assignments, class learning activities, instructional arrangements, and resources.

For example, one sixth-grade mathematics teacher presented a unit about statistics as a way to investigate their community. Data were collected about student preferences, hobbies, favorite music, and community events. These data were analyzed with options for paper and pencil calculations and graphing or computer-based spreadsheets and graphing programs. Student groups had choices for presenting their findings as videos, models, booklets, computerized slideshows, and oral reports supported with charts. They could create questions for exploration and select types of graphs. This class even designed their own assessment options when the teacher asked how she should measure what they'd learned about statistics. This type of learning engaged all students in the classroom and was carefully planned and orchestrated by the teacher. Mathematics is not information presented by the teacher or textbook, but interesting and relevant concepts to explore, ask questions about, investigate more, and use to answer further questions.

2. *Provide Multiple Means of Representation.* The representation guideline refers to how content is presented to the learner. Teachers should consider all the means for transmitting content, including written, oral, concrete, pictorial, graphic, experiential, video, and computer-based representations. Further, teachers should make connections between representations, such as linking a word problem with a diagram or model. The representation guideline asks teachers to consider options for perception (visual, auditory), language (vocabulary, symbols, syntax), and comprehension (prior knowledge, big ideas, connections, scaffolding, and promoting transfer of learning) when planning instructional delivery.

A third-grade classroom is working on word-problem solving involving multiplicative problem types (multiplication and division). In small groups, students are locating the critical parts of a word problem and either underlining those or paraphrasing the parts. Within groups students have the option of making a diagram of the problem, representing the problem with another model (e.g., interconnecting cubes, paper strips), or developing an equation. Each group's reporter explains their work to the class, including what was difficult and what strategies were used. The teacher asks critical questions to guide thinking and self-reflecting. Next, each student is given a new problem of the same type to solve using their choice of representation. Students volunteer to share their results and what they learned.

When planning to present mathematics concepts, teachers should consider the best representations for students' levels of learning. Teachers should emphasize the big ideas, connect new learning to prior knowledge, and consider vocabulary, reading demands, and overall organization of printed materials. If alternative forms of written material are not provided with the textbook, teachers can adapt those to reduce vocabulary load, items per page, or readability level using assistive technology.

3. *Provide Multiple Means of Action and Expression.* The third guideline addresses the expression of learning by students, including demonstration, writing, drawing, and speaking, as well as the actions students make while learning. The mathematics process standards are a good guide for student engagement. Students should use tools, make representations, communicate with others, and solve problems with multiple methods. What options can teachers provide for students constructing geometric objects or making graphs? What are the options for students to explain their reasoning and to question the reasoning of others? How can students' executive functions be supported for self-monitoring, strategy development, organizing information, and managing memory tasks?

A seventh-grade lesson is about constructing angles with various attributes. Tools for these constructions include rulers, compasses, and protractors, and their virtual alternatives on classroom tablet computers. Students have access to a range of papers and rulers, compasses, and protractors are available with adapted capabilities (e.g., three-sided ruler, safety compass, and alternative-form protractor). When asking students to discuss the difference between an acute and obtuse angle, the teacher provides graduated questions rather than asking for the entire explanation at once. Can you name these two parts of the angle (pointing to an arm and vertex)? What units do we use to measure angles? How are these two angles different? Can you use your angle vocabulary to describe this one? For assignments and assessments on angle concepts, students have options such as reading and solving word problems, making constructions, demonstrating to others, and giving an oral explanation or report. This classroom provides multiple means of expression for students to demonstrate what they have learned about angles. Table 6.1 illustrates mathematics examples and enhancements using UDL guidelines.

UDL and variants, such as universal design of instruction, have the potential to reduce or eliminate barriers to learning for students with learning differences. However, actual research on the effectiveness of UDL has been scant (King-Sears, 2014; Rao et al., 2014; Smith et al., 2019). Most articles about UDL provide suggestions for curriculum design and teacher training, or investigate perceptions about UDL, but lack evidence regarding implementation effectiveness. Lack of definitional consensus for what constitutes a universally designed intervention and lack of agreement on the combinations and intensity of UDL elements has hampered research as well (Rao et al., 2014; Smith et al., 2019). In two literature reviews, Rao and colleagues (2014, 2020) located one study out of 20 that included mathematics, finding that applying UDL design in a preschool setting improved learning outcomes in literacy, mathematics, and social skills (Lieber et al., 2008). Root and colleagues (2018, 2020) investigated the use of UDL mapping onto interventions for students with intellectual disability or autism. The guidelines assisted the interventionists in removing learning barriers and supporting learning so these students could be successful with real-life mathematics problems.

Researchers have also investigated an observation instrument for evaluating the implementation of UDL in classrooms and schools, the *UDL Observation Measurement Tool* (Basham et al., 2020) and universally designed technology hardware and software to support science and mathematics (STEM) learning (Izzo & Bauer, 2015). Other researchers have demonstrated that special and general education teachers can develop lesson plans using UDL guidelines to address learner variability. Spooner and colleagues (2007) compared experimental and control teachers' lesson plans after a one-hour training session on UDL. This simple introduction to the guidelines enabled teachers in designing lesson plans accessible for all students.

Table 6.1 Mathematics Examples for UDL Framework

Multiple Means of Engagement	Multiple Means of Representation	Multiple Means of Action and Expression
Options for Recruiting Interest	*Options for Perception*	*Options for Physical Action*
Build in choice within activities and assignments	Display worked examples via projector or on desks	Act out math word problems to demonstrate the situation
Plan relevant activities based on student interest and backgrounds	Use multiple models such as number lines, diagrams, and concrete objects	Employ virtual manipulatives and tools
Build in motivating aspects within math instruction	Use alternate representations for measurement, data analysis, and geometry	Work with a range of math models, in physical and virtual formats
Options for Sustaining Effort & Persistence	*Options for Language & Symbols*	*Options for Expression & Communication*
Establish a classroom climate that allows for errors and risk-taking	Use clear mathematics vocabulary and notations, supporting those with visuals and examples	Promote student-to-student communication, justification, and critique
Keep guided instruction at a challenging level with supports	Explicitly teach vocabulary and symbols	Vary individual and group responses during practice
Encourage student self-charting progress	Represent information from word problems in diagrams	Provide alternative response modes (oral, written, demonstrations)
Options for Self-Regulation	*Options for Comprehension*	*Options for Executive Functions*
Encourage self-assessment	Teach for transfer by incorporating context-based word problems	Activate prior knowledge through review, concept mapping, other strategies
Guide goal setting	Highlight patterns and connections (e.g., multiples, factors, exponents)	Promote strategy development for memory and organizational tasks
Teach self-monitoring skills	Scaffold student understanding through think-alouds, questions, dialog, and demonstrations	Facilitate information and resource organization

There is convincing evidence about specific options for learners within a UDL framework, such as the use of concrete models to support abstraction (Witzel et al., 2001), the use of graphic organizers to support executive processes of learning (Dexter & Hughes, 2011), and encouraging student verbalizations, such as think-alouds, during instruction with immediate corrective feedback (Tournaki, 2003). These elements and others verified in effective-instruction literature for students with mathematics learning difficulties (MD) can be incorporated with some confidence as options or adaptations within a UDL curriculum. It may be that UDL remains a broad, curriculum-design framework with research focused on effective elements for specific student characteristics. Additionally, some studies, such as those for multi-tiered systems of support programs, may not explicitly use UDL language when describing the design of core curricula and instruction (e.g., Tier 1), but are essentially implementing universally available differentiations and options when possible, and individualized adaptations when necessary.

TRY THIS

Apply the UDL framework to one of the fraction lessons in Chapter 4. What elements could be added to the lesson to provide multiple means of engagement, representation, and expression?

Differentiated or Specially Designed Instruction

Differentiated instruction is certainly not a new concept but is increasingly important as classrooms become more diverse. Drawing from previous concepts of individual differences and individualized instruction, differentiated instruction describes instruction that is planned and delivered so that all students within the classroom benefit. As with UDL, differentiation provides multiple means of access and support for learning with options or variations for providing content, practice, feedback, and assignments within the same grade-level topics. Differentiation should be planned when diverse students are instructed in the same venue, including core mathematics classes, intensive-intervention groups, and special education settings.

The term *differentiation* with respect to classroom instruction can be traced to Great Britain's Department of Education and Science's surveys of the late 1970s and the subsequent documents on curriculum, such as *Mixed Ability Work in Comprehensive Schools* (DES, 1978). The report summarized:

> Where the level, pace, and scope were inappropriate, this was mainly because teachers failed to provide sufficient differentiation, or sometimes any differentiation at all, in the work required of pupils of different abilities, whether they were using class teaching methods, "individual" assignment methods, or both. It was surprising to find that in a large number of cases mixed ability classes were taught as though they were homogeneous groups.

(DES, 1978, p. 49)

Differentiated instruction involves understanding student strengths and needs when planning instruction and implementing instruction that differs based on those student profiles (Tomlinson, 1999). Tomlinson identified three components of instruction that should be differentiated: content, process, and products. An example for differentiating mathematics content would be to provide options within class practice such as: *Simplify one of the following expressions:* A) $3x + 4x$, B) $6 + x - \frac{1}{2}x$, or C) $9 - 6(b^2 - b - 1) - 8b^2$. Differentiating the process of instruction might include tiered or parallel activities, individual learning contracts, flexible grouping, and peer tutoring. Product differentiation includes allowing students to demonstrate their learning through options for products and assessments.

Key components of a differentiated classroom have been identified as student choice, flexibility, on-going assessment, creativity in planning instruction, and a strong understanding of student prior knowledge, abilities, and learning needs (Anderson & Algozzine, 2007). Tomlinson (2003) described differentiated instruction as designed to provide various learning opportunities to students who differ in their readiness levels, interests, and profiles. By differentiating instruction, Tomlinson argued, teachers can challenge all students by providing a range of levels of difficulty, vary the degree of scaffolding required, and provide variety in the ways students work.

Small (2009) emphasized two primary strategies for differentiating instruction in mathematics: open questions and parallel tasks. An open question is broad, related to a big idea, and allows for a range of responses. A closed question example: *What is half of 20?* An open question: *Show the number 20 in as many different ways as you can.* Parallel tasks are sets of two or three tasks about the same important concept at different developmental levels but allow for whole-group discussion. Examples of parallel tasks for the concept of number patterns: A. *The sum of the first 10 terms of a pattern is a multiple of 3. What could the pattern be?* B. *The sum of the first 10 terms of a pattern is 195. What could the pattern be?* (Small, 2009, p. 142). Tasks and questions like these promote deeper, more engaged thinking while allowing options and differing levels of entry.

TRY THIS

Find a closed question in a mathematics text and turn it into an open question. For example: *Which fraction is larger, $\frac{5}{6}$ or $\frac{7}{8}$?*

Specially designed instruction (SDI), a term referring to special education under the Individuals with Disabilities Improvement Act (IDEA, 2004), goes beyond access to core classroom instruction. It is instruction based on individual needs and is different from the general curriculum (Rodgers & Weiss, 2019). SDI also addresses content, methods, and instructional delivery, and will look different for each student (Riccomini et al., 2017). Rodgers and Weiss (2019) articulated SDI for secondary mathematics co-taught classes when individual students may lack prerequisite skills, require different instructional delivery to master skills, need more examples or smaller instructional units, have difficulty with abstract representations, and need cognitive strategy instruction, among other individualized concerns.

Some teachers will resist specially designed or differentiated instruction and continue to teach to one level of students within the classroom, leaving some students unchallenged while others continue to struggle. Often-cited barriers include time for evaluating student needs and planning instructional options, lack of professional development and support for teachers, the view that differentiation is simply another fad, concern about classroom management issues, and lack of expertise in formative assessment strategies (Logan, 2011). Some teachers view differentiation as applicable only to students who have learning deficits, when those who are performing above grade-level expectations also need differentiation. Many teachers point to high-stakes tests that all students are expected to take at grade level as driving curriculum decisions, in effect standardizing both curriculum and instruction. Lynch and colleagues (2018) challenged middle-school mathematics teachers to differentiate instruction while maintaining productive struggle to learn mathematics meaningfully. The authors warned about pitfalls of differentiation: providing hints or formulas to students removes the cognitive demands of tasks, focusing on procedures rather than concepts will also inhibit student growth. Teachers should examine the underlying mathematics concept, prerequisite skills, and potential barriers for individual students. Then they can develop appropriate supports and structures (questions, feedback, discussions, options for engagement) that will promote meaningful learning.

Research on the effectiveness of differentiated instruction within core mathematics classes is scarce. As with research on UDL, differentiation has been called by many descriptors over the years so identifying research with specific and limited instructional characteristics is difficult. Do we include all accommodation, enrichment, response-to-intervention, and inclusion studies in diverse core mathematics classrooms? How do the studies measure actual differentiation decisions and their impact on student achievement and control those differentiations in comparison classrooms? Not easily accomplished. Differentiation in the classroom will occur regardless of what the teacher does (Ollerton, 2014). Students respond at different rates, with different levels of confidence, different outcomes, and different perspectives. Effective differentiation is actually planned implementation of the principles of effective instruction: know your students, provide instruction about important concepts in ways that connect with prior knowledge, provide learning supports such as scaffolding and feedback, continuously assess student understanding, and engage student thinking. Effective differentiation is not simply providing choices for students, as many would choose those within their comfort zones and remain at those levels. Teachers must challenge students based on their instructional levels and scaffold student learning to the next levels.

A word about student *profiles* is warranted. As Landrum and McDuffie (2010) reminded, matching instruction with students' strengths and needs is a hallmark of special education and this differentiation based on student-learning characteristics is necessary. Student profiles include information about academic progress, prior knowledge, background, interests, behavioral and emotional concerns, and strategy use. However, many teachers confuse the concept with *learning styles* (modality-based instruction), an approach that has little research support but continues to receive "gut or intuitive" support from those less informed. In a meta-analysis of studies of modality-based instruction, Kavale and Forness found effect sizes of 0.144 (slightly higher than chance) in a 1987 review of 39 studies. The primary research on learning styles, conducted by Rita and Kenneth Dunn and their colleagues and students, has been criticized as lacking rigor, requisite description, and independent replication (Kavale et al., 1998; Kavale & LeFever, 2007). Even Tomlinson (2009), who is most prominent among educators promoting differentiation, cautioned that learning profiles (including interests and preferences) should not take precedence over students' readiness for determining instructional levels.

What is problematic about using a learning styles (LS) approach to instruction? First, the approach is based on a faulty premise that the cause of learning difficulties is a mismatch between the student's learning style and the instructional method. Second, the LS approach assumes that information is processed in isolated parts of the brain and that processing remains in only one sensory modality. In fact, separate structures in the brain are highly interconnected and there is enormous cross-modal activation and transfer of information between modalities during learning (Gilmore et al., 2007). Third, LS advocates claim that students can be assessed for preferred learning styles and should be taught in a matching teaching method. For example, Çarbo (1988) suggested that global/tactile/kinesthetic learners who were poor readers should be taught with holistic reading methods and not phonics, that phonics instruction is for students who can already discriminate easily among sounds. Emphasizing holistic methods with younger and remedial students is contrary to virtually all other academic-intervention research. Fourth, other instructional methods, such as explicit instruction, cognitive-strategy training, and mnemonic strategies, have achieved much higher gains (effect sizes from 0.75 to 1.62) than learning-styles matching. Finally, attempting to implement instruction matched to each student's preferred style would be difficult at best. When little or no achievement gains could be expected, it would be hard to justify the extra effort. Kavale and Forness (1987, p. 237) reminded educators that, "all modalities are strongly involved in the learning process, and academic achievement without the inclusion of all modalities is virtually impossible (for a child without sensory deficits)." A learning-styles approach to teaching and learning underestimates the significance of learning difficulties and offers only surface-level, simplistic instructional interventions.

It's difficult to convince parents and the general public that something so intuitively appealing as "I'm a visual learner" or "I'm an auditory learner" has no basis in educational research related to achievement. Even teachers with professional training often hold to brain-based learning myths such as learning-styles approaches. In a study of 242 teachers in the United Kingdom and the Netherlands, Dekker et al. (2012) found more than 90% believed the myth, "Individuals learn better when they receive information in their preferred learning style (e.g., auditory, visual, kinesthetic)," the *neuromyth* with the greatest percentage of incorrect responses of 15 myths about learning and the brain. The researchers found that teachers were interested in brain research but had a difficult time distinguishing fact from myth. They recommended more collaboration and communication between neuroscientists and educators, especially within professional development experiences.

To simulate a mathematics task requiring executive-process resources, try the following exercise:

> You will be asked to recall a list of eight mathematical terms. You can view the list for 15 seconds, then please cover and write them out. You may use only *one* of the following memory methods: visualizing the term's concept, saying the term to yourself, imagining the term in a context, making a connection with another term, making an image in your mind of the word, or tracing the letters of the word. Ready? Go! Terms: median, scalar, coefficient, trapezoid, sine, vector, quartile.

Were you able to use only one modality for this memorization task? Most likely, when you read the term *coefficient* you visualized the length of the word, said it to yourself, noted the prefix co-, recalled related concepts of constant and variable, and attended to the spelling. You may have even thought of an example from previous experience (e.g., the 3 in 3*x*). Skilled learners bring multiple resources to a learning situation and that's what we should teach our students.

Accommodations

Accommodations for instruction are changes in the way a student accesses learning, without changing curriculum standards. (Adaptations that change standards are *modifications*.) Accommodations for instruction and assessment are necessarily interconnected. For instruction, accommodations can be classified as presentation modes, response options, material formats, and setting characteristics. Many accommodations can be designed universally, allowing all students the option to use them for learning. Some are specific to student needs, such as large print for students with visual impairments or the use of a notetaker for students with fine-motor difficulties. Box 6.1 provides examples of instructional accommodations typical for mathematics classes. Accommodations should be student- and setting-specific, considering the student's profile, classroom, and mathematics content.

Individual accommodations should be developed and documented during the individualized planning process for students with disabilities (SWD), for the individualized education program (IEP) or a 504 plan. For students with IEPs, those documents should provide substantial information about the student's present level of performance and how the disability affects his or her progress in the general education curriculum. The IEP must include statements "of measurable annual goals ... to enable the child to be involved in and make progress in the general-education curriculum" and "a statement of the special education and related services and supplementary aids and services, based on peer-reviewed research to the extent practicable, to be provided for the child to advance appropriately toward attaining the annual goals" (IDEA, 2004, 20 USC § 1414 (d)(1)(A)). Section 504 requirements are even more compelling:

> No otherwise qualified individual with a disability in the United States ... shall, solely by reason of her or his disability, be excluded from the participation in, be denied the benefits of, or be subjected to discrimination under any program or activity receiving Federal financial assistance.

(Rehabilitation Act of 1973, 29 USC§ 794 (a))

Box 6.1 Instructional Accommodations for Mathematics Classes

Presentation
- Desk copies of projected materials
- Simplified language
- Practice pages with fewer items per page
- Partner review and reciprocal tutoring
- Word problems read aloud (in person or via computer)
- Vocabulary highlighted, pretaught
- Written directions read aloud
- Pop-up glossary
- Graphics or images to supplement concepts
- Prompts and encouragement to maintain focus
- Smaller segments of instruction
- More frequent feedback
- Advance outline

Setting
- Assigned seating, grouping
- Reduction in distractions
- Self-regulation prompts, timers
- Time of the day/schedule adjustments

Responses
- Oral explanations
- Demonstration
- Multiple choice rather than open end
- Computer-based responses
- Recordable white board or tablet
- Speech-to-text software
- Note-taking technology
- More time to complete assignments
- Choice for assignments and modes of response

Materials★
- Virtual manipulatives
- Safety tools
- Enlarged materials (blocks, grid paper)
- Calculator use
- Extra textbook (print or home version)
- Visual aids (highlighters, templates)
- Tables, counters, 100s chart
- Graph paper or specially lined paper
- Schematic diagrams

Note: ★ See Chapter 7

Since these regulations have been enacted, there have been a number of court cases addressing the free and appropriate public education (FAPE) for SWD (see Yell et al., 2017 for discussion). One of the most persuasive was *Endrew F v. Douglas County School District* (2017), in which the Supreme Court rejected the appeals court ruling that slightly more than *de minimis* benefit of the provided education under an IEP was sufficient. The Court's opinion stated, "To meet its substantive obligation under the IDEA, a school must offer an IEP reasonably calculated to enable a child to make progress appropriate in light of the child's circumstances…" (137 S. Ct. 999). "…for most children, a FAPE will involve integration in the regular classroom and individualized special education calculated to achieve advancement from grade to grade" (137 S. Ct. 1000). Students with disabilities not only have protected access to the general education curriculum, but also to the supports and services needed to be successful within that curriculum. If teachers do not implement the provisions of IEPs or 504 plans, the school district can be found in noncompliance with the respective federal laws.

Students and their parents or caregivers should play a significant role in selecting and evaluating instructional accommodations. Students should have a role in selecting accommodations, to understand those, and be able to self-advocate for needed accommodations. Teachers who have not had training in legal and instructional aspects of classroom accommodations will need professional development and mentoring. Any assessment accommodations for mandated testing programs should also be used for other assessments and related instruction within the classroom.

All classroom adaptations, whether designed universally, provided as accommodations for individuals, planned as differentiated instruction, or part of instruction provided within a

MTSS, have the common goal of making instruction accessible and effective for all students. Mathematics, special education, and instructional-support teachers should work together to tackle time-consuming tasks of developing student profiles, planning adaptations within the core curriculum, and assessing student progress. Administrators should support professional development and time for necessary planning and collaboration.

Student Grouping

Beyond the instructional planning discussed in Chapter 4, teachers in core and specialized mathematics settings must plan special instructional arrangements. These include grouping students within the classroom based on curriculum goals, student-learning needs, teacher resources, and lesson-based activities. Mathematics topics lend themselves to various student groupings for learning activities. Students solve problems, conduct investigations, and present data-analysis findings in groups. When engaged in problem solving, they justify their reasoning and challenge the reasoning of others within small groups. Students may even provide segments of guided instruction and practice for their peers. The most common peer-mediated instructional approaches are cooperative learning and peer tutoring, both of which take on several versions.

Cooperative Learning

Cooperative learning involves small, heterogeneous groups of students working together on common learning objectives and is based on social interdependence theories (Johnson & Johnson, 1992), as well as theories of motivation, cognitive development, and cognitive elaboration (Slavin, 2017). Cooperative-learning tasks can be short (a few minutes), long term (over the course of a unit), or somewhere in between. Johnson and Johnson posited that five variables mediate the effectiveness of cooperative-learning approaches: "positive interdependence, individual accountability, promotive interaction, the appropriate use of social skills, and group processing" (2009, p. 366). In describing a unified theoretical model of cooperative learning, Slavin (2017) outlined elements that contribute to achievement: group goals that require members to encourage others, individual accountability built into group tasks, elaborated explanations among all group members, and interdependent roles in structured cooperative models. Some examples of cooperative-group activities in mathematics include solving a problem, conducting a survey for data analysis, contrasting various geometric properties using objects, preparing statistical results for the class, developing an alternative algorithm, and creating a new word problem.

Cooperative learning has one of the most extensive research bases of any instructional approach, dating back to the 1970s. It offers a classroom approach for very diverse groups of students that addresses academic achievement as well as interpersonal skills (O'Connor & Jenkins, 2013). However, research involving SWD has been mixed (McMaster & Fuchs, 2002) and its effectiveness may well depend on other factors such as other instructional elements (e.g., strategy instruction, modeling), setting, group size, progress monitoring, teacher skill in scaffolding and guiding group work, and prerequisite skills. Further, although academic gains of low-achieving students have been reported in research on cooperative learning (Slavin, 1995), those gains may not be commensurate with that of nondisabled peers (McMaster & Fuchs, 2005). Many research-based interventions for students with learning needs include structured group tasks, group activities with concrete materials and representations, and even group gaming, for brief episodes within a longer intervention session. These are essentially cooperative or peer-mediated sessions, structured and guided by the teacher, that afford increased practice and response, communication about concepts, and a change in the lesson's format to boost interest and maintain attention. Often these segments are not extracted from the overall intervention effects within research studies.

Some of the potential benefits of cooperative learning include increasing levels of student response, improving communication skills with regard to mathematics, developing appropriate social skills, building confidence and positive dispositions with mathematics, increasing achievement levels, and improving problem-solving abilities (Bertucci et al., 2010; Tarim, 2009). A large-scale study of the benefits of cooperative learning in mathematics found a small positive effect on achievement and significantly more positive attitudes about mathematics among eighth graders (Smith et al., 2014). Potential benefits for SWD include increased self-esteem, social skills, achievement, and participation levels (O'Connor & Jenkins, 2013). Students with disabilities may need special partner or group assignments, preinstruction in cooperative skills, task modification, additional teacher guidance, and systematic monitoring to be successful in cooperative groups (McMaster & Fuchs, 2005; O'Connor & Jenkins, 2013).

Formats of cooperative-learning models vary and include the jigsaw approach, group investigations, competitive teams, academic controversy, team-accelerated instruction, complex instruction, and structured dyadic methods (Johnson & Johnson, 2009; O'Connor & Jenkins, 2013; Olsen & Platt, 1992; Slavin, 1995). In each model, the teacher assigns student roles, specifies learning objectives, structures the group process, facilitates student interactions, and evaluates group and/or individual performance. If students are not experienced in working in groups, teachers may decide to begin with pairs before attempting groupings of three or four students (Bertucci et al., 2010). Topping and colleagues (2017) identified *constructive interactions* as the central element of cooperative-learning methods and encouraged teaching students cooperative skills and requiring students to reflect on group processing. They also encouraged generic or content-free cooperative-group structures, so that the episodes use familiar procedures.

There are many ways group efforts fail when teachers do not structure them carefully and provide assistance as needed. Slavin (1995) warned about the *free rider* effect in which some group members do all the work. This problem can be addressed with diffusion of responsibility strategies. Others are concerned about overemphasis on competition, content integrity, and stagnant student groupings—all of which should be addressed with careful planning and knowledge of students. Low task engagement in cooperative-group arrangements can be the result of teacher guidance that lacks clear and explicit assistance (Cohen, 1994). Setting up, monitoring, and evaluating cooperative-group work is a complex endeavor (Todding et al., 2017), but with careful planning and guidance, with attention to individual-student needs, this instructional approach has the potential to benefit all the students involved. Further, the ability to work with others to solve problems and create solutions is a hallmark of today's workplace.

Peer Tutoring

Peer-tutoring programs, also termed *peer-assisted* and *peer-mediated instruction*, is a cooperative and structured dyadic method involving one student assisting another in the acquisition or reinforcement of specific skills. Students may be of the same or different ages and students with disabilities can be the tutor, tutee, or take turns in each role (reciprocal tutoring). Peer tutoring may be an informal pairing of students for a one-session skill review or a highly structured class-wide system. Peer tutoring has been demonstrated to be effective with SWD in increasing academic performance and improving social acceptance, school behaviors, and mathematics self-concept (Alegre et al., 2019; Leung, 2015; Moliner & Alegre, 2020). Fundamentally, peer tutoring seeks to increase the amount of academic learning time, active responding, and guided practice with feedback in a teacher-monitored setting.

Alegre and colleagues (2019) examined peer tutoring in mathematics Grades 1 to 6 reported in 51 studies primarily addressing arithmetic skills. The average effect size was large ($d = 0.89$), half the studies examined the effects for MD, and average effect sizes were very similar for MD and non-MD students. Morano and Riccomini (2017) examined peer-tutoring interventions involving SWD at middle- and high-school levels. The researchers targeted higher-order learning (HOL) objectives because much of the research has been with lower-order skills such as vocabulary, math facts, and computation. Of the nine studies examining HOL using peer tutoring, only three targeted HOL objectives for mathematics—for problem-solving skills and mathematics concepts and applications. Of those only one reported positive effects on learning as compared with traditional instruction (Roach et al., 1983). The researchers cautioned practitioners about using peer tutoring with HOL in mathematics for secondary SWD due to lack of research support and heavy demands on time and other resources for preparation and implementation. In fact, the tutoring interventions may have overlooked students' lower-order deficits that would have benefited more from this approach.

Class-wide peer tutoring (CWPT) has its roots in the Juniper Gardens Children's Project (University of Kansas), which is an intra-class, reciprocal peer-tutoring model for an inclusive elementary setting (Delquadri et al., 1983). This model includes peer tutoring three to five days a week for 30 minutes each in reading, spelling, and mathematics. CWPT is characterized by its highly structured procedures, weekly competing teams, daily point posting, and direct practice of target instructional objectives. Effective CWPT programs require planning and resources: teacher and student training, appropriate curriculum materials, and an ongoing monitoring system, but the benefits of high levels of student engagement and increased academic achievement for both partners are compelling for teachers in diverse classrooms (Maheady et al., 2003).

Other CWPT models have been developed since the Juniper Gardens project, including The Ohio State Peer Tutoring Program, Class-wide Student Tutoring Teams, Reciprocal Peer Tutoring, and Peer-Assisted Learning Strategies (Maheady & Gard, 2010; Morano & Riccomini, 2017). Peer-Assisted Learning Strategies (PALS), developed by Doug and Lynn Fuchs and colleagues at Vanderbilt University, includes a component for K-6 mathematics that supplements the existing mathematics curriculum. The researchers were concerned with the effects of a class-wide tutoring program on all students in the classroom, not just those identified as low achieving, and the limitations of some programs to basic-fact and computation tutoring. In the PALS model pairs of students are matched based on their skill profiles from frequent curriculum-based measures. The tutor in each pair models and fades a verbal rehearsal delineating the steps for solving a problem. Then the tutor provides elaborated feedback for the tutee on each step of their practice, with both confirmation feedback and additional modeling as needed. Both the tutor and tutee are engaged in a high level of interaction during these sessions. In a study of PALS on first-graders' mathematics achievement, 20 general classroom teachers implemented the program three times a week for 16 weeks, with student pairs changing every three weeks and student roles within pairs (tutor and tutee) changing midway through each session (Fuchs et al., 2002). Content for tutoring was taken from the district's curriculum and included number knowledge, simple addition and subtraction problems and stories, place value, and two-digit addition and subtraction. This highly structured program resulted in stronger achievement for low, average, and high-ability students than for students in the contrast classrooms. More information about the PALS model can be found at the Fuchs Research Group website, Vanderbilt University.

Dozens of empirical studies have been conducted on CWPT, with similar outcomes for academic and behavioral effects by a number of different researchers across a wide range of content areas and educational settings (Alegre et al., 2019; Leung, 2018; Maheady & Gard, 2010; Topping et al., 2017). Students with low ability, high ability, average ability, disabilities, and English language learners have benefitted from these programs. However, limitations and potential problems with peer tutoring must be acknowledged. Most studies targeted lower-level skills for tutoring, so more study on concept understanding is in order. For mathematics, most research has been conducted at the elementary level. Teachers need to monitor the achievement of all students participating in CWPT programs, because some studies have indicated stronger effects for students beginning with lower achievement levels (e.g., VanDerHeyden & Codding, 2015). Other concerns include overuse of some students as tutors, lack of training of tutors in appropriate techniques, using peer tutoring for activities and instruction that should be provided by the teacher, and failing to monitor the success of the peer-tutoring effort (Olson & Platt, 1992). For CWPT to be successful, students should be trained and know what is expected, materials should be based on curriculum standards, tutoring sessions should be highly structured, teachers should monitor student progress closely, and parents should understand the benefits of the sessions for their children.

Challenges in Secondary Settings

Secondary settings, especially those for Grades 9 to 12, pose special challenges for students who struggle in mathematics. Issues such as credit units for graduation, scheduling options, and college and career preparation present challenges for teachers and students unique to this level. Students who have persistent difficulties in mathematics may have even wider gaps in mathematics learning and a longer experience with frustrations and failures. Weighing on students and their parents are the grade-point-average mandates, end-of-course exams, credit hours, and college-entrance exams that often dictate curriculum design.

Since the publication of *A Nation at Risk* in 1983 (NCEE), high-school requirements for mathematics have increased to three credits in most states, with four credits the trend since the *Common Core State Standards for Mathematics* (CCSSM) were introduced (NGA & CCSSO, 2010). By 2018, 17 states required four mathematics units and 27 states required three units, with others depending on local requirements or proficiency measures (Macdonald et al., 2019). The CCSSM recommended preparing students to take Algebra I by eighth grade, but allowing for Algebra I, Geometry, and Algebra II, or their integrated three-course counterparts, during high school. States that require four units typically include an elective, an upper-level mathematics course such as statistics, precalculus, or trigonometry, an integrated course such as STEM studies, or an applied course such as computer science or engineering. Each state's high-school requirements are different. Teachers, students, and parents should begin planning no later than seventh grade for a sequence of courses that will allow students to master concepts needed for success in their high-school studies. The CCSSM standards actually pack the middle-school grades (6–8) with "the highest priority content for college and career readiness" such as applying ratio and proportional reasoning, computing fluently with the range of rational numbers, and solving real-world geometric problems (CCSSM, 2010, p. 84).

Simply increasing the number of courses in mathematics has not increased students' preparation for college and careers (Buddin & Croft, 2014). There has been little effect on student achievement or college enrollment from the increased course requirements, especially for those students with lower abilities and less motivation. "For students to succeed in these [high school] courses, they must possess the necessary prerequisite skills to then take advantage of the advanced material. Efforts should be focused on early preparation [prior to high school]

... and targeted remediation" for students already in high school (Buddin & Croft, 2014, pp. 30–31). In a study of PISA data, Cogan et al. (2019) found the coherence and rigor of the mathematics curriculum across the grade levels and the opportunity to work with challenging mathematics problems in applied contexts was significantly related to students' mathematics literacy and preparation for college or career.

Alternative Scheduling

Another challenge for high-school students is the block schedule with alternating semesters or alternating days of longer class periods, sometimes termed 4 × 4 or A/B schedules. Rather than the traditional 50-minute periods, these classes span 85 to 90 minutes. The 4 × 4 schedules may result in students having Mathematics I in fall one year, and Mathematics II in the spring of the following academic year, a gap of 12 months. Students who take Algebra I one year may have a two-year gap before taking Algebra II. Another variation is the Copernican block, with longer classes for some subjects such as mathematics and English and then shorter class periods for other courses, switching schedules every month or two. The longer class periods within these block schedules have resulted in some teachers lecturing too long during the period or providing too great a portion of class time for homework. Further, research has been unconvincing regarding the merits of block scheduling for student achievement.

Most block-scheduling schemes were designed for management reasons—teacher preparation time, reduction in misbehavior during class changes, special coursework that requires longer periods (e.g., labs and field work), and increased credits earned per year. There is some evidence that secondary students should be involved in continuous development in mathematics, rigorous coursework over four years without major gaps, in order to be prepared for college. One study, analyzing National Assessment of Educational Progress data, found that students enrolled in 90-minute blocked classes scored two-thirds of grade-level lower in achievement than students enrolled in 50-minute yearlong courses (Zelkowski, 2010). The study also determined that college-bound students who did not take mathematics all four years in high school diminished their odds of completing a bachelor's degree by 20 percent.

Other studies on the effectiveness of block schedules on student performance have been mixed. Hackney (2013) found no differences in student performance on the Prairie State Achievement Exam (Illinois) between schools on a block schedule as compared with those on a traditional schedule. Bottge and colleagues (2003) found no differences in achievement (all content areas) for students with or without disabilities for traditional or block schedules. Murray and Moyer-Packenham (2014), in studying Algebra I achievement scores, found that 7th through 9th grade students performed better with 3/3 schedules (daily 70-minute classes over three trimesters) while 10th and 11th graders performed better with daily traditional (50 minute) schedules, as compared with block schedules. The researchers emphasized the need for daily, in-person mathematics classes for the "gateway" Algebra I course. There has been little research in the past decade on these scheduling options and their effects on achievement. It may be more important for teachers to understand how to design effective instruction for the longer class periods because what is accomplished within classrooms has the most impact on student achievement.

There are disadvantages for students with MD with block schedules, including long gaps between courses, the requirement to focus attention for long periods of time, organization and planning issues for students with confusing schedules, problems with transfers from other schools, and the effects of school absences (one day of absence is essentially two days of class). With careful planning, there can be benefits for these students. Additional special education or support classes could be provided for 40- to 45-minute sessions on alternate days or semesters, sometimes called a flex period. Co-teachers could staff the longer class period and segment

Table 6.2 Segmenting a Block Schedule Mathematics Class

Time	Grouping	Mathematics Teacher	Special Educator
10:00–10:05	Individual seats	Problem on board Distribute materials	Check student notes
10:05–10:15	Larger group + smaller group	Review previous algebra concepts with larger group, solve one problem together	Review previous algebra concepts with small group, emphasis on strategies, notation, vocabulary
10:15–10:30	Whole group, team-teaching	Introduce new algebra concept, model examples	Model additional examples using think-aloud and strategies
10:30–10:45	Small groups	Assign small group problems, move among groups for assistance	Work with same small group on problem, scaffold understanding
10:45–10:55	Small groups— report outs for discussion	Prompt discussion and questions	Clarify language and scaffold understanding as needed
10:55–11:10	Whole group, team-teaching	Introduce related geometry topic, model examples	Model additional examples using think-aloud and strategies
11:10–11:25	Small groups or stations— applications with geometry concept	Assign small group applications, move among groups for assistance	Work with a different small group on application, scaffold understanding
11:25–11:30	Whole group	Provide closure, assignments	Provide quick review

the period with review, direct instruction, problem solving, practice, small-group work, and re-teaching or extension portions, taking advantage of every instructional minute but with alternating types of engagement.

Table 6.2 illustrates a 90-minute mathematics class for ninth graders, with a mathematics teacher and special educator co-teaching one group of 25 students. This classroom should be set up with flexible seating and materials at hand. It is important for teachers and students to remember that the block class is equivalent to two regular class sessions, with the same work-load and content requirements. These teachers could decide to set aside 20 to 30 minutes of the 90-minute period for extension lessons for some students and review and reinforcement lessons for others, as long as the curriculum standards are addressed across all class sessions for the year.

High-school settings provide many challenges, including tough academic standards and difficult parameters for coursework. But high schools are rewarding environments if teachers plan together and join resources to serve students' individual needs.

Mathematics Textbooks

Textbooks have long been extremely influential on the mathematics content that students learn as well as how teachers plan and provide instruction. In fact, many teachers are reluctant to modify textbook-based instruction, opting to follow texts as they are written (Lee, 2016). While curriculum standards have garnered political attention, those are the goals for instruc-tion at each grade level, not instructional guides like textbooks. Textbook selection was an increasingly centralized decision with the advent of state curriculum standards and testing requirements of legislation (Mark et al., 2010), but textbook adoption rules have loosened recently, providing districts more flexibility (Thompson, 2017). In 2017, there were 19 states, primarily in the south and west, that had state-wide adoptions, with some customized to the state. In many of those states, districts can still make off-list purchases, adopt digital textbooks, or use open educational resources (OER). The Every Student Succeeds Act (2015) encouraged the use of funds for OER for digital learning and open courseware and the US Department of Education established the #GoOpen campaign (https://tech.ed.gov/open). The Council of Chief State School Officers reported that 20 states had joined this initiative by 2018, purporting to increase equity, keep content relevant, save money, and empower teachers (Tepe & Mooney, 2018). With this flexibility comes the risk of schools and teachers using unvetted and low quality materials to guide classroom instruction. It will become increasingly important for districts to evaluate the effectiveness of alternative curricula with the range of students they serve.

Most textbook and other curriculum material selection criteria address the support of diverse students' learning. For students who struggle with mathematics, have specific disabilities, or are English language learners, this support is usually in the form of suggestions for building vocabulary, adaptations or adjustments for activities within lessons, and "differentiation" boxes within teachers' guides. They may lack essential criteria such as explicit instruction, scaffolding, progress monitoring, and judicious review. Often the research evidence for curriculum materials under consideration does not include specific student subgroup results, so educators must examine materials with student differences in mind and monitor student progress once the programs are in use.

The What Works Clearinghouse (WWC) was established by the US Department of Education in 2002, a consolidation of the previous ERIC (Educational Resources Information Center) clearinghouses and an attempt to provide a central source of scientific evidence of what works in education (https://ies.ed.gov/ncee/wwc). One portion of the WWC is concerned with conducting systematic reviews of K-12 interventions (programs, practices, and policies) and disseminating information about specific interventions. The WWC has review teams for primary and secondary mathematics. Among several core mathematics programs found to have positive effects were: *Everyday Mathematics* (Grades 3–5); *Investigations in Number, Data, and Space* (Grades 1–5); and *University of Chicago School Mathematics Project* (Grades 7–10, algebra). However, the WWC intervention reports on these mathematics programs do not include information on student subgroups, so educators must examine the supporting research studies.

Another curriculum-review center, EdReports (www.edreports.com), is a nonprofit organization that solicits educators who will collaborate on reviews. Materials for each grade level are reviewed with an emphasis on alignment with CCSSM or state standards. For example, the Math K-8 rubric includes three indicators: focus and coherence, rigor and mathematical practices, and instructional supports and usability. The first two indicators, with their subindicators, focus on standards alignment, processes for developing concepts and procedures, and mathematical practices. Only the third indicator (usability) addresses facilitating student learning through sequencing, practice, using representations, employing technology, working with parents, research-based approaches, student misconceptions, scaffolding, feedback, differentiation, and assessing student performance. Educators must read the third indicator carefully, with all the pull-down notes, to understand how student learning is actually addressed. As one of several such examples, the *Utah Middle School Math Project* (2019, grades 6–8) scored high for the first two indicators, but only partially met standards in the third. The reviewers stated that there were no explanations or identification of research-based instructional strategies in the materials, no information for gathering students' prior knowledge, no supports or accommodations for EL or special needs students, and merely suggested teachers differentiate as needed. This indicator, of major importance for students with learning difficulties, is outweighed by standards-based indicators.

Until studies of mathematics curricula are subjected to broader quality criteria and their field studies include more diverse student populations, educators have two strategies for evaluating curriculum materials:

- evaluate curricula using additional criteria such as well-established effective instructional practices; and
- carefully monitor the implementation of any curriculum for effectiveness using frequent curriculum-based measures.

Even evidence-based curricula can be ineffective with some students or not used in ways that maintain fidelity with program design. It is the quality of instruction, not the textbook or program, that makes a difference in students' mathematics learning.

Core Mathematics Textbooks

A number of special education researchers have investigated the content and instructional methods of core mathematics textbooks. Most of these studies evaluated programs for a limited number of grade levels so the authors could examine and compare the extent to which research-based instructional practices were incorporated or how specific student learning characteristics were addressed.

Doabler et al. (2012) took an instructional-design analysis approach to evaluate three mathematics programs at Grades 2 and 4 for the presence of eight instructional principles within three topics per grade level—prerequisite skills, math vocabulary, explicit instruction, selection of instructional examples, multiple and varied practice and review opportunities, teacher-provided academic feedback, and formative feedback loops. For the textbooks examined at second grade, 44% of scores met criteria, but only 33% met criteria at fourth grade. Prerequisite skills were addressed only inconsistently, most lacked consistent opportunities for explicit and systematic instruction, and none of the textbooks offered procedures for linking formative assessments with instructional decision-making. The researchers recommended that these core curricula be enhanced by teachers for students who struggle with mathematics by making instruction more explicit and systematic, clarifying teacher-to-student interactions, checking for student understanding, and improving student practice opportunities.

Investigating sixth- and seventh-grade textbooks and their supplemental materials, van Garderen and colleagues (2012) focused on making and using representations for learning. The researchers sought to find whether textbooks provided explicit instruction about representations, math tasks that encouraged the use of representations, and instructional information about representations for teachers. Most of the instructional representations in the middle-grades texts related to content being taught rather than helping students understand representations. Two texts had no support at all for understanding representations. Most representations were related to applications rather than for instruction. In most cases representations were provided rather than supporting students in generating their own. Out of 168 lessons reviewed across eight textbooks, only 42 lessons included information for teachers for additional instructional support for representations.

Hord and Newton (2014) examined three mathematics curricula across Grades K to 5 for how they provided support for students with working-memory difficulties who are learning to solve multistep problems, focusing on finding the area of composite shapes (geometry). All three programs provided strategies for reducing working-memory difficulties, such as visual representations, activities to develop conceptual understanding, information-storage strategies, and opportunities to develop cognitive and metacognitive skills. However, teachers should be aware that these core curricula may need to be supplemented or adapted to be effective. One program didn't provide explicit instruction of thinking processes required to solve problems or direct instruction for conceptual learning early enough in the instructional sequence. Many activities in another program were too open-ended for some students, requiring more instruction on storing and organizing information. In the third program, teachers would need to supplement the curriculum with additional experiences to make connections and develop concepts at a deeper level.

Core textbooks are often inadequate for typically achieving students as well. Rohrer and colleagues (2020) examined six representative texts Grades 6–8 for blocked practice opportunities (practice on the same skill or concept) versus interleaved practice (mixed skills or concepts) that has been correlated with higher test scores. Only 9.7% of 13,505 practice problems in the texts were interleaved, requiring teachers to modify practice sets for optimal achievement. The authors emphasized that long blocks of practice problems should be avoided and that teachers could create an interleaved set from a dozen problems from various blocked sets. In studying how fraction arithmetic is represented in middle-school textbooks, Braithwaite and Siegler (2018) termed the textbook presentation spurious, with mathematically irrelevant associations. When presented with fraction arithmetic problems, students tended to base their procedure on the features of the operands because the textbook examples are associated with those procedures. For example, when presented with two equal-denominator fractions, students expected to add or subtract; when presented with mixed numbers, students expected to multiply or divide. The researchers recommended improving students' conceptual understanding of fraction operations and providing a better mix of practice problems so students will not rely on faulty associations for solution strategies.

Clearly the textbooks used in core mathematics classrooms must be adapted and enhanced to meet the learning needs of students who struggle. It may be optimal for a mathematics teacher to plan with a special educator or other support teacher in making these adaptations, drawing on the expertise of each professional. The next section describes an approach for this planning.

Adapting Mathematics Textbooks

Teachers' guides for mathematics textbooks typically offer brief suggestions for differentiating instruction for students with individual learning needs, within lessons or at the unit level. Common suggestions for students with learning difficulties include creating an atmosphere of high expectations but support and acceptance, creating consistent routines, making assessment criteria clear, considering multiple approaches to problem solving, providing additional practice sets, and occasionally offering specific suggestions for alternative instruction and re-teaching. Some teachers' manuals, curiously, emphasize addressing different learning styles where teachers are encouraged to allow students to learn math in the most comfortable style for each student—visual, auditory, kinesthetic, linguistic, logical, social, or individual—a seemingly confused mix of multiple intelligences and learning styles theory, neither of which has research support for effective mathematics intervention. Special education researchers recommended approaches to evaluating existing textbooks, emphasizing research-based instructional practices—especially the explicit elements of instructional scaffolding, student practice, and judicious review (Doabler et al., 2012; Doabler et al., 2018). They emphasized that teachers should locate where a critical topic is introduced within the text and trace it across developmental lessons for a full review.

For cooperating teachers, an effective approach for planning the use of textbook-based lessons is to consider the strengths and needs of a specific group of students while studying the elements of the lesson. Box 6.2 provides a list of questions to guide this type of lesson planning.

Note the connections between these questions and the principles for instruction discussed in Chapter 4. For student characteristics, keep in mind prior knowledge, vocabulary skill, cognitive and metacognitive abilities, need for scaffolding and practice, abilities related to making and using representations, dispositions about mathematics. Review the lesson or unit in the textbook for elements such as review of prior knowledge, sufficient examples, adequate practice, vocabulary demands, and methods for teachers to check student understanding. Create an adapted lesson plan that addresses student needs and specific lesson elements.

Box 6.2   Textbook Adaptation Planning Guide

When planning for a unit of study or a specific lesson within a mathematics textbook, consider the following.

All students

1. Is this topic important, a key idea for future mathematics learning?
     If not, move on to another topic or lesson.
2. What is the primary instructional objective?
3. What prior knowledge (informal or formal) do my students have about this topic?
4. What prior knowledge is essential for success with this objective?
5. What future concepts and skills will be developed based on this learning?

Students with learning difficulties

6. What vocabulary will be challenging?
7. What concepts might be confusing?
8. Are the examples appropriate for my students?
9. Do the representations enhance learning?
10. Is the sequence of instruction appropriate?
11. Is practice with new concepts and skills sufficient?
12. What adaptations and differentiations are required for my students?
13. Does this lesson require supplemental material?

Future lessons within the text

14. Has judicious review been planned in future lessons?
15. Do extensions and applications in future lessons have concept fidelity?

Consider the textbook lesson titled *Powers of 10* depicted in Figure 6.1. This lesson is a simulated example based on a review of fifth-grade curriculum standards and typical mathematics texts. This lesson will illustrate the questions teachers should ask while planning for a more effective lesson for all the students in the classroom.

1. Is this topic important, a key idea for future mathematics learning? This question reminds teachers that textbooks are often crammed with too many topics. Curriculum standards urge focusing on a few key topics per grade level. Teachers should be able to reorder topics and delete those not essential so that deep learning for a few topics can be planned. For fifth grade, the critical concepts include fractions, fluency with arithmetic operations, development of decimal concepts while expanding place-value understanding, and three-dimensional concepts. Powers of 10 is an important topic within place value as well as for strengthening multiplicative reasoning. In future grades students will use positive and negative exponents with other base numbers within mathematics and science.

2. *What is the primary instructional objective?* Most textbooks have clearly stated objectives; however, teachers should check for too many objectives for the span of a lesson. The *Powers of 10* lesson includes a purpose statement: This lesson is about expressing numbers as multiples of 10 using an exponent to show the power of 10. The lesson is not about all exponents and it does not include negative powers of 10. It does not include powers of other numbers, such as 4^3, but more advanced students may wonder about other applications so teachers should be prepared to support those questions. Teachers may want to turn the purpose statement into a learning objective: Students will demonstrate understanding of powers of 10 by expressing multiples of 10 in standard, expanded, and exponential forms.

3. *What prior knowledge (informal or formal) do my students have about this topic?* This question is specific to a group of diverse students. In previous lessons in this textbook, teachers may have noted students who could easily multiply or divide by multiples of 10 and students with strong place-value understanding. Some students may have had an exposure to powers of 10 within science investigations.

4. *What prior knowledge is essential for success with this objective?* What concepts must be understood and skills mastered before beginning this lesson? Some students may need a review of multiples of 10 and reading larger numbers using the names for each period (thousands, millions). This review could occur within a small group in a preteaching model or with the entire class as a review segment before beginning the lesson.

5. *What future concepts and skills will be developed based on this learning?* This question reminds teachers to extend the trajectory of learning while planning. Trace the concept through the curriculum standards in future grade levels. The powers of 10 concepts in this lesson form a foundation for reasoning about other exponential expressions within algebra, geometry, statistics, and functions.

6. *What vocabulary will be challenging?* Make a list of all vocabulary within the lesson and group the terms by familiar and new vocabulary. For the *Powers of 10* lesson, familiar vocabulary likely includes: kilometer, distance, zero, ten, hundred, thousand (and other number words), product, expanded form, standard form, whole number, factors, period (some students may have called these number groups *classes*), equivalent, cube, meter, liters, cubic centimeters. New vocabulary includes *power* (familiar word with new meaning), *exponent* (and exponential), *base* (familiar word with new meaning), and *shares of stock* (not critical for understanding lesson concept). In this text, the "Math Connections" piece is related to a previous unit on three-dimensional shapes and measurement. Additional vocabulary is found in the "Extend Your Thinking" optional section: Bolivia, Angel Falls, mm (millimeters). This section's concepts go beyond the lesson's objectives and should be used only with those students who need to extend their understanding. In this lesson for most students there are three new vocabulary words related to the concepts of the lesson: power, exponent, and base. That is a reasonable number of new words to support the lesson concepts. Some students will be confused by multiple meanings of these words and some students may need a review of previously learned vocabulary.

7. *What concepts might be confusing?* This question challenges teachers to anticipate misconceptions and areas of difficulty. The major concepts in this lesson are multiples of 10 (a prior knowledge

Lesson 6-5

Powers of 10

Our galaxy is 1,000,000,000,000,000,000 kilometers long. The distance from Earth to Mars is 10,000,000 kilometers. These numbers have a lot of zeros! Have you seen numbers like these in science books?

How many zeros are in the first number above?

This number can be written a shorter way: 10^{18}

We read this number "ten to the power of 18."

PURPOSE: This lesson is about expressing numbers as multiples of ten using an exponent to show the power of 10.

Study this pattern:

$10^1 = 10$
$10^2 = 10 \times 10 = 100$
$10^3 = 10 \times 10 \times 10 = 1000$
$10^4 = 10 \times 10 \times 10 \times 10 = 10,000$
$10^5 = 10 \times 10 \times 10 \times 10 \times 10 = 100,000$

What did you notice about the number of zeros in the products? Each zero represents one power of ten.

In the following number, the 10 is called the base and the 7 is the exponent.

$$\underset{\text{base}}{\nearrow} 10^{\overset{7}{\underset{\text{exponent}}{\uparrow}}}$$

We can write this number in expanded form using factors:

$10 \times 10 \times 10 \times 10 \times 10 \times 10 \times 10$

Or we can write it in standard form as a whole number: 10,000,000. The number of zeros gives us the exponent. A power is the product of multiplying a number by itself. 10^7 means "10 to the 7th power."

PRACTICE:

Write the numbers in exponential form:

1. $10 \times 10 \times 10 \times 10 \times 10 \times 10$ 2. $10 \times 10 \times 10$

Write in expanded form (showing factors):

3. 10^4 4. 10^8

Figure 6.1 Sample Textbook Lesson

Place commas in these numbers to distinguish each period. Read the numbers and then write them in exponential form.

5. 10000000

6. 1000000000

Write in whole-number form (standard form):

7. 10^9

8. 10^{12}

PROBLEM SOLVING:

9. A company has 10,000,000 shares of stock for sale. How can they write this number in a shorter form?

10. The first three powers of ten are 10^0, 10^1, and 10^2. List the next five powers.

EXTEND YOUR THINKING:

11. In Bolivia's Angel Falls, the water falls almost 1000 meters. If a sheet of paper is 0.1 mm thick, how many sheets of stacked paper would it take to equal the height of the falls? Write your solution two ways.

12. Explain why 10^4 is equivalent to 10,000.

Mental Math Review: Find the product in your head.

1) 12 x 10 =   2) 35 x 100 =   3) 157 x 100 =

4) 4.2 x 10 =   5) 5.6 x 100 =   6) 9.13 x 10 =

MATH CONNECTIONS:

This cube represents a cubic meter. We can write this $1m^3$ or m^3. This cube is 1 meter on each side: length, width, and height. A cubic meter holds 1000 liters. What is another way to express 1000 liters? One cubic meter equals one million cubic centimeters. How can that number be expressed?

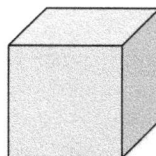

ON YOUR OWN:

Write in expanded form and standard form:

1. 10^3

2. 10^5

3. 10^{10}

Write in exponential form:

4. 10,000,000,000   5. 100,000   6. 1,000,000,000,000

Lesson 6-5   Powers of 10

Figure 6.1 Continued

concept), power of a number, exponent, and various methods for writing the same number (an equivalence concept). The term *exponent* in mathematics means a quantity representing the power to which a number or expression is raised, so it depends on understanding power. In the case of positive whole-number exponents for the base of 10, this is repeatedly multiplying by 10. The term *power* means the result of raising a base by an exponent. Most elementary and middle-school texts define power as the number of times the base is multiplied by itself. Use consistent and clear vocabulary in this lesson and minimize alternative terms for new learning. Post an example labeled with base and exponent on the wall (10^7) and refer to that while using the terms. Keep in mind that these explanations will break down in later grades with decimal, fraction, and zero exponents. At that point students should have more understanding of the full number system and functions for a more complex understanding of exponents.

Consider possible misconceptions students might have during instruction. Most common for students who struggle with mathematics is the application of additive rather than multiplicative reasoning. Having students count zeros reinforces this misconception. They are actually multiplying by 10 and counting the multiples. Create or locate learning strategies that could be taught to support student memory and organization. One student may understand that power means strength, so powers make the number stronger and stronger, expanding it by multiples. Other students may benefit from a vocabulary concept map, a diagram showing the vocabulary word, definition, characteristics, examples, and non-examples. Another student may realize that exponential notation is like using an abbreviation, but for mathematics.

8. *Are the examples appropriate for my students?* Good lessons should have carefully sequenced and sufficient examples for new concept learning. In the *Powers of 10* lesson, we see one example in the introductory box, about our galaxy. Next we see a number pattern with powers of 10 listed and their expanded and standard forms. That is followed by an example with base 10 and exponent 7. That is a total of seven examples of the power of 10. The teacher should supplement these with other examples if students do not follow the pattern or have difficulty recognizing the base and exponent. Advanced students may ask about exponents of zero and negative integers so teachers should be prepared to support those questions and differentiate examples.

9. *Do the representations enhance learning?* Representations are multiple ways to show, symbolize, and describe a concept. Representations can be symbols, words, diagrams (pictures), models (manipulatives), oral, or experiential (application). In this lesson we see two pictorial representations—the picture of the galaxy and the diagram of the cube. We also have numbers represented in multiple forms—as standard, expanded, and exponential forms—all at the symbolic level. For students who struggle with mathematics, concrete materials or diagrams depicting the base-10 system would support their understanding of multiples of 10 and powers of 10. The galaxy picture is a way to illustrate a very large number. The cube diagram should assist most students in making the connection to geometry concepts, but some students may need concrete objects for those connections. Teachers should locate these additional representations prior to instruction.

10. *Is the sequence of instruction appropriate?* Students with learning difficulties in mathematics need systematic, explicit instruction in order to grasp concepts and develop skills. This lesson begins with examples of very large numbers and demonstrates how to write and say the numbers in a shorter way. The lesson continues by showing a pattern of multiples of 10 using factors and linking those to the exponential and standard forms, introducing the concepts of base, exponent, and power. That is followed with eight problems of mixed practice and two word problems, for guided practice. Optional sections are provided for extended thinking, mental math, and math connections. The lesson concludes with six independent practice items.

Examples of adaptations to this lesson include: beginning with review of multiples of 10 and the mental math section, reviewing reading larger numbers, asking students to extend the pattern of powers of 10 on the first page of the lesson, requiring students to verbalize all the expressions using precise vocabulary, demonstrating the concepts using manipulatives or additional diagrams, using whiteboards for guided practice so all students are responding and receiving feedback, and providing additional guided practice for some students before independent practice. During these phases of instruction the teacher is modeling, providing feedback, and scaffolding student understanding. The word problems require reading comprehension so teachers may prompt, "What is the problem asking?" for students to paraphrase the question. The teacher may also decide to delay the math connections piece until the next day, when the powers of 10 concepts can be reviewed and evaluated again. Only those students with very strong concept understanding should attempt extension item #11. Item #12 is a good example of practice with reasoning and verbalization.

11. *Is practice with new concepts and skills sufficient?* Teachers who are skilled at monitoring student responses during guided practice will add practice items for the sets in 1 through 9 as needed. Teachers who plan carefully will record additional practice items in the teacher's manual or on their lesson plans. Students should achieve levels of success during guided practice between 80% and 90% before being assigned independent practice. If some students are not successful on all six independent practice items (at or above 90%), areas of confusion should be probed and retaught.

   Additional practice activities could be planned, including games, small-group work, and computer-based practice. For example, a good practice activity for this lesson would be a card game matching forms of numbers (e.g., match 10^4 to two other equivalent forms). Practice can also be differentiated, based on students' levels of understanding.

12. *What adaptations and differentiations are required for my students?* This question prompts a summary of adaptations thus far: ways to teach challenging vocabulary, additional representations for concepts, additional examples for guided practice, and a re-ordering of lesson content. Some adaptations will be planned for definite use during the lesson while others will be "on standby" in case they are needed.

   This question also focuses on specific student needs. How can the lesson elements be differentiated during instruction to meet those needs? Some students will need more frequent responding and feedback, others may need a more gradual sequence from reading large numbers to writing numbers in expanded form and then finally to exponential form. More advanced students will need more challenging items. Different examples, scaffolds, and practice items can be incorporated for different students or student groups. Students can be assigned parallel tasks, tasks that explore the same concept but at different levels. For example, two groups could work on this problem: *Create a word problem that involves multiplying 100 by 10,000. Use exponential representations and explore other solution methods.* Another group's problem: *Find powers of 10 in the science newsletter and record those in another form.* A fourth group is given problem #11 in the lesson. The first two groups demonstrated good concept understanding during the initial portions of the lesson rather quickly and their group assignment requires them to use that understanding to extend their thinking. The third group needed more practice with recognizing powers of 10 and recording those in expanded and standard forms. The fourth group came into the lesson fluent with number representations and was ready for a decimal number extension.

13. *Does this lesson require supplemental materials?* Locating additional materials before beginning instruction is critical for a successful lesson. Review your annotated lesson and make a list of needed materials. Base-10 blocks, place-value charts or mats, number grids, and interactive apps can provide additional representations for concepts, but choose the best representation for your students. Whiteboards provide for group responses in writing. Additional examples for practice sets are also needed for this lesson. Some of the differentiated elements will need

additional or adapted materials. Be sure to use precise and accurate symbols for exponents in all written materials—use 10^5 and not 10^5.

14. *Has judicious review been planned in future lessons?* This lesson about powers of 10 is a good example of a topic that appears to be taught in isolation. There are no other lessons in the fifth-grade text that deliberately apply this concept, so teachers must make an effort to teach the concept with understanding and plan for review and application. Skim through the textbook for review of concepts and skills from this lesson. If those aren't present make a note in two or three future lessons to do a quick review and check. Plan to use exponential notation in science or social studies lessons with large base-10 numbers, with a brief review at that time. Some teachers even chart key concepts across the school year, documenting when those are introduced, practiced, reviewed, and applied.

15. *Do extensions and applications in future lessons have concept fidelity?* Check future lessons in the mathematics text as well as those in other subject areas for applications with powers of 10. Are the concepts explained in a way that aligns with this lesson or are those misleading or simplistic? Make a note to use the same terminology and symbols for representing numbers within these lessons. Make explicit connections with squared and cubed units in geometry lessons. Explain to students that in sixth grade they will be using exponents with other base numbers and variables when they study algebraic expressions.

This textbook-based lesson, as adapted for a diverse group of students, is illustrated in Box 6.3. Some teachers prefer to make notes directly on the lesson in the teacher's manual, while others generate a new lesson plan with references to the text or make notes in a sharable, digital format. After instruction, teachers should reflect on student responses for each segment of the lesson, misconceptions and how those were addressed, and changes needed in future lessons. Making notes on lesson plans will ease planning the following year, although a different group of students must be considered when recycling the lesson plan. When this author field tested the sample lesson depicted in Figure 6.1, she found that one student with a mathematics learning disability resisted thinking in multiples of 10, preferring to count zeros (additive reasoning). This student also, unexpectedly, placed commas within long numbers from left to right beginning after the 1 (e.g., 1,000,00), which required some re-teaching.

Box 6.3   Adapted Textbook Lesson: Powers of 10

Lesson Objective: Students will demonstrate understanding of powers of 10 by expressing multiples of 10 in standard, expanded, and exponential forms.
Curriculum Standard: (5A) Use whole-number exponents to denote powers of 10.
Assessment Method: Independent practice items on second page of lesson (6/6).

Sequence of Instruction

1. Review using mental math items on second page of lesson. Provide more whole number review before attempting decimal numbers. Practice reading large numbers using billions, millions, and thousands for periods.
2. Introduce concept of large numbers that are multiples of 10. Use the galaxy example plus additional examples.
3. Explain that in science we use shorthand, like an abbreviation, to write these long numbers. This shorthand uses the base of 10 and a special number called an exponent

to show how many times the ten is multiplied by itself. Model writing 10^{18}. Why is this way of writing the number preferred by scientists? Model 10^5 and 10^7 with student assistance. Listen to student reasoning and vocabulary.

4. Now let's examine the patterns involved. Does anyone see a pattern on the first page of your lesson? Yes, those are factors of the numbers and the number of zeros shows the number of times the 10 is multiplied by itself. Demonstrate with place-value blocks. Let's extend this pattern on the chart. Who can do 10^6? (Extend as needed).

5. We call the 10 the **base** of this number and the 7 that is the smaller, raised portion of the number is called the **exponent**. The exponent tells us the **power** of 10 or the number of times 10 is multiplied by itself. (Point to chart with parts labeled.)

6. Now we have three ways to write these numbers: standard form, expanded form using factors, and exponential form using an exponent to show the power of 10. Model 10^7.

7. Distribute white boards and practice using items 1–8, with supplementary examples as needed. Then do items 9, 10, and 12 as a group.

8. Now in your problem-solving groups, each group has a power of 10 task. You have group roles and five minutes. I'll be circulating and asking questions about your thinking.

9. Check each group's work and assign independent practice (items 1 to 6 at end of lesson for most students, differentiate for others).

10. Tomorrow we will review this information and connect it with our geometry and measurement concepts.

Materials

Charts for patterns and power-of-10 model example, base-ten blocks, white boards and markers, task cards for group work, alternative independent practice items.

Special Provisions

Small groups have pre-assigned roles and tasks are differentiated.
Provide additional guided practice for individual students as needed.
Differentiated independent practice.

Anticipated Misconceptions

The concept of number: 10^6 is a number with two parts, base and exponent (not a 10 and a 6). This number can be written as 1,000,000.

Additive thinking rather than multiplicative: For 10^6, the exponent 6 represents the number of times 10 is multiplied by itself, the power of the 10. Counting zeros is counting the number of multiples.

Some students attempting 10^0 may confuse that with 10×0. (It is $= 1$.)

Ensure students aren't using the item numbers in the text as part of numbers.

Lesson Vocabulary

New: power, base, exponent
Review: names of numbers, whole number, standard form, expanded form, factors, period (or class), equivalent

Additional Examples

Review: 24 × 1000   102 × 1000   217 × 10,000   Read: 10,000   10,000,000

Introduce concept

It takes 10,000 hours of practice to be an expert in a field.
A library had 100,000 pages to digitize.
A flashlight has 10,000,000 candela (candlepower).
Austin (Texas) is home to about 1,000,000 people.

Additional guided practice items by set

Write numbers in exponential form: 10 × 10   10 × 10 × 10 × 10   10

Write in expanded form: 10^5 10^1 10^3

Place commas and read numbers:
1000   100000   100000000

Write in standard form:
10^7 10^8 10^1 10^{10}

Small group tasks

1 and 2: Create a word problem that involves multiplying 100 by 10,000. Use exponential representations and explore other solution methods.
3: In the science newsletter on your desks, locate three powers of 10 and write those in expanded and standard forms.
4: Can you write 1000 and 0.1 in exponential notation? Now attempt item #11 in your text. Use diagrams (place-value grids) as needed and write your solution in two ways.

More challenging independent practice items

2. 10^{-4} 3. 10^{-7} 5.   0.0001   6.   0.0000001

Teachers using core mathematics textbooks often forget to examine the many ancillary materials that accompany the text. These materials sometimes offer suggestions for representations of concepts, additional practice sets, resources for frequent assessment, and other supplements for lessons and units. In addition, textbook lessons can be supplemented with materials from the web, commercial vendors, or professional organizations, if those are evaluated carefully. If the text is presented digitally, some students may need a print version (as an accommodation) or explicit guidance on how to use digital text, apps, quizzes, and videos. Teachers should monitor how students use digital materials and make adjustments if students are struggling.

TRY THIS

Practice lesson plan adaptions with an actual textbook lesson using the questions from Box 6.2.

Models of Collaboration

Today's mathematics classrooms include diverse student populations with a range of learning needs. Mathematics teachers will need the support and collaboration of other professionals in order to successfully meet the needs of all students in the classroom. These professionals may range from special educators to mathematics support specialists and EL teachers. Further, the nature of collaboration among professionals varies considerably. Table 6.3 illustrates some collaboration models that can support mathematics learning in the general curriculum within inclusive classrooms. These models differ in their planning requirements, contact with students, and instructional arrangements, but they all contribute to understanding student needs and planning instruction that will be effective. The small-group arrangements within Table 6.3 could be used for Tier 2 and Tier 3 intensive interventions. Tier 1, core mathematics instruction, could be supported by consultation, co-teaching, or small group within class. Professionals such as curriculum administrators or mathematics coaches would not fit within this collaboration model because those professionals have a supervisory (unequal) role and typically do not have a caseload of students.

Core mathematics teachers may have no or very little training in methods for teaching students with MD. Despite decades of inclusive education and differentiated instruction, there is evidence that many general educators still use whole-group, undifferentiated instruction in both elementary and secondary levels (Swanson, 2008). In a review of observational research in classrooms that included students with learning disabilities in mathematics (MLD), McKenna and colleagues (2015) found that several instructional practices that are associated with improved outcomes for students with MLD were used only infrequently. These included checking for student understanding, providing explicit instruction, promoting student verbalizations and discussions of reasoning, and employing visual representations of concepts. The use of cognitive-strategy instruction was largely absent from these core mathematics classrooms. However, core mathematics teachers should have a deep understanding

Table 6.3 Collaboration Models

Model	Student Contact for Mathematics Instruction	Teacher Roles for Planning and Instruction
Consultation	Core teacher only for mathematics instruction	Core teacher and support teacher(s) share student information and resources for instruction and assessment. Special educator may observe core class to determine setting demands or to collect student data, but not provide instruction.
Support Class (Pull-out, Resource)	Both teachers but in different settings and different times (2 to 5 times a week)	Teachers consult so the support class supplements students' core mathematics instruction.
Small Group Within Class	Both teachers within same setting, but separate instruction during portion of class time for a small group (2 or 3 times a week)	Teachers plan together so the small group sessions support core mathematics instruction and take place during appropriate time in class.
Co-teaching (Collaborative Teaching)	Both teachers: mathematics educator and special educator assigned to one class and share responsibility for the class	Teachers plan, deliver, and evaluate instruction using one or more structures: team-teaching, alternative teaching, interactive teaching, parallel teaching, station teaching, or one teaching-one assisting.

of mathematics concepts across grade levels, the connections among concepts and trajectories students follow when developing these concepts. They should have a repertoire of methods for introducing concepts and providing examples and practice for new concepts as well as assessing student understanding within mathematics topics.

Many special education teachers and other support staff have had no training in mathematics instruction or the training was limited to an elementary-level mathematics-education course. These teachers' understanding of mathematics concepts in depth or across the curriculum may be likewise limited or just developing. These professionals should have greater knowledge of student learning differences, how to design learning strategies and accommodations, and methods for conducting formative assessments and collecting progress data. One of the most compelling arguments for collaboration and, specifically, co-teaching is that two pedagogical experts can blend their expertise so that all students have access to high-quality instruction that is differentiated to meet individual learning needs.

Collaboration for mathematics educators is often described in broader terms, as working within professional-learning communities, participating in school-university partnerships, engaging in coaching or mentoring activities, or as serving on school committees. To improve student learning in mathematics there should be

> a change from individual teachers working in isolation while striving to make sure each and every standard is "covered" and taught well, to a collaborative culture in which teams of teachers work to ensure that standards are, in fact, *learned*—by each and every student.
> (Eaker & Keating, 2012, p. 17)

In all professional-collaboration efforts, whether school-wide, between co-teachers, or among other learning communities, several essential elements have been noted:

- *The culture of a school or other organization will influence the nature of collaborations.* Cultures that have a shared purpose and collective responsibility will have stronger professional collaborations.
- *Everyone is a learner.* The goals for professional collaborations don't end with student learning, they include ongoing learning by teachers and other professionals.
- *Professional collaboration activities include seeking out best practices from research and evaluating the results of their efforts.* Professionals in these collaborations are decision-makers, using performance data to make adjustments to instruction and other elements of the learning context.
- *Collaborators take advantage of resources to achieve goals*, including parent and community partnerships, professional development, and joint planning and reflection.

When special educators are collaborating with core mathematics teachers, they should schedule time for working together at the beginning of the school year and at regular intervals throughout the year. Handing a math teacher a list of student names with their accommodations is not collaboration. Catching another teacher in the hall to discuss a problem with a student is not collaboration. Partners should acknowledge that everyone is busy, but they have common goals and mutually beneficial resources and time spent working together is part of their professional responsibility. They should make a point of outlining goals for their partnership and evaluating their efforts.

Co-Teaching

Co-teaching is a collaboration model in which two certified teachers, a general educator and a special educator, "share responsibility for planning, delivering, and evaluating instruction for a diverse group of students, some of whom are students with disabilities" and others who may be at risk of learning difficulties (Kloo & Zigmond, 2008, p. 13). Co-teaching, previously termed cooperative teaching, developed during the mainstreaming and inclusion movements of the 1970s and 1980s as a way to support students with disabilities in general education classes, a special education service-delivery model (Bauwens et al., 1989). By 1995 co-teaching was the most prevalent model for inclusive education and has grown in use with federal mandates for highly qualified teachers, high-stakes assessment, and access to the general curriculum for all students (B. G. Cook et al., 2011).

A number of structures have evolved for co-taught classrooms, based on teacher expertise and preferences, student characteristics, and content pedagogy. The most prevalent structure is *one teach, one assist* where the core content teacher takes the lead for instruction and the special educator (or other support professional) assists students during the instructional period. Assistance levels vary from unplanned "assist as needed" for monitoring student attention and providing help during practice to well-planned preview, review, or intensive intervention for one student or small groups. The *one teach, one assist* structure also includes alternating roles as lead teacher, but that form is less common. For example, a special educator has been assigned to Joseph Lopez's second period geometry class where he has five students with identified disabilities and five others who have been struggling with mathematics. At the beginning of the year the special educator tended to stand near the rear of the classroom and monitor the students on her caseload, providing those students assistance when they asked for help. After a professional development session on co-teacher roles, the special educator asked Joseph if they could plan together and co-teach the geometry class. She said:

> If I know the goals of the lesson, the vocabulary and the challenges ahead of time, I'll be a more productive co-teacher. Plus, if we alternate some of the teaching roles you will have a chance to work with individual students as well. I could lead on the review and preview segments and you could lead with the new concepts.

Another familiar form of cooperative teaching, *team teaching*, offers the most parity for teachers. Both teachers plan and deliver instruction during the class period, frequently alternating the role of primary instructor. Both teachers deliver critical content and check for student understanding with the entire group of students. A combination of team teaching and one teach, one assist, called *interactive teaching*, provides lead roles for each teacher in 5–10-minute segments while the other teacher assists, then the teachers switch roles (Walther-Thomas et al., 2000). An example for a third-grade mathematics class teams the third-grade teacher and a special educator assigned to that class. The teachers plan mathematics lessons together and map out each teacher's role during portions of the lessons. Additional flex time for mathematics will allow the special educator to provide Tier 2 interventions with third graders while the third-grade teacher will offer extensions and applications for the other students. Since the special educator is a full-time co-teacher for one third-grade class, he knows the curriculum and issues students are having within the core classroom.

There are several structures within co-teaching that involve student-grouping patterns. *Station teaching* involves two or more stations within the classroom and students rotate among stations during the class period. Typically each teacher directs a station and other stations may be planned for computer-based learning, independent practice, or application activities. *Parallel teaching* involves joint planning, but each teacher delivers instruction simultaneously to half the class, with each half a heterogeneous group of students. *Alternative teaching* is a configuration of one large group and a smaller group for intensive instruction on the same topic. An example of station teaching for a seventh-grade class with a mathematics teacher and a special educator planning together involves a unit about statistics. The mathematics teacher provides instruction about random sampling at one station and the special education teacher provides review instruction connecting data analysis with computation (mean, median, mode) at another station. Students also rotate to a station with newspapers for reading and answering questions about population data and to a computer-based station where students input data and generate graphs.

Advantages of co-teaching arrangements include access to the general curriculum with support for students with difficulties, reduced student–teacher ratio for more individualized adaptations and supports, reduction of the stigma of attending special-education classes, and providing assistance for general educators in meeting the needs of diverse students. Co-teaching offers a blending of teacher expertise that may be required for students with disabilities and other learning difficulties to be successful in the core curriculum. Stefanidis et al. (2018) found that co-teachers with higher levels of co-planning—with regular planning time used well—and those with high-quality professional relationships, viewed the benefits of co-teaching more favorably. However, there have been issues with implementing co-teaching. In some classrooms the support teacher's role is that of an assistant rather than an equal professional. The role of the special educator is too often unplanned, not providing elaborated explanations, portions of instruction, or re-teaching. In other classrooms the support teacher has been relegated to assisting only identified students, essentially separate instruction within a core classroom. Some schools, seemingly stretched for resources, assign co-teachers for limited days of the week, not a true co-teaching model.

There has been inconclusive research on the effectiveness of co-teaching for improved learning for students with disabilities. Co-teaching cannot be considered an evidence-based practice due to the lack of high-quality studies (B. G. Cook et al., 2011; S. C. Cook et al., 2017). Co-teaching holds promise and is recommended as an inclusive education model as long as the essential tenants of special education are included: explicit, systematic instruction; flexibility; intensive, individualized instruction in small groups; progress monitoring; and the use of research-based practices. Without those elements it is unlikely that students with MD will achieve within the core mathematics curriculum. Because of the dearth of quality research and the difficulty of designing research about the effectiveness of co-teaching, S. C. Cook and colleagues (2017) emphasized the importance of monitoring student progress toward educational goals and evaluating the effectiveness of implemented co-teaching models within a classroom.

Some recommendations for successful co-teaching include (e.g., B. G. Cook et al., 2011; S. C. Cook et al., 2017; Rexroat-Frazier & Chamberlin, 2019; Treahy & Gurganus, 2010):

- Co-teachers must plan instruction together and have dedicated time for that planning. Planning should occur at least weekly for lesson planning and student progress monitoring, quarterly for unit planning as well as for parent reports and conferences, and with grade-level teams as those are typically scheduled.
- Roles for both teachers should be clearly delineated, professional, and active.
- Both teachers must know the learning characteristics of all students in the classroom.
- Co-teachers should discuss their educational philosophies, approaches to instruction and assessment, disciplinary expectations, and communication skills. In some cases teachers' philosophies and approaches will be so different that a co-teaching arrangement cannot be successful.
- The core mathematics teacher should develop a repertoire of skills for differentiating instruction.
- The special educator should develop mathematics content and pedagogical knowledge.
- Teachers should work toward a common instructional language. For example, the term *accommodation* is used differently by different disciplines, but in special education has a

specific, legal meaning. Further, the mathematics language both teachers use should be accurate and consistent.

- Special educators should continue to adhere to the principles of special education instruction while co-teaching (intensive, individualized instruction; progress monitoring; flexibility; and research-based practices).
- Co-teaching works best with supportive administrators who schedule common planning times, provide adequate classroom space, and support professional development.
- Unanticipated problems will arise in any co-teaching arrangement. Those should be expected and resolved with effective communication and professional processes for problem solving.
- Co-teachers should jointly set the classroom climate for learning, including interest in mathematics, appropriate expectations, respect, and positive attitudes. They should model communication, problem solving, and working together for their students.
- Teachers should monitor and deliberately evaluate their model(s) of co-teaching, types of student groupings and engagements, and instructional and assessment strategies with their effects on student learning, making adjustments based on those analyses.

Although not well-researched, co-teaching and other collaboration models for supporting student learning within core classrooms are prevalent and offer many advantages if implemented with careful planning, flexibility, specific professional roles for both teachers, and research-based instruction. However, if core mathematics classes maintain traditional roles and instructional approaches such as whole-group instruction with little differentiation, regardless of the number of teachers, outcomes for students who struggle to learn are not likely to improve (B. G. Cook et al., 2011). For students with IEPs, special education continues to be a legally mandated service regardless of the setting in which it is provided (Zigmond & Kloo, 2017). With the inclusion movement and collaborative models, it is important to remember that students with disabilities still require specialized, individualized, and intensive instruction in order to make expected educational progress. Occasional assistance in the core mathematics classroom will not be adequate.

Collaboration with Parents

Parents and other caregivers should be partners with teachers and other school professionals for their children's education. They offer insight into students' learning strengths and needs, background knowledge, interests and preferences, language and communication skills, and community and cultural backgrounds. They can support learning outside the classroom setting if they have personal resources and partnerships with teachers. Parents, like their children, will differ considerably in their abilities and interests for partnering with teachers to support their children's education, but it is incumbent on teachers to facilitate the partnership. Successful parent-teacher collaborative relationships are especially important for the success of students with identified disabilities (Solone et al., 2020).

Teachers should communicate with parents, at a minimum, the curriculum standards and grade-level expectations as well as students' progress within those expectations at frequent intervals across the school year. Students who have IEPs or 504 plans will have specific, documented requirements for communication with parents. Optimally, parents should have a good understanding of the curriculum expectations at the specific grade level, assessment methods, and their role in the partnership for their children's education.

For mathematics classes, schools or groups of teachers often provide evening orientation seminars for parents about grade-level curriculum standards and assessments. Many teachers maintain websites for parents to learn about mathematics topics, view schedules for assessment, and to ask questions. Other teachers send home parent-involvement kits with activities parents and their children can do together, with the necessary materials. Some teachers open their classrooms for parent visits so they can witness and participate in class activities. All of these approaches have a goal of increasing parent engagement with their children's learning. Particularly with mathematics, some parents may feel unprepared or inadequate in assisting their children or even following their progress. Teachers should find strategies for bridging this gap, to promote positive attitudes about mathematics at home.

Jay et al. (2018) described school-centered and parent-centered approaches to parental involvement in children's mathematics learning, as two ends of a continuum. School-centered activities are focused on curriculum goals, homework practice, and other attempts to assist parents in being involved in school-directed mathematics learning. Parents are often unprepared and lack confidence for this role. Parent-centered approaches take advantage of parental resources to promote mathematics thinking and activities within natural home situations and everyday problems. While parent-centered activities hold promise for bridging the school-family gap and build positive attitudes about mathematics, they require much more time, communication, parent guidance, and planning.

TRY THIS

Create a parent involvement packet that could be sent home with a student about one of the following topics. Include clear, parent-friendly directions and any needed materials.

a.   Conduct a survey about vehicles and record the results in a bar graph.
b.   Draw at least three different rectangles on graph paper with areas of 36 sq cm.
c.   Find the value ½ in ordinary objects and in magazines. Make a list and cut out pictures.

Mathematics homework has been one of the flash points of the CCSSM, as parents often struggled to assist their children with these assignments. "This isn't the way we did math when I was in school" is a frequent complaint. The purpose of homework for mathematics is to provide additional practice and applications for skills and procedures already introduced and practiced to the independent level within the classroom. Too often homework is assigned at an incorrect level or students struggle with the directions. Some teachers take an extra step when assigning homework to go over the instructions and allow students to complete one or two examples before leaving class, ensuring student understanding. Other teachers plan options within homework so that students can work at their level. Homework, if carefully designed, can also facilitate partnerships with parents and more positive attitudes about mathematics (Van Voorhis, 2011). Homework that is frequent, short, purposeful, adjusted to the student's ability, and includes some degree of choice was positively related to homework effort and academic achievement, while long, repetitive assignments were negatively related (Cunha, 2019).

In a study of the role of homework on mathematics achievement, Kitsantas et al. (2011) analyzed *Programme for International Student Assessment* (PISA) data for US students in Grades 9 to 11 (n = 5200). The researchers found a reduction in the achievement gap between White and minority students correlated with two variables: resources for completing homework and higher levels of mathematics self-efficacy. Homework resources included a quiet study environment with materials such as calculators, dictionaries, tools, and computers. The proportion of homework-support resources was related to higher mathematics achievement for all student groups. However, increasing homework time actually decreased mathematics achievement.

In a study of parental involvement with mathematics homework, O'Sullivan et al. (2014) found that getting parents involved with their children's homework in the home setting, especially by providing structure for doing homework, was beneficial for mathematics learning among low-income families. Structure included setting aside quiet time for homework and providing incentives for completion. The researchers noted that some parents do not realize that providing the structure for homework can assist their children in mathematics even if parents are not able to assist with content learning.

Some recommendations for mathematics homework include:

- Assign homework only if that contributes to instructional goals for the student. It should be accessible, at the student's instructional level.
- Know student circumstances for homework resources. Don't assign homework that will penalize students who lack requisite resources.
- Homework should be time-limited based on grade-level guidelines. Cooper et al. (2006) recommended a limit of 10–20 minutes for first graders with an additional 10 minutes per grade level (all subjects combined).
- Vary homework types and options to maintain student interest and promote generalization. One interesting problem is better than a 30-item page of rote practice.
- Design homework that encourages mathematics discussion at home. Tasks can be accessible and engaging, promoting positive attitudes among all family members. This author overheard a middle-grades student challenge her dad while they were waiting in line at a baseball game, "Hey dad, I bet you can't solve a problem we did in math class today! A bat and ball together cost 1 dollar and 10 cents. The bat costs one dollar more than the ball. How much does the ball cost?" (This is a classic problem attributed to Kahneman, 2002.)
- Plan homework follow-up (Rosário et al., 2019). Often teachers check for completion but may not provide feedback, not understanding homework's role in the learning process. Some teachers lead group homework checking practices that can lead to whole-group discussion or reteaching, while others check individual work and provide oral or written feedback.

Best practices for family-educator collaboration include communication, equity, trust, and respect (Solone et al., 2020). Communication focuses on quality and quantity, is honest, avoids jargon, and is tailored to the needs of each family. Equity acknowledges that families are experts on their children. Educators are aware that some families perceive administrators and teachers as having power and authority and make an extra effort to share decision making. Trust is built through actions of protecting private information and communicating well. Respect for a family's language, background, culture, and other commitments is also essential for effective collaboration. Respectful educators set high expectations for students and model professional behaviors such as being reliable and on time. Some parents, possibly due to previous experiences, may be confrontational and overly demanding. Educators must listen with empathy and reset dialog and trust with these parents, devoting more time to establishing their collaborative relationship.

Most students who struggle with mathematics receive instruction in core mathematics classes. Many of those students will not be successful in learning mathematics without specially designed instruction, accommodations, or other adaptations based on their specific learning needs. Professional and parental collaborations can support more effective planning and effective instruction for these students.

References

Alegre, F., Moliner, L., Maroto, A., & Lorenzo-Valentin, G. (2019). Peer tutoring in mathematics in primary education: A systematic review. *Educational Review (Birmingham)*, *71*(6), 767–791. doi: 10.1080/00131911.2018.1474176

Algozzine, B., & Anderson, K. M. (2007). Tips for teaching: Differentiating instruction to include all students. *Preventing School Failure*, *51*(3), 49–53. doi: 10.3200/PSFL.51.3.49-54

Basham, J., Gardner, J., & Smith, S. (2020). Measuring the implementation of UDL in classrooms and schools: Initial field test results. *Remedial and Special Education*, *41*(4), 231–243. doi: 10.1177/0741932520908015

Bauwens, J., Hourcade, J., & Friend, M. (1989). Cooperative teaching: A model for general and special education integration. *Remedial and Special Education*, *10*, 17–22. doi: 10.1177/074193258901000205

Bertucci, A., Conte, S., Johnson, D. W., & Johnson, R. T. (2010). The impact of size of cooperative group on achievement, social support, and self-esteem. *The Journal of General Psychology*, *137*, 256–272. doi: 10.1080/00221309.2010.484448

Bottge, B. J., Gugerty, J. J., Serlin, R., & Moon, K.-S. (2003). Block and traditional schedules: Effects on students with and without disabilities in high school. *NASSP Bulletin*, *87*, 2–14. doi: 10.1177/019263650308763602

Braithwaite, D., & Siegler, R. (2018). Children learn spurious associations in their math textbooks: Examples from fraction arithmetic. *Journal of Experimental Psychology. Learning, Memory, and Cognition*, *44*(11), 1765–1777. doi: 10.1037/xlm0000546

Buddin, R., & Croft, M. (2014). Do stricter high-school graduation requirements improve college readiness? *ACT Working Paper Series*. (ED560234) ERIC. https://files.eric.ed.gov/fulltext/ED560234.pdf

Çarbo, M. (1988). Debunking the great phonics myth. *Phi Delta Kappa*, *70*, 226–240.

Center for Applied Special Technology (CAST) (2011). *Universal design for learning guidelines version 2.0.* CAST. www.udlcenter.org

Cogan, L., Schmidt, W., & Guo, S. (2019). The role that mathematics plays in college- and career-readiness: Evidence from PISA. *Journal of Curriculum Studies*, *51*(4), 530–553. doi: 10.1080/00220272.2018.1533998

Cohen, E. G. (1994). *Designing group work: Strategies for the heterogeneous classroom* (2nd ed.). Teachers College Press.

Cook, B. G., McDuffie-Landrum, K. A., Oshita, L., & Cook, S. C. (2011). Co-teaching for students with disabilities: A critical analysis of the empirical literature. In J. M. Kauffman & D. P. Hallahan (Eds.), *Handbook of special education* (pp. 147–159). Routledge. doi: 10.4324/9780203837306

Cook, S. C., McDuffie-Landrum, K. A., Oshita, L., & Cook, B. G. (2017). Co-teaching for students with disabilities: A critical and updated analysis of the empirical literature. In J. M. Kauffman, D. P. Hallahan, & P. C. Pullen (Eds.), *Handbook of special education* (2nd ed., pp. 233–248). Routledge. doi: 10.4324/9781315517698

Cooper, H., Robinson, J. C., & Patall, E. A. (2006). Does homework improve academic achievement? A synthesis of research, 1987–2003. *Review of Educational Research*, *76*, 1–62. doi: 10.3102/00346543076001001

Cunha, J., Rosário, P., Núñez, J., Vallejo, G., Martins, J., & Högemann, J. (2019). Does teacher homework feedback matter to 6th graders' school engagement?: A mixed methods study. *Metacognition and Learning*, *14*(2), 89–129. doi: 10.1007/s11409-019-09200-z

Dekker, S., Lee, N. C., Howard-Jones, P., & Jolles, J. (2012). Neuromyths in education: Prevalence and predictors of misconceptions among teachers. *Frontiers in Psychology*, *3*, 429, 1–8. doi: 10.3389/fpsyg.2012.00429

Delquadri, J., Greenwood, C. R., Stetton, K., & Hall, R. V. (1983). The peer tutoring spelling game: A classroom procedure for increasing opportunity to respond and spelling performance. *Education and Treatment of Children*, *6*, 225–239.

Department of Education and Sciences (1978). Mixed ability work in comprehensive schools. *HMI Series: Matters for Discussion No. 6.* London: Her Majesty's Stationery Office. www.educationengland.org.uk/documents/

Dexter, D. D., & Hughes, C. A. (2011). Graphic organizers and students with learning disabilities: A meta-analysis. *Learning Disability Quarterly*, *34*(1), 51–72. doi: 10.1177/073194871103400104

Doabler, C. T., Fien, H., Nelson-Walker, N. J., & Baker, S. K. (2012). Evaluating three elementary mathematics programs for presence of eight research-based instructional design principles. *Learning Disability Quarterly, 35*, 200–211. doi: 10.1177/0731948712438557

Doabler, C., Smith, J., Nelson, N., Clarke, B., Berg, T., & Fien, H. (2018). A guide for evaluating the mathematics programs used by special education teachers. *Intervention in School and Clinic, 54*(2), 97–105. doi: 10.1177/1053451218765253

Eaker, R., & Keating, J. (2012). Improving mathematics achievement: The power of professional learning communities. In J. M. Bay-Williams & W. R. Speer (Eds.), *Professional collaborations in mathematics teaching and learning: Seeking success for all* (pp. 3–18). NCTM. www.nctm.org

Endrew F. v. Douglas County School District RE-1, 15-827, 137 U. S. 988 (2017). www.supremecourt. gov/opinions/16pdf/15-827_0pm1.pdf

Every Student Succeeds Act, 20 USC § 6301 (2015). www.congress.gov/114/plaws/publ95/PLAW-114publ95.pdf

Fuchs, L. S., Fuchs, D., Yazdian, L., & Powell, S. R. (2002). Enhancing first-grade children's mathematical development with peer-assisted learning strategies. *School Psychology Review, 31*, 569–583. doi: 10.1080/02796015.2002.12086175

Gilmore, C. K., McCarthy, S. E., & Spelke, E. S. (2007). Symbolic arithmetic knowledge without instruction. *Nature, 447*(7144), 589–592. doi: 10.1038/nature05850

Hackney, J. (2013). *The impact of high-school schedule type on instructional effectiveness and student achievement in mathematics* [Doctoral dissertation]. Sycamore Scholars at Indiana State University. http:// hdl.handle.net/10484/8147

Higher Education Opportunity Act of 2008. 20 USC § 1003 (2008). www.govinfo.gov/content/pkg/ PLAW-110publ315/pdf/PLAW-110publ315.pdf

Hord, C., & Newton, J. A. (2014). Investigating elementary mathematics curricula: Focus on students with learning disabilities. *School Science and Mathematics, 114*(4), 191–201. doi: 10.1111/ssm.12064

Individuals with Disabilities Education Improvement Act, USC 20 §1400 *et seq.* (2004). www.congress.gov/108/plaws/publ446/PLAW-108publ446.pdf

Institute of Education Sciences (2018). *Digest of education statistics.* National Center for Education Statistics. https://nces.ed.gov/programs/digest/d18/

Izzo, M., & Bauer, W. (2015). Universal design for learning: enhancing achievement and employment of STEM students with disabilities. *Universal Access in the Information Society, 14*(1), 17–27. doi: 10.1007/ s10209-013-0332-1

Jay, T., Rose, J., & Simmons, B. (2018). Why is parental involvement in children's mathematics learning hard? Parental perspectives on their role supporting children's learning. *SAGE Open, 8*(2), 1–13. doi: 10.1177/2158244018775466

Johnson, D. W. & Johnson, R. T. (1992). Implementing cooperative learning. *Contemporary Education, 63*, 173–180.

Johnson, D. W., & Johnson, R. T. (2009). An educational psychology success story: Social interdependence theory and cooperative learning. *Educational Researcher, 38*, 365–379. doi: 10.3102/ 0013189x09339057

Kahneman, D. (2002, December). Maps of bounded rationality: A perspective on intuitive judgment and choice. In T. Frangsmyr (Ed.). *Les prix Nobel* [Nobel Prize Lecture] (p. 451). http://nobelprize.org

Kavale, K. A., & Forness, S. R. (1987). Substance over style: Assessing the efficacy of modality testing and teaching. *Exceptional Children, 54*(3), 228–239. doi: 10.1177/001440298705400305

Kavale, K. A., Hirshoren, A., & Forness, S. R. (1998). Meta-analytic validation of the Dunn and Dunn model of learning-style preferences: A critique of what was Dunn. *Learning Disabilities Research & Practice, 13*, 75–80.

Kavale, K. A., & LeFever, G. B. (2007). Dunn and Dunn model of learning-style preferences: Critique of Lovelace meta-analysis. *The Journal of Educational Research, 101*, 94–97. doi: 10.3200/joer.101.2.94-98

King-Sears, P. (2014). Introduction to special series on universal design for learning: Part one of two. *Learning Disability Quarterly, 37*, 68–70. doi: 10.1177/0731948714528337

Kitsantas, A., Cheema, J., & Ware, H. W. (2011). Mathematics achievement: The role of homework and self-efficacy beliefs. *Journal of Advanced Academics, 22*(2), 310–339. doi: 10.1177/1932202x1102200206

Kloo, A. & Zigmond, N. (2008). Coteaching revisited: Redrawing the blueprint. *Preventing School Failure, 52*(2), 12–20. doi: 10.3200/psfl.52.2.12-20

Landrum, T. J., & McDuffie, K. A. (2010). Learning styles in the age of differentiated instruction. *Exceptionality, 18*(1), 6–17. doi: 10.1080/09362830903462441

Lee, K.-H. (2016). Changes in attitudes towards textbook task modification using confrontation of complexity in a collaborative inquiry: Two case studies. In G. Kaiser, H. Forgasz, M. Graven, A. Kuzniak, E. Simmt, & B. Xu (Eds.), *Invited lectures from the 13th International Congress on Mathematical Education* (pp. 343–361). Springer. doi: 10.1007/978-3-319-72170-5

Leung, K. C. (2015). Preliminary empirical model of crucial determinants of best practice for peer tutoring on academic achievement. *Journal of Educational Psychology, 107*(2), 558–579. doi: 10.1037/a0037698

Leung, K. C. (2018). An updated meta-analysis on the effect of peer tutoring on tutors' achievement. *School Psychology International, 40*(2), 200–214. doi: 10.1177/0143034318808832

Lieber, J., Horn, E., Palmer, S., & Fleming, K. (2008). Access to the general education curriculum for preschoolers with disabilities: Children's school success. *Exceptionality, 16*(1), 18–32. doi: 10.1080/09362830701796776

Logan, B. (2011). Examining differentiated instruction: Teachers respond. *Research in Higher Education Journal, 13*, 1–14.

Lynch, S. D., Hunt, J. H., & Lewis, K. E. (2018). Productive struggle for all: Differentiated instruction. *Mathematics Teaching in the Middle School, 23*(4), 194–201. doi: 10.5951/mathteacmiddscho.23.4.0194

Macdonald, H., Zinth, J. D., & Pompelia, S. (2019). *50-state comparison: High school graduation requirements.* Education Commission of the States. www.ecs.org/high-school-graduation-requirements/

Maheady, L., & Gard, J. (2010). Class-wide peer tutoring: Practice, theory, research, and personal narrative. *Intervention in School and Clinic, 46*(2), 71–78. doi: 10.1177/1053451210376359

Maheady, L., Harper, G. F., & Mallette, B. (2003). *Class-wide peer tutoring.* (Current Practice Alert Series, 8). Division for Learning Disabilities and Division for Research, Council for Exceptional Children. www.TeachingLD.org/alerts/

Mark, J., Spencer, D., Zeringue, J. K., & Schwinden, K. (2010). How do districts choose mathematics textbooks? In B. J. Reys, R. E. Reys, & R. Rubenstein (Eds.), *Mathematics curriculum: Issues, trends, and future directions* (pp. 199–211). NCTM.

McKenna, J. W., Shin, M., & Ciullo, S. (2015). Evaluating reading and mathematics instruction for students with learning disabilities: A synthesis of observation research. *Learning Disability Quarterly, 38*(4), 195–207. doi: 10.1177/0731948714564576

McMaster, K. N., & Fuchs, D. (2005). *Cooperative learning for students with disabilities* (Current Practice Alert Series, 11). Division for Learning Disabilities and Division for Research, Council for Exceptional Children. www.TeachingLD.org/alerts/

McMaster, K. N., & Fuchs, D. (2002). Effects of cooperative learning on the academic achievement of students with learning disabilities: An update of Tateyama-Sniezek's review. *Learning Disabilities Research and Practice, 17*(2), 107–117. doi: 10.1111/1540-5826.00037

Moliner, L., & Alegre, F. (2020). Effects of peer tutoring on middle school students' mathematics self-concepts. *PloS One, 15*(4), e0231410–. doi: 10.1371/journal.pone.0231410

Morano, S., & Riccomini, P. (2017). Reexamining the literature: The impact of peer tutoring on higher order learning. *Preventing School Failure: Alternative Education for Children and Youth, 61*(2), 104–115. doi: 10.1080/1045988X.2016.1204593

Murray, G. V., & Moyer-Packenham, P. S. (2014). Relationships between classroom schedule types and performance on the Algebra I criterion-referenced test. *Journal of Education, 194*(2), 35–43. doi: 10.1177/002205741419400205

National Commission on Excellence in Education (NCEE). (1983). *A nation at risk: The imperative for educational reform.* www2.ed.gov/pubs/NatAtRisk/risk.html

National Governors Association Center for Best Practices & Council of Chief State School Officers. (2010). *Common core state standards for mathematics.* www.corestandards.org/Math/

O'Connor, R. E., & Jenkins, J. R. (2013). Cooperative learning for students with learning disabilities: Advice and caution derived from the evidence. In H. L. Swanson, K. R. Harris, & S. Graham (Eds.), *Handbook of learning disabilities* (2nd ed., pp. 507–525). The Guilford Press. www.guilford.com

Ollerton, M. (2014). Differentiation in mathematics classrooms. *Mathematics Teaching, 240*, 43–46.

Olson, J., & Platt, J. (1992). *Teaching children and adolescents with special needs.* Merrill/Macmillan.

Orkwis, R., & McLane, K. (1998). *A curriculum every student can use: Design principles for student access* (ED423654). ERIC. https://files.eric.ed.gov/fulltext/ED423654.pdf

O'Sullivan, R. H., Chen, Y.-C., & Fish, M. C. (2014). Parental mathematics homework involvement of low-income families with middle-school students. *School Community Journal, 24*, 165–187. https://files.eric.ed.gov/fulltext/EJ1048611.pdf

Rao, K., Ok, M., & Bryant, B. R. (2014). A review of research on universal design education models. *Remedial and Special Education, 35*(3), 153–166. doi: 10.1177/0741932513518980

Rao, K., Ok, M., Smith, S., Evmenova, A., & Edyburn, D. (2020). Validation of the UDL reporting criteria with extant UDL research. *Remedial and Special Education, 41*(4), 219–230. doi: 10.1177/0741932519847755

Rehabilitation Act of 1973. 29 USC § 701 et seq. (1973). www2.ed.gov/policy/speced/leg/rehab/rehabilitation-act-of-1973-amended-by-wioa.pdf

Rexroat-Frazier, N., & Chamberlin, S. (2019). Best practices in co-teaching mathematics with special needs students. *Journal of Research in Special Educational Needs, 19*(3), 173–183. doi: 10.1111/1471-3802.12439

Riccomini, P., Morano, S., & Hughes, C. (2017). Big ideas in special education: Specially designed instruction, high-leverage practices, explicit instruction, and intensive instruction. *TEACHING Exceptional Children, 50*(1), 20–27. doi: 10.1177/0040059917724412

Roach, J. C., Paolucci-Whitcomb, P., Meyers, H. W., & Duncan, D. A. (1983). The comparative effects of peer tutoring in math by and for secondary special needs students. *Pointer, 27*(4), 20–24.

Rogers, W. J., & Weiss, M. P. (2019). Specially designed instruction in secondary co-taught mathematics courses. *TEACHING Exceptional Children, 51*(4), 276–285. doi: 10.1177/0040059919826546

Rohrer, D., Dedrick, R., & Hartwig, M. (2020). The scarcity of interleaved practice in mathematics textbooks. *Educational Psychology Review, 32*(3), 873–883. doi: 10.1007/s10648-020-09516-2

Root, J. R., Cox, S. K., Hammons, N., Saunders, A. F., & Gilley, D. (2018). Contextualizing mathematics: Teaching problem solving to secondary students with intellectual and developmental disabilities. *Intellectual and Developmental Disabilities, 56*, 442–457. doi:10.1352/1934-9556-56.6.442

Root, J. R., Cox, S. K., Saunders, A. F., & Gilley, D. (2020). Applying the universal design for learning framework to mathematics instruction for learners with extensive support needs. *Remedial and Special Education, 41*(4), 194–206. doi: 10.1177/0741932519887235

Rosário, P., Cunha, J., Nunes, A., Moreira, T., Núñez, J., & Xu, J. (2019). "Did you do your homework?" Mathematics teachers' homework follow-up practices at middle school level. *Psychology in the Schools, 56*(1), 92–108. doi: 10.1002/pits.22198

Slavin, R. E. (1995). *Cooperative learning: Theory, research, and practice.* Allyn & Bacon.

Slavin, R. E. (2017). Instruction based on cooperative learning. In R. E. Mayer & P. A. Alexander (Eds.), *Handbook of research on learning and instruction* (2nd ed., pp. 388–404). Routledge. doi: 10.4324/9780203839089.ch17

Small, M. (2009). *Good questions: Great ways to differentiate mathematics instruction.* Teachers College Press and NCTM. www.tcpress.com

Smith, S., Rao, K., Lowrey, K., Gardner, J., Moore, E., Coy, K., Marino, M., & Wojcik, B. (2019). Recommendations for a national research agenda in UDL: Outcomes from the UDL-IRN preconference on research. *Journal of Disability Policy Studies, 30*(3), 104420731982621–185. doi:10.1177/1044207319826219

Smith, T. J., McKenna, C. M., & Hines, E. (2014). Association of group learning with mathematics achievement and mathematics attitude among eighth-grade students in the US. *Learning Environment Research, 17*(2), 229–241. doi: 10.1007/s10984-013-9150-x

Solone, C. J., Thornton, B. E., Chiappe, J. C., Perez, C., Rearick, M. K., & Falvey, M. A. (2020). Creating collaborative schools in the United States: A review of best practices. *International Electronic Journal of Elementary Education, 12*(3), 283–292. doi: 10.26822/iejee.2020358222

Spooner, F., Baker, J. N., Harris, A. A., Ahlgrim-Delzell, L., & Browder, D. M. (2007). Effects of training in universal design for learning on lesson plan development. *Remedial and Special Education, 28*(2), 108–116. doi: 10.1177/07419325070280020101

Stefanidis, A., King-Sears, M. E., & Brawand, A. (2018). Benefits for coteachers of students with disabilities: Do contextual factors matter? *Psychology in the Schools, 56*(4), 539–553. doi: 10.1002/pits.22207

Swanson, E. A. (2008). Observing reading instruction for students with learning disabilities: A synthesis. *Learning Disability Quarterly, 31*(3), 115–133. doi: 10.2307/25474643

Tarim, K. (2009). The effects of cooperative learning on preschoolers' mathematics problem-solving ability. *Educational Studies in Mathematics, 72*(3), 325–340. doi: 10.1007/s10649-009-9197-x

Tepe, L., & Mooney, T. (2018). *Navigating the new curriculum landscape: How states are using and sharing open educational resources.* Council of Chief State School Officers. https://ccsso.org/resource-library/navigating-new-curriculum-landscape

Thompson, G. (2017, 20 October). The shifting textbook adoption market. *Victory Productions.* https://victoryprd.com/blog/shifting-textbook-adoption-market/

Tomlinson, C. A. (1999). *The differentiated classroom: Responding to the needs of all learners.* Pearson.

Tomlinson, C. A. (2003). *Fulfilling the promise of the differentiated classroom.* ASCD. www.ascd.org/publications/books/103107.aspx

Tomlinson, C. A. (2009). Learning profiles and achievement. *School Administrator, 66*(2), 28–34. www.aasa.org/schooladministratorarticle.aspx?id=3460

Treahy, D. L., & Gurganus, S. P. (2010). Models for special needs students. *Teaching Children Mathematics, 16*(8), 484–490.

Topping, K., Buchs, C., Duran, D., & van Keer, H. (2017). *Effective peer learning: From principles to practical implementation.* Routledge. doi: 10.4324/9781315695471

Tournaki, N. (2003). The differential effects of teaching addition through strategy instruction versus drill and practice to students with and without learning disabilities. *Journal of Learning Disabilities, 36*(5), 449–458. doi: 10.1177/00222194030360050601

van Garderen, D., Scheuermann, A., & Jackson, C. (2012). Developing representational ability in mathematics for students with learning disabilities: A content analysis of grades 6 and 7 textbooks. *Learning Disability Quarterly, 35*(1), 24–38. doi: 10.1177/0731948711429726

Van Voorhis, F. L. (2011). Adding families to the homework equation: A longitudinal study of mathematics achievement. *Education and Urban Society, 43*(3), 313–338. doi: 10.1177/0013124510380236

VanDerHeyden, A. M., & Codding, R. S. (2015). Practical effects of classwide mathematics intervention. *School Psychology Review, 44*(2), 169–190. doi: 10.17105/spr-13-0087.1

Walther-Thomas, C. S., Korinek, L., McLaughlin, V., & Williams, B. (2000). *Collaboration for inclusive education: Developing successful programs.* Allyn & Bacon.

Witzel, B., Smith, S. W., & Brownell, M. T. (2001). How can I help students with learning disabilities in algebra? *Intervention in School and Clinic, 37*(2), 101–104. doi: 10.1177/105345120103700205

Yell, M. L., Crockett, J. B., Shriner, J. G., & Rozalski, M. (2017). Free appropriate public education. In J. M. Kauffman, D. P. Hallahan, & P. C. Pullen (Eds.), *Handbook of special education* (2nd ed., pp. 71–86). Routledge. doi: 10.4324/9781315517698

Zelkowski, J. (2010). Secondary mathematics: Four credits, block schedules, continuous enrollment? What maximizes college readiness? *The Mathematics Educator, 20,* 8–21. http://math.coe.uga.edu/tme/Issues/v20n1/20.1_Zelkowski_p.8-21.pdf

Zigmond, N. P., & Kloo, A. (2017). General and special education are (and should be) different. In J. M. Kauffman, D. P. Hallahan, & P. C. Pullen (Eds.), *Handbook of special education* (2nd ed., pp. 249–261). Routledge. doi: 10.4324/9781315517698

7 Resources to Support Mathematics Instruction and Integration

Chapter Questions

1. How are manipulatives and other hands-on materials best selected and used for mathematics instruction?
2. How can technology enhance mathematics learning?
3. What are benefits and concerns about integrating mathematics with other content areas?
4. What mathematics skills are required for home, community, and work settings?

The entire staff of Pinetops Elementary School, including Chris Johnson, the mathematics support teacher, is attending a January workshop on integration of content areas across the elementary curriculum. Segments of the inservice training address the integration of mathematics and science standards, the use of children's literature across subject areas, writing across the curriculum, implementing technology across the curriculum, and using the arts with other content areas. Chris discusses her concerns with the other teachers at her table before the first session begins, "I feel so pressured to cover the math standards in time for spring testing, I can't imagine how we can fit in more content."

"I've had concerns for several years about curriculum connections," offers Mei Wang. "My fifth-grade students just aren't able to use their math skills in social studies or science. They act like they've never heard of fractions or measurement with those applications, even though we completed units in both areas. It seems to me there are many commonalities among the content areas that we don't stress enough. My concern is having enough planning time, especially with new technologies."

"My students need integrated mathematics, but for skills they will need in everyday life," comments Robin Small, a special educator. "How can I address grade-level standards with my students when they are two and three grade levels behind? They love to use the computer, but they tend to select programs that are easy for them."

Teachers of mathematics, whether they are special educators, core mathematics teachers, or other support professionals, often seek supplementary materials, representations, tools, and technology to enhance their students' learning. They also must understand learning goals across the curriculum and integrate those where possible. This chapter provides an orientation to resources for the mathematics classroom but is not intended to be comprehensive. Chapter

DOI: 10.4324/9781003096733-7

topics include manipulatives and other representations for mathematics concepts, technology for instruction, and integration with other content areas. The chapter closes with a section on life-skills mathematics, those skills that are not explicitly described in standards but are replete with mathematics concepts and applications.

Selecting, Creating, and Using Manipulatives

The use of manipulatives, or concrete objects that can be handled by students in a sensory manner to foster mathematical thinking, has been researched for decades with mixed results (Sowell, 1989; Swan & Marshall, 2010). In a meta-analysis of concrete manipulative use for mathematics, Carbonneau et al. (2013) identified 55 studies that compared instruction involving manipulatives with abstract-symbol instruction across grade levels. The researchers concluded that manipulative use produces small- to medium-sized effects on student learning. The strength of the effect, however, was dependent on other instructional variables such as perceptual richness of the object, mathematics topic, instructional time, level of guidance offered to students, and the developmental status of the learner. Some of the research on these variables was mixed. For example, perceptual richness was determined as too distracting in some studies while others found that perceptually rich manipulatives enhanced transfer-of-learning. Uribe-Flórez and Wilkins (2017) analyzed longitudinal student data across Grades K to 5 from a large, national database and found that manipulative use decreased as grade level increased, perhaps due to curricular emphases or teacher preferences. Although manipulative use was not correlated with achievement at a specific grade level, there was a positive relationship between manipulative use and student learning across grade levels. Students demonstrated an increased rate of learning with regular and ongoing manipulative use, a statistically significant 2.1 points per year for medium use and 5.09 for high manipulative use in their classrooms.

Research on manipulative use with preschool children has typically evaluated two instructional elements: level of teacher guidance and perceptual qualities of manipulatives. Carbonneau and Marley (2015) found, in a study with preschool children developing early number sense, that participants who had higher levels of instructional guidance outperformed those with low guidance. When teachers made explicit links between manipulatives and symbols, children demonstrated higher procedural and conceptual knowledge as well as greater transfer. For conceptual knowledge, children who used bland manipulatives (e.g., blocks or sticks) outperformed those who used rich, realistic ones (e.g., frogs or play money). The researchers hypothesized that realistic manipulatives tended to distract learners from the abstract mathematical concepts while bland ones were understood to be a referent for the concepts. Clements and Sarama reinforced these findings by reminding us that "manipulatives to not 'carry' mathematical ideas" (2018, p. 3). Children need teachers to help them reflect on their actions with manipulatives and help them move to more sophisticated mathematical representations. Teachers should move away from manipulatives for a given concept when students are able to use drawings and verbal or written symbols.

Studies have indicated moderate to strong positive results for manipulative use with students with learning difficulties and disabilities. One of the recommendations of the expert panel report *Assisting Students Struggling with Mathematics* (Gersten et al., 2009) was that mathematics "materials should include opportunities for students to work with visual representations of mathematical ideas…" (p. 6). The evidence included studies with systematic use of visual representations and manipulatives, including research on the concrete-representational-abstract (CRA) sequence. Bouck and Park (2018) conducted a systematic review of literature on manipulative use to support students with disabilities (SWD). Most of the 36 studies employed the CRA framework and were implemented with elementary and secondary students across many mathematics topics. Teachers and researchers effectively used both teacher-made and commercial manipulatives to enhance student learning. In a review of the CRA approach for students with specific mathematics learning disabilities (MLD), Bouck, Satsangi, and Park (2018) identified 20 studies since the 1970s and, based on the quality indicators and results of eight high-quality studies, the researchers concluded that CRA is an evidence-based practice for students with MLD, especially for basic-operation interventions. Peltier et al. (2020) conducted a meta-analysis of single-case research using mathematics manipulatives for students with mathematics difficulties (MD, 48 studies). The researchers calculated an omnibus effect size of 0.91 (strong) and concluded that positive effects were consistent across age of participants, implementer (teacher vs. researcher), and for both mathematics processes and solution strategies. It is important to note that all except one study employed explicit instruction combined with manipulative use and a variety of teacher-made and commercial manipulatives were effective representations.

Manipulatives have been demonstrated to improve learning by students with MD for place-value concepts (Peterson et al., 1988, using cubes and paper strips), fact mastery and fluency for the four arithmetic operations (*Strategic Math Series*, Mercer & Miller, 1992), for algebra word problems using algebra tiles (Maccini & Hughes, 2000); and for teaching fraction concepts using construction paper models (Jordan et al., 1999). Marsh and Cooke (1996) demonstrated the effectiveness of using Cuisenaire® Rods to support problem-solving instruction with students with MLD. Cass et al. (2003) used geoboards to teach perimeter and area concepts to students with MLD. Milton et al. (2019) demonstrated that the multiplication and division modules of the CRA-based *Strategic Math Series* (Mercer & Miller, 1992) could be alternated for better concept understanding of the relationship between the operations. Flores et al. (2020) applied the CRA framework to a Tier 2 fraction intervention (17 fifth-graders) using area and linear models of fractions. Students in the CRA intervention group demonstrated significantly greater learning ($\eta^2 = 0.67$) and improved from below 50% to 89% on fraction measures as compared with a control comparison using small-group interventions with model and guided practice. The linear models of fractions (e.g., fraction towers and number lines) provided students a greater sense of magnitude than area models (e.g., fraction circles).

In general, concrete manipulatives have the potential to support the development of abstract reasoning, simulate real-world knowledge, promote nonverbal coding to support learning, provide motivating activities, and enhance transfer of learning and language development (Bouck & Park, 2018; Carbonneau & Marley, 2015; Peltier et al., 2020). Since mathematics is essentially a science of number and space and their interrelationships, concrete materials support student understanding of these relationships. However, the effective use of concrete materials takes time, reflection, and practice. Manipulatives themselves do not cause concept understanding or transfer. Teachers must understand concepts at deeper levels and select materials that enhance the understanding of that content where it is most appropriate—for concept introduction, guided-learning experiences, small-group activities, individual practice, and applications. Most importantly, manipulative use should be accompanied by explicit teacher guidance through modeling, feedback, scaffolding, and language, and should be faded as soon as possible to more abstract representations (Clements & Sarama, 2018; Fyfe et al., 2014; Gersten et al., 2009; Morin & Samelson, 2015). The next two sections address concrete manipulatives, those that are teacher-made or purchased for hands-on manipulation. Virtual manipulatives are computer-based simulations of the concrete versions and are discussed in a later section.

Creating Materials

The most effective concrete and representational materials for mathematics are those that are:

• valid representations of important concepts;
• inexpensive enough for each student to have adequate materials;
• durable enough for repeated student use (unless consumable); and
• age appropriate.

Teacher- or student-created materials and inexpensive everyday items can meet all of these criteria. Additional requirements may include the ease of creating larger or projected versions for demonstration, safety with specific students, the ability to send materials home with assignments, and storage capability. This section will provide useful examples of simple materials and should spark the reader's imagination for additional adaptations or creations. A list of free and inexpensive items for a mathematics classroom is featured in Box 7.1. Give teachers a pile of construction paper and graph paper and they can teach most mathematics concepts!

Box 7.1   Free and Inexpensive Materials for the Mathematics Classroom

Paper Products	Cloth	Other Household Items
paper rolls	ribbon	shoe boxes
envelopes	string	paper clips
construction paper	yarn	paper brads
graph paper	felt	small mirrors
library card pockets	cloth scraps	file folders
cake/pizza circles		sponges
paper cups and plates	**Print Materials**	shower curtains
paper bags	menus	tape measures
cardboard inserts	pamphlets	poker chips
boxes	newspapers	playing cards
	magazines	dice
Plastics	calendars	tiles
clean food trays	maps	children's blocks
coffee stirrers		clocks
straws	**Kitchen Items**	coins
buttons	scales	game boards and pieces
clear plastic cups	thermometers	dominoes
beads	measuring cups, spoons	Lego blocks
	empty food containers	plastic bins
Wood	egg cartons	stickers
popsicle sticks	dry beans	pipe cleaners
toothpicks	cupcake pans	play dough
paint sticks	bottle caps	needlepoint plastic
yard and meter sticks	pasta shapes	rubber bands
wooden dowels		common objects

For the early grades, simple counting objects such as blocks, buttons, bottle caps, crayons, game-board objects, and other everyday objects with similar characteristics can be used to develop whole-number concepts and place value for numbers up to 20. Smaller or edible objects (e.g., beans, macaroni, candy) should be used with care, as will be discussed in the next section. Younger children developing number concepts to 10 may find using 5- or 10-frames easier to visualize than scattered objects. These simple frames may be created from boxes or line drawings. Frames encourage children to name the total without counting each item. A 10-frame with round disks as a transition between concrete and representational is depicted in Strand A, Figure A.8. To demonstrate the properties of sets and groupings, objects can be placed on paper plates or sheets of construction paper, in clear plastic cups or sections of cupcake pans, or in other containers that are easy to view and manipulate.

Concepts of larger numbers, requiring place-value representations, can be developed using bundled straws, wooden sticks or coffee stirrers, or, in more representational forms, coins (penny, dime, dollar) or sections of graph paper (units, strips of 10, blocks of 100, cube with 100 on each face). Place-value mats (see Appendix II) are useful devices for organizing objects by place, engaging in trading (regrouping) activities, and transfer into symbol (number) form. Teachers and students can represent decimal forms of number using these same objects if the one's place is the larger grouping (i.e., the bundle, cube, or dollar is one unit) that can be divided into 10 parts to represent tenths and 100 parts to represent hundredths. The place-value mat in this case would include a decimal point.

Many free and inexpensive materials can be adapted to represent fractions, percents, decimals, ratios, and probability. Although teachers often reach first for the pizza example, strips of construction paper (e.g., 3 in. by 12 in.) are better representations for initial learning, with one intact strip representing one whole and folded or cut sections representing parts. These strips are easily folded, torn, labeled, and glued. Fraction concepts can be extended with representations for mixed numbers and with more than one strip representing the "whole." Students can explore the answers to questions such as, "What are other names for 1/2?" and, "How do we divide a fraction by a fraction?" using paper strips.

Because percents are units per hundred, materials representing percents should clearly depict that base. Examples include pennies and dollars, 100s grids or graph-paper squares, or meter sticks with centimeters labeled. Ratio comparisons can be made within sets of similar items or between different items, so materials representing ratios should include those with similar (cubes, crayons, coins) or different (collections of objects, pictures) characteristics. Standard ratios (rates; e.g., miles per hour, price per yard) should be represented with real-world materials such as maps or advertisements. Probability of events can be represented with many concrete objects: dice, spinners, coins, cards, cubes, and semi-concrete grids and patterns. It is important that students have experience with a variety of objects for the generalization of concepts. Even data sets (surveys, weather data, baseball statistics), magazine clippings (ads, photographs), and game boards (checkers) can be adapted for probability experiments.

For geometry and measurement concepts, two- and three-dimensional figures and their features come to life when constructed by students. Construction paper, graph paper, straws and string, paper brads, clay, and food packaging are inexpensive and flexible construction materials. Paper-folding activities have been used to develop many geometric concepts such as angles, shapes, symmetry, lines, rotations, and other transformations. Cubes and pyramids can be constructed with thick paper using simple directions that actually demonstrate the concepts of face, vertex, and volume. Straws strung together with string can create open three-dimensional shapes with the straws representing edges. These activities can also be used to check concept and math-language understanding, develop general spatial sense, and promote overlapping fraction concepts (e.g., Wheatley & Reynolds, 1999).

Measurement tools can be standard devices or alternatives. For measuring distance, paper tape measures and free or inexpensive rulers and yardsticks are used for standard units while any object with a constant length can make alternative measures (e.g., shoe, block, wooden dowel, vinyl floor tile, unsharpened pencil, paper clip). Dollar store measuring sets for liquid volume sometimes include both metric and customary units. Other everyday containers are marked with units (pint of milk, liter of soda, 32 oz of juice). Teachers often use water with a drop of food coloring for easier reading, or, when liquids are not feasible, inexpensive dry goods such as rice. Measures of nonliquid volume and capacity (cubic units) can be made with the same materials using three-dimensional geometric objects.

Safety and Individual Concerns

Teachers should consider safety issues before using certain manipulatives with students. A number of publications promote edible manipulatives (e.g., m & m's, pretzels, jellybeans, chocolate bars) for mathematics instruction. Edible manipulatives can be extremely motivating but teachers should be aware of their disadvantages. Many students have severe allergies to food items, especially peanuts, tree nuts, and products containing milk, eggs, or wheat. The use of food for working with math concepts is usually unsanitary, unless carefully controlled. Rice or corn used for measurement activities should be composted, not cooked. "We're going to put the ones we've used in the compost and cook a fresh batch to eat afterward." Some food items, such as beans and corn kernels, are too small for safe use with younger children.

Other math tools pose safety hazards, especially for students with behavioral and self-control issues. Students should be taught how to safely use the rubber bands with geoboards or, if not practical, substitutes such as virtual manipulatives, string, or dot paper should be used. Compasses with metal points are especially dangerous and safety compasses, offering the same construction capabilities, should be used instead. Scissors are available in safer versions with blunt tips. Other construction supplies, such as glue, crayons, markers, toothpicks, building clay, and straight edges, should be checked for toxicity or sharpness and monitored for safe use.

Although not safety issues, some manipulatives are not appropriate for use with some students because of individual needs or some objects may need adaptations. Students who do not discern between specific colors will not be successful with color-coded manipulatives unless the colors are selected with care or also have patterns. Students with fine-motor issues may need larger versions of manipulatives or even virtual simulations. Children who cannot use vision for learning or have low vision will need adaptations to manipulatives and related diagrams. If these students are served under an individualized education program (IEP), those adaptations should be specified on the IEP and provided. Other students with disabilities can access adaptations under *504 plans*.

Representational Materials

Materials at the representational level also bridge mathematics understanding to the symbolic. In the CRA sequence, representational refers to the iconic, or pictorial, a two-dimensional figure that represents mathematics concepts. Representational materials are also called semi-concrete because in some forms they can be handled, cut, marked, and folded. Number lines, among the most commonly used mathematics tools, are flexible representations of the sequence of whole numbers and can be adapted for negative integers and other rational numbers (fractions and decimals). Number lines can be drawn in small versions for students' individual use or in large formats across the classroom floor or wall. Hundred charts also have powerful applications for a range of number concepts including sequencing, count-bys, multiples, and factors. Hundred charts can be depicted in 0 to 99 or 1 to 100 formats, in either right-to-left or left-to-right sequences. They can be teacher made or downloaded from a number of websites (see Appendix II). Larger pocket versions allow for removal and manipulation of numbers and number patterns. Individual hundred charts can be used with plastic or paper discs or drawn on directly.

Graph paper, also readily available and mathematically powerful, is available in many formats: linear, metric (10, 100), isometric, Cartesian, single quadrant, logarithmic, geometric, dots, and polar. These papers can be cut, marked, folded, and compared. They can be used for solving problems and creating demonstrations. Various grids are also available on student-size white boards. Number tiles and fraction shapes are created by copying templates onto sturdy paper and cutting out the shapes (see Appendix II). Other representations can be created from photographs, diagrams, drawings, and calendars. Students can assist with making mathematics materials. They can cut out number tiles and fraction shapes, create number lines, and use *nets* to create three-dimensional objects. Individual sets of materials can be stored in easy-close plastic bags or bins.

Commercial Materials

Catalogs and websites for teachers offer dozens of concrete materials to support mathematics instruction. This section will describe select commercial materials that have the power to demonstrate big ideas and connect critical concepts. These materials are only a sampling of what is available and were selected because they frequently appear in classroom kits that accompany textbooks. Swan and Marshall (2010) found, in a survey of teachers, that the most commonly used commercial manipulatives were pattern blocks, attribute blocks, base-10 blocks, geoshapes, Unifix® and other linked cubes, square tiles, and Cuisenaire® Rods.

Cuisenaire® Rods, invented by Emile-Georges Cuisenaire in 1930 in Belgium to convey critical mathematics concepts to his students, were described in the collaborative book *Mathematics with Numbers in Colour* in partnership with Caleb Gattegno (Cuisenaire & Gattegno, 1954). The rods, now available in wood or plastic, were designed in specific lengths (1 to 10 units) and colors to be visual representations of numbers. Cuisenaire® Rods are continuous representations (not scored or separable) of number and therefore have arithmetic and algebraic properties (see Figure 7.1). For example, the white rod is 1 unit, the red 2 units, the green 3 units, and the purple 4 units. Four whites equal one purple. Two reds equal one purple. The purple is twice as long as the red. The white is one-fourth the length of the purple. If the

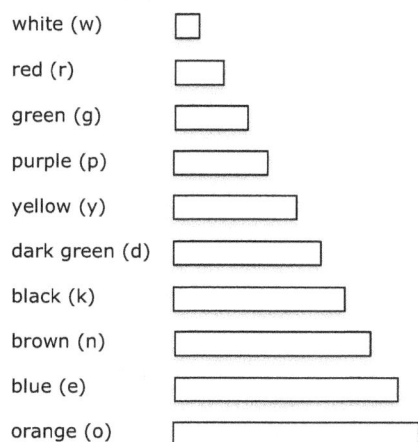

white (w)

red (r)

green (g)

purple (p)

yellow (y)

dark green (d)

black (k)

brown (n)

blue (e)

orange (o)

Figure 7.1 Cuisenaire® Rods

Source: Cuisenaire® Rods is a registered trademark of hand2mind, Inc. and is used with permission from hand2mind, Inc.

white represents 8 fish, then the red would represent 16 fish. The orange rod could represent a 1, 10, 100 or even 0.5, depending on the context.

Cuisenaire® Rods have powerful applications in developing number and spatial sense, as well as geometry, measurement, probability, and statistics concepts. Students with MD need explicit orientation to the value of the rods with each activity. Rods are typically called by their color names, but their values are intentionally variable. Students who cannot discriminate colors may need their rods coded with letters. For a full-color view and example activities and demonstrations, visit the website www.hand2mind.com.

Geoboards also have a direct connection to Caleb Gattegno, who referred to the *Gattegno Geoboard* in a 1954 article. Geoboards, available in wood and plastic in many sizes, have pins or nails arranged in rectangular arrays or circular patterns. Geometric figures are created on geoboards using rubber bands. Painted-surface geoboards can be marked with chalk. Clear plastic geoboards are useful for overhead demonstrations and comparing patterns by overlay. Geoboards were developed to demonstrate geometric concepts such as perimeter, area, circumference, line, angle, symmetry, and transformations. Additional mathematics concepts, such as those of rational numbers, multiplication, and graphing, also can be modeled with geoboards.

Unit cubes are available in many sizes, colors, and materials. The smallest commercial cubes are 1 cm^3 and weigh 1 gm, too small for younger children, but perfect for developing base-10 and decimal-system concepts. The most commonly used cubes for younger children measure 2 cm^3, 3/4 in ^{3}, and 1 in^3, appropriate for exploring a range of number and spatial concepts. Many commercial-brand cubes (e.g., Unifix®, Multilink, Snap Cubes®, and PopCubes®) have connection devices on two or more sides, for creating cube towers and other shapes. Some cubes are plain, while others feature dots, numbers, and other symbols on the sides (like dice), or even write-on surfaces. Fraction Tower® Equivalency Cubes are interconnecting cubes labeled with fractions, decimals, or percents for comparison of these rational number forms. Cubes are useful for developing number and spatial sense through activities such as counting, sorting, comparing, measuring, estimating and constructing.

Base-10 blocks were most likely developed by the Hungarian Mathematician Zoltan Paul Dienes (Picciotto, n.d.), and are sometimes called Dienes Blocks. Base-10 sets are available in wood and plastic versions in many colors and sizes. However, since color is not the attribute of concern in most base-10 applications, single-color sets are recommended, especially for students with MD. Each set should include unit blocks, rods (a connected bar of 10 units), flats (a connected square of 10 rods or 100 units), and cubes (representing 1000 units, sometimes called decimeter cubes). Sets are also available in clear and soft versions and sets with holders (Digi-Blocks). Useful supplements are the overhead sets, stamps, and place-value mats. A place-value mat with base-10 blocks is depicted in Strand B, Figure B.5.

Attribute blocks are common in early childhood classrooms but also have applications for more advanced mathematics topics. Attribute blocks are proportionally sized blocks with five shapes (circle, square, rectangle, triangle, hexagon) in two sizes (small and large), three colors (yellow, red, and blue), and two thicknesses. They can be purchased in wood, plastic, or foam or made by the teacher using patterns, templates, or stamps. Attributes are characteristics of objects that can be compared, sorted, identified, and used for object description. The concept of attributes is important for understanding geometric properties, logical reasoning, problem solving, probability, and set theory. Students with MD sometimes have trouble distinguishing more than one attribute at a time or identifying the important attributes for creating shapes or finding members of sets.

Pattern blocks are often confused with attribute blocks but are distinct—they come in six specific polygons (square, triangle, trapezoid, hexagon, rhombus, and parallelogram) that can form fractional parts of each other. Pattern blocks can be obtained in a single-color set or mixed colors in many sizes and materials. Also available are pattern block overhead sets, templates, stamps, paper repeating patterns, and die cutters. Pattern blocks can be used throughout the mathematics curriculum for concepts such as patterns, geometry, fractions, tessellations, problem solving, estimation, and probability. Commercially available representations of rational numbers include *fraction pieces* or *tiles* such as fraction tiles and fraction circles. When using these concrete representations, the teacher should make explicit references to the specific rational-number types, not just parts-of-whole concepts (see Strand C).

Tangrams and *pentominoes* can extend students' concepts of spatial relationships such as area, flips, turns, perimeter, ratio, and properties of triangles and rectangles. Tangrams, most likely an ancient Chinese game, are actually squares that have been divided into seven pieces: five triangles, one square, and one parallelogram. By turning and flipping pieces, different shapes can be created with just seven pieces. Pentominoes are five connected square blocks, and, although they may be older, the name is attributed to Solomon Golomb from a lecture in 1953 to the Harvard Mathematics Club (Bhat & Fletcher, 1995). A set of 12 different shapes are identified by the letters which assist with identification: T, U, V, W, X, Y, Z, F, I, L, P, and N. Challenges for pentominoes include creating 6×10, 5×12, 4×15, and 3×20 rectangles using all 12. Students may also enjoy creating full sets of triominoes or tetrominoes.

Algebraic concepts were first modeled using base-10 blocks by Dienes and later, using multi-base blocks, by Laycock (Picciotto, n.d.). Rasmussen developed a tile form to represent algebra concepts and added color on each side of the tiles for positive and negative values as well as a frame with a corner. When selecting the appropriate model for students, teachers should consider the range of concepts to be addressed, students' cognitive flexibility, and the limitations of some visual models. These geometric-based models of algebraic concepts have the additional advantage of connecting math concepts and prior student knowledge.

Algebra Tiles and other xy tiles are inexpensive and widely available. Algebra Tiles include three pieces—the small square, the larger square, and a rectangle. The small and larger squares are not proportionately related. Colored sides represent positive values while the red sides represent negatives. Algebra Tiles are limited to representing two dimensions and the blocks can represent units (whole numbers) or variables but cannot represent whole numbers and more than one variable at a time. The representation of negatives and positives with color also limits their use and may cause misconceptions if extended too far. Algebra Tiles can assist initial algebra concepts such as modeling polynomials, performing the four operations with polynomials (including the distributive property and factoring), and modeling quadratic equations. Figure 7.2 illustrates multiplication of the polynomials $(x + 2)(3x + 1)$.

Algeblocks provide a three-dimensional view of algebraic equations that include units and two variables. Algeblocks offer a quadrant map to assist with more accurate representations of minus. They are also limited in representing minus in operations but can expand polynomial concepts to two variables with units and extend the distributive property to include examples such as $(x^2)(x + 5y)$. The *Lab Gear* (Picciotto, n.d.) algebra model involves two different representations at once, positions related to the mat and blocks on top of each other, which allows for a geometrically sound representation of minus but may be initially more difficult to understand. The Lab Gear model was developed for high-school algebraic applications and includes models for 5 and 25 per the original Rasmussen and Laycock versions and can be extended with base-10 blocks. The Lab Gear model includes a unit block with scored flats for 5, 25, $5x$, and $5y$; a rod representing x; a longer rod representing y; flats representing x^2, y^2, and xy; and cubes representing x^3, y^3, x^2y, and xy^2.

Other commercial manipulatives include the *abacus*, most likely dating back to Sumeria between 2700 and 2300 BCE (Ifrah, 2001) but found in different formats today, and is useful for developing number sense, the four operations, square roots, and cubic roots. The *geared clock*

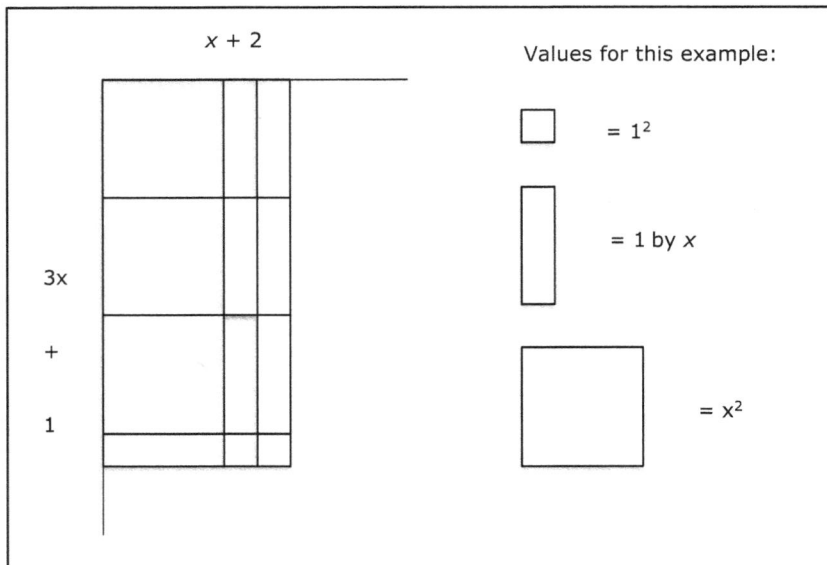

Figure 7.2 Algebra Tiles: $(x + 2)(3x + 1) = 3x^2 + 7x + 2$

may have advantages over a teacher-made clock because hidden gears maintain correct hour and minute relationships. A *math balance* has arms with pegs for hanging weighted labels, useful for equivalence, operation, property, and measurement concepts in arithmetic and algebra. *Two-color counters*, with a different color on each side of a disk, are useful for developing number sense and operations with positive and negative integers as well as probability concepts and relationships on hundred charts. For older students, *hands-on trigonometry* sets include 20 foam pieces in four colors for demonstrating proofs or relations such as the Pythagorean identity, a sum formula, a double-angle formula, and sum-to-product formulas.

Teachers should spend some time exploring the use of these concrete models before introducing them to students. Consider what might be confusing or supportive during student instruction. If introduced gradually, with clear and explicit references to numbers and variables depicted and deliberate transfer activities to other representations such as graphs and equations, these models have the potential to unlock critical concepts for students. No representative model will be a perfect match for the more abstract forms of mathematics, but they offer sound conceptual visualizations that can scaffold, or support, fragile and superficial understanding.

TRY THIS

Select one of the following mathematics *big ideas* and explain how concrete manipulatives can introduce the concept:

place value (whole number)	proportional reasoning
linear equation	factoring
perimeter related to area	fraction to decimal equivalence

Technologies for Mathematics

Technology means the practical application of knowledge to a specific art or domain. For mathematics, technologies have come to include the range of tools teachers and students use to explore and connect concepts and facilitate problem solving. Pencil and paper are technologies that assist learning, as are the abacus, slide rule, and ruler. These technologies, often termed *low-tech*, have assisted learners with mathematics for centuries. With electricity and mass production capabilities in the 20th century, more technologies for teaching were developed, including lighted projectors, audio- and videotape, early computing machines, and handheld calculators. Relatively recent technologies—personal computing devices, the internet, virtual tools, augmented reality, cloud computing, screen casting, and 3D printing—will continue to evolve and expand, challenging teachers to consider effective applications.

Technologies, in themselves, are not "better" or even "effective." The effectiveness of *any* tool for learning depends on a number of factors, including the teacher's knowledge and skill, student characteristics, accessibility features, resource budget, and nature of the content. Digital technologies, used with planning and teacher guidance, have the potential to enhance learning by providing motivating interactions with mathematics content, virtual representations of difficult concepts, multiple means of access to content and teachers, and multiple tools for understanding and solving problems as well as for organizing and expressing ideas. Technology enhancements also have the potential to influence the ways in which students think about mathematics (Young, 2017). However, technologies also have serious issues. These include hampering communication and face-to-face interactions, cost for each student to take advantage of access and interactions, teacher training to keep up with the most effective technologies for mathematics applications, and the risk of viewing the technology as the teacher rather than as a tool for the student in the learning process. Some students have not been mentally challenged or engaged by technology, others may be too distracted or select components at the independent level rather than their instructional level. Further, teachers may be overwhelmed by available technology and not be trained in its best, most effective use for the mathematics classroom.

The National Council of Teachers of Mathematics recommended *strategic* use of technology for teaching and learning of mathematics "in thoughtfully designed ways and at carefully determined times so that the capabilities of the technology enhance how students and educators learn, experience, communicate, and do mathematics" (NCTM, 2011, p. 1). The NCTM position paper on technology referred to content-specific technological tools such as dynamic geometry environments and handheld computation devices as well as content-neutral tools such as adaptive technologies and web-based digital media.

In a *second-order meta-analysis* for 30 years of technology-enhanced mathematics instruction, Young (2017) identified 19 prior meta-analyses including 663 studies. Dividing the research into four categories, he found mean effect sizes of 0.47 for computation enhancement studies (e.g., calculators), 0.42 for instructional delivery studies (curriculum-assisted or curriculum-based instruction), and 0.36 for studies using presentation and modeling technology for conceptual understanding (e.g., virtual manipulatives, dynamic geometry, and learning software). A fourth category included studies using combinations of technologies, with a mean effect size of 0.21. Overall, prior meta-analyses reported small to large effect sizes (0.16 for intelligent tutoring systems to 1.02 for dynamic geometry software), with overall statistically significant moderator effects on student achievement in mathematics.

Researchers such as Young (2017) and Drijvers (2015) considered technologies for mathematics as three didactical types: tools to outsource mathematics (computation enhancement), tools for practicing skills, and tools for concept development. There is, of course, overlap across these categories. For example, virtual manipulatives could be categorized within all three groups, depending on the use for the student. This section provides an overview of common digital technologies for mathematics teaching and learning as well as guidelines for teachers implementing technologies.

Virtual Manipulatives

Among the most powerful digital applications for mathematics are *virtual manipulatives* (VM). These interactive, internet-based or downloadable app-based tools can be found with most mathematics textbook supplementary materials, on mathematics websites such as NCTM's *Illuminations*, and in mobile tablet and phone app forms. NCTM *Illuminations* features interactive VM, data graphers, calculus tools, and other web-based digital tools (http:// illuminations.nctm.org). Some of the NCTM tools allow students to explore geometric figures and simulate constructions. Others provide practice with equivalent fractions, factoring, and making change, to name a few. Some apps can be downloaded to mobile devices, while others require Java (*Java applets*). NCTM also offers *Core Math Tools*, a downloadable suite of interactive software tools for high-school topics (e.g., computer algebra systems, geometry platform, vertex-edge graphs). Other sources of virtual manipulatives for mathematics include *GeoGebra*, a site with various calculators, graphing apps, and a geometry simulator (www.geogebra.com); *The Math Learning Center*, a math learning site with free apps (web based or downloadable) for clocks, number lines, geoboards, and other mathematics manipulatives (www.mathlearningcenter.org); and *Fun Fraction*, an interactive site about multiplying fractions (www.funfraction.org*)*. The National Library of Virtual Manipulatives at Utah State University began developing a collection of VM in 1999 for K–12 mathematics applications using Java applets. The collection is not available for mobile formats but still can be accessed with some computer operating systems (http://nlvm.usu.edu).

Virtual manipulatives offer students a way to interact with mathematics concepts, communicate about those concepts, and solve problems. They are a blending of concrete and representational levels for learning and also provide accessibility for students who have difficulties with other representations.

TRY THIS

Explore the virtual manipulatives on one of the websites described above. Consider how the VM would be used for a specific mathematics concept.

Virtual manipulatives include all the concrete manipulatives and pictorial representations that are commonly used for mathematics instruction, such as attribute blocks, geoboards, clocks, money, tangrams, algebra tiles, spinners, and fraction bars. They also include function machines, games, puzzles, problems, and fractals. A simulated example of a VM geoboard is shown in Figure 7.3. VMs can be used individually, with small groups, or for larger groups by projecting content. Some have features that allow teachers to monitor student progress or take quizzes. For applications such as attributes of solids and volume, three-dimensional representational apps or online modeling are available (e.g., *SketchUp* by Trimble, www.sketchup. com). Two- and three-dimensional simulations and interactive tasks have applications across the curriculum.

In comparison studies, VM were as effective as concrete manipulatives with third- and fourth-grade students for fraction and symmetry concepts (Burns & Hamm, 2011), with

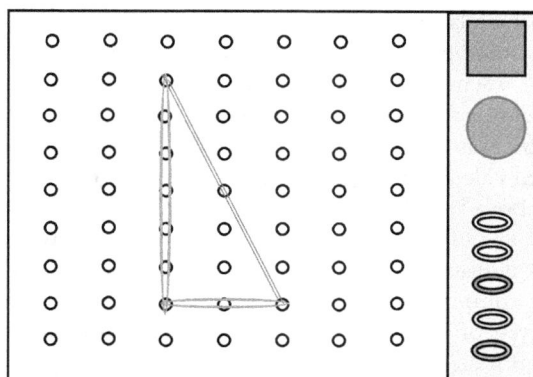

Figure 7.3 Virtual Manipulative Example of the GeoBoard

high-school students with MLD for perimeter and area concepts (Satsangi & Bouck, 2015) and algebra equations (Satsangi et al., 2016), with middle-school SWD for fraction addition (Bouck, Shurr, et al., 2018), and with elementary and secondary students with developmental disabilities on multidigit subtraction, multiplication, and division (Bouck et al., 2020; Park et al., 2019). VM have advantages for SWD including accessible features, motivating aspects, and the range of possible tasks. These tools can support students' mental images of mathematics concepts across the grade levels and topics (Shin et al., 2017). However, students should be evaluated for fine-motor ability and interaction with VM tools to ensure those don't present barriers. Some students may perform better with screen touch, pen, mouse, or various track pads. Further, careful selection of specific VM to match concepts, teacher guidance, and scaffolding are still required in order for students to achieve. Teachers should monitor students' use of VM and have them verbalize their understanding, as with concrete manipulatives.

Data Analysis Tools

The most common data analysis and graphing tools for mathematics are calculators, spreadsheets, and dedicated apps. Handheld calculators are the most common calculators for mathematics classrooms and can be as simple as four-function (arithmetic) calculators or as complex as scientific and graphing calculators. The functions of these calculators are also found in device-based apps and online applications. Teachers should be familiar with provisions for calculator use on state- and district-tests and ensure students have practice with the permitted types. Calculator use has been questioned for students in the initial stages of developing concepts and skills. For example, when and to what extent should elementary students use calculators for basic operations? In a position statement, the NCTM (2015) recommended that calculators at these grade levels be used strategically and with a pedagogical purpose, to support and advance learning. Calculators are essential elements of mathematics classrooms for learning, problem solving, working together, and building a positive view of mathematics. Some students will need more explicit instruction in the most effective and efficient use of calculators, how various types of numbers are displayed, how to check the results, how to estimate using calculators, and the meaning of mathematics symbols with the order of operations. After all, calculators are employed extensively in work and everyday life situations so students should be fluent users.

There is little research on the efficacy of calculators for students who struggle to learn mathematics, particularly concerning because calculators are a common accommodation. Bouck et al. (2013) examined calculator use by middle-school students with and without disabilities on their scheduled assessments across a school year. Both groups of students were more accurate on mathematics assessments when they used calculators, although the proportion of students who reported they used calculators was low. The finding that only 8% to 19% of sixth and seventh graders with MLD employed calculators suggests that students may need more explicit instruction in calculator use. Yakubova and Bouck (2014) studied the effects of calculator types on the mathematical performance of five fifth-grade students with mild intellectual disabilities. Both scientific and graphing calculators were effective in increasing accuracy and efficiency of all students on mathematics computation and word-problem solving. However, some students performed better on one type of calculator than the other, indicating the need for matching individual student ability and preference to type of calculator.

Recent enhancements for calculators may benefit some students, including the talking calculator and enlarged key calculators for students with limited fine-motor skills. Calculators can also enhance word-problem exploration by reducing the calculation demands on students with working-memory issues while they are attempting to understand problems and explore various solution strategies. The voice-input, speech-output (VISO) calculator was developed for students with visual impairments or visual-processing difficulties. In a study with three high-school students with visual impairments, Bouck et al. (2011) found that the students were more efficient with solving calculations using the VISO calculators, taking less time on mathematics assessments. The students were more independent in mathematics class, not depending as much on other students for calculations. However, these students needed time to develop their use of the voice-input capabilities.

Spreadsheets and other dedicated data tools are used to collect and analyze data and produce graphs and conversions. Although spreadsheets are not mentioned in the *Common Core State Standards for Mathematics* (CCSSM, NGO & CCSSO, 2010) until high school, many teachers incorporate their use in elementary mathematics. Spreadsheets, such as *Microsoft Excel* or *Google Sheets*, use cells within a grid for entering data and formulas. Other functions are used to analyze data, such as sorting and creating graphs. Students may need explicit, step-by-step instruction to use spreadsheets for mathematics goals. Spreadsheets use notations, abbreviations, equations, and syntax that sometimes differ from those commonly used in mathematics, such as the caret (^) for exponents, ★ for some multiplication tasks, =ROUND for the function of reducing the number of decimal places, and nested functions that are performed in a specific order. Each cell is referenced by its coordinate pair from the upper left-hand corner without a comma (e.g., B7). Since spreadsheets have extensive uses for work and real-life applications, they should be used as tools for mathematics. Other dedicated data-analysis tools include measurement converters, amortization tables, graphing calculators (computer algebra systems), probability simulators, and statistical analysis programs.

Interactive Applications

The term *app* refers to application programs that are downloaded to a computer or other electronic device or used online, as with virtual manipulatives or a calculator app. Online apps are generally less expensive than those that are downloaded, although many mobile and computer-based apps are free. Apps for mathematics applications include spreadsheets, sketchpads, graphing tools, calculators, calculation programs, and algebra systems. Apps include word-processing programs and games. Apps include mathematics software programs for concept development, practice, and assessment. Apps are available through textbook packages, commercial companies, government and university projects, online app stores, and through educational websites and range from free to quite expensive. Bouck, Working, and Bone (2018) described app-based manipulatives as similar to VM, but existing on a mobile device, not requiring plugins or even the internet. Some examples of mobile apps are base-10 blocks, geoboard, number line, and 3D geometry, and are free or inexpensive to download. One extensive source for mathematics and science computation, representations, graphing, solving, artificial intelligence, data collections, and even problem creation is WolframAlpha®, a "computational knowledge engine" (www.wolframalpha.com). This powerful engine defies categorization and will be useful for students through college and beyond.

A group of technological tools for geometry is termed *dynamic geometry software* or *environment* (DGE), which can be used on calculators, computers, and handheld devices. DGEs, such as *The Geometer's Sketchpad*, *Cabri*, and *GeoGebra*, provide ways of representing and manipulating geometric objects that are not possible with paper, pencil, compass, and straightedge alone (Hollebrands & Lee, 2012). These environments allow students to engage and reason with geometric objects to develop stronger understandings of concepts, properties, and theorems. Although their features vary, most DGEs have a set of basic objects (points, lines, rays, etc.); provide for the construction of lines and objects; and have the ability to make transformations, perform calculations, hide objects, and create procedures.

General computer-based apps also can assist mathematics learning, including word processors with speech-to-text capabilities and enhanced text, on-line dictionaries, drawing and graphics programs, and graphic-organizer generators. For example, the graphic organizer is a pedagogical tool across the curriculum, for initial learning, practice, and review of concepts and procedures. In mathematics, teachers and students can create their own devices using free or commercial apps. Mathematics organizers can be created for new vocabulary, comparing and contrasting concepts, outlining the steps of a process, and organizing categories of concepts. An example of a graphic organizer is illustrated in Figure 7.4. Unfortunately, there has been scant research on the effects of assistive technologies specifically for mathematics learning of students with disabilities. Teachers must implement their use with effective instructional strategies and monitor their effectiveness.

A personal interactive device, the smart digital pen, allows the user to record notes and drawings as they write them and transmit the information to a computer (e.g., Livescribe pen, FLY Fusion pen). Writing mathematics notation is easier with a digital pen than on a computer's word-processing program. Teachers can use digital pens during tutoring sessions so that students have a record of examples. Small working groups can take notes for distribution and study. Students with writing, reading, listening, and memory issues can capture their notes,

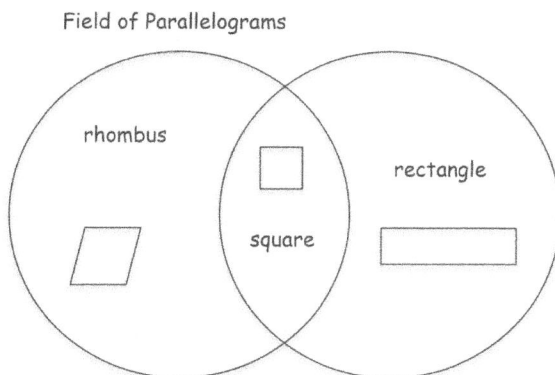

Figure 7.4 Graphic Organizer of Geometric Concepts

from written or audio sources, and later organize those on the computer. Research on digital-pen interfaces shows promise for increasing engagement and decreasing working-memory load. For example, Oviatt and Cohen (2014) investigated the use of digital pens with high- and low-performing high-school students. Six student triads were presented 16 geometry and algebra problems via computer screen in two sessions. During the problem-solving sessions, 80.3% of written material was nonlinguistic (numerical, symbolic, diagrams, and markings) and these notations increased with problem difficulty. Higher-performing students had lower written and spoken "disfluencies" and were more adept at constructing representations. They increased their use of representations with the most difficult problems. Lower-performing students put more effort into easy and moderate problems, indicating a frustration-level break-down on difficult problems, similar to problem-solving research using pencil and paper (e.g., Rosenweig et al., 2011). Other digital pens are available for reading sentences or words (as with word problems or new vocabulary) and scanning short excerpts of text (optical character recognition). Digital-pen technology has the promise of increasing nonlinguistic recording activity, such as making diagrams and symbol notations, but lower-performing students clearly need explicit instruction in problem-solving strategies regardless of the technology used.

Technology-Based Learning Programs

Technology-based learning programs are those that support the learning process for academic goals, either through a software package or interactive website. They include computer-assisted instruction (CAI), computer-based instruction (CBI), and integrated learning systems. Cheung and Slavin (2013) identified three major categories for this type of instruction: computer-managed learning (e.g., *Accelerated Math)*, comprehensive models that include non–computer components (e.g., *Cognitive Tutor*), and supplemental CAI (e.g., *Schoolhouse Math Lab*). In a meta-analysis of 74 studies, Cheung and Slavin (2013) found that technology-based programs produced positive, but small effects (0.16) on mathematics achievement, with supplemental CAI having the largest effect (0.19). It is not surprising that supplemental CAI had the strongest effects as these programs are designed to supplement regular instruction, so teachers select levels for the additional practice. The primary problem with computer-managed learning systems was the low level of implementation by teachers—less time was devoted than was recommended—because teachers attempted to use those as supplements that were not well integrated. The researchers also concluded that achievement has not improved over time with advances in technology. In another meta-analysis, Higgins et al. (2019) examined student outcomes, motivation, and attitudes when using technology with mathematics instruction, locating 24 studies with these variables, across all mathematics content domains. The weighted Cohen's *d* effect sizes were 0.68 for achievement, 0.30 for motivation, and 0.59 for attitudes, with both technology-assisted and technology-based instructional approaches positively influencing achievement and attitudes.

In a synthesis of technology-mediated mathematics (TMM) interventions for K-12 students with MLD, Kiru et al. (2018) located 19 studies since 2000 and concluded that TMM interventions had mainly positive outcomes for these students. The interventions included CAI or anchored instruction packages such *Math Explorer, Math Drills, Texas Instruments InterActive,* or researcher-designed programs. Most TMM interventions applied one or more principles of explicit instruction, specifically practice and feedback. However, most did not provide overt demonstrations or modeling and feedback tended to be minimal (correct or incorrect). In a summary of meta-analyses on the effects of CAI on elementary mathematics learning, Räsänen (2015) found that two groups tended to benefit more from CAI instruction: younger students and students with special needs. Further, supplemental CAI was more effective than programs that replaced instruction.

Technology for Teachers

Teachers often use technology for group instruction and preparing classroom materials. Interactive whiteboards offer the potential for engaging group lessons and activities related to mathematics concepts. Teachers and students can interact with the information much like they would with a touchscreen computer—manipulating apps, videos, images, spreadsheets, and other tools and information—while also having the ability to write on the surface, click on embedded links, and drag or draw using objects. Teachers creating classroom materials can use mathematics notation tools embedded in word-processing programs. For example, the following expression was created using *Microsoft Word*'s document *insert equation* option: $\sqrt{a^2 + b^2}$. Also available for teacher use are materials downloaded from websites, such as graph paper, patterns for geometric shapes, sample word problems, video demonstrations, and activities. Teachers can also access videos and other materials that explain mathematics concepts and connections, as long as those are credible sources.

Some guidelines for implementing technologies within mathematics teaching and learning:

- Consider technologies like any other tool in the classroom. Does the technology support concept understanding and enhance learning?
- Start planning with the curriculum goal, not the technology. Make a decision about the best activities and tools for the concepts and skills, not the other way around.
- Consider possible disadvantages for using technologies for specific lessons or with individual students. Do those outweigh the benefits?
- Teach students how to use technologies and understand related vocabulary and notations.
- Take advantage of the capabilities of technology, to extend and enhance what could be achieved without technology.
- When using videoclips as *anchors* or to illustrate concepts or procedures, keep those short (30 s to 2 min). Longer videoclips should be interrupted with questions or discussion. Many students with MD or ADHD will have difficulty viewing long video assignments, maintaining attention and self-monitoring.
- Use technology as a bridge—between concrete and representational depictions, between concepts in two or more domains, between home and school.
- Take advantage of teacher collaborations, professional development, and other resources to learn about key features and applications of technologies. Teachers should be learners as well.

Technologies with applications for teaching and learning mathematics will continue to evolve and have the potential to have an impact on student achievement. However, technology has not yet met many predictions related to student learning. For students with disabilities, it appears that there is higher use of assistive technologies with low-incidence disabilities than those of high-incidence such as learning disabilities, with the most prevalent being calculators and computers (Bouck, 2016). One issue has been a disconnect between technology developers and classroom practice. The context of the classroom, with considerations for the teacher's role in providing and assessing instruction, the nature of student interactions, and the need for differentiations and scaffolding must be considered when planning the incorporation of technologies within instruction. Another issue for many schools has been resource commitment for teacher training, time for implementation, funding for classroom devices and other supports, and home-school connections. We know that technologies are pervasive in the home, community, and in work settings so it is important that students understand their use for specific applications and are technologically literate across the curriculum.

Integrating Mathematics

Mathematics standards such as the NCTM standards (2000), state standards, and the CCSSM (NGA & CCSSO, 2010) are not just for mathematics teachers. English and language arts standards such as the CCSS for English Language Arts Literacy in History/Social Studies, Science, and Technical Subjects (CCSS/ELA, NGA & CCSSO, 2010) are not just for English or language arts teachers. Other content areas have curriculum standards as well, developed by states or professional organizations. All teachers should be familiar with these standards and support their implementation. The goal of these global standards is to prepare PreK–12 students for the demands of college and entry-level careers, focusing on communication, critical-thinking, problem-solving, and analytical skills. This section addresses the integration of mathematics concepts and skills with other content areas with a few examples to illustrate the nature of content integration.

STEM Studies

Since the 1990s, the acronym STEM has been used to encourage the integration of science, technology, engineering, and mathematics, an interdisciplinary approach to education that focuses on real-world applications and prepares students for careers in STEM fields. Although other acronyms have been used by various organizations for similar interdisciplinary studies (e.g., STEMTEC, GEMS, STEMM, and STREM), STEM is used by federal agencies in the United States, such as the National Science Foundation. The range of fields represented within STEM varies, for example the NSF also includes the fields of information technology and social sciences for scholarships and grant programs.

An emphasis on STEM education through an interdisciplinary approach is supported by workforce projections. The US Department of Labor projected 6 million new jobs between 2019 and 2029 (BLS, 2020). The occupations with the greatest projected increases included wind turbine technicians (61%), nurse practitioners (52% increase), solar photovoltaic installers (51%), occupational therapy assistants (35%), statisticians (35%), and other health-care positions (31%–34%), all requiring STEM competencies. Careers that are high-paying as well as fast growing include software developers, management analysts, financial managers, and health services managers, all of which require education beyond high school. However, only 40% of 2018 high-school graduates were prepared for college-level mathematics and only 36% were prepared for college-level science (ACT, 2018). Only 20% of 2018 high-school graduates met the STEM Readiness Benchmark, a combination of math and science scores and success in higher-level coursework (e.g., calculus, physics). Underserved student populations, such as minorities, those from low-income households, and students with low-education parents, represent only 3%–15% of students interested in STEM careers.

In elementary and secondary schools, STEM education best practices include providing hands-on learning experiences where students collaborate, problem solve, explore, and engage in critical thinking and reasoning. Teachers should ensure that STEM curricula focus on the most important topics for each discipline and relate those to students' daily lives (CADRE, 2012; Making Sense of Science, 2016). Students should be engaged in science and mathematics investigations from the earliest grades, to increase their interest and motivation in these areas. A deeper understanding of the most important topics, as well as an emphasis on the interrelated nature of the STEM fields is essential if students are to be successful in preparing for STEM-related careers (Hanover Research, 2011).

The focus on STEM studies has not been without criticism, primarily for promoting a narrow focus on technical skills while de-emphasizing other skills and concept understanding required by today's employees: communication, facility with other languages and cultures, arts, and humanities. "A broad general education helps foster critical thinking and creativity. Exposure to a variety of fields produces synergy and cross fertilization" (Zakaria, 2015). Moreover, students in today's classrooms are likely to change careers often and beyond traditional boundaries. According to the Bureau of Labor Statistics (2020), workers between the ages of 25 and 34, reporting for the years 2010–2020, remained in one position an average of 2.8 years. Teachers must understand the current generation's likelihood to change careers several times and encourage a well-rounded education.

Some educators promote the Engineering Design Process (EDP) as a framework for STEM lessons. This approach is similar to the scientific approach and the problem-solving process in mathematics. The EDP begins with defining a problem and doing background research, followed with specifying the requirements of the problem, brainstorming solutions, selecting the best solution, doing development work, building a prototype, and finally testing and redesigning as needed (Ertas & Jones, 1997). For younger students, this process is often simplified as Ask (about the problem), Imagine (some solutions), Plan, Create, Improve, with cycling back to Ask (Museum of Science, 2016). The design process is iterative as designers and problem-solvers seek ever more refined solutions through collaboration and investigation.

An example of integrating science with mathematics that employs the EDP is an integrated unit for kindergarten students about forces and interactions. The *Next Generation Science Standards* (NGSS) include alignments with CCSS standards for each of its standards (NGSS Lead States, 2013). The science standard for a sequence of lessons, within the physical science domain for kindergarten, is: "Plan and conduct an investigation to compare the effects of different strengths or different directions of pushes and pulls on the motion of an object" (K-PS2-1). Related CCSSM standards: (K.MD.A.1) Describe measurable attributes of objects, such as length or weight. Describe several measurable attributes of a single object. (K.MD.A.2) Directly compare two objects with a measurable attribute in common, to see which object has "more of/less of" the attribute and describe the difference. Related ELA standards are about asking and answering questions and participating in shared research projects.

In the unit, students have previously read the book *Push and Pull* by Charlotte Guillain (2008) and kicked soccer balls on the playground to observe their motion when kicked, rolled, and bumped. Vocabulary words such as push, pull, motion, direction, strength, distance, and mass are introduced through these activities. In this three-day lesson, students plan and conduct an investigation to compare the effects of different strengths or directions of pushes and pulls on the motions of objects, using ramps and measuring devices, and analyze data to test a design solution for changing the speed or direction of an object (Figure 7.5). Teacher questions and guidance prompt students to: *Ask* questions to fully understand the problem; *Imagine* some solutions and brainstorm a hypothesis; *Plan* an investigation of the hypothesis; *Create* the model for testing and record data during testing; and *Improve* and retest as needed.

Figure 7.5 Exploring with Ball and Ramp in EDP Investigation

Students share their findings and make conclusions or recommend further testing. The teacher assesses student understanding through data charts, oral descriptions, and responses to clarifying questions.

Science and mathematics integrations are, perhaps, the most common due to the comparable problem-solving process, similar tools for inquiry, and related investigations of the world around us. When the NGSS were developed, the writing teams collaborated with CCSS developers to assist with alignment of skills and concepts. For example, if first-grade students are not yet measuring with standard units, science standards do not require that type of measurement, just the comparison of lengths.

Writing in Mathematics

The mathematics process standards of communication and reasoning are skills that are also developed within language arts and are language-based. Mathematics tasks require students to communicate their work with others, to record solution strategies, to use notation systems that are standard, and to justify their reasoning orally and in writing. Having students explain their reasoning orally and in writing also benefits mathematics learning by offering teachers a view of student thinking and requiring students to consider their responses. However, some students with disabilities or language differences may not be able to adequately express their mathematics reasoning in writing.

Consider the following students' written explanations of the same mathematics problem. Al has a mathematics disability but above average language skills, Tami has mathematics and language disabilities, and Ciara has been learning English for only six months but was on grade level in mathematics in her native China. Jen is gifted in mathematics and sometimes leaves out steps. The problem: Convert this expression into its simplest form: $5(4x + 2y^2) + 2x + y$.

> Al: I followed the steps in "Please Excuse My Dear Aunt Sally" and did the parentheses first getting $20x + 10y^2 + 2x + y$ and then I couldn't do anything with the exponent and there was no multiplication or division so I then I added the like terms, getting $22x + 10y^2 + y$. I rearranged that into $22x + 10y^2 + y$.
>
> Tami: The () things first and so the answer is $20x + 10y^2 + 2x + y$ because that's the answer.
>
> Ciara: The 5 many the $4x$ and the $2y^2$ and the $20x$ and the $2x$ so $= 22x + 10y^2 + y$.
>
> Jen: I can distribute the 5 over $4x + 2y^2$. Next, I start combining like terms, so we have $22x + 10y^2 + y$. A step further would factor the y terms for $22x + y(10y + 1)$. Is that the correct form for what you are asking?

All four students were observed beginning their work the same way and both Al and Ciara simplified the expression completely. Al was able to provide an understandable written explanation with the protocol for his final result, but his reasoning was purely procedural rather than conceptual. Both Tami and Ciara struggled with accurate vocabulary. Jen used correct vocabulary for her actions and skipped writing out some of her thinking. She noticed something in the expression the others didn't see, and she wanted to make sure her answer was in the correct form. A more conceptual written description would have been:

> *The distributive property should be applied first because of the parentheses, distributing the 5 to the terms $4x$ and $2y^2$ by multiplication. No other division or multiplication can be performed, so look for common terms to add. We can combine the $20x$ and $2x$ into $22x$. No other terms can be combined, so put the terms in order of degree for each variable: $22x + 10y^2 + y$.*

For students with limited vocabulary or written-language skills, mathematics teachers should allow alternative means of expression, such as oral explanations or diagrams, while building writing skills. Mathematics vocabulary can be developed through activities similar to those in language arts: charts of new terms, morphological analysis, mnemonic devices, repeated use of new words by students in both oral and written forms, and personal mathematics dictionaries or word banks. Producing quality written descriptions of mathematics reasoning also requires instruction, practice, and reinforcement. Students should share their descriptions with others for clarification, view and discuss model descriptions, develop descriptions in cooperative groups, and practice critiquing descriptions for elements of logical reasoning. Like language-arts writing activities, mathematics writing can take many forms, such as reflections, descriptions, captions, word problems, and technical reporting. Students should have grade-appropriate practice with many forms and specific, corrective feedback on their efforts.

There is research support for the effects of writing activities on learning mathematics. Writing contributes to concept and procedural understanding (Stonewater, 2002), to the development of problem-solving skills (Borasi & Rose, 1989) and metacognitive skills (Pugalee, 2005), and to mathematics vocabulary development (Urquart, 2009). Writing fosters communication among students. Further, students' writing cues teachers to student understanding, allowing for instructional adjustments (Pugalee, 2005). In a meta-analysis of the effects of writing in the content areas across Grades 1 to 12, Graham et al. (2020) reported that the mean effect size of 21 studies on writing for mathematics achievement was 0.32, with effects ranging from (−0.41 to +1.26). Most mathematics writing took the form of journal writing, analysis and interpretation, and informative writing. Most interventions using mathematics writing involved metacognitive prompting, encouraging students to reflect on the process, content, or achievement when writing. Overall, writing in mathematics can enhance learning, improve comprehension and application of content knowledge. However, most mathematics teachers do not integrate writing or recognize its importance (Atieri, 2010).

Some recommendations for writing in mathematics include:

- Mathematics writing should be simple, clear, uncluttered, and well organized. It should not be flowery or full of forced synonyms. Use complete sentences to enhance clarity. Many times the same phrases will be repeated as with, "To solve for …," and, "we see that," and, "we can show … by …," and "given the value …." It is not considered redundant to repeat these phrases.
- Clearly label all charts, tables, graphs, and other representations using words. Define all variables and formulas.
- Avoid using words that can cause confusion such as time, center, and first. If necessary, use these multiple-meaning words very precisely with modifiers. For example, "t stands for the time in seconds" and "the first term of the four terms in this polynomial."
- Use symbols within sentences for clarity. The following sentence is accurate for mathematics writing: If A = area, L = length, and W = width, then we can find the area of a rectangle given the formula A = LW if we know the values of L and W. But avoid other abbreviations and symbols that are not clearly mathematical or accurate for the situation (e.g., etc., re., OK, #, ea, &).
- Tables, graphs, or diagrams should be clearly labeled and referenced within the text of the student's narrative. They should not stand alone but be explained in terms of how they were created and what they mean.
- If students use word processors for writing mathematics, teach them to use equation editors to make the actual mathematics symbols. This expression does not mean the square root of 5: √5, while this expression does: $\sqrt{5}$.

- Have students write for a specific audience, such as their peers, rather than writing for the teacher. Collect "quick-write" prompts for journal writing and use a rubric for evaluation. For example, *"Explain why 50 percent is equivalent to half."*
- Students who struggle with writing will need more explicit instruction. Provide instruction, modeling, and prompts for different types of mathematics writing, such as journal writing or writing about an incorrect worked example.
- Older students should read mathematics articles on their concept level for exposure to mathematics writing. Good sources include *Plus Magazine, Pi in the Sky, Discover,* and Wolfram *MathWorld.*

Writing is a critical communication skill across curriculum areas but has specific applications for mathematics that merit development. Good writing instruction for mathematics is not simply telling students to "write in your math journal" but involves planning, appropriate instruction, and ongoing, standards-based assessment as with any other curriculum goal.

TRY THIS

This is Abe's work finding the unknown value in the proportion $2: x = 3: 9$. Explain in writing: 1) What was Abe's error? 2) How should you solve for x?

$$\frac{2}{x} = \frac{3}{9} \quad \rightarrow \quad x = \frac{6}{9} \quad \rightarrow \quad x = \frac{2}{3}$$

Literature

One of the most popular curriculum integrations has been that of children's literature with mathematics concepts. There is research support for the use of literature with typically achieving or at-risk young children to enhance mathematics concepts such as counting and number knowledge (Anderson et al., 2004; Young-Loveridge, 2004), measurement (Van den Heuvel-Panhuizen & Iliada, 2011), geometry (Skoumpourdi & Mpakopoulou, 2011), and mathematics vocabulary, language skills, and discourse (Hassinger-Das et al., 2015; Purpura et al., 2017). In a rare study with young children with disabilities, Green et al. (2018) examined the effects of a shared storybook-reading intervention with 50 preschool children. The children, with a range of developmental disabilities, were engaged with interactive storybook reading in small groups three times a week for six weeks. Scripted questions were posed during the reading (e.g., *Did the caterpillar eat more strawberries or more oranges?*) with comprehension and mathematical questions balanced. Related mathematics activities followed each reading (e.g., comparing amounts) using systematic, explicit instruction and inexpensive manipulatives. The intervention groups demonstrated significantly higher scores on total mathematics ability, a quantity comparison task, one-to-one correspondence skills, and oral counting than the control groups with small-group story time with the same books. The researchers recommended selecting books intentionally and noted that quality books can address mathematics implicitly or explicitly. Careful planning of questions during reading and related mathematics activities after reading can promote greater achievement.

Research on the use of literature to promote mathematics learning with older children and adolescents is scarce. Most publications on the topic are anecdotal, noting that using short stories, as with children's story books, can be a real-life application that motivates students to explore mathematical concepts. Literature can be a vehicle, like a problem situation, for introducing a concept or applying a math skill. White (2016) encouraged elementary teachers to create a problem-solving community that emphasizes the standards for mathematics practice by using children's literature regularly, providing examples of thinking aloud, prompting problem-solving steps, and scaffolding understanding with specific texts. Boerman-Cornell et al. (2017) described using the Harry Potter series, using well-chosen excerpts or film clips, to bridge literature and high-school mathematics such as moving between two and three dimensions and even higher dimensions.

Literature for children and adolescents that is appropriate for making mathematics connections are of three categories: literature developed specifically for mathematics connections, literary works that have the potential for strong mathematics connections, and literature that has only incidental or minor connections to mathematics. Books and stories in each of these categories range from high quality to downright terrible and include both fiction and nonfiction. Examples of high-quality fiction are *Grandfather Tang's Story* (Tompert, 1990), developed to demonstrate the origin and use of tangrams; *Alice's Adventures in Wonderland* (Carroll, 1865/1941) that includes many complex mathematics concepts such as shape changes and proportion, logical argument, time, space, and puzzles; and *A Year Down Yonder* (Peck, 2000) with its incidental references to train travel in 1937, seasonal changes on a farm, and depression-era wages. Special mathematics series include *Stories to Solve: Folktales from Around the World* by Shannon (1985), and the *Sir Cumference* series by Neuschwander (1997). Nonfiction examples include *Fermat's Enigma* (Singh, 1977) for high-school students, *Crafts around the Ancient World* (Jovinelly, 2002) for middle-school students, and *Food Chains in a Meadow Habitat* (Nadeau, 2002) for elementary students.

Teachers who want to integrate literature with their mathematics activities should consider these guidelines:

- Consider the mathematics concepts that the selection supports. They should be aligned with grade-level mathematics standards and offer the opportunity for deeper exploration and understanding rather than just a bit of humor or simplistic treatment.
- Use reference books, journals, and websites to locate and read reviews of appropriate literature.
- Read and reread the book or story before using it with students. A book with a great title and wonderful illustrations can still be poorly written, inappropriate, or inaccurate regarding mathematics concepts.
- Decide whether the selection is appropriate for the age, interests, and reading levels of students. A work written for younger children can be used by older students for deeper analysis and extensions as long as students understand the purpose. The length of the work, vocabulary needs, and relationship to other subject areas also should be considered.
- Plan whether to use a good literary work to introduce a mathematics concept, to develop the concept, or to reinforce the concept after instruction. For example, the book *How Big is a Foot?* (Myller, 1962) serves as a good introduction to the concept of units of measure. *The Doorbell Rang* (Hutchins, 1986) helps to develop the concept of partitive division. After much practice with two-digit numbers and their meaning, the book *17 Kings and 42 Elephants* (Mahy, 1987) brings together many mathematics concepts and skills such as place value, number sense, combining, grouping, and comparing.
- Select works or portions of works that increase in value with repetition. Literature selections are typically reread and explored rather than given a "once-through."
- Plan mathematics extensions for the literary selection such as creating similar problems or researching related data.

TRY THIS

Select one book (fiction or nonfiction) for children or adolescents and analyze its possible use for teaching or reinforcing mathematics concepts.

Social Studies

Social studies are

> the integrated study of the social sciences and humanities to promote civic competence.... The primary purpose of social studies is to help young people make informed and reasoned decisions for the public good as citizens of a culturally diverse, democratic society in an interdependent world.
>
> (NCSS, 2010)

Within the social studies domain are many disciplines, including anthropology, archaeology, history, law, philosophy, economics, geography, political science, psychology, religion, and sociology, although K-12 studies focus primarily on history, political science, geography, and economics. The National Council for the Social Studies developed the *National Curriculum Standards for Social Studies* (NCSS, 2010) as a guide for states and school districts. A perusal of these standards across the grade levels highlights many mathematics applications, concepts, and processes. Making connections in social studies includes those across time and cultures, patterns in history, the effects of events on people, the production and use of goods and services, and applications of science and technologies. Problem-solving skills can be applied to simple or complex problems such as developing systems for food and water in third-world countries or studying the impact of political change. Communication skills are critical for studying and understanding interpersonal and intercultural relationships, the effects of media, and researching topics of interest. Reasoning is essential for solving problems, making connections, applying theory, and making predictions. Representations are used in social sciences to convey ideas through charts, graphs, maps, and models.

The *C3 Framework* (College, Career, & Civic Life) *for Social Studies State Standards* (NCSS, 2013, updated 2017) was developed to assist states in upgrading social studies standards based on an Inquiry Arc. This Arc focuses on the nature of inquiry and the pursuit of knowledge through questions across four dimensions: developing questions and planning inquiries, applying disciplinary tools and concepts (such as those from mathematics, economics, geography, history), evaluating sources and using evidence, and communicating conclusions and taking informed action. This approach to social studies is intended to add rigor and better prepare students for college, careers, and civic life. An example of a compelling question—a question that is academically rigorous yet student relevant—that might involve mathematics tools and concepts is, "What should be done about the gender wage gap?"

Study of history and cultures involves a mental number line that represents time and patterns of events. Study of geography requires spatial sense, measurement, and proportions. Anthropology, economics, sociology, and political science all depend on data analysis, probability, and statistical representations. Archaeology combines most other disciplines. The content strands in this book conclude with cultural connections that relate specific mathematics concepts to applications in other cultures. Some examples of social studies and mathematics integration include:

- developing surveying skills using maps and compasses;
- comparing wages, taxes, and prices among several countries after converting currencies, comparisons can also be made for one country across time;
- collecting population data for analysis and prediction using spreadsheets;
- researching how mathematics differs in other cultures;
- interpreting and creating charts and graphs to depict economic, political, geographic, cultural, and sociological phenomena;
- engaging in service-learning projects that meet community-level needs while addressing social studies, mathematics, and other content standards and cross-curricular processes (e.g., environmental surveys, community transportation studies, children's education projects, and parks and recreation projects for diverse citizens).

TRY THIS

Select one of the following social studies topics and describe a companion mathematics unit for the same grade level:

Fourth grade: Resources of various regions of the US and their impact on occupations of citizens.

Seventh grade: Economic relationships of trading partners such as the US, European Union, and China.

The Art of Mathematics

The entire mathematics curriculum could be taught through an artist's point of view. From early childhood concepts of two-dimensional symmetry and visual patterns to the dynamical systems and chaos theory of advanced mathematics, art can be a vehicle for making sense of the numerical, functional, and spatial properties of mathematics. Mathematics is also a tool for creating and understanding art.

Many intriguing aspects of the history of mathematics involve art and architecture. The ancient Egyptians used mathematics in architecture but did not demonstrate perspective in their drawings because those were not intended to represent reality, rather project a symbolic representation of the world based on proportions (University College London, 2002). Understanding the construction of the pyramids (e.g., the Great Pyramid of Khufu, *c.*2694 BCE) depends on geometric concepts such as surface area and volume of the pyramid, slope, and cross-section analysis of solids (Figure 7.6). There is a lot of disagreement about whether the architects employed mathematical theories—were the measurements' relationship to the golden ratio (Φ) a coincidence or a deliberate application of mathematical formulae? The golden ratio is $\frac{1+\sqrt{5}}{2}$ or 1.61803 ... and an angle based on this number will have the measure 51°49'38. The sides (lateral surfaces) of the Great Pyramid have gradients of 51° 50' 40, found by computing the lower angle of a right triangle with the height of the pyramid at its center on one side and the distance from the base to middle and top forming the other two sides (Bartlett, 2014). Lengths were measured in cubits based on seven palms (of the Royal hand), and one palm was four digits (fingers) of approximately 7.5 cm. The side lengths at the base of the Great Pyramid average 230.478 m (440 cubits) with only 4.4 cm maximum difference, remarkable precision for a structure built over 4,700 years ago. Many ancient temples, burial

Figure 7.6 Great Pyramid

The Metropolitan Museum of Art, New York: *Vue de la grande pyramide (Chéops) prise à l'angle S.E.*, Maxime Du Camp (French), December 1849, photograph. www.metmuseum.org.

chambers, and depictions of human forms reflect the golden ratio, however, there is no evidence that the structures were based on that design. Bartlett (2014) hypothesized that ancient Egyptians had no implicit intention to incorporate mathematical or geometrical theories, but structures such as the Great Pyramid were based on "religiously significant idealism of human proportions used to depict the gods and pharaohs," thus applying the proportions that had been used for millennia (Bartlett, 2014, p. 309).

The early Greeks contributed the mathematical descriptions of Phi (Φ) and proportion concepts that have permeated art, architecture, photography, design, music, and even poetry. The golden mean (Phi, Φ), with a proportion of 1:1.618 ..., appears in the proportions of the Parthenon in Athens and a construction for the golden section point was found in Euclid's *Elements* from 300 BCE, "A straight line is said to have been cut in extreme and mean ratio when as the whole line is to the longer part, so is the longer part to the shorter" (as cited by Foutakis, 2014, p. 71). The Italian mathematician Pacioli (1509) is credited for connecting the Euclidean proposition to the golden (or divine) proportion, and the German mathematician Maestlin (1597) made the first calculation of the number (Foutakis, 2014). Nature most likely provided the inspiration for artists and mathematicians: the patterns in seashells (the nautilus shell), the spiral growth patterns of plant stems (phyllotaxis) and pinecones, human proportions, and even the rotation of hurricanes. The golden ratio is also related to the numerical sequence known as the Fibonacci sequence: 0, 1, 1, 2, 3, 5, 8, 13, 21, 34, ... that is created by adding the two previous terms in the sequence. Students of all ages enjoy explorations and applications with these universal concepts.

Another major application of mathematics for art was the development of perspective during the Renaissance. Brunelleschi is credited for making the first correct formulation of linear perspective in 1413 with the concepts of vanishing point (and the convergence of parallel lines toward that point) and that of scale as the relation between the measurement of an actual object and the object in the picture (O'Connor & Robertson, 2003). Teachers can demonstrate these mathematics principles by having students study works of art (e.g., Alberti's *Vanishing Point*) or produce their own scale-perspective drawings. Perspective is actually an optical illusion that uses mathematical principles to "fool the eye" into viewing what has been drawn on a two-dimensional surface as three-dimensional space. Leonardo da Vinci (around 1490) furthered the work on perspective, writing about related mathematical formulae and two types of perspective: artificial (with foreshortening) and natural (with consistent relative sizes; O'Connor & Robertson, 2003). Dürer (around 1525) extended theories of perspective to shadows cast by objects (see Figure 7.7). Hogarth (1697–1764) and Escher (1898–1972) are

Figure 7.7 Mathematics Concepts in Art: Perspective and Shadowing
The Metropolitan Museum of Art, New York: *Self-portrait, Study of a Hand and a Pillow,* Albrecht Dürer (German), 1493, pen and ink. www.metmuseum.org.

known for deliberately misusing perspective in their art. Escher's work has been the topic of many mathematics classes, especially those in hyperbolic geometry.

There are dozens of excellent resources on the integration of mathematics and art for PreK-12 settings. For example, the North Carolina Museum of Art (NCMA) has one of the most diverse programs for educators and their students in the country. On the *NCMA Learn* website educators will find teacher-created lessons for integrating art across the curriculum referenced to specific art within the museum's collections and content standards (https://learn.ncartmuseum.org). The museum's special exhibits always include powerful educational connections. NCMA programs can be found at ncartmuseum.org. The NCMA, as well as other museums, schools, and community art programs across the country, such as The Kennedy Center's *ArtsEdge*, the Carnegie Museum of Natural History, the Seattle Art Museum, and the Smithsonian museums, emphasize the importance of integrating the arts within STEM disciplines, now called STEAM initiatives (STEM to STEAM, 2016). First championed by the Rhode Island School of Design, STEAM collaborations and projects encourage the integration of the arts within education at all levels and promote hiring artists and designers to drive innovation and creativity within businesses and other organizations. STEAM goals include preparing students for the challenges and careers of the 21st century. STEAM studies that are well designed are motivating, engaging, focused on important concepts and their connections, and allow for student differentiation. Many professional journals, such as *Art Education* and NCTM's *Mathematics Teacher: Learning & Teaching PK-12*, have special issues or columns addressing STEAM education.

Activities that integrate mathematics concepts with art include:

- Enhance the study of patterns, tillings, and tessellations by viewing how various cultures create fabric (rugs, quilts, wall hangings, etc.), architecture (windows, floors, roofs, fences, etc.), and decorations for baskets, pottery, wallpaper, and many other applications. These patterns have been made popular by the work of Escher, who was inspired by the art of 13th–15th-century Alhambra, Spain (of Islamic influence).

- Create polyhedra using repeated two-dimensional patterns of regular polygons: triangles for tetrahedrons (4 faces), octahedrons (8 faces), and icosahedrons (20 faces); squares to create cubes; and pentagons for dodecahedrons (12 faces). These are the five platonic solids. There are 13 Archimedean solids that use two or more types of regular polygons.
- Explore the properties of Möbius (Moebius) strips, discovered simultaneously by Möbius and Listing in 1858 (Derbyshire, 2003). To create a Möbius strip take a long strip of paper and give it a half twist. Tape the two ends together. Some activities with the strip include determining how many sides there are by drawing a line down the middle of the strip, cutting the strip along the line to see what happens, and predicting what will happen with a second cut along the midline. The mathematics related to the Möbius strip depends on trigonometric functions. The Möbius strip and its extensions (double Möbius, surfaces achieved by attaching disks, etc.) have been applied to architecture, physics, and even symbolism.
- Create mathematical mobiles after studying the work of Alexander Calder. The concepts of balance, fulcrum, symmetry, and distribution are critical for creating a mobile that is balanced yet interesting in motion. The National Gallery of Art's website for teachers features lesson plans connecting mobile making and mathematics with connections to an online gallery of Calder's work (www.nga.gov).
- Study the architectural features of buildings in the community. Look for examples of the golden proportion, other proportions, the Pythagorean triangle (ratio 3:4:5), and arches. Students can also draft floor plans of classrooms or public spaces to apply measurement and geometric concepts.
- Use origami to explore mathematics concepts. In Japanese, the word "origami" refers to any type of paper folding. The traditional methods typically involved a single square of paper. But origami applications have expanded to modular forms with several sheets and differing dimensions of paper. Huzita formulated a set of six basic axioms for origami including the most basic "given two points P_1 and P_2, we can fold a line connecting them" (Frigerio & Huzita, 1989, p. 144). The rest develop systematically through two points and two lines with mapping points onto lines. Other mathematicians have proposed origami theorems. For example, Kawasaki's Theorem (1989) states that if the angles surrounding a single vertex in a flat origami crease pattern are $a_1, a_2, a_3, \ldots a_{2n}$, then:
  - $a_1 + a_3 + a_5 + \ldots + a_{2n-1} = 180$, and
  - $a_2 + a_4 + a_6 + \ldots + a_{2n} = 180$.

If you add up the angle measures of every other angle around a point, the sum will be 180. Origami applications can be as simple as paper folding to gain experience with basic shapes and lines of symmetry or as advanced as topology and combinatorics.

TRY THIS

Explore the website of a local or regional art museum for STEAM connections.

Integration of mathematics concept development with other content areas has the potential to boost student interest in mathematics, create deeper understanding, scaffold generalization of new mathematics skills, and prepare students for the next levels of learning and careers.

Integration with integrity requires fidelity with all content area standards, careful research and planning, and the continued application of effective instructional methods including the ongoing assessment of student learning.

Many teachers place content integration in the "extras" category of activities to fill time once the content required for the next test has been covered and the weeks of testing have been completed for the year. However, isolating these activities will reduce their effectiveness and make actual concept connections less viable. Successful integration should be planned to introduce new concepts, provide real-world situations, assist with concept development, or serve as culminating experiences to units of study that will connect and extend new learning. Content integration is only as powerful as the planning invested and integrity with which it has been designed. The internet offers excellent sources for high-quality mathematics integration, including the NCTM's *Student Explorations in Mathematics*, the American Association for the Advancement of Science's *Project 2061*, as well as the websites of many other organizations and university projects.

Life-Skills Mathematics

Mathematics study prepares students to solve problems, reason about real situations, collect and analyze data, use tools for solving problems, communicate with others about concepts and processes and make connections among concepts. All of these skills are critical for life, for real-world situations at home, in the workplace, and in the community. All students should be challenged with real-world problems and develop skills in solving and communicating about those problems. Can I afford to buy and maintain a car? How can I change this recipe for more people? Can I take this medication with my other supplements? How can I improve my running skills? Why is my computer dropping its internet connection? How should I wash this new pair of slacks? What will the weather be for the weekend? Daily life is full of such problems.

In the United States, life-skills mathematics has historically been designed for students who are not college bound and students with moderate to severe disabilities. Sometimes called functional or consumer skills, life skills are those that are needed to work and live independently. The *Programme for International Student Assessment* (PISA) measures not only students' success in mathematics curricula but whether students can apply mathematics concepts outside the classroom. For example, one test item asked students to interpret a graph showing the amount of space used on a memory stick by various applications; another involved the renovation of an ice cream shop. Fifteen-year-olds in the United States scored lower than the international average in mathematics, lower than 36 other countries including Slovenia, Portugal, the UK, and Poland (OECD, 2018). Implications for these students include not being able to compete for entry-level positions and live independently.

In a survey of adult skills, the Organisation for Economic Co-operation and Development (OECD, 2020) collected information from 39 countries about employment needs and skills shortages of adult workers and identified a mismatch of skill demand and supply internationally. Approximately 60% of workers identified a skills mismatch, with either under- (12%) or over-qualified (22%) or field-of-study mismatch (40% of workers internationally, 48% in the US). At the same time, 83% of employers in the US reported difficulties in finding employees with the required skills for specific positions (SHRM, 2019). Beyond the narrow job skills described in the US Bureau of Labor Statistics publications, job seekers need employability skills such as communication, teamwork, analysis and investigation (problem solving), self-motivation, planning and organization, flexibility, time management, global languages and cultures, literacy, numeracy, and computing skills.

Life-Centered Education

The Council for Exceptional Children offers the *Life Centered Education Transition Curriculum* (LCE, CEC, 2013), which is a framework of instructional goals, with lesson plans and measures of student progress in three life domains: daily living skills, self-determination and interpersonal skills, and employment skills. Originally developed by Donn E Brolin as *Life Centered Career Education*, LCE offers a framework for ensuring that students with disabilities have appropriate instructional goals that address skills for independence beyond school. Many of the competencies within the LCE are related to mathematics skills, as are illustrated in Table 7.1.

Mathematics teachers should make a point to incorporate real-life problems within their instruction and provide real-world simulations, such as collaborative problem solving with planning and communication demands. Some students will require more explicit connections with applications. For example, a unit about percentages could include problem situations such as: *Molly wants to purchase a new notebook that is on sale. The original price was $12.85, but it is marked as 25 percent off. How much will Molly pay?* A unit on geometry could include problem examples from carpentry, sewing, and sports. Even the highest-achieving students need instruction in life skills, such as the management of money, time, transportation, and living arrangements, whether that be through integrated coursework, parental guidance, or workshops (Moran, n.d.). Tasks such as doing the laundry, shopping for groceries, and house cleaning are often overlooked. Other skills such as setting goals and self-advocacy should be taught and practiced throughout school.

Students with moderate to severe developmental disabilities should learn a progression of foundational mathematics skills while applying these skills to grade-aligned content in elementary school (Spooner et al., 2019). Secondary students may benefit from instruction on simplified grade-aligned content, such as a smaller portion with smaller numbers in real-life contexts. Evidence-based practices for this modified mathematics instruction across 36 studies include systematic, explicit instruction; manipulatives, technology assisted instruction, graphic organizers, and task analysis (Spooner et al., 2019). Authentic stories can be used to present problem-solving situations, with a careful task analysis of the steps that will be required. An example of a task analysis for computing a tip can be found in Box 7.2.

Box 7.2   Task Analysis of Computing a Tip

To Compute a 15% Tip:

1.	Find the total.	$ 24.80
2.	Move the decimal one position to left to find 10%.	$ 2.480
3.	Take half of that number.	$ 1.240
4.	Add that to the 10%.	$ 3.720
5.	Drop the 0 or round up to have tip amount.	$ 3.72
6.	You may decide to leave an amount a bit higher.	**$ 4.00**

To Compute an 18% Tip with a Calculator:

1.	Find the total and enter that.	$ 18.47
2.	Hit the × key.	
3.	Enter .18 and hit =.	$ 3.3246
4.	Hit the + key and enter the original total.	$21.7946
5.	Drop zero(s) or round up for total and tip.	**$21.80**

Table 7.1 Life-Centered Education with Mathematics Examples

Domain	Competencies	Mathematics Examples
DAILY LIVING SKILLS	Managing personal finances	Understand credit card fees and interest.
	Selecting & managing a household	Compute apartment costs with utilities and renter's insurance.
	Caring for personal needs	Plan appropriate exercise.
	Demonstrating relationship responsibilities	Monitor a sick child and administer medications.
	Buying, preparing, and consuming food	Budget food purchases for a week.
	Buying and caring for clothing	Estimate the sales price with a percentage reduction.
	Exhibiting responsible citizenship	Study a bond issue on the costs of new roads for an upcoming vote.
	Utilizing recreational facilities and engaging in leisure	Engage in carpentry as a hobby.
	Choosing and accessing transportation	Understand bus schedules and routes.
SELF DETERMINATION AND INTERPERSONAL SKILLS	Understanding self-determination	For options identified during problem solving, identify potential consequences.
	Being self-aware	Identify one's educational needs and develop a plan to meet those.
	Developing interpersonal skills	Listen to others and respond.
	Communicating with others	Provide one's reasoning to another person.
	Good decision making	Collect data for making decisions.
	Developing social awareness	Understand the rights of others, perspective taking.
	Understanding disability rights and responsibilities	Problem solve about needed disability services.
EMPLOYMENT SKILLS	Knowing and exploring employment possibilities	Understand employment remuneration.
	Exploring employment choices	Understand one's aptitude for various positions.
	Seeking, securing, and maintaining employment	Take a math test as part of an employment application.
	Exhibiting appropriate employment skills	Solve job-related problems.

Karl et al. (2013) demonstrated that students aged 15–18 with moderate intellectual developmental disabilities (IDD) could compute percentages in cooking activities as well as read age-appropriate content and demonstrate the effects of force on motion (science). For the mathematics objective, the teacher constructed a task analysis for computing the sale price of groceries needed to make a cake using calculators, with simultaneous verbal prompts during training. The students mastered the mathematics goal of the intervention with 8–10 trials, in partial preparation for their state's alternate assessment. Further, the students gained a skill that will continue to be useful in daily living. Root et al. (2018) taught mathematics problem solving involving percent of change word problems (with multiplication and subtraction) to secondary students with IDD, important personal-finance skills. The researchers employed a community-theme menu (local locations on a grid such as car wash, coffee shop); video anchors (30–40 sec) for applying skills in each location; worksheets with word problems, a six-step task analysis, and graphic organizer; and an Excel workbook for the iPad for self-graphing progress. Instructional practices included task analysis, least-intrusive prompting, self-monitoring instruction, explicit instruction with modeling and feedback on practice, and goal setting with self-graphing. The research demonstrated a functional relationship between the modified schema-based intervention and problem-solving skills, with a large effect size (Tau-U = .87). Students were also able to compare the final discounted price of an item or activity with a given amount of money and generalize skills to real-world stimuli.

Transition Services

The continued development of mathematics processes and skills will contribute to high-school completion and successful transition to adult life for students with disabilities. Students with disabilities must have transition services included in their IEPs before they turn 16 years old. Transition services are:

> a coordinated set of activities for a child with a disability that (A) is designed to be a results-oriented process, that is focused on improving the academic and functional achievement of the child with a disability to facilitate the child's movement from school to post-school activities, including post-school education, vocational education, integrated employment (including supported employment), continuing and adult education, adult services, independent living, or community participation; (B) is based on the individual child's needs, taking into account the child's strengths, preferences, and interests; (C) includes instruction, related services, community experiences, the development of employment and other post-school adult living objectives, and, when appropriate, acquisition of daily living skills and functional vocational evaluation.
>
> (IDEA, 2004; 34 CRF 300.43)

John will be a student at Pine Road High School in the fall and is attending an IEP meeting in the spring of eighth grade with middle- and high-school teachers and his parents. In the meeting John, his parents, and the teachers discuss John's postsecondary goals, based on John's skills, interests, and preferences:

- Complete a certification program at a local culinary school.
- Live with a roommate while in culinary school.
- Work part-time in the food industry while in culinary school.
- Use public transportation for travel to work, school, and within the community.
- Monitor own medications and health appointments, with assistance as needed from his parents.

These post-secondary goals were based on John's interests and skill with cooking, his desire to live more independently after high school, his inability to drive due to epilepsy, and his academic strengths and needs. These goals will be reviewed and updated annually through his exit from high school. Examples of transition services for the IEP (including coursework) include:

- Work with guidance counselor to identify culinary schools and their information.
- High school *Culinary Arts I* and *II*.

- *Life-skills Mathematics* for an elective.
- *Employment Skills* course that will include a community transportation component and internship within a restaurant or cafeteria.

In addition to these specific transition services, John's IEP will include annual, measurable goals to be addressed within his classes, along with any additional services and accommodations he needs to be successful. IEP goals shouldn't repeat the grade-level curriculum goals; they are more individualized. For example, IEP goals for John's *Culinary Arts I* class include the following:

1. Using a phone-based app, John will convert metric to standard measures for 19 of 20 provided.
2. Given a recipe and measurement tools (cups, scales, and spoons), John will measure liquid and dry items with 100 percent accuracy.
3. In small-group activities, John will ask questions and provide answers within the group, demonstrating three of each during a one-hour activity for five consecutive sessions.

What mathematics concepts and skills are evident in these goals? Other goals for the *Culinary Arts* classes also reflect mathematics content, such as converting recipes, budgeting for meals, understanding food storage and cooking temperatures, managing time, communicating about concepts and processes, and solving problems.

TRY THIS

Locate an occupation, such as *cook* or *grounds maintenance*, in the Bureau of Labor Statistics' *Occupational Outlook Handbook* (www.bls.gov/ooh) and evaluate its mathematics requirements.

Life skills—for work, home, and community—are typically integrated skills rather than discrete content areas. "Shopping for a meal with a specified budget and designated nutritional values" requires mathematics, science, reading, oral communication, notetaking, and social skills. "Scheduling and participating in a job interview" requires mathematics, social studies, reading, written language, oral communication, and social skills. The nature of integrated skills for life underscores the importance of all teachers addressing core content standards in mathematics and other areas, as well as broader communication, problem-solving, self-determination, and social skills necessary to function as an adult.

References

ACT (2018). *The condition of college & career readiness 2018*. www.act.org

Anderson, A., Anderson, J., & Shapiro, J. (2004). Mathematical discourse in shared storybook reading. *Journal for Research in Mathematics Education, 35*, 5–33. doi: 10.2307/30034801

Atieri, J. (2010). *Literacy + Math = Creative connections in the elementary classroom*. International Literacy Association. www.readwritethink.org

Bartlett, C. (2014). The design of the Great Pyramid of Khufu. *Nexus Network Journal: Architecture and Mathematics, 16*(2), 299–311. doi: 10.1007/s00004-014-0193-9

Bhat, R. & Fletcher, A. (1995). *Pentominoes*. www.andrews.edu/~calkins/math/pentos.htm

Boerman-Cornell, W., Klanderman, D., & Schut, A. (2017). Using Harry Potter to bridge higher dimensionality in mathematics and high-interest literature. *Journal of Adolescent & Adult Literacy, 60*(4), 425–432. doi: 10.1002/jaal.597

Borasi, R., & Rose, B. J. (1989). Journal writing and mathematics instruction. *Educational Studies in Mathematics, 20*(4), 347–165. doi: 10.1007/bf00315606

Bouck, E. C. (2016). A national snapshot of assistive technology for students with disabilities. *Journal of Special Education Technology, 31*(1), 4–13. doi: 10.1177/0162643416633330

Bouck, E. C., Flanagan, S., Joshi, G. S., Sheikh, W., & Schleppenbach, D. (2011). Speaking math: A voice input, speech output calculator for students with visual impairments. *Journal of Special Education Technology, 26*(4), 1–14. doi: 10.1177/016264341102600401

Bouck, E. C., Joshi, G. S., & Johnson, L. (2013). Examining calculator use among students with and without disabilities educated with different mathematical curricula. *Educational Studies in Mathematics, 83*(3), 369–385. doi: 10.1007/s10649-012-9461-3

Bouck, E., & Park, J. (2018). A systematic review of the literature on mathematics manipulatives to support students with disabilities. *Education & Treatment of Children, 41*(1), 65–106. doi: 10.1353/etc.2018.0003

Bouck, E., Shurr, J., Bassette, L., Park, J., & Whorley, A. (2018). Adding it up: Comparing concrete and app-based manipulatives to support students with disabilities with adding fractions. *Journal of Special Education Technology, 33*(3), 194–206. doi: 10.1177/0162643418759341

Bouck, E., Shurr, J., & Park, J. (2020). Virtual manipulative-based intervention package to teach multiplication and division to secondary students with developmental disabilities. *Focus on Autism and Other Developmental Disabilities, 35*(4), 195–207. doi: 10.1177/1088357620943499

Bouck, E., Satsangi, R., & Park, J. (2018). The concrete–representational–abstract approach for students with learning disabilities: An evidence-based practice synthesis. *Remedial and Special Education, 39*(4), 211–228. doi: 10.1177/0741932517721712

Bouck, E., Working, C., & Bone, E. (2018). Manipulative apps to support students with disabilities in mathematics. *Intervention in School and Clinic, 53*(3), 177–182. doi: 10.1177/1053451217702115

Bureau of Labor Statistics (2020). *Median years of tenure with current employer for employed wage and salary workers by age and sex, selected years 2010–2020.* US Department of Labor. www.bls.gov

Bureau of Labor Statistics (2020). *Occupational employment projections to 2029.* US Department of Labor. www.bls.gov

Burns, B. A., & Hamm, E. M. (2011). A comparison of concrete and virtual manipulative use in third- and fourth-grade mathematics. *School Science and Mathematics, 111*(6), 256–261. doi: 10.1111/j.1949-8594.2011.00086.x

CADRE (2012). *Improving STEM curriculum and instruction: Engaging students and raising standards* (Smart Brief). Education Development Center. http://cadrek12.org

Carbonneau, K. J., & Marley, S. C. (2015). Instructional guidance and realism of manipulatives influence preschool children's mathematics learning. *The Journal of Experimental Education, 83*(4), 495–513. doi: 10.1080/00220973.2014.989306

Carbonneau, K. J., Marley, S. C., & Selig, J. P. (2013). A meta-analysis of the efficacy of teaching mathematics with concrete manipulatives. *Journal of Educational Psychology, 105*(2), 380–400. doi: 10.1037/a0031084

Carroll, L. (1865/1941). *Alice's adventures in Wonderland.* Macmillan.

Cass, M., Cates, D., Smith, M., & Jackson, C. (2003). Effects of manipulative instruction on solving area and perimeter problems by students with learning disabilities. *Learning Disabilities Research & Practice, 18*(2), 112–120. doi: 10.1111/1540-5826.00067

Cheung, A. C. K., & Slavin, R. E. (2013). The effectiveness of educational technology applications for enhancing mathematics achievement in K-12 classrooms: A meta-analysis. *Educational Research Review, 9*, 88–113. doi: 10.1016/j.edurev.2013.01.001

Clements, D., & Sarama, J. (2018). Myths of early math. *Education Sciences, 8*(2), 1–8. doi: 10.3390/educsci8020071

Council for Exceptional Children (2013). *Life centered education transition curriculum.* www.exceptionalchildren.org

Cuisenaire, G., & Gattegno, C. (1954). *Numbers in colour: A new method of teaching the processes of arithmetic to all levels of the primary school.* Heinemann.

Derbyshire, J. (2003). *Prime obsession: Bernhard Riemann and the greatest unsolved problem in mathematics.* Joseph Henry Press. doi: 10.17226/10532

Drijvers, P. (2015). Digital technology in mathematics education: Why it works (or doesn't). In S. Cho (Ed.), *Selected regular lectures from the 12th international congress on mathematical education* (pp. 135–151). Springer. doi: 10.1007/978-3-319-17187-6

Ertas, A., & Jones, J. C. (1997). *The engineering design process* (2nd ed.). John Wiley & Sons.

Flores, M., Hinton, V., & Meyer, J. (2020). Teaching fraction concepts using the concrete-representational-abstract sequence. *Remedial and Special Education, 41*(3), 165–175. doi: 10.1177/0741932518795477

Foutakis, P. (2014). Did the Greeks build according to the golden ratio? *Cambridge Archaeological Journal, 24*(1), 71–86. doi: 10.1017/s0959774314000201

Frigerio, E., & Huzita, H. (1989). Axiomatic development of origami geometry. In H. Huzita (Ed.), *Proceedings of the 1st international meeting of origami science and technology* (pp. 143–158). Universita di Padova. www.origamiusa.org

Fyfe, E. R., McNeil, N. M., Son, J. Y., & Goldstone, R. L. (2014). Concreteness fading in mathematics and science instruction: A systematic review. *Educational Psychology Review, 26*(1), 9–25. doi: 10.1007/s10648-014-9249-3

Gattegno, C. (1954). The Gattegno geoboards. *Bulletin of the Association for Teaching Aids in Mathematics, 3.*

Gersten, R., Beckmann, S., Clarke, B., Foegen, A., Marsh, L., Star, J. R., & Witzel, B. (2009). *Assisting students struggling with mathematics: Response to intervention (RtI) for elementary and middle schools* (NCEE 2009–4060). National Center for Education Evaluation and Regional Assistance, Institute of Education Sciences, US Department of Education. https://ies.ed.gov/ncee/wwc/PracticeGuide/2

Graham, S., Kiuhara, S. A., & MacKay, M. (2020). The effects of writing on learning in science, social studies, and mathematics: A meta-analysis. *Review of Educational Research, 90*(2), 179–226. doi: 10.3102/0034654320914744

Green, K., Gallagher, P., & Hart, L. (2018). Integrating mathematics and children's literature for young children with disabilities. *Journal of Early Intervention, 40*(1), 3–19. doi: 10.1177/1053815117737339

Guillain, C. (2008). *Push and pull (Investigate!)*. Heinemann Library.

Hanover Research (2011). *K-12 STEM education overview*. www.hanoverresearch.com

Hassinger-Das, B., Jordan, N. C., & Dyson, N. (2015). Reading stories to learn math: mathematics vocabulary instruction for children with early numeracy difficulties. *The Elementary School Journal, 116*(2), 242–264. doi: 10.1086/683986

Higgins, K., Huscroft-D'Angelo, J., & Crawford, L. (2019). Effects of technology in mathematics on achievement, motivation, and attitude: A meta-analysis. *Journal of Educational Computing Research, 57*(2), 283–319. doi: 10.1177/0735633117748416

Hollebrands, K., & Lee, H. (2012). *Preparing to teach mathematics with technology: An integrated approach to geometry*. Kendall-Hunt.

Hutchins, P. (1986). *The doorbell rang*. Greenwillow Books.

Ifrah, G. (2001). *The universal history of computing: From the abacus to the quantum computer*. Wiley.

Individuals with Disabilities Education Improvement Act, USC 20 §1400 *et seq.* (2004). www.congress.gov/108/plaws/publ446/PLAW-108publ446.pdf

Jordan, L., Miller, M. D., & Mercer, C. D. (1999). The effects of concrete to semiconcrete to abstract instruction in the acquisition and retention of fraction concepts and skills. *Learning Disabilities: A Multidisciplinary Journal, 9*, 115–122.

Jovinelly, J. (2002). *Crafts of the ancient world* (series). Rosen.

Karl, J., Collins, B. C., Hager, K. D., & Ault, M. J. (2013). Teaching core content embedded in a functional activity to students with moderate intellectual disability using a simultaneous prompting procedure. *Education and Training in Autism and Developmental Disabilities, 48*(3), 363–378.

Kawasaki, T. (1989). On the relation between mountain-creases and valley-creases of a flat origami. In H. Huzita (Ed.), *Proceedings of the 1st international meeting in origami science and technology* (pp. 229–237). Universita di Padova. www.origamiusa.org

Kiru, E. W., Doabler, C. T., Sorrells, A. M., & Cooc, N. A. (2018). A synthesis of technology-mediated mathematics interventions for students with or at risk for mathematics learning disabilities. *Journal of Special Education Technology, 33*(2), 111–123. doi: 10.1177/0162643417745835

Maccini, P., & Hughes, C. A. (2000). Effects of an instructional strategy incorporating concrete representations on the introductory algebra performance of secondary students with learning disabilities. *Learning Disabilities Research & Practice, 15*(1), 10–21. doi: 10.1207/sldrp1501_2

Mahy, M. (1987). *17 kings and 42 elephants.* Dial Books for Young Readers.

Making Sense of Science (2016). *Planting the seeds for a diverse US STEM pipeline: A compendium of best practice K-12 STEM education programs.* www.makingsciencemakesense.com/static/documents/Resources/K-12-STEM-edu-programs.pdf

Marsh, L. G., & Cooke, N. L. (1996). The effects of using manipulatives in teaching math problem solving to students with learning disabilities. *Learning Disabilities Research & Practice, 11*, 58–65.

Mercer, C. D., & Miller, S. P. (1992). Teaching students with learning problems in math to acquire, understand, and apply basic math facts. *Remedial and Special Education, 13*(3), 19–35, 61. doi: 10.1177/074193259201300303

Milton, J., Flores, M., Moore, A., Taylor, J., & Burton, M. (2019). Using the concrete–representational–abstract sequence to teach conceptual understanding of basic multiplication and division. *Learning Disability Quarterly, 42*(1), 32–45. doi: 10.1177/0731948718790089

Moran, K. (n.d.). *Life skills for happy, successful teens.* We Are Teachers. https://s18670.pcdn.co/wp-content/uploads/Allstate-Foundation-SEL-Parent-Guide-072220.pdf

Morin, J., & Samelson, V. M. (2015). Count on it: Congruent manipulative displays. *Teaching Children Mathematics, 2*(6)1, 362–370. doi: 10.5951/teacchilmath.21.6.0362

Museum of Science, Boston (2016). *Engineering is elementary* (Project). www.eie.org

Myller, R. (1962). *How big is a foot?* Dell.

Nadeau, I. (2002). *Food chains in a meadow habitat.* Rosen.

National Council for the Social Studies (2010). *National curriculum standards for social studies: A framework for teaching, learning, and assessment.* www.socialstudies.org

National Council for the Social Studies (2013, 2017). *College, career, and civic life (C3) framework for social studies state standards: Guidance for enhancing the rigor of K-12 civics, economics, geography, and history.* www.socialstudies.org

National Council of Teachers of Mathematics (2000). *Principles and standards for school mathematics.* www.nctm.org/Standards-and-Positions/Principles-and-Standards/

National Council of Teachers of Mathematics (2011). *Strategic use of technology in teaching and learning mathematics* (Position statement). www.nctm.org/Standards-and-Positions/NCTM-Position-Statements/

National Council of Teachers of Mathematics (2015). *Calculator use in elementary grades* (Position statement). www.nctm.org/Standards-and-Positions/NCTM-Position-Statements/

National Governors Association Center for Best Practices & Council of Chief State School Officers (2010). *Common core state standards for English language arts & literacy in history/social studies, science, and technical subjects.* www.corestandards.org/ELA-Literacy/

National Governors Association Center for Best Practices & Council of Chief State School Officers (2010). *Common core state standards for mathematics.* www.corestandards.org/Math/

NGSS Lead States (2013). *Next generation science standards: For states, by states.* The National Academies Press. www.nextgenscience.org

Neuschwander, C. (1997). *Sir Cumference and the first round table.* Charlesbridge Publishing.

O'Connor, J. J., & Robertson, E. F. (2003). *Mathematics and art: Perspective.* MacTutor History of Mathematics. www-history.mcs.st-andrews.ac.uk

OECD (2018). PISA *2018 results: What students know and can do.* www.oecd.org/pisa

OECD (2020). *Employment outlook 2020.* https://doi.org/10.1787/1686c758-en

Oviatt, S., & Cohen, A. (2014). Written activity, representations and fluency as predictors of domain expertise in mathematics. In A. A. Salah, J. Cohen, & B. Schuller (Eds.), *Proceedings of the 16th international conference on multimodal interaction* (pp. 10–17). Association for Computing Machinery. doi: 10.1145/2663204

Park, J., Bouck, E., & Smith, J. (2019). Using a virtual manipulative intervention package to support maintenance in teaching subtraction with regrouping to students with developmental disabilities. *Journal of Autism and Developmental Disorders, 50*(1), 63–75. doi: 10.1007/s10803-019-04225-4

Peck, R. (2000). *A year down under.* New York, NY: Dial Books.

Peltier, C., Morin, K., Bouck, E., Lingo, M., Pulos, J., Scheffler, F., Suk, A., Mathews, L., Sinclair, T., & Deardorff, M. (2020). A meta-analysis of single-case research using mathematics manipulatives with students at risk or identified with a disability. *The Journal of Special Education, 54*(1), 3–15. doi: 10.1177/0022466919844516

Peterson, S. K., Mercer, C. D., & O'Shea, L. (1988). Teaching learning disabled students place value using concrete to abstract sequence. *Learning Disability Research & Practice*, 4, 52–56.

Picciotto, H. (n.d.). *Algebra manipulatives: Comparison and history.* www.mathedpage.org/manipulatives/ alg-manip.html

Pugalee, D. K. (2005). *Writing to develop mathematical understanding.* Christopher-Gordon.

Purpura, D. J., Napoli, A. R., Wehrspann, E. A., & Gold, Z. S. (2017). Causal connections between mathematical language and mathematical knowledge: A dialogic reading intervention. *Journal of Research on Educational Effectiveness*, 10(1), 116–137, doi: 10.1080/19345747.2016.1204639

Räsänen, P. (2015). Computer-assisted interventions on basic number skills. In R. Kadosh & A. Dowker (Eds.), *Oxford handbook of numerical cognition* (pp. 745–766). Oxford University Press. doi: 10.1093/oxfordhb/9780199642342.013.63

Root, J., Cox, S., Hammons, N., Saunders, A., & Gilley, D. (2018). Contextualizing mathematics: Teaching problem solving to secondary students with intellectual and developmental disabilities. *Intellectual and Developmental Disabilities*, 56(6), 442–457. doi: 10.1352/1934-9556-56.6.442

Rosenzweig, C., Krawec, J., & Montague, M. (2011). Metacognitive strategy use of eighth-grade students with and without learning disabilities during mathematical problem solving: A think-aloud analysis. *Journal of Learning Disabilities*, 44(6), 508–520. doi: 10.1177/0022219410378445

Satsangi, R., & Bouck, E. C. (2015). Using virtual manipulative instruction to teach the concepts of area and perimeter to secondary students with learning disabilities. *Learning Disability Quarterly*, 38(3), 174–186. doi: 10.1177/0731948714550101

Satsangi, R., Bouck, E., Taber-Doughty, T., Bofferding, L., & Roberts, C. (2016). Comparing the effectiveness of virtual and concrete manipulatives to teach algebra to secondary students with learning disabilities. *Learning Disability Quarterly*, 39(4), 240–253. doi: 10.1177/0731948716649754

Shannon, G. (1985). *Stories to solve: Folktales from around the world.* HarperCollins.

Shin, M., Bryant, D., Bryant, B., McKenna, J., Hou, F., & Ok, M. (2017). Virtual manipulatives: Tools for teaching mathematics to students with learning disabilities. *Intervention in School and Clinic*, 52(3), 148–153. doi: 10.1177/1053451216644830

Singh, S. (1977). *Fermat's enigma: The epic quest to solve the world's greatest mathematical problem.* Walker & Co.

Skoumpourdi, C., & Mpakopoulou, I. (2011). The prints: A picture book for pre-formal geometry. *Journal of Early Childhood Education*, 39(3), 197–206. doi: 10.1007/s10643-011-0454-0

Society for Human Resource Management (2019). *The skills gap 2019.* www.shrm.org

Sowell, E. J. (1989). Effects of manipulative materials in mathematics instruction. *Journal for Research in Mathematics Education*, 20(5), 498–505. doi: 10.5951/jresematheduc.20.5.0498

Spooner, F., Root, J. R., Saunders, A. F., & Browder, D. M. (2019). An updated evidence-based practice review on teaching mathematics to students with moderate and severe developmental disabilities. *Remedial and Special Education*, 40(3), 150–165. doi: 10.1177/0741932517751055

STEM to STEAM (2016). *What is STEAM?* Rhode Island School of Design. http://stemtosteam.org

Stonewater, J. (2002). The mathematics writer's checklist: The development of a preliminary assessment tool for writing in mathematics. *School Science and Mathematics*, 102(7), 324–334. doi: 10.1111/j.1949-8594.2002.tb18216.x

Swan, P., & Marshall, L. (2010). Revisiting mathematics manipulative materials. *Australian Primary Mathematics Classroom*, 15(2), 13–19.

Tompert, A. (1990). *Grandfather Tang's story.* Crown.

University College London (2002). What is ancient Egyptian art? *Digital Egypt.* www.ucl.ac.uk/museums-static/digitalegypt/

Uribe-Flórez, L., & Wilkins, J. (2017). Manipulative use and elementary school students' mathematics learning. *International Journal of Science and Mathematics Education*, 15(8), 1541–1557. doi: 10.1007/s10763-016-9757-3

Urquhart, V. (2009). *Using writing in mathematics to deepen student learning* (ED 544239). ERIC. Mid-continent Research for Education and Learning. https://files.eric.ed.gov/fulltext/ED544239.pdf

Van den Heuvel-Panhuizen, M., & Iliada, E. (2011). Kindergartener's performance in length measurement and the effect of picture book reading. *ZDM Mathematics Education*, 43(5), 621–635. doi: 10.1007/s11858-011-0331-8

Wheatley, G. H., & Reynolds, A. M. (1999). "Image maker": Developing spatial sense. *Teaching Children Mathematics, 5*, 374–378.

White, J. (2016). *Using children's literature to teach problem solving in math: Addressing the standards for mathematical practice in K-5.* Taylor & Francis. doi: 10.4324/9781315527536

Yakubova, G., & Bouck, E. C. (2014). Not all created equally: Exploring calculator use by students with mild intellectual disability. *Education and Training in Autism and Developmental Disabilities, 49*, 111–126.

Young, J. (2017). Technology-enhanced mathematics instruction: A second-order meta-analysis of 30 years of research. *Educational Research Review, 22*, 19–33. doi: 10.1016/j.edurev.2017.07.001

Young-Loveridge, J. M. (2004). Effects on early numeracy of a program using number books and games. *Early Childhood Research Quarterly, 19*, 82–98. doi: 10.1016/j.ecresq.2004.01.001

Zakaria, F. (2015, 26 March). Why America's obsession with STEM education is dangerous. *The Washington Post.* www.washingtonpost.com

Introduction to Content Strands

The following strands offer a more in-depth study of the big ideas of PreK–12 mathematics content. These key concepts are not only foundational to mathematics understanding; they provide the framework for most other concepts and skills throughout the curriculum. If teachers develop good understandings of these concepts, they will be able to develop curriculum connections, plan applications and problem situations, prepare students for the next levels of mathematics learning, and have confidence in their ability to adapt the mathematics curriculum to meet students' needs. Each strand provides an overview of key concepts, challenges for students with learning difficulties, and recommendations for instruction and intervention. Each strand includes a box with an example intensive intervention for the content. The strands are not organized by national standards; rather, the concepts featured in each strand are important for more than one domain.

Content Strands:

A. Number Sense, Place Value, and Number Systems
B. Whole-Number Relationships
C. Rational Numbers
D. Spatial and Geometric Reasoning
E. Measurement, Data Analysis, and Probability
F. Algebraic Reasoning

DOI: 10.4324/9781003096733-8

Strand A

Number Sense, Place Value, and Number Systems

After studying this strand, the reader should be able to:

1. discuss number sense and its importance for the range of mathematics topics;
2. implement several strategies with PreK–12 students that promote stronger number sense;
3. use teacher language to boost students' concept understanding of number;
4. determine student place-value understanding and plan activities to build those concepts;
5. discuss challenges and strategies for teaching about negative integers.

Emily, a student in Pinetops Elementary School's fifth-grade mathematics support class, was asked by Ms. Johnson to estimate and then quickly count a group of 217 unit cubes. Emily estimates 100 cubes but cannot explain how she selected that estimate. Then she counts each cube, stumbling over the numbers after 30, 80, and 110.

MS. JOHNSON: Let me ask you a question. When you are counting, let's say you counted to 90, tell me the number before that and the number after that.
EMILY: 89.
MS. JOHNSON: And the number that comes after 90?
EMILY: 100.
MS. JOHNSON: OK, now I'd like to hear you count by twos, starting with zero up to 20.
EMILY: 2, 4, 6, 8, 9, 10, 12, 14, 15, 18, 20.
MS. JOHNSON: Can you count by fives beginning with zero?
EMILY: 0, (slowly counting in head) 9, 14, 19.
MS. JOHNSON: Thank you. Now I'd like to see you write a number. Can you write the number three hundred two?
EMILY: Writes 31002
MS. JOHNSON: Can you read that number?
EMILY: Three hundred and two.

Emily demonstrates significant problems with numbers including patterns of numbers, place-value knowledge, and estimation. As a fifth grader, she is still struggling with recalling basic facts, applying operations to numbers, and solving problems involving whole numbers. Ms. Johnson noted that Emily works well with spatial concepts, such as describing features of geometric figures and interpreting graphs.

Virtually all mathematics in the elementary and secondary curriculum is strongly grounded in number, even extensions of geometry and measurement. This strand explores the importance of understanding number, ways of representing number, number sense, number relationships, number values, number systems, and how teachers can promote the development of these understandings throughout the mathematics curriculum.

DOI: 10.4324/9781003096733-9

Number Sense

Number sense is a concept that has evolved over several decades and still is not defined the same way by researchers in mathematics, psychology, and mathematics education. Mathematician Tobias Dantzig, in his groundbreaking 1954 book on number, used the term *number sense* to describe the ability that allows us to recognize that something has changed in a small collection of objects when an object is removed or added without our direct knowledge. This number-sense ability comes before counting and is shared with some animals.

Cognitive and developmental psychologists continued to explore young children's concepts of number, how those develop, and how certain concepts contribute to formal mathematics learning (Dehaene, 1997; Piaget, 1965; Sekuler & Mierkiewicz, 1977). For example, an innate ability termed the approximate number system allows for humans (and a range of animals) to discriminate between quantities without counting or using numbers (Berteletti et al., 2010). This ability to approximate representations of numerosities provides the foundation for the development of symbolic representations (Malone et al., 2020). In a longitudinal study of kindergarteners across a year, Malone and colleagues (2020) found that nonsymbolic numerosity discrimination was a predictor of early arithmetic ($r = .60$).

Psychologists have started to identify the contributions of basic number intuitions to formal school mathematics, and evidence suggests that the quality of one's approximate, nonsymbolic number system in preschool is correlated with more success in school mathematics (Chen & Li, 2014; Feigenson et al., 2013; Malone et al., 2020; Mazzocco et al., 2011; Schneider et al., 2017). Students with disabilities in mathematics (MLD) tend to have less accurate numerical approximation than their peers (Piazza et al., 2010). This approximate system of numerosity continues to develop and sharpen through early adulthood (Halberda & Feigenson, 2008), providing an important cognitive tool for determining the reasonableness of mathematics calculations involving symbols.

Mathematics educators have a related history of research on children's number concepts. Brownell wrote, "To be intelligent in quantitative situations children must see sense in the arithmetic they learn. Hence, instruction must be meaningful and must be organized around the ideas and relations inherent in arithmetic as mathematics" (1954, p. 5). Swedish mathematics educator Ekenstam (1977) studied how students with difficulties in mathematics (MD) developed their understanding of number concepts. He found that many middle-school students could calculate quite mechanically but failed to grasp what fraction and decimal numbers actually mean. Other mathematics-education researchers in the 1980s were concerned with how concepts of number and related abilities, such as estimation and mental calculation, contributed to students' higher-order conceptual understanding, or sense-making, of mathematics (Berch, 2005). This view of number sense was a more complex, broader concept than the intrinsic sense of number, with applications to the range of domains within mathematics.

Sowder (1988) described number sense as a well-organized conceptual network that enables students to relate number and operation properties and to solve number-based problems in creative ways. Abilities in students that indicate good number sense include inventing procedures for conducting operations, representing a number in several ways, recognizing number patterns, recognizing errors of magnitude, and discussing general properties of a numerical expression without depending on precise computation. This broad view of number sense was introduced to teachers and curriculum developers by the editors of *The Arithmetic Teacher* who proposed that "good number sense involves the development of understanding of … number meanings and relationships, the relative magnitudes of numbers, the relative effects of operations on numbers, and referents for quantities and measures as numbers used in everyday situations" (Thompson & Rathmell, 1989, p. 2). The National Council of Teachers of Mathematics' 1989

Curriculum and Evaluation Standards for School Mathematics described number sense as "an intuition about numbers that is drawn from all varied meanings of number" (NCTM, p. 39).

The 2000 NCTM standards and 2010 *Common Core State Standards for Mathematics* (CCSSM, NGA & CCSSO) reinforced the importance of number sense throughout the mathematics curriculum. "All the mathematics proposed for prekindergarten through grade 12 is strongly grounded in number" (NCTM, 2000, p. 32).

> During the years from kindergarten to eighth grade, students must repeatedly extend their conception of number… In high school, students will be exposed to yet another extension of number, when the real numbers are augmented by the imaginary numbers to form the complex numbers.
>
> (NGA & CCSSO, 2010, p. 58)

Both sets of standards include deep understanding of numbers and number systems throughout the grade levels. Selected examples of number-sense applications in the CCSSM:

- Kindergarteners know number names, use numbers to represent quantities, and compare numbers.
- Second graders can skip-count by 5s, 10s, and 100s. They can read and write numbers to 1000 using base-10 numerals, number names, and expanded form.
- Third graders understand place value of whole numbers and the properties of operations.
- Sixth graders understand that positive and negative numbers describe quantities that have opposite directions or values.
- Eighth graders can use rational approximations of irrational numbers to compare size and estimate the value of expressions.
- High-school students extend the properties of arithmetic to exponents and complex numbers.

The range and categories of the real-number system, including rational and irrational numbers, are depicted in Figure A.1.

The construct of number sense has evolved since the 1950s and is applied in often confusing ways in research and mathematics curricula. In noting the different uses of the construct *number sense* across fields of study (psychology, mathematics education, special education), Whitacre and colleagues (2020) analyzed 141 seminal research articles and clarified, for the first time, the researchers' use of one term (number sense) for three distinct constructs. *Approximate number sense* (ANS) is the inborn ability to perceive and discriminate magnitudes. *Early number sense* (ENS) includes learned skills such as number knowledge, counting, and comparing and is applied primarily to children ages 4–7 and those at risk of MD. ENS overlaps with ANS and *mature number sense* (MNS) that includes multidigit and rational-number sense, applied more to upper-elementary and middle-school students. The ANS is innate and most research on this construct is conducted by cognitive psychologists on children, adults, and animals and often uses brain imaging. The three neurological abilities of the ANS are perceptual subitization (rapid identification of three or four objects), magnitude discrimination (determining which of two sets includes more or less objects), and the mental number line (nonsymbolic but sequential and analogical representation of number; Dehaene, 2001; Whitacre et al., 2020). Friso-van den Bos et al. (2018), in noting the importance of mapping nonsymbolic number understanding (ANS) onto symbolic representations for later mathematics performance, demonstrated the effectiveness of training children (5 years old) in counting skills (through games and other counting activities) to advance symbolic-number understanding (ENS) and early arithmetic skills. Training had no effect on nonsymbolic processing (ANS).

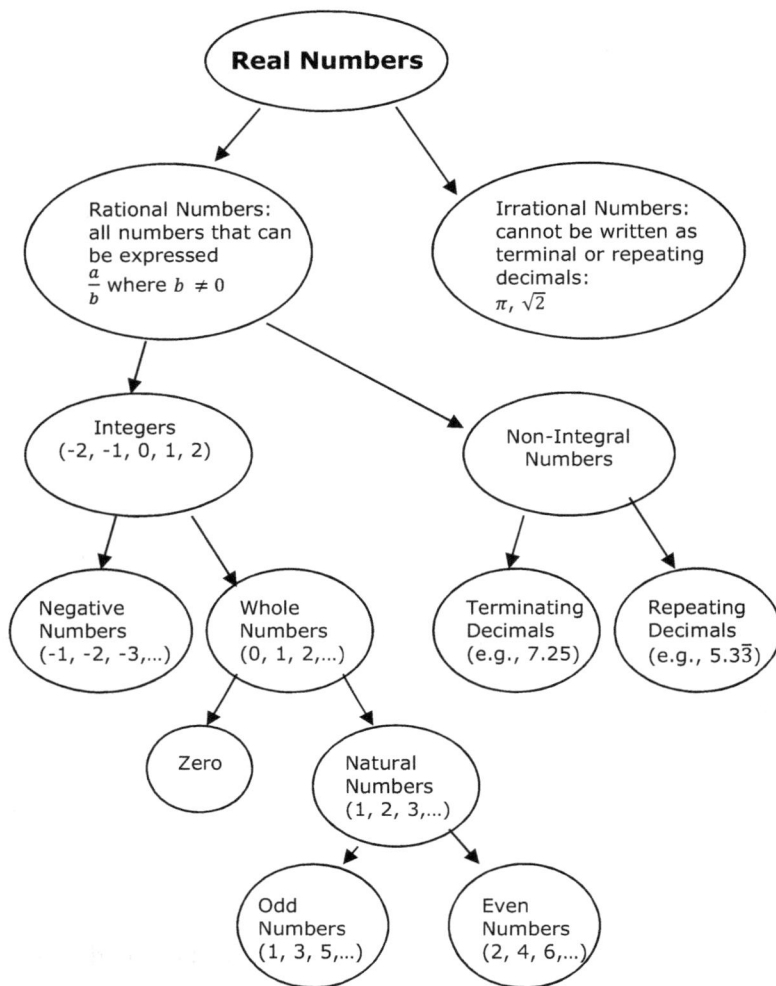

Figure A.1 The System of Real Numbers

Early Number Sense

Educational research on the development of early number sense (ENS) and its impact on later mathematics achievement has produced significant results in recent years. Special education researchers Gersten and Chard (1999) proposed that early number sense in mathematics may be analogous to phonemic awareness in reading development for early mathematical experiences. They built a strong case for the importance of number-sense development in early, preformal schooling experiences as a foundation for later mathematics achievement. Number sense was identified as the missing component of early math-facts learning, the reason rote drill and practice does not lead to significant improvement in mathematics ability. Their line of research, and that of others, articulated key components of ENS, or early numerical proficiency, in young children (Andrews & Sayers, 2015; Baroody et al., 2012; Dyson et al., 2011; Gersten et al., 2007; Soto-Calvo et al., 2015; Ulrich & Norton, 2019). The components most frequently identified are:

- numeral recognition (connecting names for numbers with what they represent and their symbols);
- counting efficiently and strategically (one-to-one correspondence, fixed order, cardinality, count-ons);
- recognizing and working with number patterns (pattern copying, missing number identification);
- comparing number magnitudes (pairs and on number line);
- transformation of sets (add-ons and take-aways as operations, verbal word-problem solving);
- fluent retrieval of basic number combinations (addition and subtraction); and
- estimation (magnitude estimation including using a number line to approximate number location).

Geary and colleagues, in assessing 12 quantitative competencies in preschool children, identified cardinal principle knowledge (CPK)—understanding of the quantities represented by number words—as the key predictor of later mathematics outcomes (Geary et al., 2019). The achievement of CPK predicts young children's readiness for formal mathematics at school entry (Geary et al., 2018). The five measures of quantitative abilities foundational to CPK are the ANS, sensitivity to more and less, learning number words, verbal counting, and recognizing numerals (symbols). Four-year-old children who do not know the count-word list and cannot enumerate up to four are at risk for delayed CPK and positive mathematics outcomes (Geary et al., 2019). However, these children may also have deficits in executive functions, such as attention and inhibition control, and letter recognition (visual-symbol learning). Some children may need more explicit, direct instruction on core numerical relations, enumeration, and cardinality to be successful in mathematics.

There is also evidence that kindergarten and early first-grade number-sense ability is a predictor of later mathematics achievement. Locuniak and Jordan (2008) found that some aspects of number sense in kindergarteners predicted their calculation fluency in second grade, a stronger predictor than memory, oral vocabulary, or reading ability. Especially important predictors were number knowledge (magnitude judgments) and experience with kindergarten-level addition and subtraction number combinations. Jordan et al. (2010) later used a number-sense screening battery (*Number Sense Screener™*) to measure number-sense ability in kindergarten and early first-grade children. The battery included counting (sequences and sets), magnitude judgments (comparing amounts), simple addition and subtraction problems (nonverbal set transformations, verbal story problems, and verbal number combinations). The number-sense measures were strong predictors of mathematics achievement at the end of first and third grades for these children, over and above age and cognitive factors. Early number-sense competencies predicted applied problem-solving abilities even stronger than calculation skills in the third grade. The finding that ENS as a predictor became stronger over time suggested that the effects of weak ENS may be cumulative.

Malone and colleagues (2020) assessed a range of potential predictors of early mathematics development upon school entry (age 5) and 12 months later. Symbolic-number knowledge (of Arabic digits, $r = .68$) and nonsymbolic-numerosity discrimination ($r = .60$) were unique and distinct predictors of arithmetic performance, even controlling for executive function. The researchers emphasized that, while number knowledge is often assessed during early childhood screening, tests of nonsymbolic-numerosity discrimination are not frequently used but could help identify children at-risk of arithmetic difficulties. Wong and Chan (2019) also identified both symbolic and nonsymbolic tasks as predictors of persistent low achievers across Grades 1 to 5. Measures of mapping of number symbols to magnitude (e.g., number-line tasks) and understanding the structure of the symbolic-number system (e.g., strategic counting of 1s, 10s, and 100s represented by squares) should be included in screenings.

Researchers in Belgium measured ENS abilities in kindergarten, first-, and second-grade children during spring of the school year and assessed their mathematics achievement a year later (Sasanguie et al., 2012). The number-sense measures included number priming (considering whether number pairs or sets of dots were larger or smaller than 5), number comparison (selecting the larger of two numbers or sets), and number-line estimation (positioning numbers on 0 to 10 and/or 0 to 100 number lines). Children who performed well in number comparison and number-line estimation had higher scores on mathematics achievement a year later than those with weaker abilities, even controlling for mathematics achievement over time. Especially predictive were children's abilities to connect number symbols with their meanings and mapping numbers onto space (number lines). Early screening tools that include number-sense components are described in Chapter 3. As Gersten et al. (2007) emphasized, there are measurable differences among children in kindergarten and first grade with respect to their early number sense and quick-screening tools can help teachers identify gaps in this foundational knowledge of number. Early targeted instruction to develop stronger number understanding may lessen the need for more formal interventions in later grades.

Early number sense is foundational for mature number sense (MNS), which "refers to a person's general understanding of number and operations along with the ability and inclination to use [MNS] in flexible ways to make mathematical judgments and to develop useful strategies for handling numbers and operations" (McIntosh et al., 1992, p. 3). Components of MNS include understanding of the meaning and size of numbers, using equivalent expressions of number, understanding the meaning and effect of operations, using equivalent expressions for other terms, applying flexible computing and counting strategies, judging the reasonableness of numerical results, and using benchmarks for measurement (Lin et al., 2016; Whitacre et al., 2020). MNS represents a more sophisticated and flexible use of number when solving problems and varies even more widely among students. Students with MD often lack ENS abilities, hindering their progress into more mature number sense. They often demonstrate overreliance on memorized procedures, lack basic conceptual understanding, demonstrate misconceptions about number and operations (e.g., longer numbers are larger), and cannot develop strategies for solving novel problems.

Promoting Number-Sense Development

Given the importance of number sense for developing deep mathematical understanding and promoting continuous achievement through the grade levels, what can teachers do to develop this ability in students with learning difficulties? Some interventions should target specific students while others can include small groups or the whole class.

A few caveats about number-sense development should be noted. First, promoting early number sense in young children (before formal schooling or in kindergarten) will look different than interventions in later grades and may have more impact over time. Young children in the verbal stage, just approaching symbolic understanding, are beginning to connect numbers to objects, by name and symbol (Jordan et al., 2010). Young children's delays may be due to lack of experience with number or a fundamental disability in language, working memory, approximate number system, visuospatial skills, or a combination of cognitive systems. Shown a group of five blocks, the child should say "five" and connect that with the numeral "5." Shown a dot card with four dots, the child should match it with four objects and say "four." When shown one more object and asked, "What is the number now?" the child should say "five" and match that with the five-dot card. Number-sense activities with young children are less formal, draw on everyday objects and experiences, and always make connections between what the child knows and is attempting to grasp. Researchers have demonstrated successful interventions with young children with delays in number sense using dot cards, dominoes, 5- and 10-frames, cubes, number charts, and other concrete or iconic materials, making explicit, language-supported connections between numbers, symbols, and their meanings (Dyson et al., 2011; Dyson et al., 2015; Sood & Jitendra, 2011).

Second, regardless of the grade level, teachers should make connections between form and meaning, between process and concept. Use verbal scaffolds, demonstrations, hands-on objects, and experiences to make connections. Use what students already know to promote connections. Without connections the "sense" in number sense will not take hold and be useful in later applications. For example, when helping students struggling to count beyond 20 (or some other stumbling point), use number lines, number charts, and other representations so students can visualize the relationships of numbers to each other, their relative magnitudes, and can map numbers spatially. Use language and encourage students to use their language to describe number relationships.

Finally, don't assume that students in upper elementary and even middle school have firm basic number understanding. As with Emily at the beginning of this strand, many students still get lost counting or comparing numbers in the 80s and 90s. Some cannot compare number magnitudes (which is larger: 400 or 389) or locate positions on a number line or coordinate plane. This author conducted think-aloud assessments with a number of third- through sixth-grade students with MLD (Gurganus & Shaw, 2006). One task required students to estimate the number of cubes in a clear container, with a range of 100 to 150 being reasonable for a container with 123 cubes. Most students estimated too low (20 to 50) or too high (200 to 1000). When asked to count the cubes, most students counted them one-by-one, not being confident in more efficient count-by strategies. Several stumbled on 89 to 90, 99 to 100, and 110 to 111 (decade transitions). Quick screening, such as with simple think-aloud tasks, and review at the beginning of the school year can help teachers identify which students need some gaps addressed. Critical transition points in number topics, such as moving from whole numbers to fractions or from whole numbers to integers including negative numbers, may need extra support through concrete materials, careful mathematics language, and sufficient, varied practice.

This section focuses on four areas critical for developing number sense (ENS and MNS) within any mathematics curriculum—mental mathematics, number-line applications, estimation skills, and number-meaning development.

Mental Mathematics

All engagement with mathematics is, of course, mental. But mental mathematics denotes "doing math" in your head without paper and pencil or other assistive devices. It forces students to visualize numbers and their relationships. Many teachers view *counting exercises* before a math lesson as coaches view stretches before a workout. These quick drills can strengthen students' sense of numbers and their patterns.

Begin with counting up, beginning at zero, to 30 or 40 by 1s, 2s, 5s, and 10s, then have students count back down again. On other days challenge students to count by 3s, 4s, 6s, 7s, 8s, and even 9s. With practice, students can continue patterns such as 1, 2, *skip 2*, 5, 6, *skip 2*, 9, 10, *skip 2*…, or 1000, 1050, 1100, 1150…, or create their own patterns (Gurganus, 2004).

If students who struggle with mathematics have trouble with counting exercises, try some of the following modifications:

- Start with a few numbers and practice until firm: 0, 5, 10, 15, 20, 20, 15, 10, 5, 0.
- Allow students to count using a number line or chart at first, then fade its use.
- Use familiar objects associated with the number (coins for 5s, 10s, 25s; shoes for 2s; clover for 3s; pictures of dog's feet for 4s; etc.) then fade their use.

- Have students respond chorally until comfortable before calling on individual students to count alone.
- Individualize the counting challenge for each student and scaffold students to the next level without causing frustration.
- Have students toss bean bags or bounce balls while counting.
- Play games that involve active counting (e.g., board games, skipping rope).

Visualizing numbers, objects represented by numbers, and number patterns can also strengthen number sense. Ask students to name their personal associations with specific numbers. For example, 12 is typically associated with a carton of eggs or a child's age. Younger students tend to view numbers as labels, like names, rather than the total number of items in a set. While the nominal form is common for addresses and other means of identification, mathematics uses primarily *cardinal* and *ordinal* forms \ in problem solving. Ask students about the relationships between numbers: Which number is larger, 47 or 52? How many whole numbers fall between 39 and 44? Just don't overburden students' capacities for mental math by presenting problems better computed by pencil or calculator. Many students with MLD have working-memory difficulties, as discussed in Chapter 2.

Another type of visualization involves the presentation of objects (concrete) or diagrams (representational) and asking students to enumerate or compare sets (see Figure A.2). Research on instant enumeration of objects in a set (small number system, in the ANS) demonstrated that humans could quickly identify three objects but started making errors with four (Mandler & Shebo, 1982). The first three objects do not have to be counted one-by-one. With larger numbers human brains have a hard time instantaneously identifying five objects from four $(n - 1)$ or six $(n + 1)$, especially when spacing and configuration are varied. It takes practice and some cognitive scaffolding to increase one's speed and accuracy with rapid enumeration and comparison.

Teachers can also promote number sense by setting up *number games* that challenge visualization and mental manipulation. Variations on dominoes and dice assist with rapid enumeration of whole-number sets and set comparisons. Simple games with playing cards or teacher-made flash cards can build number concepts and associations. Magic squares promote mental problem solving with number patterns. Magic squares (also magic triangles and sudoku) have some numbers missing from the grid and all number relationships are based on a rule. For example, the rule for the rows, columns, diagonals, and the four central squares in Figure A.3 is the sum of four numbers is 34. The rule can be found by placing letters in unknown boxes and solving simultaneous equations or through guess and check methods.

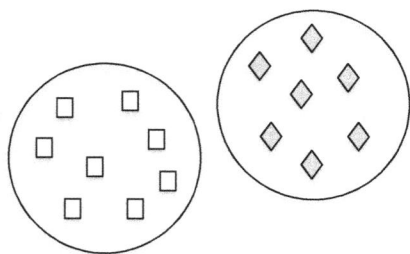

Figure A.2 Comparing Sets

	15		12
8	10		13
		16	
14		9	7

Figure A.3 Magic Square

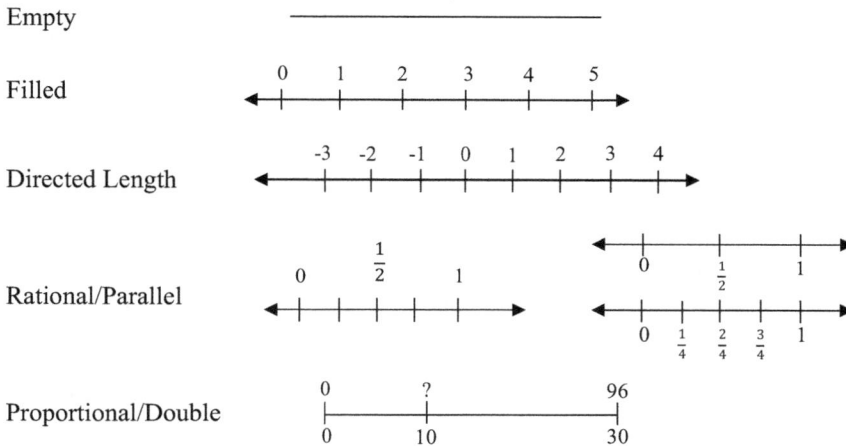

Empty

Filled

Directed Length

Rational/Parallel

Proportional/Double

Figure A.4 Types of Number Lines

Number-Line Applications

Number lines are relatively recent devices for teaching and learning about number magnitude, relationships, and operations. Although mathematicians have depicted number lines in their writings since the 17th century (e.g., Napier, 1616 for logarithms; Wallis, 1685 for addition and subtraction of negative numbers; and Chambers, 1728 for the coordinate system; Lemonidis & Gkolfos, 2020), and number lines occasionally appeared in mathematics textbooks around the turn of the 20th century (e.g., Wentworth's *Practical Arithmetic*, 1897, a line segment with equal parts to explain common fractions), it was not until the mid-20th century that number lines became common pedagogical tools and were widely depicted in mathematics textbooks (Wessman-Enzinger, 2018).

Diezmann et al. (2010) classified two major types of number lines: *structured number lines* with mathematical markings of proportional segments and *empty number lines* that are blank lines. Structured number lines can be further classified as *filled* (with equidistant points of rational numbers), *directed-length* (measured from zero to support arithmetic operations), *rational* (units are divided into equal intervals), and *proportional* (or double) number lines (Teppo & Van den Heuvel-Panhuizen, 2014). Number lines have been used to understand and convey number magnitude and relational concepts of negative integers, arithmetic operations, fractions, proportions, and algebra. Various types of number lines are depicted in Figure A.4. Structured number lines typically illustrate the one-to-one correspondence of the real numbers to points on the line. Their attributes include *ordinality* (increasing order),

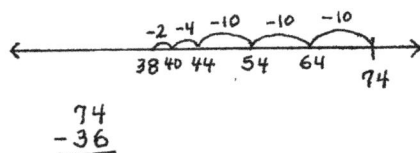

Figure A.5 Two-Digit Subtraction on Empty Number Line

directionality (conventionally positive is to the right, infinity in both directions), *relativity* (points are relative to others), and *density* (infinite set of real numbers on the number line and between any two real numbers on the number line; Wessman-Enzinger, 2018). These attributes are not intuitive for learners.

Number lines can serve as devices for understanding and manipulating number relationships, however teachers should be aware that number lines are not natural tools or depictions for students. When number lines are introduced for a specific concept, those should be carefully designed and introduced explicitly. Teachers should plan how to guide students through well-chosen examples to develop accurate concepts. For example, a teacher planning to use number lines for fraction concepts should begin with a number line labeled with 0 and 1, with 3 equidistant points in between, as with the rational number line in Figure A.4. Careful use of language and the emphasis on distance between points rather than the point will assist students in developing accurate concept understanding of common fractions.

Teachers using the empty number line with students with MD should also model accurate uses. For example, when using empty number lines to understand two-digit subtraction, students should already be familiar with structured number lines and some place-value concepts, such as counting by 10s. For the problem 74 − 36, the teacher models placing the 74 in an approximate location on the line, as in Figure A.5. He then demonstrates jumping in a negative direction, while labeling, each −10. Finally, he can jump −4 and −2 to arrive at the difference. The jumps illustrated above the line represent −36. This use of the empty number line allows the student to represent relative number magnitudes and the subtraction operation without using zero as a reference. Empty number lines offer flexibility in solution strategies but are more cognitively demanding than labeled lines.

Number lines also can be used to translate word-problem types onto a familiar representation, as with the schematic approach discussed in Chapter 5. For example, in change problems there is a beginning amount, a change amount, and an ending amount, with one of those unknown. *We have 10 bricks in our wheelbarrow, but it is too heavy to push, so we remove 2. How many bricks remain?* Find 10 on the number line, then move 2 units to the left, landing on the 8. Students who have difficulties with the transition from concrete objects to number symbols and equations can be assisted through the use of number lines.

There are common difficulties with using number lines. Younger children tend to overestimate small numbers and underestimate larger numbers, a logarithmic representation (e.g., number placement: ←--0----10----20----30---40---50--60--70-80-90-100--→). Children tend to make the shift to a more linear conception about ages 7 or 8 for the 0–100 number line and not until age 9 for the 0–1000 number line (Booth & Siegler, 2006). For students with learning delays in mathematics, their conception of number positions may continue to be compressed (Geary et al., 2012).

Luwel and colleagues (2019) found that for the 0–1000 number line with no other points labeled, third and fifth graders tended to self-generate a mid-point, however fifth graders used quartile strategies to a greater extent than third graders. Fifth-grader estimation ability improved when quartile benchmarks were marked with points on the number line (from 30%

with no benchmarks indicated to 89%), but third graders' accuracy increased only slightly (8%–34%) with quartile benchmarks probably because they had difficulty assigning values to those points. The researchers also found that more accurate number-line estimation was associated with better mathematics achievement in general. Teaching students how to locate and label benchmarks, such as 250, 500, and 750 on the 0–1000 number line, can support their number sense. When using number lines for instruction, labeled benchmarks along the number line may help support student learning.

Another common problem for some students is counting the marks along a number line rather than recognizing the distance between points as the value. For example, in viewing the difference between 5 and 8 on a 0–10 number line, a student might count the 2 dots in between. Another student begins counting at 1 rather than 0. Other students may have difficulty locating a point, for example the value 18, on a number line with the points 10 and 20 labeled, without reference to the 0 and an endpoint. One day they are working with a 0–100 number line, the next day a line with points 20 and 80 indicated, the following day a number line with 50 and 60 or even negative integers. When the spans change, the value of units between points change, an adjustment that is difficult for many students. Teachers should explicitly orient students to the beginning and endpoint values indicated and model locating other points. Strategies, such as locating a midpoint, quartile, or decile, should be modeled and practiced.

Other difficulties with using number lines have been noted with representing the subtraction and division of negative integers, without resorting to pure rule instruction. Students may also have difficulty with decimal fractions (e.g., 2.5, 1.3, 0.8), especially when a number line includes only whole numbers. Orienting students to decimal benchmarks within whole units on the number line using familiar benchmarks first (e.g., 1.25, 1.5, 1.75) can scaffold their understanding of in-between values in decimal form.

In a meta-analysis of 41 articles (ages 4–14), Schneider et al. (2018) concluded that number-line estimation correlates at a medium level with mathematics competencies ($r = .443$). Number-line estimation ability increased with age, was higher for fractions than for whole numbers, and demonstrated a stronger correlation with mathematics competencies than magnitude comparison for students 6–9 years old. The authors described number-line estimation as drawn from a wide range of abilities including magnitude processing, proportional reasoning, rounding strategies, spatial visualization, visual-motor integration, measurement, and counting.

Structured number lines should be introduced with explicit references to the points and relationships between points. They should be drawn with arrows on each end and clear indications of at least two reference points. For example, students in kindergarten should work with number lines labeled with 0 and 5 (or 10) with points indicated between, as in the filled number line in Figure A.4. Create large number lines across the floor using tape. Have students stand on reference points and compare distances, as the number line is a measurement model. This is a great way to model rounding numbers to the nearest 10 and simple addition and subtraction operations. Ask students to explain their thinking while working with number lines for insight into misconceptions about number. Older students should be challenged with segments of number lines depicting decimals, fractions, negative numbers, and very large numbers. Students working with ratios and proportions could benefit from experience with a double number line, a visual model of the equivalence of values, as in Figure A.4.

Estimation Skills

Until the 1990s, mathematics texts rarely emphasized estimation skills. Occasionally there would be a special box on a page with one or two estimation practices. More frequently students were encouraged to round and then compute. Estimation skills are important for mathematics and other areas of life. Some things cannot be exacted—the measurement of a coastline, the number of cells in your body. Some offer no practicality for exactness—there are about 150 oysters in a bushel, it will take a little more than 2 gallons of paint to paint this porch, we expect about 40 people at our party. The importance lies in good estimation based on a sense of number, space, and time.

Estimation is a critical part of problem solving. Using estimation assists students in selecting operations and judging the relative sense of their answers. Good problem solvers begin a problem with a sense of the range of reasonable answers and complete the problem-solving sequence by asking themselves, "Does my answer make sense given what I've established as the important parts of the problem?" These students tend to notice number relationships and make connections between past experiences and the current problem.

Estimation is developed over time through practice with different problem types. Students with difficulties in this area may be anxious about not having an exact answer, may not have estimation strategies such as rounding, or may not have a sense of the approximate size of an expected number as with the mental number line. Students should first have experience estimating relatively small amounts or sizes. About how many students are seated at each table in the cafeteria rather than about how many students are in the cafeteria. About how long is this bulletin board? Students often mistake estimation for guessing and should be taught to consider a reasonable range and begin relating their experiences with familiar distances, numbers, or areas with more challenging ones. Examples of problems without possible precise answers but requiring reasonable ranges:

- It is 20 miles between our school and the mall. If you could run about 6 or 7 miles per hour, about how long would it take to run to the mall?
- We know it is 100 miles from Charleston to Columbia because we took a field trip there last week. If this distance on your map is 100 miles, about how far is Washington?

Teachers should model strategies such as using referents and chunking so that students have estimation tools (Lang, 2001). Using referents involves finding number or amount benchmarks with which the student is already familiar and using those to estimate unknowns. For example, the student may be familiar with the length of a baseball bat and be able to use that in estimating someone's height. Chunking is like sampling in statistics—breaking a total into parts, estimating a part and then applying that back to the whole. This section of the auditorium has about 25 students so there must be about 200 students in all. Estimation with larger numbers can develop a stronger sense of number (MNS) and place value.

Number Meaning

Students who have more and varied experiences with numbers develop a deeper understanding of numbers and number structures. They develop a sense that numbers represent other things and can themselves be represented in various ways. In addition to the number line and number-game activities described in the previous sections, teachers can promote deeper understanding of number meaning in the following ways:

- Guiding students through 100 charts explorations: Using both 1–100 and 0–99 charts (Figure A.6; see Appendix II), students cover up various number patterns and see relationships such as factors, doubles, and primes.
- Demonstrating the use of objects or pictures to represent numbers: Interlocking cubes can represent various number patterns or groupings. Graph paper is a wonderful, inexpensive tool for showing the relationship between number and space.
- Incorporating number experiences into other content areas: Teachers should make a point to use number and estimation skills in social studies, reading, science, and other areas to promote generalization and reinforce number concepts.

1	2	3	◯	5	6	7	◯	9	10
11	◯	13	14	15	◯	17	18	19	◯
21	22	23	◯	25	26	27	◯	29	30
31	◯	33	34	35	◯	37	38	39	◯

Figure A.6 Use of a 100 Chart, Multiples of 4

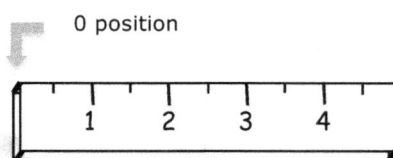

Figure A.7 Using Ruler with Concept of Zero

- Teaching the concept of zero: Many children think zero means nothing. But in the system of real numbers, zero holds a position on the number line. In computation, zeros have place value. Children often have trouble positioning a ruler because of their lack of understanding of zero as a starting position (see Figure A.7). To promote better understanding, use games with negative and positive numbers, incorporate zero into the number line, and include zero in counting exercises.
- Representing numbers in other forms: When familiar numbers take on unfamiliar positions such as in the denominator of a fraction or to the right of the decimal point, many students with learning difficulties are confused or attempt to maintain a whole-number concept. Begin these new concepts at the concrete level with familiar examples. Use language and modeling to make each numeral's role explicit.
- Connecting number meaning with number functions: From the number line, hundred charts, graph paper, objects, and other number experiences, students are better able to conceptualize the actions, relationships, and properties represented by addition, subtraction, multiplication, division, and more advanced operations (e.g., exponentiation).
- Using teacher and student language to support number concepts: Teachers scaffold student understanding using language. Students use language to explain their work, allowing for teachers to glimpse misconceptions and firm understandings. Without practice and feedback, students with learning difficulties tend to offer irrational or simplistic justifications such as, "It was the best number," or "It was the biggest number because it was the biggest of all." Sometimes their explanations will unearth true misconceptions, such as a fifth-grade student asked to arrange the digits 9, 8, 6, 4, and 1 into the lowest possible five-digit number and to explain which digit should be in the one's place (Yang, 2019). The student responded, "The 1 because 1 is the smallest," revealing serious place-value misconceptions.

In an effort to promote number-sense understanding among teachers and to assist teachers in implementing instruction that will improve number-sense development across grade levels, Cain et al. (2007) proposed the *Components of Number Sense* framework to represent the discussions and connections to be made in *every* math lesson throughout the curriculum (Faulkner & Cain, 2009). The concepts to be connected are quantity and magnitude, numeration, equality, base-10, form of a number, proportional reasoning, and geometric and algebraic thinking. Rather than teaching hierarchically, or one step at a time, teachers should support students' attempts to make simultaneous connections among concepts, to create an interconnected network for deeper number understanding. For example, the middle-school teacher attempting to teach percent of a whole number (e.g., 38% of 90) is faced with teaching a rote procedure that students may not be able to apply in a real-world context if they don't understand the concepts. Drawing on all parts of the framework, the teacher would guide a discussion and hands-on demonstrations to illuminate the meaning of 38% as equivalent to 0.38 or 38 parts of a 100. Students computing 38% of 90 would estimate that the result could not be greater than 90, would be less than half of 90 (< 45) and more than a quarter of 90 (> 23), so an answer of about 25 to 40 would be in the realm of reasonable. Answers of 3,000 or 300 or 3 would not be reasonable. Concepts of quantity, numeration, equivalence, place value (base 10), forms of the same number, proportional reasoning (38 is to 100 as x is to 90), and possibly geometric (using graph paper) and algebraic representations of the problem all connect for deeper understanding.

All students need to develop a strong sense of number and number relationships throughout school, but students with learning difficulties may need more explicit strategies to develop these fundamental understandings. Number sense may be key to most other mathematics learning. Box A.1 illustrates an intensive intervention (Tier 2) for three kindergarten students with early number-sense deficits. The next section extends number-sense development to the big-idea concept of place value.

Box A.1 Intensive Intervention for Early Number Sense

Target: Three kindergarten students demonstrate difficulty with counting, comparing number magnitudes (0 to 10), and naming the total of sets (cardinality).

Intervention Schedule: Daily 10-minute sessions for 6 weeks. (Or 15-minute sessions 3 times a week.)

Nature of Intervention: Each session follows the format:

1. Open with quick review of one skill from previous session (< 1 min).
2. a. Object-centered instruction (cubes, discs, markers, etc.), choral counting up and back, naming total of set, naming one more or one less than a number ($n + 1$, $n - 1$), comparing two numbers; then moving to
   b. Number-line based instruction, counting, naming points on the line with the number word, placing number cards under points, comparing two points, moving one point right or left from a number (3–4 min).
3. Game involving counting, comparing, and/or taking one more or one less from a set (3–4 min), card games, board games. Alternative: children's book activity with same skills.
4. Quick practice of skill(s) from lesson (< 1 min).

Begin with values of 0 to 4, move to 10, then to 15 and 20 if students make sufficient progress.

Data Collection (Twice Weekly): After an intervention session or at another point in the day, each child is asked to count from a given number (forward and backward), to name the number of objects in four sets (or $n + 1$, $n - 1$), compare numbers in four pairs, for 10 possible points. Results are graphed.

Place Value

As children enter formal mathematics instruction, place value is an essential concept that evolves in complexity throughout the curriculum. Place value is not innate; it is a cultural invention. The concept of place value—the quantity that a digit represents varies depending on the place it occupies within a number and is multiplicatively related to its neighbors—is one of the most fundamental for mathematics. Many teachers, regardless of grade level, begin the school year with place-value review and extension activities. Students are also fascinated to learn that our base-10 system of number is not the only system.

Before Place Value

The earliest number systems were developed as people needed to express larger numbers and make computations such as those for astronomy. If each real number had its own name and symbol, one couldn't get much past 33 distinct digits, as in the New Guinea system in the mid-20th century, before losing track of the amount (Dehaene, 1997, p. 94). Consider the challenge of memorizing each digit (that they assigned to body parts) and performing addition and subtraction by moving forward and backward. Go beyond the body positions and one's ability to understand larger number concepts is restricted.

Early systems for numbers included grouping for ease in managing larger numbers. The Egyptians used symbols for numbers and a special symbol for 10 as early as 3400 BCE. The 3000 BCE Sumerians (modern southern Iraq) used 1s, 10s, and 60s (the first positional number system—a combination decimal and sexagesimal system), using symbols pressed into clay. However, there was no zero. Even the Greeks, those mathematical marvels, had no positional number system or zero.

Babylonians (about 2000 BCE) continued to use the base-60 system and used the first indicator of a special space for "nothing in this place" between 700 and 300 BCE (Kaplan, 1999). The zero marker was used only in the middle of a number, not at the end. (Think about it—when someone says a car costs 24, you know they mean 24,000.) Base 60 has advantages because it allows for the expression of the factors of 2, 3, 5, 6, 10, 12, and 15, and simplified the most common fractions. Even today we use 60 for seconds and minutes in telling time, 12 months of the year, and 360 degrees in a circle.

The first written appearance of a circular symbol for zero with its current meaning was in Hindu India between 200 and 400 CE, in the Bakshali manuscript, with the first known definition of zero and its operations by Hindu astronomer and mathematician Brahmagupta in 628 CE (Szalay, 2017). Indian mathematicians, such as Brahmagupta and Aryabhata, also contributed rules of arithmetic including work with zero, place value, and negative numbers. Simultaneously, but independently, the Mayans (and possibly the Olmecs as early as 31 BCE) developed a place-value system (base 20) by 655 CE. Interestingly, they had a symbol for zero (a shell) along with counting numbers before developing a place-value system (O'Connor & Robertson, 2000). The Mayan's system was influenced by their work with calendars and astronomy.

Al'Khwarizmi, the Islamic mathematician whose name gives us the term *algorithm*, wrote *Al'Khwarizmi on the Hindu Art of Reckoning* (about 825 CE) describing the Indian place-value system based on the numerals 1, 2, 3, 4, 5, 6, 7, 8, 9, and 0, introducing it to Islamic and Arabian mathematicians. We refer to this system as the Hindu-Arabic number system. In Moorish Spain, approximately 950 CE, the use of the base-10 place-value system is documented. Small dots above numerals denote their value for that place, for example 4⋯ 6·

would mean 400 + 6 or 406. Al'Khwarizmi's book on the Indian place-value system was translated into Latin in the 12th century, and thus introduced into Europe (O'Connor & Robertson, 2000).

The Value of Place Value

It doesn't matter whether our place-value system uses 2, 5, 10, or even 20 digits, any system using zero can become operational. Twenty might be too many to recall, especially for math facts. Computer systems use binary systems (only two digits), represented in code with 0 and 1 for the circuit off and on positions. However, a binary system results in very long number representations. Some historians have posited that a 10-digit system is natural for humans because we have 10 fingers, harking back to body-based counting (Barrow, 2000).

Having a decimal (base-10) number system with a manageable number of digits, meaningful relative positions, and zero serving as a placeholder allows us to express very large and small numbers, make complex computations, and express rational numbers in different forms. We can read and understand the size of a number such as 50,002,600. We can use exponential notation such as 2.15×10^5 to represent 215,000 (the exponent 5 means 10 is multiplied by itself 5 times; the decimal is moved 5 places to the right) or .00043 can be expressed as 4.3 $\times 10^{-4}$ (10^{-4} is the same as $\frac{1}{10^4}$ or 0.0001, moving the decimal 4 places). With a place-value system we can compute using the operations (and the properties of the operations) for addition, subtraction, multiplication, and division for the entire set of rational numbers. We can express fractions in decimal form so comparisons are more easily made, measurements are more precise, and technological tools (calculators and computers) more useful. For example, consider which fraction is larger, $\frac{4}{15}$ or $\frac{6}{16}$. A difficult comparison! But it is easy to compare .2667 with .375. If students have a deeper understanding of place value, they can devise (or at least understand) nonroutine algorithms (see Strand B).

Researchers in mathematics education have noted that teachers rarely move beyond the surface-level understanding of place value, that a digit has a certain value in a column. "There is a near total neglect of effort to assist the student in developing a variety of conceptualizations of place value" for topics such as the basis for estimating, the foundation for alternative representations of numbers (e.g., 348 represented as 300 + 40 + 8), the foundation of the base-10 system including decimals, a way to construct alternative algorithms, the conservation of number embedded within alternative representations (e.g., four 10s and sixteen 1s), and a way to interpret the oral and written number system (Cawley et al., 2007, p. 22). Further, students who experience difficulty with place value tend to struggle with algorithms (Moeller et al., 2011).

Teachers should understand that our place-value system involves two inseparable principles: a *positional principle* (the position of each digit corresponds to a unit) and the *decimal principle* (each unit is equal to ten units of the immediately lower order; Houdement & Tempier, 2019). For example, in the number 473, the digit 7 represents 7 tens and its position's value is 10 times that of the position of the 3 (one-unit position). Numbers can be represented in three systems: written number (473), spoken number (four hundred seventy-three), and in numeration units (4 hundreds, 7 tens, 3 ones). The standard form (473) can be written in expanded form (400 + 70 + 3) to encourage students to understand arithmetic operations (Howe, 2019). Adding multiplicative reasoning deepens place-value understanding: $4 \times 100 + 7 \times 10 + 3 \times 1$ and also $4 \times 10^2 + 7 \times 10^1 + 3 \times$

10^0. Understanding the overall multiplicative structure of the base-10 place-value system (including decimal fractions) is essential for teachers.

Teaching Place-Value Concepts

Most young children view numbers as labels. As they learn to count, the numbers are names. I'm three years old. We live in apartment 209. It's much harder to add to or take away from the amount labeled with a rigid number name. Even when children count up to 12, numbers have discrete names (in English and most European languages). For young children counting by rote, the numbers up to 20 remain discrete. It's only when they attempt 21, 22, and so forth that the place-value pattern is apparent. The numbers from 13 to 19 are particularly tricky. "Thir" in thirteen and "fif" in fifteen may not sound like third or fifth to a child. The -teen suffix represents "plus ten" and is reversed when we reach the twenties and thirties and so forth. It is interesting to investigate how other languages incorporate the decimal system in verbal form. For example, in German the verbal form of 24 is *vierundzwanzig*—four and twenty, naming the one's place first (also the earlier English form, as in "four and twenty blackbirds"). In Chinese, the verbal form of 24 is translated as "two-ten four," literally naming the structure.

Consider the following examples of two students whose teachers did not understand the importance of place-value understanding. One student was working with subtraction of whole numbers and had been taught to subtract single digits using a number line. His teacher continued to use the number line when she taught him to perform two-digit minus one-digit computations (e.g., $17 - 5$). Perhaps she could not imagine that he would ever subtract larger numbers such as $398 - 127$. Another student had been taught two-digit addition with small digits, not requiring regrouping (e.g., $25 + 43$). Her teacher was careless and never explained that the 2 was actually 20 and the 4 was 40. The student added $2 + 4$ for 6 and $5 + 3$ for 8, with the sum of 68. Her answer was correct. However, when larger digits were required and the teacher tried to explain regrouping and the value of each column, this student was extremely resistant to a different procedure. It took three times as long to reteach addition from a place-value perspective. As illustrated in Strand B, students can perform alternative algorithms (such as those from left to right) as long as they understand the values of the digits and conserve number.

There have been disagreements among researchers regarding the appropriate age to begin place-value instruction (e.g., Clements & Sarama, 2014; Fuson, 1990; Kamii, 1985; McGuire & Kinzie, 2013). Developing place-value concepts, such as understanding the value of numerals in the ones, tens, and hundreds places, requires children to have conservation of number (decomposing one ten into ten ones doesn't change the value) and multiplicative reasoning with multiples of tens and hundreds. They must also have skills in composing and decomposing values, recognizing the difference between the face value of a number (numeral) and its complete value within a place, and language for naming numbers and places of digits. However, delaying place-value instruction until two-digit addition and subtraction instruction may result in misunderstandings that can persist until fifth or sixth grade in multidigit operations (McGuire & Kinzie, 2013).

Byrge et al. (2014) found that most preschoolers (4–6 years old) understood larger three- and four-digit numbers, probably from world experience, but when asked to write those numbers they tended to write them as the language sounds, in an expanded form. For example, 642 was written as 60042 or 600402. Their research suggested that not only do young children have experience with larger numbers and the language for those numbers, but that expanded form may be a foundational approach for teaching place value. There is evidence that young children process symbolic and verbal numbers differently (MacDonald et al., 2018). Younger children developing place-value concepts need time and planned activities to transition from concrete to abstract, from low numbers to high numbers, and to develop mental conceptions of the structure of numbers.

The CCSSM outline the following progression for place-value instruction:

- Kindergarten: Counting to 100 by 1s and 10s, composing and decomposing numbers from 11 to 19 into 10s and 1s (with objects or drawings).
- First grade: In the range 0 to 120, read and write numerals and represent objects with written numerals, add two-digit plus one-digit numbers using concrete models or drawings and place-value strategies, add and subtract multiples of 10 (from 10 to 90).
- Second grade: Read and write numbers to 1000 (numerals, number names, and expanded form); skip-count by 5s, 10s, and 100s; add and subtract within 1000 using models, drawings, or place-value strategies.
- Third grade: Round whole numbers to nearest 10 or 100, fluently add and subtract within 1000 using algorithms based on place value, and multiply one-digit numbers by multiples of 10 (10 to 90).

These grade-level benchmarks don't reveal the intermediate, more nuanced competencies that students with MD will need to develop, such as number to symbol mapping, unit coordination (construction and reorganization of units and unitizing composite units such as tens and hundreds), mental actions on number structures, understanding the zero concept, fluency with number combinations 0 to 10, strategic counting, comparing number magnitude based on structure, millennial transitions (across number centuries, e.g., from 198 to 203), and extending number-structure understanding to two, three, and more digits. Chan and colleagues (2014), in studying children from kindergarten through second grade, found that low-mathematics achievers at the end of second grade were behind their peers in strategic counting—counting taking advantage of groups of 10 and 100—since early first grade. These students had not grasped the structure of the place-value system, even with two–digit numbers. Many students with MD have weak place-value concepts into the upper elementary grades because they lack fundamental composing/decomposing skills and regrouping conceptual understanding (Lambert & Moeller, 2019; MacDonald et al., 2018). These students continue to have difficulty with multidigit arithmetic (even two–digit addition) because of their impaired understanding of place-value structure, lack of fact fluency, less efficient strategies for operations, and working-memory deficits (Lambert & Moeller, 2019).

Howe (2019) argued that early instruction should begin gradually, with learning addition and subtraction facts to 10 very well and flexibly (through activities such as "making 10"), learning the teen numbers as a 10 and some 1s (with explicit attention to language), and learning higher addition and subtraction facts by making and unmaking a 10. Equations involving "making 10," such as $3 + \_\_ = 10$ and $10 = \_\_ + 4$, should be supported with manipulatives (e.g., 10 frames, unit blocks). Instruction in addition should emphasize that two–digit numbers are a sum of a certain number of tens and a certain number of ones, with converting ones to tens and combining tens practiced with manipulatives and using mental math. Fraivillig (2018) demonstrated, in a two-year longitudinal study with first graders, that place-value instruction could be incorporated within the daily circle time "Days in School" counting routine. By using Digi–Blocks and stacked place-value cards ($100 + 20 + 4$ overlap to show 124), teachers practiced place-value with first graders daily, resulting in three times the knowledge of place-value concepts for the treatment group over a control group, with no additional classroom time.

Some researchers suggest that students spend sufficient time with two-digit place-value experiences before adding additional places (Fuson, 1990; McGuire & Kinzie, 2013). Effective teachers must understand what their students already conceptualize and scaffold that understanding using students' prior knowledge. Teachers need to demonstrate two-digit numbers using multiple representations and questioning so that students can verbalize their thinking. McGuire and Kinzie (2013) recommended using snap cubes, number charts, children's literature that involves counting, and games in which students count groups of 10, name leftovers, and identify the positional meaning of numerals. Students need multiple and varied experiences to be firm in place-value understanding to two digits.

Additional strategies for teaching place-value concepts:

- Use a single 10-frame to assist students' automatic recognition of zero to 10 objects (Figure A.8) (attributed to Wirtz, 1974). Adding 4 + 5 will leave one spot open on the frame. Adding 7 + 4 will fill the frame with one left-over object.
- Employ a two-box frame (for two-digit representations) with two-color disks. On one side of each disk is the face value (8) and on the other side the complete value (80). This method offers a support step between objects and symbols and connects with the expanded form (Varelas & Becker, 1997).
- Provide place-value mats and number cards for practice reading numbers and constructing numbers read aloud (Figure A.9). Longer mats can include a decimal point. These mats can be extended to form grids for addition and subtraction.
- Stress the patterns in reading larger numbers. One strategy is the concept of three houses in each block on a street, always ones, tens, hundreds, and a comma separating each block indicating where students should say the name of the block. For example: two hundred

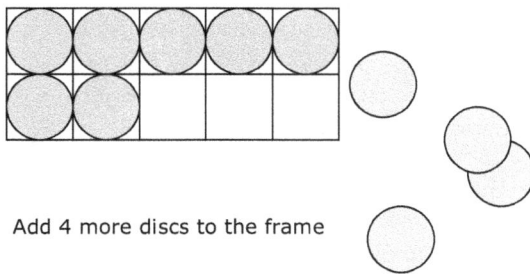

Add 4 more discs to the frame

Figure A.8 Ten-Frame to Demonstrate Place-Value Conversion

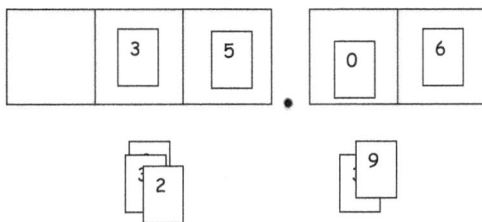

"Thirty-five and six one-hundredths."

Figure A.9 Place-Value Mat with Number Cards

thirty-five *million*, three hundred fifteen *thousand*, six hundred forty-seven. Place-value exercises are also excellent for mental-math activities. Reading large numbers, adding on in different places, rounding numbers to a given place, and writing dictated numbers are some of the activities that promote facility with place value. These activities can continue throughout the elementary and middle-school years with the inclusion of decimal numbers and expanded notation.

- Engage students with strategic counting. Use manipulatives such as a ten-block and 25 unit-blocks and encourage students to group units into tens and combine tens as a more strategic way of counting than counting units one-by-one. Practice with two- and three-digit representations.

- Present students with manipulative collections that need regrouping and combining. For example: 1 hundreds block, 15 tens blocks, 45 ones blocks. Have students verbalize their actions on the manipulatives and record the final number.

- Have students represent numbers in expanded notation before adding or subtracting as a concept builder. For example, 56 + 28 would be expanded into 50 + 6 over 20 + 8. The ones column would result in 10 + 4 and the tens column 70, requiring additional combining.

- Reinforce place value concepts with money. It is motivational! Using dollars for hundreds, dimes for tens, and pennies for ones, demonstrate the trading procedures. Money can also be used to demonstrate decimal positions (dollar is ones, dime is tenths, and penny is hundredths). Younger children may be confused with money because it is not physically proportional. Base-10 blocks, Digi-blocks, or other proportional materials would be more effective.

- Back up and work on simpler trading activities with students who may not have a solid conservation of number. Show one nickel for five pennies, one solid color Cuisenaire® rod for the same value of discrete cubes, items with marked prices (two 5¢ pencils for one 10¢ pen), and number cards with numbers on one side and circles or tallies on the other.

- Practice adding on by 10s, 100s, or even greater multiples using calculators. Call out a number (2,315), and have students enter it. Then say, "Add 100 and try to predict." Then have students push the equal button and read the result.

- Use lined paper sideways or graph paper to assist older students computing multidigit problems to keep columns lined up. Alternative algorithms such as those described in Strand B can lessen the memory load and are based on solid place-value concepts. If students are taught procedures by rote rather than through understanding, they will experience difficulty when faced with difficult situations such as zeros, decimals, or numbers that don't need regrouping. Other strategies for older students include decomposing/expanding numbers (3,127 = 3000 + 100 + 20 + 7) and estimating answers before computing: 58.12 divided by 2.3 will be close to 60 divided by 2 or 30.

Place value is what gives our number system its power. This big idea of mathematics is developed throughout the curriculum, with some concepts rather simple, others quite complex.

Negative Integers

Negative numbers were first recorded in the Chinese text *The Nine Chapters on the Mathematical Arts* (approximately 200 BCE), with rules for addition and subtraction of positive and negative numbers (O'Connor & Robertson, 2003). In seventh-century India, negative numbers were used to represent debts and positive numbers, assets. In *The Opening of the Universe* (628

CE), Indian mathematician Brahmagupta discussed the zero and negative numbers along with rules for operations with negative numbers. The Arabs and Europeans were much slower to accept negative numbers. These numbers were referred to as absurd, strange, fictitious, and impossible, probably due to the mathematicians' need to connect numbers with objects or geometric figures (Bishop et al., 2014). However, driven by other types of problems (quadratic equations and, more practically, accounting ledgers) many mathematicians in 17th- and 18th-century Europe finally (and reluctantly) began to use negative numbers with exponents and equations. By the 19th century a formal view of number (rather than the limited view of quantity and magnitude) was accepted and extended to include negative, rational, irrational, and imaginary numbers. Over the millennia, symbols for negative numbers have evolved, from slashing through the final digit of a number to the letter *m* and finally to the "−" symbol we use today.

In modern mathematics, numbers called integers include both positive and negative whole numbers and zero. The NCTM standards introduce negative integers in grade five: Students should "explore numbers less than 0 by extending the number line and through familiar applications" (NCTM, 2000, p. 148). In middle school, students develop a deeper understanding of integers and compare quantities, compute, and simplify expressions with the range of integers. The CCSSM introduce negative integers in the sixth grade, along with the full system of positive and negative rational numbers, absolute value, and the four-quadrant coordinate plane.

Some researchers promote an earlier introduction to negative numbers, or incorporating some transitional experiences earlier (as early as first grade), so that students in fifth or sixth grade facing negative numbers for the first time won't have to unlearn rules and concepts (e.g., "You can't subtract a larger number from a smaller one.") and relearn number relationships, representations, addition, and subtraction with a different set of numbers (Bofferding, 2014a). An earlier introduction to negative integers could provide deeper meaning for elementary-level mathematics topics such as subtraction and the meaning of zero. Aqazade et al. (2016) used number paths with second graders who had firm whole number sequence knowledge to explore negative number sequences and operations. The researchers used game boards with a range of integers for solving simple problems such as, "A gingerbread man is moving up and down a hill (labeled with a number path −10 to 10). Can you show 5 − 3? What about −5 − 3?" These second graders' abilities to correctly extend their whole number sequence into the negatives increased. Boffering (2019) recommended beginning the process of learning negative numbers at least by second grade because these children are able to understand initial concepts and the development of negative number concepts, such as those for operations and equations takes several years.

What makes negative integers so challenging for students? First, students may have been taught vocabulary, concepts, and rules that don't hold up when facing negative integers and their operations. For example, numbers that are longer are bigger. Or "more" means moving right on the number line (Bofferding, 2014b). Second, the lack of tangible, concrete, and countable representations for negative numbers creates an obstacle to concept acceptance for some students, not unlike the views of medieval mathematicians (Bishop et al., 2014). Even when representations, such as number lines or red and black chips are used, students often have difficulty depicting various problems when negative numbers are included. Third, the symbolic representation of negative numbers may be confusing or contradict some students' number and symbol paradigms. How could you possibly solve $5 + \underline{\quad} = 3$? What does −3 −−5 = ___ mean?

Students should understand that the minus sign has three meanings: subtract (binary function, 8 − 5), negative (unary function, −4), and opposite (multiply by −1, −(−6) = 6, symmetric function; Vlassis, 2008). Students will move through developmental stages in forming mental models of these concepts—from initial and often faulty concepts (e.g., viewing negative integers as the same as positive ones) to synthetic and unstable concepts (e.g., including some new number forms but with errors) and finally to a formal, accurate understanding (Bofferding, 2019). Experiences with good models such as number lines (directional) and number paths (count model) as well as careful language and symbol use with scaffolding can support students' negative-number understanding. In studies of middle-school students, Booth and Davenport (2013) found that conceptual knowledge of the minus sign was a strong predictor of success in solving equations and Young and Booth (2020) found that the knowledge of negative numbers on a bidirectional number line predicted improvement in overall algebra knowledge. Booth et al. (2014), in a study of conceptual errors in Algebra I, identified negative-sign errors as the most prevalent error type across six topics of study (e.g., solving equations, inequalities).

Promoting Negative-Number Sense

For students in first through fourth grades, transitional experiences can help students establish a stronger and more valid sense of integers before formal introduction in fifth or sixth grade. Bofferding (2014a) recommended using the term number *values* instead of number *quantities* in earlier grades. She argued that values and order of numbers are foundational for solid computational understanding. In conducting research with first graders, Bofferding (2014b) found that these young students had many intuitions about negative numbers that could be harnessed, such as intuition that the number line could keep going below zero, and personal experiences with negative representations such as with temperature.

When working with number patterns and counting, students should be encouraged to include zero as a number and continue counting below zero, or counting backward by a few integers, to develop a sense for the values and order of negative and positive integers and zero. Teachers should be aware of students' understanding of the minus sign and its multiple meanings. Students should be engaged in activities, such as memory or matching games, where they can discriminate between expressions (e.g., $4 - 1$, $-6 + -2$, $8 --4$, etc.) without actually computing. They can label horizontal and vertical number lines that represent real-life situations involving negative numbers—above and below water, floors above and below ground, yards rushing or lost on a football field, and temperatures—and play games manipulating various values on the number lines (Boffering, 2014a). Negative integers should be incorporated into an inclusive number system whenever possible and practicing with negative-number sequences (counting up and back) can improve students' performance.

When negative integers are introduced formally, in fifth or sixth grade, teachers should informally assess students' negative-number concepts. Simple tasks such as putting integers in order $(4, -9, 3, 0, 5, -7)$, filling in a number line with missing integers, comparing the values of two integers (which number is greater? -2 or -7), and simple addition and subtraction items can illuminate students' foundational concepts and misconceptions. Rather than teaching rules rotely, such as adding two negatives gives a positive, introduce bidirectional number lines for practice with simple equations (e.g., $-6 - -3$, $-5 + 8$, $-7 - 6$). Use spatial cues so students don't confuse the negative symbol with subtraction by moving the negative sign closer to the numeral and encouraging students to write symbols this way or use parentheses for clarity (e.g., $-6 - (-3) = x$). Provide guided practice with activities that involve thinking about integers as directed quantities such as number sentences and word problems involving realistic contexts supported with number paths or bidirectional number lines. Listen to student explanations to uncover misconceptions that require reteaching.

Some contexts for studying negative integers include:

- Money: borrowing and repaying money as a concept of debt (Whitacre et al., 2014).
- Local contexts and names: negative temperatures for northern areas (the temperature was −15 degrees in Chicago), building levels above and below ground, mining operations levels above and below ground, or water levels above and below a reference point (Beswick, 2011).
- Varying number-line formats and pairing those with their context: horizontal (distance on a football field), vertical (as in a thermostat), and circular (dial) (Figure A.10).
- Other content areas: examples from social studies, science, literature books from the same grade level. For example, use a timeline showing BCE to CE dates of Egyptian history or the boiling temperature points of various gasses.
- Simple board games: roll the dice (one red and one black) and move backward or forward, keeping track on paper through positive and negative notations.
- Discrete counters: red and black chips, dice, and cards can be used to manipulate integers (represent zero as well) and to create integer games.

The integrated theory of numerical development (Siegler et al., 2011; Siegler, 2016) posits that the magnitudes of all rational numbers are represented on a mental number line that begins with small whole numbers and expands, over the course of development, to include larger whole numbers, negative numbers, and interstitially to include fractions and decimals. Numerous processes, such as association and analogy, influence the development of number-magnitude knowledge, which is central to numerical development. Interventions that help students understand number-system structures, map numbers onto their magnitudes, and learn which properties apply to which numbers have the potential to improve students' mathematical outcomes. Properties of number and operations are addressed in the next chapter.

Number sense is foundational for all mathematics; therefore, teachers should create opportunities for students to strengthen their number concepts regardless of grade level or subject area. Numbers are part of the language of mathematics. For students to be proficient in mathematics they should be fluent with numbers and their meanings.

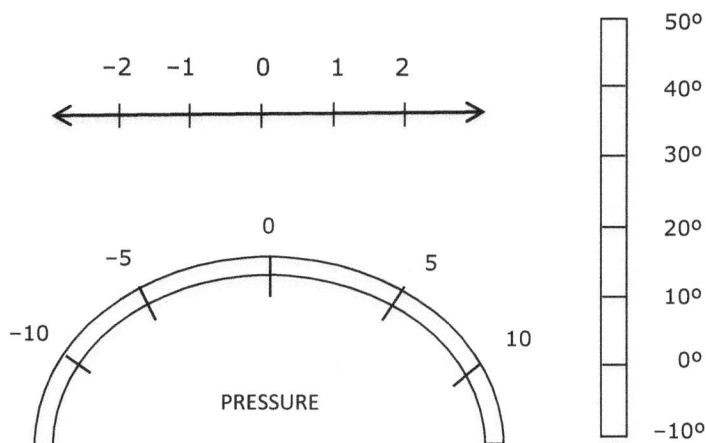

Figure A.10 Number Line Variations

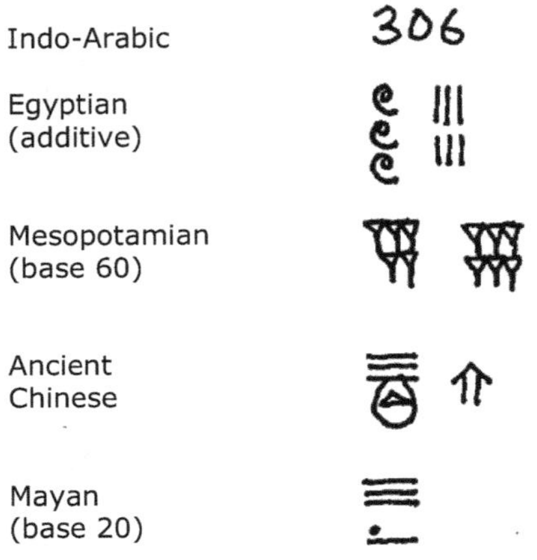

Indo-Arabic	306
Egyptian (additive)	
Mesopotamian (base 60)	
Ancient Chinese	
Mayan (base 20)	

Figure A.11 Multicultural Connection: Place-Value Systems

Multicultural Connection

The Indo-Arabic system of place value uses a base-10 system of numbers with the zero symbol, as is depicted for the number 306 in Figure A.11. There are three hundreds, no tens, and six ones in the number 306. Early number systems depicted in Figure A.11 did not use a zero as a place holder, except for the Mayans for some situations (a shell symbol). Preservice teachers and elementary students should work with Egyptian or Mayan number systems to represent and operate with two-, three-, and four-digit numbers as a way of understanding grouping structures, especially the value of place value and its multiplicative structure (see for example Thanheiser & Melhuish, 2019; find systems at www.dcode.fr/en).

References

Andrews, P., & Sayers, J. (2015). Identifying opportunities for grade one children to acquire foundational number sense: Developing a framework for cross cultural classroom analyses. *Early Childhood Education Journal, 43*(4), 257–267. doi: 10.1007/s10643-014-0653-6

Aqazade, M., Bofferding, L., & Farmer, S. (2016). Benefits of analyzing contrasting integer problems: The case of four second graders. In M. B. Wood, E. E. Turner, M. Civil, & J. A. Eli (Eds.), *Proceedings of the 38th annual meeting of the North American Chapter of the International Group for the Psychology of Mathematics Education* (pp. 132–139). The University of Arizona. www.pmena.org/proceedings/

Baroody, A. J., Eiland, M. D., Purpura, D. J., & Reid, E. E. (2012). Fostering at-risk kindergarten children's number sense. *Cognition and Instruction, 30*(4), 435–470. doi: 10.1080/07370008.2012.720152

Barrow, J. D. (2000). *The book of nothing: Vacuums, voids, and the latest ideas about the origins of the universe.* Pantheon Books.

Berch, D. B. (2005). Making sense of number sense: Implications for children with mathematical disabilities. *Journal of Learning Disabilities, 38*, 333–339. doi: 10.1177/00222194050380040901

Berteletti, I., Lucangeli, D., Dehaene, S. Piazza, M., & Zorzi, M. (2010). Numerical estimation in preschoolers. *Developmental Psychology, 46*(2), 545–551. doi: 10.1037/a0017887

Beswick, K. (2011). Positive experiences with negative numbers: Building on students' in- and out-of-school experiences. *Australian Mathematics Teacher, 67*(2), 31–40.

Bishop, J. P., Lamb, L. L., Philipp, R. A., Whitacre, I., Schappelle, B. P., & Lewis, M. L. (2014). Obstacles and affordances for integer reasoning: An analysis of children's thinking and the history of mathematics. *Journal for Research in Mathematics Education, 45*, 19–61. doi: 10.5951/jresematheduc.45.1.0019

Bofferding, L. (2014a). Order and value: Transitioning to integers. *Teaching Children Mathematics*, *20*(9), 546–554. doi: 10.5951/teacchilmath.20.9.0546

Bofferding, L. (2014b). Negative integer understanding: Characterizing first graders' mental models. *Journal for Research in Mathematics Education*, *45*, 194–245. doi: 10.5951/jresemathheduc.45.2.0194

Bofferding, L. (2019). Understanding negative numbers. In A. Norton & M. W. Alibali (Eds.), *Constructing number: Merging perspectives from psychology and mathematics education* (pp. 251–277). Springer. doi: 10.1007/978-3-030-00491-0_12

Booth, J. L., Barbieri, C., Eyer, F., & Paré-Blagoev, E. J. (2014). Persistent and pernicious errors in algebraic problem solving. *Journal of Problem Solving*, *7*, 10–23. doi: 10.7771/1932-6246.1161

Booth, J. L., & Davenport, J. L. (2013). The role of problem representation and feature knowledge in algebraic equation-solving. *The Journal of Mathematical Behavior*, *32*, 415–423. doi: 10.1016/j.jmathb.2013.04.003

Booth, J. L., & Siegler, R. S. (2006). Developmental and individual differences in pure numerical estimation. *Developmental Psychology*, *42*, 189–201. doi: 10.1037/0012-1649.41.6.189

Brownell, W. A. (1954). The revolution in arithmetic. *The Arithmetic Teacher*, *1*(1), 1–5. doi: 10.5951/at.1.1.0001

Byrge, L., Smith, L. B., & Mix, K. S. (2014). Beginnings of place value: How preschoolers write three-digit numbers. *Child Development*, *85*, 437–443. doi: 10.1111/cdev.12162

Cain, C., Doggett, M., Faulkner, V., & Hale, C. (2007). *The components of number sense*. NC Math Foundation Training, Exceptional Children's Division of the North Carolina Department of Public Instruction.

Cawley, J. F., Parmar, R. S., Lucas-Fusco, L. M., Kilian, J. D., & Foley, T. E. (2007). Place value and mathematics for students with mild disabilities: Data and suggested practices. *Learning Disabilities: A Contemporary Journal*, *5*(1), 21–39.

Chan, W. W. L., Au, T. K., & Tang, J. (2014). Strategic counting: A novel assessment of place-value understanding. *Learning and Instruction*, *29*, 78–94. doi: 10.1016/j.learninstruc.2013.09.001

Chen, Q., & Li, J. (2014). Association between individual differences in non-symbolic number acuity and math performance: A meta-analysis. *Acta Psychologica*, *148*, 163–172. doi: 10.1016/j.actpsy.2014.01.016

Clements, D. H., & Sarama, J. (2014). *Learning and teaching early math: The learning trajectories approach* (2nd ed.). Routledge. doi: 10.4324/9780203520574

Dantzig, T. (1954). *Number: The language of science*. Macmillan.

Dehaene, S. (1997). *The number sense: How the mind creates mathematics*. Oxford University Press.

Dehaene, S. (2001). Precis of the number sense. *Mind & Language*, *16*(1), 16–36. doi: 10.1111/1468-0017.00154

Diezmann, C. M., Lowrie, T., & Sugars, L. A. (2010). Primary students' success on the structured number line. *Australian Primary Mathematics Classroom*, *15*(4), 24–28.

Dyson, N., Jordan, N. C., Beliakoff, A., & Hassinger-Das, B. (2015). A kindergarten number-sense intervention with contrasting practice conditions for low-achieving children. *Journal for Research in Mathematics Education*, *46*(3), 331–370. Doi: 10.5951/jresemathheduc.46.3.0331

Dyson, N. I., Jordan, N. C., & Glutting, J. (2011). A number-sense intervention for low-income kindergarteners at risk for mathematics difficulties. *Journal of Learning Disabilities*, *46*(2), 166–181. doi: 10.1177/0022219411410233

Ekenstam, A. A. (1977). On children's quantitative understanding of numbers. *Educational Studies in Mathematics*, *8*(3), 317–332. doi: 10.1007/bf00385928

Faulkner, V., & Cain, C. (2009). The components of number sense: An instructional model for teachers. *Teaching Exceptional Children*, *41*(5), 24–30. doi: 10.1177/004005990904100503

Feigenson, L., Libertus, M. E., & Halberda, J. (2013). Links between the intuitive sense of number and formal mathematics ability. *Child Developmental Perspectives*, *7*(2), 74–79. doi: 10.1111/cdep.12019

Fraivillig, J. L. (2018). Enhancing established counting routines to promote place-value understanding: An empirical study in early elementary classrooms. *Early Childhood Education Journal*, *46*(1), 21–30. doi: 10.1007/s10643-016-0835-5

Friso-van den Bos, K. (2018). Counting and number line trainings in kindergarten: Effects on arithmetic performance and number sense. *Frontiers in Psychology*, *9*, 975–975. doi: 10.3389/fpsyg.2018.00975

Fuson, K. C. (1990). Conceptual structures for multiunit numbers: Implications for learning and teaching multidigit addition, subtraction, and place value. *Cognition and Instruction, 7*, 343–403. doi: 10.1207/s1532690xci0704_4

Geary, D. C., Hoard, M. K., Nugent, L., & Bailey, D. H. (2012). Mathematical cognition deficits in children with learning disabilities and persistent low achievement: A five-year prospective study. *Journal of Educational Psychology, 104*(1), 206–223. doi: 10.1037/a0025398

Geary, D. C., vanMarle, K., Chu, F. W., Hoard, M. K., & Nugent, L. (2019). Predicting age of becoming a cardinal principle knower. *Journal of Educational Psychology, 111*(2), 256–267. doi: 10.1037/edu0000277

Geary, D. C., vanMarle, K., Chu, F. W., Rouder, J., Hoard, M. K., & Nugent, L. (2018). Early conceptual understanding of cardinality predicts superior school-entry number-system knowledge. *Psychological Science, 29*, 191–205. doi: 10.1177/0956797617729817

Gersten, R., & Chard, D. (1999). Number sense: Rethinking arithmetic instruction for students with mathematical disabilities. *The Journal of Special Education, 33*(1), 18–28. doi: 10.1177/002246699903300102

Gersten, R., Clarke, B. S., & Jordan, N. C. (2007). *Screening for mathematics difficulties in K-3 students.* RMC Research Corporation, Center on Instruction. www.centeroninstruction.org

Gurganus, S. P. (2004). 20 ways to promote number sense. *Intervention in School and Clinic, 40*(1), 55–58. doi: 10.1177/10534512040400010501

Gurganus, S. P., & Shaw, A. (2006, February). *Think-alouds for informal math assessment.* Presentation at the meeting of the South Carolina Council for Exceptional Children, Myrtle Beach.

Halberda, J., & Feigenson, L. (2008). Developmental change in the acuity of the "number sense": The approximate number system in 3-, 4-, 5-, and 6-year-olds and adults. *Developmental Psychology, 44*, 1457–1465. doi: 10.1037/a0012682

Houdement, C., & Tempier, F. (2019). Understanding place value with numeration units. *ZDM, 51*(1), 25–37. doi: 10.1007/s11858-018-0985-6

Howe, R. (2019). Learning and using our base ten place value number system: theoretical perspectives and twenty-first century uses. *ZDM, 51*(1), 57–68. doi: 10.1007/s11858-018-0996-3

Jordan, N. C., Glutting, J., & Ramineni, C. (2010). The importance of number sense to mathematics achievement in first and third grades. *Learning and Individual Differences, 20*(2), 82–88. doi: 10.1016/j.lindif.2009.07.004

Kamii, C. K. (1985). *Young children reinvent arithmetic.* Teachers College Press.

Kaplan, R. (1999). *The nothing that is: A natural history of zero.* Oxford University Press.

Lambert, K., & Moeller, K. (2019). Place-value computation in children with mathematics difficulties. *Journal of Experimental Child Psychology, 178*, 214–225. doi: 10.1016/j.jecp.2018.09.008

Lang, F. K. (2001). What is a "good guess," anyway? Estimation in early childhood. *Teaching Children Mathematics, 7*, 462–466.

Lemonidis, C., & Gkolfos, A. (2020). Number line in the history and the education of mathematics. *Inovacije u Nastavi: Časopis Za Savremenu Nastavu,* [Innovation in Teaching], *33*(1), 36–56. doi: 10.5937/inovacije2001036L

Lin, Y.-C., Yang, D.-C., & Li, M.-N. (2016). Diagnosing students' misconceptions in number sense via a web-based two-tier test. *Eurasia Journal of Mathematics, Science & Technology Education, 12*(1), 41–55. doi: 10.12973/eurasia.2016.1420a

Locuniak, M. N., & Jordan, N. C. (2008). Using kindergarten number sense to predict calculation fluency in second grade. *Journal of Learning Disabilities, 41*, 451–459. doi: 10.1177/0022219408321126

Luwel, K., Peeters, D., & Verschaffel, L. (2019). Developmental change in number line estimation: A strategy-based perspective. *Canadian Journal of Experimental Psychology, 73*(3), 144–156. doi: 10.1037/cep0000172

MacDonald, B. L., Westenskow, A., Moyer-Packenham, P. S., & Child, B. (2018). Components of place value understanding: Targeting mathematical difficulties when providing interventions. *School Science and Mathematics, 118*(1–2), 17–29. doi: 10.1111/ssm.12258

Malone, S. A., Burgoyne, K., & Hulme, C. (2020). Number knowledge and the approximate number system are two critical foundations for early arithmetic development. *Journal of Educational Psychology, 112*(6), 1167–1182. doi: 10.1037/edu0000426

Mandler, G., & Shebo, B. J. (1982). Subitizing: An analysis of its component processes. *Journal of Experimental Psychology: General, 111*(1), 1–21. doi: 10.1037/0096-3445.111.1.1

Mazzocco, M. M. M., Feigenson, L., & Halberda, J. (2011). Preschoolers' precision of the approximate number system predicts later school mathematics performance. *PLoS One, 6*(9), e23749. doi: 10.1371/journal.pone.0023749.t001

McIntosh, A., Reys, B. J., & Reys, R. E. (1992). A proposed framework for examining basic number sense. *For the Learning of Mathematics, 12*(3), 2–8.

McGuire, P., & Kinzie, M. B. (2013). Analysis of place-value instruction and development in pre-kindergarten mathematics. *Early Childhood Education, 41*(5), 355–364. doi: 10.1007/s10643-013-0580-y

Moeller, K., Pixner, S., Zuber, J., Kaufmann, L., & Nuerk, H. C. (2011). Early place-value understanding as a precursor for later arithmetic performance: A longitudinal study on numerical development. *Research in Developmental Disabilities, 32*(5), 1837–1851. doi: 10.1016/j.ridd.2011.03.012

National Council of Teachers of Mathematics (1989, 2000). *Principles and standards for school mathematics.* www.nctm.org/Standards-and-Positions/Principles-and-Standards/

National Governors Association Center for Best Practices & Council of Chief State School Officers. (2010). *Common core state standards for mathematics.* www.corestandards.org/Math/

O'Connor, J. J., & Robertson, E. F. (2000). *History topic: Indian numerals.* https://mathshistory.st-andrews.ac.uk/HistTopics/Indian_numerals/

O'Connor, J. J., & Robertson, E. F. (2003). *History topic: Nine chapters on the mathematical art.* https://mathshistory.st-andrews.ac.uk/HistTopics/Nine_chapters/

Piaget, J. (1965). *The child's conception of number.* Norton.

Piazza, M., Facoetti, A., Trussardi, A. N., Berteletti, I., Conte, S., Lucangeli, D., Dehaene, S., & Zorzi, M. (2010). Developmental trajectory of number acuity reveals a severe impairment in developmental dyscalculia. *Cognition, 116*, 33–41. doi: 10.1016/j.cognition.2010.03.012

Sasanguie, D., Van den Bussche, E., & Reynvoet, B. (2012). Predictors for mathematics achievement? Evidence from a longitudinal study. *Mind, Brain, and Education, 6*(3), 119–128. doi: 10.1111/j.1751-228x.2012.01147.x

Schneider, M., Beeres, K., Coban, L., Merz, S., Susan Schmidt, S., Stricker, J., & De Smedt, B. (2017). Associations of non-symbolic and symbolic numerical magnitude processing with mathematical competence: A meta-analysis. *Developmental Science, 20*, e12372. doi: 10.1111/desc.12372

Schneider, M., Merz, S., Stricker, J., De Smedt, B., Yorbeyns, J., Verschaffel, L., & Luwel, K. (2018). Associations of number line estimation with mathematical competence: A meta-analysis. *Child Development, 89*(5), 1467–1484. doi: 10.1111/cdev.13068

Sekuler, R., & Mierkiewicz, D. (1977). Children's judgments of numerical inequality. *Child Development, 48*, 630–633. doi: 10.2307/1128664

Siegler, R. S. (2016). Magnitude knowledge: The common core of numerical development. *Developmental Science, 19*, 341–361. doi: 10.1111/desc.12395

Siegler, R. S., Thompson, C. A., & Schneider, M. (2011). An integrated theory of whole number and fractions development. *Cognitive Psychology, 62*, 273–296. doi:10.1016/j.cogpsych.2011.03.001

Sood, S., & Jitendra, A. K. (2011). An exploratory study of a number-sense program to develop kindergarten students' number proficiency. *Journal of Learning Disabilities, 46*, 328–346. doi: 10.1177/0022219411422380

Soto-Calvo, E., Simmons, F. R., Willis, C., & Adams, A.-M. (2015, December). Identifying the cognitive predictors of early counting and calculation skills: Evidence from a longitudinal study. *Journal of Experimental Child Psychology, 140*, 16–37. doi: 10.1016/j.jecp.2015.06.011

Sowder, J. T. (1988). Mental computation and number comparison: Their roles in the development of number sense and computational estimation. In J. Hiebert & M. Behr (Eds.), *Number concepts and operations in the middle grades* (pp. 182–197). Lawrence Erlbaum.

Szalay, J. (2017, 18 September). Who invented zero? *LiveScience.* www.livescience.com/27853

Teppo, A., & van den Heuvel-Panhuizen, M. (2014). Visual representations as objects of analysis: The number line as an example. *ZDM, 46*(1), 45–58. doi: 10.1007/s11858-013-0518-2

Thanheiser, E., & Melhuish, K. (2019). Leveraging variation of historical number systems to build understanding of the base-ten place-value system. *ZDM*, *51*(1), 39–55. doi: 10.1007/s11858-018-0984-7

Thompson, C. S., & Rathmell, E. C. (1989). By way of introduction [to the special issue on number sense]. *The Arithmetic Teacher*, *36*(6), 2–3. doi: 10.5951/at.36.6.0002

Ulrich, C., & Norton, A. (2019). Discerning a progression in conceptions of magnitude during children's construction of number. In A. Norton & M. W. Alibali (Eds.), *Constructing number: Merging perspectives from psychology and mathematics education* (pp. 47–68). Springer International. doi: 10.1007/978-3-030-00491-0_3

Varelas, M. & Becker, J. (1997). Children's developing understanding of place value: Semiotic aspects. *Cognition and Instruction*, *15*(2), 265–286. doi: 10.1207/s1532690xci1502_4

Vlassis, J. (2008). The role of mathematical symbols in the development of number conceptualization: The case of the minus sign. *Philosophical Psychology*, *21*(4), 555–557. doi: 10.1080/09515080802285552

Wentworth, G. A. (1897). *A practical arithmetic*. Ginn and Company.

Wessman-Enzinger, N. M. (2018). Descriptions of the integer number line in United States school mathematics in the 19th century. *Convergence*. Doi: 10.4169/convergence20180217

Whitacre, I., Bishop, J. P., Philipp, R. A., Lamb, L. L., & Schappelle, B. P. (2014). Dollars and sense: Students' integer perspectives. *Mathematics Teaching in the Middle School*, *20*(2), 84–89. doi: 10.5951/mathteacmiddscho.20.2.0084

Whitacre, I., Henning, B., & Atabaş, Ş. (2020). Disentangling the research literature on number sense: Three constructs, one name. *Review of Educational Research*, *90*(1), 003465431989970–134. doi: 10.3102/0034654319899706

Wirtz, R. (1974). *Drill and practice at the problem-solving level*. Curriculum Development Associates.

Wong, T. T-Y. & Chan, W. W. L. (2019). Identifying children with persistent low math achievement: The role of number-magnitude mapping and symbolic numerical processing. *Learning and Instruction*, *60*, 29–40. doi: 10.1016/j.learninstruc.2018.11.006

Yang, D. C. (2019). Development of a three-tier number sense test for fifth-grade students. *Educational Studies in Mathematics*, *101*(3), 405–424. doi: 10.1007/s10649-018-9874-8

Young, L. K. & Booth, J. L. (2020). Don't eliminate the negative: Influences of negative number magnitude knowledge on algebra performance and learning. *Journal of Educational Psychology*, *112*(2), 384–396. doi: 10.1037/edu0000371

Strand B

Whole-Number Relationships

After studying this strand, the reader should be able to:

1. compare the properties and operations of addition, subtraction, multiplication, and division with whole numbers;
2. describe challenges of students with learning difficulties in computing with whole numbers;
3. create example addition and subtraction problems using change, compare, and combine situations;
4. identify applications for which the concept of equivalence is critical;
5. create example multiplication and division problems using equal groups, multiplicative comparison, Cartesian product, and area or array situations;
6. demonstrate at least one alternative algorithm for each of the four operations using whole numbers.

Ciara is working on whole-number word problems with Ms. Smith in the special education classroom at Balsam Middle School. The current problem states, "There were 176 students in the sixth grade at the beginning of the year, but 19 students moved. How many are still in the sixth grade at our school?"

MS. SMITH: How do you think we should solve that problem?
CIARA: Plus it? Plus that number and that number (pointing to the 176 and the 19)?
MS. SMITH: If students have moved away, then we will have fewer students, what should we do?
CIARA: Take away.
MS. SMITH: Good, can you write the problem and solve it?
CIARA: WRITES: 176
                   $\underline{-\ 19}$
                   163
MS. SMITH: Please tell me how you worked that problem.
CIARA: First I just wrote down the one, then I said 7 take away 1 is 6. Then I did take away 6 from 9 and got 3. The answer is one sixty-three.
MS. SMITH: Think about this, if almost 20 students moved and we started with about 175, would we have about 165 still here?
CIARA: I wrote 163.
MS. SMITH: Let's add 19 to 163 and see if we get the beginning number back.

One of the most widespread myths of mathematics reform is that computation and fluency are not emphasized. Actually, these skills should be strengthened by the reorientation of the standards. The National Council of Teachers of Mathematics (NCTM) curriculum

DOI: 10.4324/9781003096733-10

standards stressed that "understanding number and operations, developing number sense, and gaining fluency in arithmetic computation form the core of mathematics education for the elementary grades" (NCTM, 2000, p. 32). The *Common Core State Standards for Mathematics* (CCSSM) outlined three aspects of rigor: conceptual understanding, procedural skills and fluency, and application. "Students must practice core functions, such as single-digit multiplication, in order to have access to more complex concepts and procedures" (NGA & CCSSO, 2010, *Key Shifts*).

What has changed is an emphasis on developing a deep understanding of number, number systems, and the meaning and interrelationships of the operations. Rote learning is out. Teaching for understanding is essential. Research on mathematics learning by students with and without learning difficulties is compelling. Students learn more and are able to maintain and generalize learning that is taught through context, developing concepts and strategies, rather than rote learning through drill and practice of isolated computations (e.g., Carpenter et al., 1998; Schneider et al., 2011). Developing fluency is still important and is emphasized across all grade levels. But fluency means more than rapid recall of facts. Computational fluency is the "connection between conceptual understanding and computational proficiency" (NCTM, 2000, p. 35). Fluency includes the deeper understanding of concepts and flexible, ready use of computation skills across a variety of applications.

The CCSSM content standards emphasized whole-number concepts, skills, and problem solving in kindergarten through fifth grade, under the domain *Number and Operations in Base 10* (NGA & CCSO, 2010). First and second graders are adding and subtracting within 100, third graders within 1000. Multiplication by one-digit whole numbers by multiples of 10 in third grade progresses to multiplication and division using one- and two-digit multipliers and one-digit divisors in fourth grade. Fifth graders should fluently compute with all four operations using multidigit numbers, demonstrating concept understanding. The NCTM standards (2000) included all numbers—including whole numbers, decimals, and fractions—within the number and operations domain. Students in kindergarten through second grade focus on addition and subtraction with whole numbers while students in Grades 3 through 5 focus on multiplication and division. By middle school, students should be fluent with basic number combinations (facts), the four operations with whole numbers, the properties of whole-number operations, and a range of methods for computing including estimation, calculator use, and paper-and-pencil approaches. For many students with mathematics learning disabilities (MLD), these skills won't be mastered in the same time frame required by some standards.

Middle-school students should maintain their facility with whole-number operations while expanding their work with rational numbers. Number topics within fractions, decimals, algebra, geometry, measurement, and data analysis are built upon a strong foundation of whole-number concepts and skills. For example, the concepts of factoring in fractions and area in geometry depend on multiplicative fluency. The properties of algebra begin with the properties of integers: identity, commutative, associative, and distributive.

Students with mathematics difficulties (MD) or MLD typically have difficulties with whole-number computation and related skills, such as procedural errors and fact retrieval (Geary, 2015). They often lag several grade levels behind their peers in number combination (fact) fluency (Rotem & Henik, 2020) and computation (Dennis & Gratton-Fisher, 2020; Geary et al., 2012). When presented with whole-number, multidigit computations, many of these students demonstrate a wide range of errors, including applying the wrong operation, poor recall and mistakes with basic facts, incorrect procedures during computation, misinterpreting symbols and other prompts, and regrouping errors (Nelson & Powell, 2018). Many errors reflect poor concept understanding, lack of strategies, and issues with executive function and working memory (McDonald & Berg, 2018; Moser Opitz et al., 2017). Difficulties with whole-number computation will continue to affect mathematics achievement as students attempt to apply multiplicative reasoning, solve problems in other domains, move into the full range of rational-number computations, and begin algebraic topics. Some students will require intensive interventions in whole-number concepts and skills in order to be successful with grade-level mathematics.

This strand explores concepts with whole numbers including properties of the four arithmetic operations, single-digit number combinations (fact fluency), multidigit computation, and alternative algorithms. Recommendations for developing concepts and skills with students with MD are provided in each section. The reader may discover that the topic of whole-number relationships—what would seemingly be the simplest mathematics concepts to teach—can be challenging to understand from the perspective of underlying constructs, interrelated attributes, and developmental progressions. It is well worth the time to study these concepts before planning instruction that develops deep understanding as well as fluency.

Understanding Single-Digit Operations

For years, teachers and parents urged children to learn their math facts—referring to addition, subtraction, multiplication, and division combinations of single-digit numbers (0 to 9). The term *fact* implies an isolated bit of information that must be memorized, so the mathematics community now uses more accurate terms such as *single-digit combinations* or *basic-number combinations* instead (Kilpatrick et al., 2001), although some mathematics materials still refer to fact learning. Extensive research into how children learn single-digit operations demonstrates that they move through individually determined sequences of progressively more abstract and more efficient methods of finding solutions. Immediate recall and rapid procedures eventually become students' primary means of computing, but students who sometimes forget a quick solution and understand the basic concepts can figure out sums, differences, products, and quotients in a number of ways.

Addition and Subtraction Concepts

The concept of addition is the earliest of the four operations developed by children and can be observed in activities such as counting discrete objects and assigning a number to a whole group of things (cardinality). Children between the ages of two and five develop the underlying concepts of number sequence, enumeration (one-to-one correspondence), cardinality, and relative magnitude (Baroody, 1987). A child counts his buttons: one, two, three, four, five, ... and announces, "I have five buttons. You have only three so I have more than you."

Early addition typically involves adding to an existing set, an often-natural progression from counting activities, or moving to the right on the number line. Addition also can be viewed as combining two or more sets (part-part-whole). Addition has a number of properties, including *commutative* (the order of addends does not affect sum) and *associative* (grouping of addends does not affect sum). Addition is also the inverse of subtraction, so addition, subtraction, and their properties should be taught simultaneously. Adding a number to a collection can be undone by subtracting the same number (*inversion principle*). Subtraction is counting back, moving left on the number line, or removing items from a set. Subtraction is a more difficult concept because it is restrictive; it does not have commutative or associative properties. We cannot reverse the order of numbers or make a regrouping of numbers to be subtracted and achieve the same solution.

Identifying the total of a set (cardinality) and comparing sets (relative magnitude) lead to other critical concepts for addition and subtraction development: equality or inequality, subitizing (automatically recognizing number of objects without counting), and conservation of number (what action effects a change in number and what does not). Preschool children should recognize that explicitly adding or taking away items from a set will change the total number of the set, but children may still be fooled when the items in a set are changed in their spacing but not number.

When children begin counting they start with the number 1. (They are actually using the positive integers but most teachers call them counting numbers.) The concepts of empty set and adding zero to an existing set, resulting in the original number (identity property), are very problematic for children (Anthony & Walshaw, 2004). If teachers use the word *nothing* for zero, confusion increases because nothing implies no importance. Some students may even confuse zero with the letter O. Concepts involving zero form a critical foundation for algebra and should be modeled through hands-on activities and discussions as the other digits and arithmetic properties are explored. Zero concepts are also critical for number theory, measurement, and geometric and statistical concepts such as computing with negative integers, measuring length, applying the concepts line and plane, and graphing data.

One problem with how beginning arithmetic is taught and depicted in textbooks is the use of symbols for addition (+), subtraction (−), and equal (=) before those concepts are fully developed. Addition and subtraction problem situations each involve three quantities—two addends and a sum—with one element unknown. Technically, addition involves two or more addends and a sum, while subtraction is the minuend − subtrahend = difference. However, mathematics standards do not emphasize the use of this nomenclature. More important are concepts such as subtraction being the inverse of addition and understanding that subtraction doesn't follow all of the properties of addition.

The situations for which addition and subtraction can be used vary conceptually by three basic types—*compare, combine,* and *change*—in two forms—static and active (Fuson, 1992). Compare and combine describe static relationships while change is active (dynamic). Research indicates that the change type (dynamic) is easier for young children while compare and combine types take more time to develop (Fuson et al., 1996). The CCSSM (2010) described the change situation as either *add to* or *take from* and combine as *put together* or *take apart.* These standards included a combine type where both addends are unknown, a precursor to algebraic thinking. For example: *If you have 8 pencils, how many could you put in the round pencil jar and how many in the pencil cube?* $x + y = 8$. Textbooks often include separate chapters on addition and subtraction before developing their connections, potentially problematic for students in understanding problem types, basic concepts, and properties.

Some children interpret the equal sign as "and the answer is," an *operator* meaning, rather than "equal to the amount," a *relational* meaning, leading to more confusion when higher-level concepts are introduced. Textbooks now introduce the horizontal equation before the vertical one and stress the use of "is equal to" for the = sign. However, textbooks tend to introduce symbols in the first chapter, before addition and subtraction concepts are developed. Long before introducing symbols, teachers should present problem situations orally and with the use of objects until children can describe and manipulate the concepts flexibly and with their own strategies. Mathematics is not always a paper-and-pencil task. For example, consider the types of addition and subtraction prompted in the following discussion.

TEACHER (TO AMY AND BILLY): I am going to give each of you five crayons. Help me count them out.

CHILDREN: One, two, three, four, five, one, two, three, four, five.

TEACHER: Good, do you both have the same number?

CHILDREN: Yes, five.

TEACHER: Amy, would you please lend Billy one crayon? *Amy hands a crayon to Billy.* Amy, do you now have fewer or more than five?

AMY: I have four.

TEACHER: OK, do you have more or does Billy have more?

AMY: He gots more cause he gots mine, too.

TEACHER: That's right, Billy has more crayons now. And you have fewer crayons because you lent him one of yours. Let's say what happened. Amy had five crayons and gave one to Billy so now she has four. Can you say that?

CHILDREN: Amy had five crayons and gave one to Billy and now Amy has four.

TEACHER: Can you tell me what happened with Billy's crayons?

AMY: Billy got six.

TEACHER: OK, but what did he start with and how does he now have six?

BILLY: I had five but now I have six 'cause Amy lended me her red one.

AMY: I gots only four because Billy gots one of mine.

This discussion illustrates the *change* situation in addition and subtraction. In the crayon exchange, the original number of five changed with the addition or subtraction of a crayon. There is a beginning amount, change amount, and ending amount. If the teacher were to ask the children next to compare the number of crayons Billy and Amy have, they would be in a *compare* situation. Billy has one *more than* Amy; Amy has one *fewer than* Billy. If the teacher asked the children to put all their crayons together into one group, a *combine* situation arises, and the children would count 10 crayons. In all three situations, the element not known determines the operation. Further, the concept of equality has been enhanced by explicitly comparing groups of crayons. But equivalence is more difficult to establish with change and combine situations because the original sets no longer exist. The use of descriptive language, diagrams, and symbols help keep track of the original and new sets.

With more experience describing and manipulating situations involving zero to five objects, young children form a stronger understanding of addition and subtraction concepts. By the time larger digits and symbols are introduced, the concepts are familiar. Seven-year-old children should be able to manipulate addition and subtraction concepts quite flexibly and fluently, adding or subtracting by one (e.g., $8 + 1$), adding or subtracting by one from known facts (if $4 + 4 = 8$, then $4 + 5 = 9$; if $1 + 4 = 5$, then $2 + 4 = 6$), doubling (e.g., $4 + 4$), applying the commutative property ($4 + 5 = 5 + 4$), and even using known combinations that add up to 10 to determine those with addends of 9 ($9 + 3$ is the same as $10 + 3 - 1$). The addition and subtraction table in Table B.1 illustrates with shading those combinations that are readily learned through strategies. Students may be amazed to see that after mastering the strategies, only five number combinations need to be memorized.

Computation tables, such as the one in Table B.1, *should not* be used to introduce or teach number-combination concepts. They are interesting to examine for patterns after students understand the concepts of addition and subtraction and have developed some computing strategies. They may help students understand how many combinations and strategies they have mastered. Students who are dependent on these tables to find answers are not thinking or strategizing, merely copying. Students learn more about numbers by counting fingers or referring to a number line than using tables.

Teachers can promote deeper understanding of addition and subtraction concepts through the following activities:

* Children between the ages of three and five should be engaged with a variety of objects for counting, comparing, and manipulating. Adults should ask prompting questions that will assist children with making connections, building vocabulary, and developing new concepts, such as those for adding onto and subtracting from sets and the equivalence of collections.
* Although parents and school boards seem to expect it, asking four- and five-year-old children to write equations may be too developmentally challenging. If children have fine-motor difficulties but can identify numbers and other symbols in print, they can work with number and symbol cards to form equations (see Appendix II).

Table B.1 Addition and Subtraction Combinations Learned through Strategies

	0	1	2	3	4	5	6	7	8	9	10
0	0	1	2	3	4	5	6	7	8	9	10
1	1	2	3	4	5	6	7	8	9	10	11
2	2	3	4	5	6	7	8	9	10	11	12
3	3	4	5	6	7	8	9	10	11	12	13
4	4	5	6	7	8	9	10	11	12	13	14
5	5	6	7	8	9	10	11	12	13	14	15
6	6	7	8	9	10	11	12	13	14	15	16
7	7	8	9	10	11	12	13	14	15	16	17
8	8	9	10	11	12	13	14	15	16	17	18
9	9	10	11	12	13	14	15	16	17	18	19
10	10	11	12	13	14	15	16	17	18	19	20

Strategies: Identity Property (+ 0, − 0), doubling (4 + 4 = 8), Inverse Property (8 − 4 = 4), adding 10, Commutative Property (4 + 2 = 6; 2 + 4 = 6), known 10s & related 9s (4 + 6 = 10 therefore 4 + 5 = 9), plus or minus 1 from known fact.

- For five- and six-year-olds, scaffold student understanding of number combinations by drawing diagrams for sets and changes, combinations, and comparisons. Here are three blocks and over here we have five blocks. If we put them all together, how many do we have? Yes, we have eight blocks. What if we moved two away? Yes, there would be six left. Let's draw what happened:

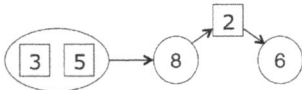

- To strengthen students' concepts of zero, use paper plates, place mats, or simple containers for grouping objects as sets to be added together or subtracted from. Randomly use an empty plate to represent zero. Three blocks in this plate and zero blocks in this plate equals how many blocks altogether?
- An activity for strengthening fluency and number concepts is *seeking sums*. Present the sum "8" and ask students for all the ways they can create addition equations that give the sum of 8 (1 + 7, 2 + 6, 3 + 5, 4 + 4, 5 + 3, 6 + 2, 7 + 1; and don't forget 0 + 8 and 8 + 0). Present equations in noncanonical (nonstandard) forms such as 8 = 5 + 3 and 2 + 3 + 3 = 8. Then do the reverse for subtraction equations with 8 as the beginning number (8 − 5 = 3; 8 − 3 = 5) or a missing term (5 =? − 3). Repeat with other sums.
- Another strategy used with early subtraction practice is *counting back*. This strategy can be used with minus one and minus two to promote fluency of those combinations. For example, in the problem 9 − 2 =?, the student would begin with the nine, then count back twice, 8, 7 to achieve the difference. Counting backwards by more than one or two is difficult for many children, so they should use counting up instead: 9 − 3 =?, starting with 3, count up to 9 and keep track of the counts with fingers or tallies.
- Teach addition and subtraction concepts within the context of realistic problems. *Outside the window I see 7 birds pecking in the grass. Two flew away. How many birds are still there?*

- Focus on operational concepts when working at the concrete, representational, and abstract levels. What do addition and subtraction mean? How can the operations of addition and subtraction be represented? Employ simple objects, 5- and 10-frames, and number lines.
- Support students' addition developmental progression from counting all, to counting on from one addend, to more sophisticated thinking strategies. Some students tend to stick with count-on strategies and need explicit instruction in decomposing and recomposing addends, forming doubles, and adding one to known sums. For example, $9 + 6$ is the same as $10 + 5 = 15$. These mental exercises will promote more flexibility in viewing equations and finding solutions.
- Remember that subtraction is a more challenging concept and is more restrictive. The developmental progression is typically removing one, removing more from a group and counting, counting up from the subtrahend (for $5 - 3$, count beginning at 3 up to 5), and relating a subtraction equation to its addition inversion ($8 - 5 =$ ___ is the same as $5 +$ ___ $= 8$). Again, thinking strategies may need explicit instruction, such as decomposing and recomposing, using 10 as a reference, and using doubles. For example, for $9 - 4$ we think of adding 1 to each for $10 - 5 = 5$. Reinforce the concept that the order of minuend and subtrahend cannot be reversed ($5 - 3$ is not the same as $3 - 5$).
- For students in first and second grade who are required to solve written equations, consider the format (horizontal, vertical, noncanonical), spacing, and numbering (or lettering) of items on a page. Some students try to figure in the item number for an equation such as the following item #3 on a worksheet: $3. 6 - 5 =$ ___.
- Varying the format of equations reinforces addition and subtraction strategies, properties, and the equivalence concept: $7 +$ ___ $= 2 + 9$ and ___ $+ 4 = 3 + 7 + 1$. These exercises can be completed in small groups, on dry-erase boards, and with number/symbol cards. Having students explain their thinking illuminates misconceptions and broadens the thinking of peers.
- *Avoid* teaching students to match cue words to operations. For example, *in 1 basket there were 12 pieces of fruit altogether, a mix of apples and oranges. If there were 4 apples, how many oranges were there?* Many students would see the word "altogether" and add $12 + 4$. Or even add the $1 + 12 + 4$. Instead, prompt: "What is the problem asking?"

There is significant research support over decades for moving through a concrete-representational-abstract (CRA) sequence of instruction with arithmetic operations if instructional time devoted to manipulatives and diagrams is carefully controlled. For example, Mercer and Miller (1992) implemented three lessons at each stage before fluency practice, with students achieving 80% concept understanding within six 30-minute lessons. In another example of a CRA sequence, Funckhouser (1995) introduced numbers and sums to 5 using beans glued to five-cell frames to represent numbers. During the first two weeks, kindergarten children built frames for the numbers 0 to 5. In the third week, combinations of number frames were explored. In the fourth week the + symbol was introduced and after the four-week intervention, all students could recognize and match numbers 0 to 5 and add sums up to 5. Five- and ten-frames bridged the concrete to representational stages of learning (see Figure A.8 in Strand A).

In a study examining instructional strategies for the inversion principle for addition and subtraction with kindergarten students, Ching and Wu (2019) found that concreteness fading (CRA sequence) with explicit guidance was a superior approach, especially for lower achieving students and for transfer tasks. It was superior to concrete-only, abstract-only, and abstract-to-concrete sequences of instruction. Simply providing "multiple representations" does not produce growth and transfer; the order of presentation matters in developing accurate concepts.

In a longitudinal study of second graders who were persistent counters (continued to use finger-counting strategies for addition and subtraction), Moser Opitz et al. (2018) implemented a well-structured and intensive cooperative-learning intervention within 38 general education classrooms over ten weeks (30 min, twice a week). The intervention, led by classroom teachers with practice in student pairs, focused on number sense and developing addition and subtraction strategies with the support of a 20 frame and number line—decomposition, counting by groups, and recognizing related problems. The cooperative-learning intervention was effective for students with greater prior knowledge (pretest scores above median), but not for those with low prior computation competence. The researchers recommended more intensive, individualized approaches for the low achievers.

Concept of Equivalence

As mentioned in the previous section, children are often introduced to the equal sign as an operator rather than as a relational equivalence, usually with children's first experiences with addition equations in kindergarten or first grade. They tend to think in terms of the "problem part" and the "answer part" for an equation such as $4 + 1 = 5$, when this equation means that the $4 + 1$ part is equivalent to the 5 part, therefore $5 = 4 + 1$ is also true. Many students in the elementary grades are entrenched in the operator view, most likely because of their limited and narrow exposure to equivalent terms in mathematics. Studies over decades have found that students with and without disabilities have misconceptions about equivalence and these often persist into the secondary grades (Knuth et al., 2006) and even college (Fyfe et al., 2020). McNeil, Hornburg, Brletic-Shipley, and Matthews (2019) assessed second graders' understanding of equivalence and found that 80% held misconceptions that were already resistant to change. These second graders' equivalence understandings predicted their third-grade mathematics achievement, reinforcing the need for early interventions. A number of studies have concluded that students who hold the relational view of the equal sign perform better in algebra. Byrd et al. (2015) found that equal-sign understanding at the beginning of the year predicted end-of-year performance on early algebra assessments for third- and fifth-grade students. Matthews and Fuchs (2020) concluded that equal-sign knowledge of second graders, measured through varying mathematics equivalence problems, was a strong predictor of fourth-grade algebraic knowledge. Equal-sign understanding (concept of equivalence) may be an important bridge between arithmetic and algebra.

Equivalence is a concept of sameness. Equivalence understanding is knowing that quantities or expressions can be interchanged without changing the value. One child has five pennies and another has a nickel. With experience, children understand that coins can have equivalent values regardless of number, appearance, or size. They understand that equivalence does not mean identical. In mathematics, equivalence refers to a state of having the same value, magnitude, weight, force, or other comparative property. Some examples: $5 = 2 + 3$; 1 pound $= 16$ ounces; the area of square A is equivalent to the area of triangle C. Table B.2 illustrates the importance of equivalence across the mathematics curriculum.

In a mathematics equation, the equal sign denotes the equivalence of the two sides. Students need to understand that the two sides are mathematical objects, abstract entities that can be manipulated (McNeil, Hornburg, Brletic-Shipley, & Matthews, 2019). Students with this more formal understanding of equivalence know that numbers and expressions can be represented in a variety of interchangeable ways. For example, in the equation $6x + 5 = 20 - x$, the two sides are equivalent, therefore we can add, subtract, multiply, divide by the same expressions on each side and not change the equivalency. We can even switch the sides and maintain equivalence. This equation is the same: $7x = 15$ because we added x to both sides and subtracted

Table B.2 Equivalence across the Mathematics Curriculum

General Domain	Illustrative Topics	Examples
Number	Number Sense Negative Integers Exponents	$\Delta\Delta\Delta = 3$ $-5 - (-3) = -5 + 3$ $5^2 = 25$ $8.3 \times 10^{-4} = .00083$
Integral (Base-10) Operations Non-Integral Operations	Arithmetic operations Arithmetic properties Simplifying fractions	$5 + 2 = 7; 27 = 9 \times 3$ Commutative: $3 + 4 = 4 + 3$ $\dfrac{5}{15} = \dfrac{1}{3}; \dfrac{6}{3} = 2$
	Converting between fractions, decimals, percents	$\dfrac{1}{2} = \dfrac{50}{100} = .50 = 50\%$
Geometry	Angles Symmetry Polygons	Sum of angles of rectangle = 360° Reflection: divisible into two mirror images Perimeter = Sum of sides Area of Rectangle = L×W
Algebraic Concepts	Algebraic properties of equality Solving equations and inequalities with one unknown Factoring polynomials	Transitive (if $a = b$ and $b = c$, then $a = c$) Substitution (if $a = b$, then a may replace b or b may replace a) $4x + 7 = 264$ $36 > 4y$ $ca + cb = c(a + b)$
Measurement and Data	Time Distance Volume Currency Graphing Least-squares regression line	60 minutes = 1 hour 1 foot = .3048 meter L × W × H = cubic volume 4 quarters = 1 dollar The same data can be represented in line, bar, and other graph formats. $y = .33x - 93.9$

5 from both. We could solve for x if we divided both sides by 7. The equation $4 = 4$ is true; the equation $9 = 10 - 1$ is true. The equation $6 + y = 24$ can be solved for y by subtracting 6 from both sides.

Because of the importance of early intervention for equal-sign understanding, Chow and Wehby (2019) conducted whole-class instruction with second graders. Twenty-one classrooms were assigned to symbolic intervention, nonsymbolic intervention, or business-as-usual control groups. The symbolic intervention taught relational understanding of the equal sign with standard and nonstandard open equations that were increasingly more difficult. The nonsymbolic group had the same instruction except with pictures (e.g., dots, turtles, flowers) instead of numbers in equations with = and + symbols. In only three 20-minute lessons the students in the intervention groups significantly outperformed the control classrooms ($g = 0.53$ for nonsymbolic and $g = 0.45$ for symbolic). Students in the nonsymbolic classrooms outperformed those in symbol-only instruction, reinforcing the need to support students' language and early symbol use.

In another study with second graders, McNeil, Hornburg, Devlin, et al. (2019) implemented an intervention of well-structured, nontraditional arithmetic practice with an emphasis on equivalence concepts. The classrooms with nontraditional practice interventions (e.g., ___ = 3

+ 4; 1 + __ = 3) plus exposure to the equal sign outside arithmetic contexts (e.g., 1 ft. = 12 in.), concreteness fading (CRA), and activities that required students to compare and explain problems and strategies outperformed classrooms with nontraditional practice alone (active control). On average 85% of students demonstrated basic understanding and 21% achieved mastery of equivalence concepts with the comprehensive intervention, compared with 30% and 2% of the active control students. The researchers concluded that practice with non-traditional equation formats alone is not enough to prevent equivalence misconceptions and resistance to change from the operational view of the equal sign. Students need well-structured, nontraditional practice with equivalent expressions and scaffolding to connect math symbols to their everyday knowledge as well as opportunities to explain and compare strategies to increase their relational thinking.

Recommendations for promoting an integrated and generalized view of equivalence:

- Expose children to the different meanings of the equal sign, including nonarithmetic contexts, emphasizing the relational model (Seo & Ginsburg, 2003).
- Be more explicit in teaching for transfer of equivalence concepts to various forms of equations.
- Use a concrete-representational-abstract (CRA) sequence to scaffold student understanding of equivalent expressions and their symbols.
- Use relational phrases when teaching about equivalence. Rather than "the answer is" use "this is equivalent to" and "the same as."
- Consider using another symbol, such as an arrow, for the operator meaning of simple add-ition and subtraction (Seo & Ginsburg, 2003).
- Use diagnostic interviews and group discussions to assess individual student understanding and changes in understanding with experience. Conduct brief assessments of equal-sign understanding at each grade level to target needed interventions.
- When children begin using written equations, build the concept of equivalence by writing equations horizontally and placing the unknown in different positions within the equation. What is the missing number in this equation? $5 +$ ___ $= 8$.
- Be careful when using calculators to emphasize the relational equivalence of input and output, not the operator view of the equal sign.
- Teach students to read mathematic equations as relational. Have them read sentences orally: "The terms $4x$ added to 6 on this side are equivalent to the 54 on the other side. I can maintain the equivalence of the expressions if I subtract the same number from each side."
- Use analogies for equivalence, such as a balance scale, also available in virtual form on the computer (e.g., the Pan Balance at NCTM's *Classroom Resources*; www.nctm.org). If 10 red cubes and 5 blue cubes are on this side of the equation, how many cubes need to be on the other side to make it balanced? $10 + 5 = 15$. If we take three cubes off this side, how many do we take off the other side? Let's write our new equation. $(10 + 5) - 3 = 15 - 3$.

A relational understanding of equivalence supports the transition from arithmetic to algebra. Asking students to explain their understanding of the equal sign, represent simple number-combination operations in different formats, solve word problems that promote a relational view, and use analogies such as the balance scale all lead to a more flexible understanding of quantitative relationships. An example of an intensive intervention with first-grade students on equivalence, addition, and subtraction concepts is provided in Box B.1.

Box B.1   Intensive Intervention for Equivalence Concept

Target: Four first-grade students still have difficulty with horizontal equations and their symbols for addition and subtraction after mid-year screening. They demonstrate weak understanding of equivalence.

Intervention Schedule: 20-minute sessions, three times a week until mastery (estimate three weeks).

Nature of Intervention: Each session follows the format:

1. Begin session with brief review or quick addition and subtraction combination check.
2. Show students one example of a nonarithmetic equivalence and discuss (e.g., feet to inches, pennies to quarters, minutes to hours, etc.).
3. Explicit lessons: Conduct early sessions using concrete objects, later with representations (e.g., flowers, dots on cards), and finally symbol-only practice. Demonstrate nontraditional equations such as 8 + __ = 10 and 15 = ___ − 4. Provide practice with feedback.
4. Provide one example for independent practice, then ask students to explain and compare their problems (alternate equations with story problems).
5. Conclude with quick review of the lesson's main concept.

See McNeil, Hornburg, Brletic-Shipley, & Matthews (2019) for complete discussion of similar methods.

Data Collection (Weekly): CBM of equivalence concept (8 equations, 2 story problems); addition & subtraction fluency check (81 random problems, 1 min).

Properties of Operations

Understanding addition, subtraction, multiplication, and division requires understanding their properties. Properties inform us about what we can and cannot do when operating on numbers and unknown values. They are the rules that guide operations. The properties include identity, inverse, commutative, associative, distributive, and closure relationships and are depicted in Table B.3.

There are also a number of properties of equality, or what the student can do to an equation and maintain equivalence on both sides of the equal sign. Rather than teach a list of these properties, work them into practice with equations and equivalence. After students are working confidently at the abstract level these properties can be formalized with symbols. For example, if $a = b$ and you add c to a, then you must add c to b to maintain equivalence: $a + c = b + c$. Related concepts, such as *negation* (a number subtracted from itself equals zero, $a - a = 0$) and *inversion* are also best taught through practice that leads to these understandings. Then the concepts can be made more explicit through rules. These concepts take time to develop and many students will need scaffolding and more explicit instruction to make connections.

In investigating understanding of these basic arithmetic properties, Robinson (2019) conducted research with students in Grades 1 to 3, asking them to solve simple problems

Table B.3 Properties of Operations

Property	Addition	Multiplication
Identity	The identity element is zero: $n + 0 = n$	The identity element is one: $n \times 1 = n$
Commutative	The order of the addends does not affect the sum: $a + b = b + a$	The order of the factors does not affect the product: $a \times b = b \times a$
Associative	The grouping of addends does not affect the sum: $(a + b) + c = a + (b + c)$	The grouping of factors does not affect the product: $(a \times b) \times c = a \times (b \times c)$
Closure	The sum of two integers equals an integer.★	The product of two integers equals an integer.★
Inverse	For every number n, $n + (-n) = 0$	For every non-zero number n, $n \times 1/n = 1$
Zero	Identity Property $n + 0 = n$	For any number n, $n \times 0 = 0$
Distributive	The product of a number and a sum is the same as the sum of two products: $a(b + c) = (a \times b) + (a \times c)$	

	Subtraction	Division
Zero	Identity Property	Zero divided by any number is equal to 0: $0/n = 0$ Division by zero is undefined.
Identity	The identity element is zero: $n - 0 = n$	The identity element is one: $n \div 1 = n$
One		A non-zero number divided by itself is equal to 1: $n/n = 1$
Inverse	Subtraction is the inverse operation to addition.	Division is the inverse operation to multiplication.

★ The four operations are closed for real numbers (except for division by zero). The property of closure is related to a specified set of numbers (e.g., whole numbers, integers, real numbers, even numbers, etc.).

mentally and explain their reasoning. Of six concepts, third graders applied accurate concept knowledge 50% of the time, second graders 40%, and first graders 20%. The identity property was the easiest for the students, followed by negation and inversion (of addition and subtraction). The concepts of commutativity, associativity, and equivalence were the weakest, with scores increasing to just over 20% for third graders for commutativity and associativity, and only 10% for equivalence.

When properties have been explored and given names (either the classroom name or the official name), then students will be ready to think more deeply about properties. One teacher provided groups in her class a word problem to solve together:

> *Jo Lynn calculated 40 × 700 mentally as follows: 4 times 7 is 28, 10 times 100 is 1000, and 28 times 1000 is 28,000. Write a series of equations that shows why Jo Lynn's method is valid. Which properties are involved? Write your equations under each other with the equal symbols lined up.*

Another teacher noticed that some students were having difficulty with fluency with sums such as 4 + 5, 5 + 6, 6 + 7, and 7 + 8. He knew that these students were fluent with doubles (4 + 4, 5 + 5, 6 + 6, 7 + 7). He pointed out a + 1 strategy, adding 1 on each side of their mental equations: 4 + 4 = 8, so 4 + 4 + 1 = 8 + 1 or 4 + 5 = 9. Which property allows students to use the plus one strategy? What allows us to substitute the 5 for the 4 + 1? What allows us to add the one to both sides of the equation? This practice encouraged students to integrate their thinking of operations and equations with the properties that allow them to solve problems in different ways.

Multiplication and Division Concepts

Multiplication literally means counting multiples. It is counting groups of the same number, or repeated addition. Instead of adding 4 + 4 + 4 + 4 + 4, we say we have 5 groups of 4, or 5 times 4 is equivalent to 20. Multiplication's identity element is the number 1. Like addition, the order or grouping of factors doesn't affect the product (commutative and associative properties). Division is the inverse operation of multiplication and should be introduced simultaneously. Like subtraction, division has no associative or commutative properties (order and grouping do matter). Division has a rule that the divisor (denominator) cannot be zero. Division begins with a total amount that is then shared or partitioned. Further, division can result in leftovers (remainders or fractional parts). For example, *a family held a yard sale and made $200 on things they no longer needed. If there are six people in this family, how much money will each receive?* Each person receives $33.33 with two cents left over.

Multiplication and division concepts are connected in the process of *multiplicative reasoning*. Multiplicative reasoning is very different than additive reasoning. In addition, the unit is one, while in multiplication the unit is "one as well as more than one" simultaneously (Carrier, 2014). There are two variables in a fixed ratio. Consider 15 divided by 3. Fifteen is to 3 as 5 is to 1. Look for the multiplicative relationships in the equations $3 \times 5 = 15$ and $15 \div 3 = 5$. Fifteen divided by 3 is equivalent to 30 divided by 6, the ratio is fixed. Multiplicative situations actually involve four numbers, not three, as the "each" or per 1 is understood. Students need to move beyond equal-group representations to develop multiplicative reasoning.

Multiplicative reasoning is a big idea of mathematics, a gateway to algebra and more complex mathematics. Understanding the whole "conceptual field of multiplicative structures" takes at least 11 years beyond first grade for typical learners (Vergnaud, 1988). Restricting the understanding of multiplicative situations to repeated addition and subtraction masks the relationships within these situations (Askew et al., 2019). One of the earliest concepts that supports multiplicative thinking is the concept of even and odd. Even numbers can be divided by two without a remainder. We can skip-count by even or odd numbers. We can share an amount evenly among several people (12 books divided among 2 friends will give each person an equal number of 6 books). Two odd numbers added will result in an even number. Even kindergarteners can benefit from activities involving equal groups, such as through stories and drawings (Cheeseman et al., 2020). Multiplication and division applications continue through fractions, ratios, rate, proportions, products of measures, and linear/nonlinear functions. These concepts also rely on the development of rational-number understanding (Strand C).

Hackenberg (2010) described a multiplicative concept hierarchy (trajectory) as emergent (counting in 1s), MC1 (counting in 2s, 3s, 4s, 5s; can establish two levels of units through counting groups), MC2 (can treat a composite unit as a unit of one and coordinate three levels of units), and MC3 (anticipates three levels of units to coordinate multiple levels). Students at MC1 solve 6×3 by counting to 18 by 3s. Those at MC2 could solve 3×16 by first finding 3×6, then counting from 18 to 48 by three 10s. Working at MC3 involves working at three levels and coordinating those, as with $3 \times 16 = (3 \times 6) + (3 \times 10)$.

Kosko (2020) pointed out that many textbooks in the US may not engage multiplicative reasoning because their illustrations allow for direct counting of 1s. In examining student interpretation of set, length, and area models of multiplication/division, he concluded that how visual models are presented will elicit student responses at different levels. Models that do not allow for counting of 1s and use larger operands (> 5) encourage multiplicative reasoning at a higher level. For example, a model with two bars and an unlabeled segmented line underneath with one bar labeled 16 and the other "?" requires coordinating three levels of units, as in Figure B.1.

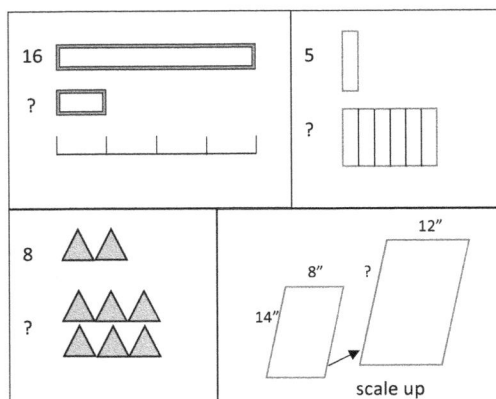

Figure B.1 Assessing Multiplicative Reasoning

Researchers investigating the development of multiplicative reasoning recommend using arrays and area models to encourage the commutative nature of multiplication and a composite-group view (Askew et al., 2019; Tillema, 2018). Other powerful models are T-tables, double number lines, and ratio diagrams. Researchers also recommend context-based problems to promote reasoning (Askew et al., 2019; Venkat & Matthews, 2019). Teachers at each grade level should assess students' multiplicative reasoning through untimed tasks at a range of levels, such as those depicted in Figure B.1 so that individualized interventions can be developed. Measures of accurate computation and fluency will not be adequate for determining needed interventions.

Basic multiplication and division can be seen in four types of situations: equal groups (also called equivalent sets, equipartitioning, vary, fair share); multiplicative comparison (restate, scaling, or compare); Cartesian product; and rectangular area (or array; Greer, 1992). Word problems with these types are discussed in Chapter 5. The types should be introduced simultaneously because problems requiring multiplication can be converted into division problems and vice versa. For example, *a classroom with 6 rows of desks, 4 in each row includes how many desks? A classroom with 24 desks in 6 rows has how many desks in each row?*

If you think about it, multiplicative reasoning, as with the examples above, actually involves four numbers—there is a "hidden" 1 or "each" in these problems (Askew et al., 2019). *You have 24 desks, 4 desks in 1 (each) row, 6 rows in all.* Therefore, these multiplication/division problems can be set up as proportions with an unknown. Teachers should be aware that specific terminology for division is dividend ÷ divisor = quotient. Multiplication's terms are multiplicand × multiplier = product. The national standards do not emphasize memorizing this terminology, but it can be useful in using precise language and comparing inverse situations for multiplication and division.

Some textbooks refer to two models of division within equal groups that help strengthen students' understanding of the relationship between multiplication and division: *partitive* and *measurement*. Partitive situations are when the whole is known (dividend) and the number of groups is known (divisor), but the number of items in each group (quotient) is missing (Jong & Magruder, 2014). *If we have 80 pennies and 20 students, how many pennies will each student have?* Measurement situations (also termed quotative) are where the number of groups is unknown. *How many 12-inch pieces of ribbon can you cut from 89 inches?*

Special education intervention research has established that explicit instruction of problem types (i.e., equal groups, multiplicative comparison, Cartesian product, array) connected with schematic diagrams that illustrate the relationships within the problems assists students with MLD to move beyond concrete stages to concept understanding and generalization (see Chapter 5, Figure 5.2). Schematic diagrams serve as an aid for organizing and representing problem information.

The following activities promote multiplication and division concept understanding:

- Expand the range of problem situations for multiplication and division beyond objects in groups. Include different types of groups (connected, such as egg cartons and soda packs; discrete, such as people, unit cubes, or pencils), rates (something per unit), multiplicative comparisons (times as many as), Cartesian products, and area models or arrays.
- Link multiplication and division concepts by using the terms *factor* and *product* to describe the elements in both operations. Also, integrate multiplication and division instruction by using problems that can be reversed to show the inverse operation.
- Encourage students to develop their own word problems to represent multiplication and division equations, beyond the simple equal-grouping situations.
- Like developing fluency with addition and subtraction combinations, students should be encouraged to develop strategies rather than viewing fact learning as a rote memory task. The most common strategies include:
  - The rule of zero—any number times zero equals zero.
  - The rule of one (the identity property)—any number times one equals the original number.
  - Using students' prior knowledge of counting by 5s and 10s with nickels and dimes.
  - Applying the inverse ($2 \times 6 = 12$; $12 \div 6 = 2$) and commutative properties ($6 \times 2 = 12$).
  - Multiplying by 2 doubles the number.
  - Nines are particularly interesting—one can use "digital computation" or the 10-finger method. Or see the pattern: any digit times nine equals a two-digit product where the first digit is one less than the multiplier and the two digits add up to nine.
    $9 \times 3 = 27$ (2 is 1 less than 3; $2 + 7 = 9$)
    $9 \times 7 = 63$ (6 is 1 less than 7; $6 + 3 = 9$)
- For products whose factors include 6, 7, 8, or 9, another finger computation method is simple and fun. Touch two fingers to multiply such as the pointing finger of one hand with the middle finger of the other hand for 7×8. Count the touching fingers and all those below as tens: 10, 20, 30, 40, 50. Next multiple the fingers above on each side: $3 \times 2 = 6$. The answer is $50 + 6 = 56$. Figure B.2 illustrates this concept first introduced by Barney (1970).
- Divisibility patterns (for division without remainders) include:
  - Even numbers are divisible by 2.
  - A number is divisible by 3 if the sum of its digits is divisible by 3.
  - A number is divisible by 4 if the last two digits are divisible by 4.
  - If the last digit is a 0 or 5, the number is divisible by 5.
  - A number is divisible by 6 if it is also divisible by 2 and 3.
  - If the last three digits are divisible by 8, the number is divisible by 8.
  - A number is divisible by 9 if the sum of digits is divisible by 9.
  - If the number ends in 0, it is divisible by 10.
  - A number is divisible by 11 if subtracting and adding digits in an alternate format (right to left) results in 0 or 11, 22, 33 (Sidebotham, 2002). For example, 3586 is divisible by 11 because $6 - 8 + 5 - 3 = 0$.
- Avoid teaching the facts in isolation or one "number family" at a time. Teach combinations in groups, such as 1 to 4.

Step 1: Count off your fingers, beginning with thumbs as 6s.

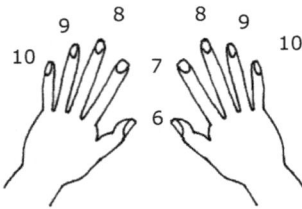

Step 2: Touch fingers to multiply. For example, 6 x 8 is touched below.

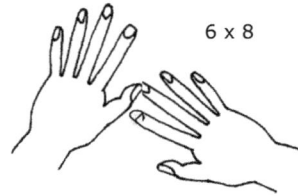

Step 3: Count touched fingers and the fingers below those as tens. (10, 20, 30, 40)

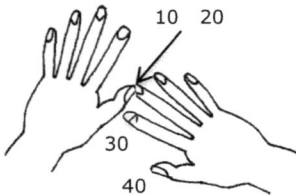

Step 4: Multiply the exact number of fingers above. (4 x 2 = 8)

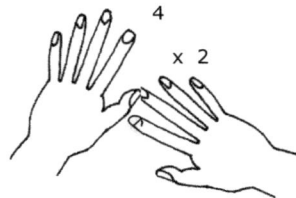

The product is 40 + 8 = 48. Note: 6 x 6 = 20 + 16, and 6 x 7 = 30 + 12.

Figure B.2 Digital Computation of 6s, 7s, 8s, and 9s

- Teach a "draw" strategy for making tallies if the product isn't fluent, at the representational stage (Mercer & Miller, 1991–1994). For example, the problem reads 4 times 3. Make four groups with circles and three tallies in each group. That reminds students of a count-by strategy.

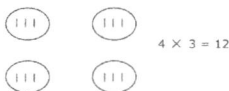

$4 \times 3 = 12$

- Be cognizant of symbol confusion by students. Various signs are used for division: $\div$, $/$, and $\overline{)}$ and the horizontal line such as in $\frac{a}{b}$. Multiplication can be expressed by $2 \times 3, 2 \cdot 3$, and $(2)(3), 5(15)$, and even $4yz$.
- Anticipate that students will tend to overgeneralize their experiences with whole numbers when they begin working with other rational numbers. Common misconceptions are the view that multiplication and addition will make the answer larger, division and subtraction will result in smaller numbers.
- Enhance multiplicative reasoning through mental-math and estimation activities.
- Use arrays to represent multiplication situations and help connect multiplication with division. These can be represented on graph paper.
- The distributive property "may be the single most powerful and useful concept related to multiplication" and should be developed with the concept of multiplication (Kinzer

& Stanford, 2014, p. 303). Understanding this property supports students' understanding of multiplication strategies, the connection between multiplication and addition, the connection of area concepts with multiplication and division, how factoring works, and how to break down complicated problems into simpler ones.

Single-Digit Computation Fluency

Fluency with basic-number combinations is important for multidigit computation, working with fractions and decimals, solving algebraic equations, applying statistics, measuring, and manipulating geometric representations. Most important is fluency with addition and multiplication, as subtraction is the inverse of addition and division is the inverse of multiplication. Fluency is more than speed; it refers to a basic skill "in carrying out procedures flexibly, accurately, efficiently, and appropriately" (CCSSM, 2010, p. 6).

Baroody and colleagues (Baroody et al., 2009; Baroody & Purpura, 2018; Baroody & Tiilikainen, 2003) hypothesized that learners move through three interdependent phases in learning a basic-number combination or family of combinations: (1) counting strategies (object or verbal counting); (2) deliberate reasoning strategies (known facts and relationships); and (3) automatic or fluent retrieval from a memory network. Students with mathematics difficulties (MD) tend to remain at phase one, overly reliant on slow counting strategies and not likely to devise their own, more efficient strategies. Research has identified differences between number-combination fluency of students with mathematics difficulties (MD) and their TA peers. In a study of second graders, de Chambrier and Zesiger (2018) found that the two groups did not differ in solving addition ties (e.g., 3 + 3, 4 + 4) but students with MD were slower than their peers in solving small and large non-tie problems (e.g., 6 + 2, 8 + 6). These non-tie problems required number comparisons and procedural strategies if they were not automatic. Rotem and Henik (2020) compared multiplication fluency of sixth and eighth graders with mathematics learning disabilities (MLD) with TA students in Grades 2–4, and 6. Sixth-grade students with MLD performed similar to TA second graders. Eighth graders with MLD performed similar to TA fourth graders, improving only on problems with ties (doubles) and involving numbers ≤5. This research demonstrates the severity of multiplication fact deficits in students with MLD. Even TA students did not reach mature fact retrieval for multiplication until sixth grade. This study highlighted the nature of building a multiplicative network mentally, from low-digit numbers and ties to medium- (4×8) and large-number problems (both digits over 5; e.g., 8×7).

Differences in fluency between students with MLD and their typically achieving (TA) peers have been attributed to problems with working memory (WM) on these tasks, a disruption in establishing an association of three numbers, two addends and their sum, in the situation of addition (Geary, 1993). Further, the executive processing of a seemingly simple task (e.g., 4 + 3) requires accessing the numbers in long-term memory (LTM), recalling the operation, holding those in WM while retrieving the sum from LTM, and possibly employing an effective strategy, all requiring attention and switching among components in WM (Berg & Hutchinson, 2010). Researchers in Belgium investigated the role of memory functions in fourth graders' fact fluency and found that students with low fact fluency experienced hypersensitivity to interference in these memory tasks as compared with TA peers (De Visscher & Noël, 2014). Since fact learning involves very similar associations, number combinations that share many features, interference from irrelevant information that was previously relevant for another task (proactive interference) has implications for fact retrieval and storing fact learning through associations (fact networks in LTM).

Fluency is improved through varied practice and problem situations, lots of opportunities to respond, efficient strategies for when combinations are not automatic, and awareness of one's fluency levels. In a meta-analysis of 17 single-case design studies of basic-fact fluency interventions with students who struggled with mathematics, Codding et al. (2011) found that the largest treatment effects came from a combination of interventions: modeling how to practice by the teacher, drill (practice of isolated items via flash cards, computer, etc.); and practice (use of learned responses in contexts).

Fluency with number combinations may be even more important for future mathematics achievement for students with MLD who have limited working-memory (WM) capacity. Effort expended on retrieving a product and sum for each number combination in a multidigit multiplication problem (e.g., 567 × 149) leaves little attention and space in WM for recalling and carrying out the procedural steps for solving the problem. Some methods for building fluency in students who struggle with mathematics include:

- For addition and subtraction combinations, teach effective counting strategies explicitly (Fuchs et al., 2010). These include + 1 and − 1 strategies, the identity property (+ 0, − 0), doubles, and + 2 and − 2 strategies. These strategies should be taught for concept understanding using manipulatives and number lines. Additional supportive strategies are *min* for addition (start with the larger number and count up by the smaller number for the minimum counts), and *minus number* for subtraction (start with the minus number and count up).

- For multiplication and division, emphasize the 2s (doubles from addition), 5s (from counting by 5), and 10s. From those known number combinations other combinations can be derived through strategies. Strother (2010) recommended calling flash cards *strategy cards* and having pairs of students describe the related-fact strategies they used to solve a combination if it wasn't automatic.

- Integrate instruction of properties into number-combination instruction. For example, the commutative property of multiplication allows a student who knows 5 × 6 to also know 6 × 5. The inverse properties allow students who know 9 − 3 = 6 to know 6 + 3 = 9.

- Vary practice using flash cards, paper-and-pencil practice, oral responses, story problems, 10-frames, dice, and computerized practice. When using flash cards, show students how to place fluent responses in one pile, correct using a strategy in a second pile, and incorrect responses in a third, as a self-evaluation step.

- Employ the research-based cover-copy-compare (CCC) practice strategy (Poncy et al., 2006). After practice with flash cards to determine fluency, prepare CCC practice sheets with nine known facts, one unknown fact, and 2–3 fluent facts in a first column, with two blank columns (Riccomini et al., 2017). Teach students to look at the first equation (e.g., 4 × 2 = 8), cover the equation, write the equation from memory, uncover and compare, and cross out/repeat if incorrect. Students should work down the list one equation at a time. Model, then monitor student practice. Build in three to four 5-minute CCC practice sessions each week with individualized practice sheets.

- Use simple, meaningful games to motivate students during practice. For example, the *Tens Go Fish* game focuses on additive doubles (Russell & Economopoulos, 2008). The game is played like Go Fish, but students are looking for combinations of tens in the cards they draw and ask for.

- When assessing fluency, assess more than speed and accuracy. Ask students what strategies they used for combinations that weren't automatic. Code and chart strategies students can use effectively and efficiently, for example M10 for making 10 or ND for the near-doubles strategy (Kling & Bay-Williams, 2014). Design assessments of number-combination knowledge so that error patterns are evident (Box B.2). For some students timed assessments aren't appropriate. These students may demonstrate less skill, fewer strategies, and lower overall progress with anxiety-provoking timed drills. Some students with greater concept understanding will actually slow down to work problems accurately.

Box B.2   Assessing Number Combination Fluency

$7 \times 2 =$ ___	$5 \times$ ___ $= 10$	$2 \times 8 =$ ____	$4 \times 2 =$ ____
$3 \times 7 =$ ___	$8 \times 3 =$ ____	___ $\times 5 = 15$	$3 \times 6 =$ ____
$4 \times 4 =$ ___	$6 \times 6 =$ ____	$3 \times 3 =$ ____	$7 \times$ ___ $= 49$

Teacher Directions: Watch the student completing the assessment and note on your copy which problems take longer and the order of responses. Point to a problem that is correct and ask how the student computed. If a problem took longer ask what strategy the student used. Point to an incorrect problem and ask the same question. Ask the student which problem was the easiest, which was the hardest, and why.

Analyze error patterns: Top row involves 2s, middle row 3s, bottom row doubles, and one diagonal has response position to the left of the equals mark. Two corners are 4s and two are 7s. Two diagonal problems involve 8s and two involve 5s.

- Even in the upper grades, teachers should assess number-combination and computation fluency and build in practice for maintenance. There is evidence that many high-school students with MLD have fluency levels of second and third graders, putting them at a disadvantage for upper-level mathematics topics (Dennis & Gratton-Fisher, 2020).

Multidigit Computation

Multidigit computation differs from single digit in that solutions cannot be automatic. Multidigit computation requires a solution strategy, how the numbers are operated on respecting the place value of digits within the problem (Hickendorff et al., 2019). Working with multidigit numbers is challenging for students with MD for a number of reasons. One is faulty conceptualization of digits within large numbers as individual numbers rather than representing a value depending on placement within the number (e.g., 78 as a 7 and an 8 rather than 70 and 8). Another, for English-speaking students, involves translating words into numbers (e.g., 1807 can be read one thousand eight hundred seven or eighteen hundred seven—not possible in many languages; 316 reads three hundred sixteen—rather than three-hundreds one-ten and six). Other common problems involve following the sequence of steps in a standard *algorithm*, recalling one-digit combinations rapidly, holding numbers and steps in working memory while carrying out other steps, overgeneralizing addition and multiplication properties to subtraction and division, and even keeping track of numbers in written form. Also, many students ready to work with larger numbers in mental computations that can enhance place-value understanding $(100 + 400; 1500 - 100)$ are forced to wait until they master their facts.

In comparing the computation errors of third-grade students with MD and their TA peers, Nelson and Powell (2018) found that using the wrong operation was the most common error for all students. Multidigit addition errors were primarily regrouping procedures and miscalculation, while multidigit subtraction errors tended toward disregarding digit placement (subtracting smallest from largest regardless of position) and wrong operation. For multidigit multiplication, the most common errors were incorrect facts and incomplete procedures, while for division items students often attempted the wrong operation. Students with MD made significantly more errors than their TA peers, and their errors were of a wider range. The researchers emphasized the difficulty of pinpointing interventions for such a wide range of errors and recommended emphasizing both concepts and procedures during practice with immediate feedback.

It is easy, and common, for teachers to overemphasize practice with computations involving multidigit whole numbers. Lengthy practice keeps students busy and is simple to check. For students with MD, practice without immediate feedback may be practicing and reinforcing errors. If students have a strong grounding in place-value and operational concepts, as well as a wide range of problem types, computational practice with larger numbers should not consume a large portion of instructional time. Multidigit computation, after all, is not the goal of mathematics instruction. Applying this computation when it is called for in a given problem situation is a goal.

It is also easy to simply leave computation to calculators. Given the readily accessible technologies today, why spend time understanding and practicing pencil-and-paper algorithms? Usiskin (1998) identified a number of reasons algorithms should be taught: they are powerful by helping us generalize the same procedures to other problems, they are reliable and accurate, they provide a written record (especially important during instruction), they help provide mental images of problem situations, they can be used in future algorithms, and they are instructive. The last point, that algorithms are instructive, may be the most critical. Working through algorithms assists students in seeing the connections among numbers, operations, and properties. This work supports concept understanding and development. New technologies will allow working with much larger (and smaller) numbers, delving deeper into mathematics topics because not as much time will be devoted to algorithm practice. But not devoting any time to paper-and-pencil algorithms would be like learning to tell time or temperature using only digital clocks and thermostats—the sense-making and concept development is lost.

Multidigit Addition and Subtraction

The first introduction of numbers with multiple digits should be accompanied by manipulatives that assist with the understanding of the values of each digit, such as base-10 blocks, and problem contexts. The basic understanding of number (with place value and magnitude knowledge) is critical for understanding computational operations. The number 245 is represented by two flats, four longs, and five units—a more efficient representation than a pile of 245 units. But students need practice with trading—10 units for one long, 10 longs for one flat (see Figure B.3).

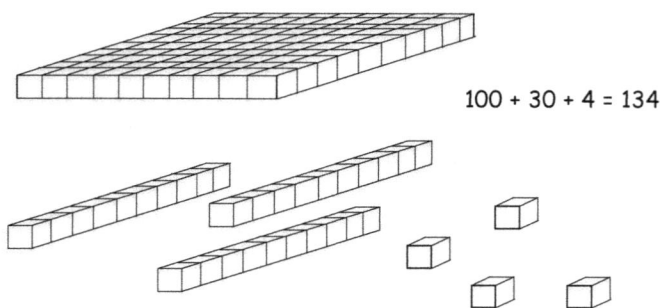

100 + 30 + 4 = 134

Figure B.3 Base-10 Blocks for Multidigit Addition and Subtraction

The number 627 is the number of students in our school and can be expressed in expanded form: 600 + 20 + 7. Later, when faced with multidigit addition and subtraction, students will be familiar with the concepts.

Transition from single-digit combinations to two-digit/one-digit computation is the first area for conceptual problems. For problems such as 17 + 4 and 25 − 3, it is tempting to continue using counting methods and number lines because the numbers aren't that large. It may be better practice to jump right into problems such as 54 + 47 or 83 − 26 so that students don't over-generalize with their comfortable counting procedures.

Once students move to three-digit computations, the same concepts apply, with a few potential difficulties. Most students are taught to perform these computations from right to left, although working left to right is not incorrect if the concepts remain intact. The most common errors are simple fact errors, using the correct operation inconsistently (in a set of mixed problems), not performing *regrouping* when needed, mistreating 0s, confusing two different algorithms, and not holding amounts or steps in working memory long enough to use them.

Addition is different than the other three operations in that it can be performed with multiple addends as in column addition. A collection of addends is common for combine types of addition problems. For example, *the students stocking the school store noticed that many boxes of black pens had been opened. They found 26 pens in one box, 43 in a second, 37 in a third box, 105 in a fourth, and a fifth box was unopened and had 120 pens. How many black pens are in stock?* The properties of addition allow us to work this problem in a number of ways. The standard textbook approach is to list all numbers in a column, taking care to line up digits by place values, then adding each column and regrouping as needed. An alternative is provided in the last section in this strand.

Research involving students with MLD and multidigit addition and subtraction verifies the use of explicit instruction of procedures grounded in concept development, cognitive-strategy instruction to support recall of procedures, a concrete-representational-abstract (CRA) sequence, student-engaged progress monitoring, and integrated ongoing practice of basic-number combinations. Mancl et al. (2012) extended the Mercer and Miller CRA research on number combinations to multidigit subtraction. After eleven 30-minute intervention lessons— five at the concrete level with base-10 blocks and place-value mats, three at the representational level with drawings of base-10 blocks, a strategy lesson for recalling the regrouping processes, and two lessons at the abstract level—the five students with MLD demonstrated mastery (at or above 80% on each lesson), with a few lessons repeated to achieve mastery. Dennis et al. (2014) used weekly curriculum-based measures (CBM) of computation with error-pattern analysis to support explicit computation instruction (CRA, cognitive-strategy instruction, and varied practice) with middle-school students with mathematics disabilities. The teachers in the study were able to target areas for individually designed remediation and their students achieved significant improvement over eight weeks with 30-minute-per-week interventions.

Thanheiser (2012) reminded us that preservice teachers often have gaps in multidigit computation concepts. Even teachers who could provide conceptually based explanations of regrouping in two-digit problems and had the knowledge base of place value and addition and subtraction algorithms had difficulty making connections with three-digit computation situations. Thanheiser posited that some concepts of teachers and students are unexpectedly weak and too isolated. In some situations, teachers and students have not explored the connections between concepts—the context in which procedures such as regrouping are applicable, the same concepts with different unit types or settings, and strategic knowledge related to the concepts.

Recommended strategies for providing instruction in multidigit addition and subtraction include:

- Ensure that underlying concepts regarding place value and operations are firm.
- Allow students to use alternative or invented algorithms as long as they are efficient and effective.
- Diagnose errors carefully by watching students work computations and conducting diagnostic interviews. Have students work back through simpler problems until the basis of the error is detected.
- Teach and encourage the use of estimation before computing. For example, the problem 578 + 213 should be about 800, not 365 or 7712. Curiously, many textbooks encourage estimation only after a strict computation algorithm has been mastered.
- For students having trouble keeping track of columns and regrouping, use 1 cm or 0.75 cm graph paper and label each column with its value (see Figure B.4).
- Use base-10 blocks or Digi-Blocks for instruction, carefully connecting what is happening with numbers to the manipulation of blocks. Have students place the manipulatives on a place-value mat (see Figure B.5 and Appendix II) and record each step with written symbols. Ask students to use think-alouds to detect strategies.
- Teach cognitive strategies to support procedures that are understood. For example, the BBB Sentence when subtracting positive whole numbers: Bigger number on Bottom means Break down and trade (Mercer & Miller, 1994).

Multidigit Multiplication and Division

Fourth- and fifth-grade teachers will attest that multidigit multiplication and division computation are not their students' favorite mathematics topics. While these are identified as critical areas (focal points) for fourth and fifth grade (CCSSM, 2010; NCTM, 2006), it may be how multiplication and division are related to each other through fluency with procedures for solving problems that is a more important goal. The other critical areas at these grade levels are dependent on strong multiplicative reasoning: equivalent fractions, decimals (as an extension of the base-10 system), area and volume (in geometry), and extending whole-number operations to the range of rational numbers.

Multidigit multiplication requires broad concept understanding—place value, multiplication and addition combinations and properties, sequencing steps to ensure all factors have been addressed, and some estimation ability. Students have the most problems with lining up numbers by place value, conducting multiplication and addition computations almost

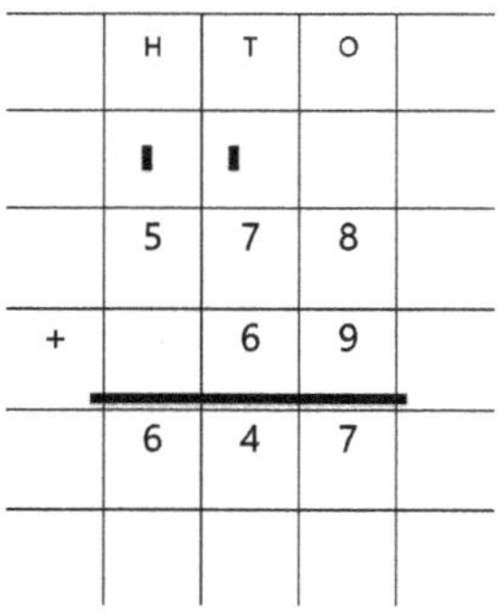

	H	T	O	
	▮	▮		
	5	7	8	
+		6	9	
	6	4	7	

Figure B.4 Using Graph Paper with Multidigit Computation

Hundreds	Tens	Ones

```
   3 0 5                    3 0 5
                           ┌─────────┐
                           │ 2 9 15  │
                           └─────────┘
  -  7 8                   -  7 8
 ─────────                ─────────────
```

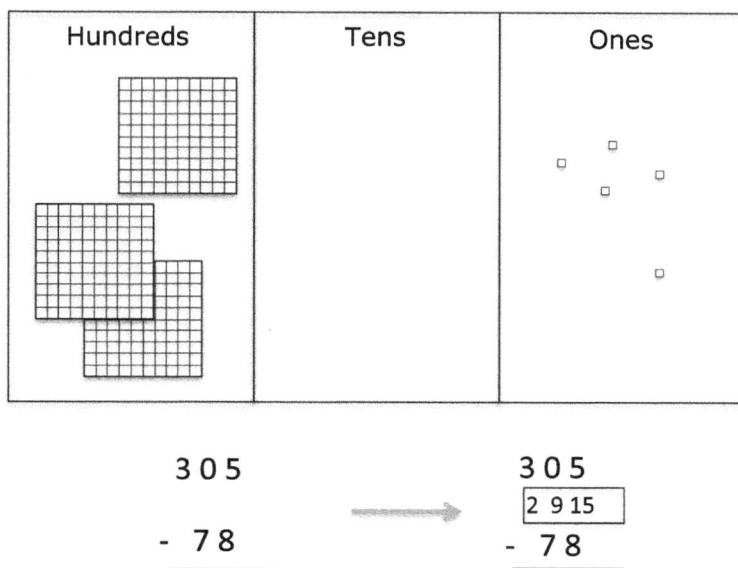

Figure B.5 A Place-Value Mat for Multidigit Subtraction

simultaneously, holding each piece of information in working memory long enough to use it in the next step, keeping track of multiple steps, and performing column addition correctly. To alleviate some of these issues, students can use alternative algorithms. When allowed to invent procedures to handle these larger numbers, students who understand the concepts involved tend to use variations on decomposition (Ambrose et al., 2004). The problem 15×8 can be decomposed into $(10 \times 8) + (5 \times 8)$ if the student understands place value and properties of multiplication and addition (a variation of the distributive property). The problem 178×56 can be decomposed into 100×56, 70×50, 70×6, 8×50, and 8×6 (finding partial products). Using multiples of 10 allows for simpler multiplication.

Explicit instruction of procedures for computation embedded in concept understanding and taught using CRA sequences and cognitive strategies are effective for students with MLD learning multidigit multiplication. Flores et al. (2014) used these approaches to teach two-digit multiplication, employing base-10 blocks and place-value mats in concrete and representational lessons. The teachers embedded the commutative property and reinforced place-value and regrouping concepts taught during addition and subtraction instruction. After ten 25-minute lessons, students demonstrated fluency in solving two-digit multiplication problems and their skills were maintained two weeks after the intervention ended. This research demonstrated that even with larger numbers, manipulatives and representations can enhance concept understanding.

Some recommended instructional strategies for multidigit multiplication:

- Allow students to use alternative strategies that are conceptually sound and efficient, such as deconstruction: $245 \times 14 = (200 + 40 + 5)(10 + 4)$. This problem can be solved through an area model (such as lattice multiplication) or using the distributive property.
- Use graphic organizers and flow charts (Ashlock, 2002).
- After a brief use of base-10 blocks for the concrete stage, use an area model (representational stage) to represent multiplication situations (see Figure B.6).

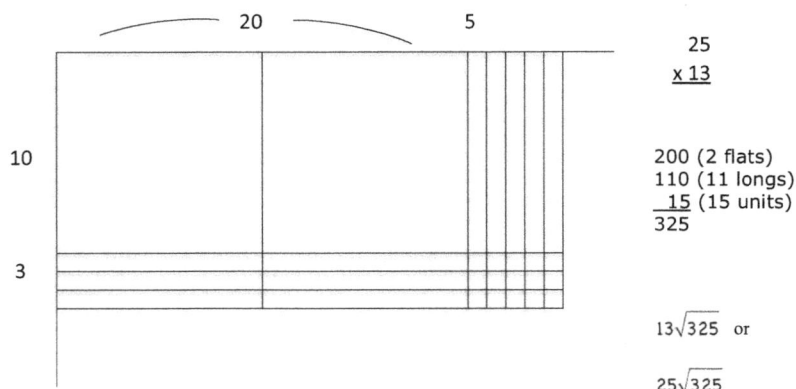

Figure B.6 Area Model of Multiplication and Division

- For practice with multistep algorithms use easy-erase surfaces such as whiteboards.
- After experience with two or more algorithms, ask individual students which they prefer and why. Make students aware of what works for them.

Long division is particularly challenging because the divisor is typically placed in front of the $)$ symbol and the dividend is underneath, a different structure than the other operations. The standard algorithm calls for working through the dividend from left to right, estimating factors and then multiplying and subtracting all the way through. It truly requires a combination of all previous arithmetic learning to accomplish. The most common errors with multidigit division include keeping numbers and symbols lined up, estimating factors closely, holding a lot of information in working memory at one time, following an effective sequence accurately through to the final step, and dealing with zeros and remainders.

Multidigit division strikes fear in the hearts of many adults, years after fifth grade. Although division is the inverse of multiplication and can be modeled in similar ways, division often involves those messy remainders (Pope, 2012). Division is sometimes introduced before the closely related concepts of fractions and factoring. Division doesn't always make things smaller (as in 3 divided by 1/2). Developing relational understanding and not just procedural competence will better prepare students for subsequent learning in mathematics. There is a paucity of research on multidigit division instruction for students who struggle with mathematics. There is some evidence that area models (at the representation level) connected with division problems can enhance conceptual understanding not only of division algorithms, but of subsequent algebraic concepts using algebra tiles (Richardson et al., 2010). See Figure B.6 for area representation of division.

Although the national standards include multidigit division, most experts emphasize that not many of us actually compute division problems by hand. More critical is an understanding of contexts in which multidigit division is required and evaluating the reasonableness of a quotient. Pose situations such as: "When would you need to divide a large number like 459 by a two-digit number such as 24?" Brainstorm situations that are relevant for students: times on base/number of home runs, students in school/classrooms, sandwiches sold/minutes, length of rope/number of sections to cut. Would a reasonable answer be 200, 70, 50, or 20? Will different situations require a remainder, a fraction, or a decimal? Practice setting up experiences with multidigit numbers that don't require calculation.

In addition to the previous instructional strategies, recommendations for multidigit division include:

- Focus on factor-estimation practice.
- For one-digit divisors, estimate whether to expect a remainder by applying the divisibility rules before computing. For example, in the problem $546 \div 5$, the dividend is not evenly divisible by 5 because it does not end with 0 or 5. It will have a remainder.
- Teach students to consider how a remainder should be expressed, depending on the context—whole number (109 R. 1), fraction ($109 \frac{1}{5}$), or decimal (109.2).
- For two- and three-digit divisors, consider making a quick list of factors before beginning the division-estimation process. For example, the problem $3984 \div 17$ is challenging. Making a list of the multiples of 17 will save trial-and-error time: 17, 34, 51, 68, … These are the most time-consuming computations and most adults simply estimate a quotient ($4000 \div 20 = 400 \div 2 = 200$) or pick up a calculator.
- Emphasize place value and use devices such as lined or graph paper to keep track of values. Particularly problematic are even products (nothing to bring down) and zeros.
- Allow the use of alternative algorithms that are conceptually sound and efficient.

Alternative Algorithms

Alternative algorithms, sometimes termed *low-stress algorithms*, refer to alternative computation procedures that are usually more meaningful or efficient than the standard, textbook approaches to multidigit computation. For centuries, mathematicians have developed and published alternative algorithms, while others have been copied from around the world (e.g., the Treviso method from at least 1478; Swetz, 1987). These algorithms have several common characteristics: they are effective computational methods, they reduce the working-memory demands on students, they are based on sound math concepts, and they can be a lot of fun. These approaches are not available as a cohesive curriculum program but can supplement any program. Research on alternative algorithms found that they reduced the time required for mastery, increased computational power, and reduced students' stress while computing (Hutchings, 1976). An additional benefit is the complete record of steps available to the student while working the algorithm. Ambrose et al. (2004) noted that standard algorithms are so compact they tend to mask the underlying principles, such as place value and properties of operations, that make them work. Standard algorithms tend to encourage thinking in digits (by rote rather than conceptually), instead of considering composite numbers (e.g., 346 = 300 + 40 + 6; 497 is 3 less than 500).

In a study on first through third graders in 82 classrooms, Sievert et al. (2019) found that strategies presented in textbooks had a substantial effect on students' adaptive expertise—the flexible use of strategies. The researchers noted that students rarely invent their own strategies and that textbooks influence the strategies teachers demonstrate and support though practice. Even textbooks based on the same curriculum standards can vary considerably in quality and opportunities to learn alternative arithmetic strategies. Alternative arithmetic strategies are effective and efficient but should be taught explicitly with their related concepts.

In a study investigating multidigit-multiplication algorithms, Speiser et al. (2012) were concerned with the working-memory and attention demands of the standard algorithm. The researchers compared the Treviso method (lattice method, see Chapter 4, Figure 4.4) with the partial-products method (see Figure B.9B). The Treviso method was superior to the partial-products method, possibly because of its spatial representations. This method was particularly advantageous for lower-performing students. Students should be encouraged to explore these

A. Standard (Textbook) Algorithm

```
  ||
  387
+ 936
 1323
```

B. Left-to-Right Method
(Pearson, 1986)

```
   387
 + 936
    12
    11
    13
  1323
```

C. Partial Sums (right-to-left), India
(Basserear, 1997)

```
   387
 + 936
    13
    11
    12
  ----
  1323
```

D. Tens Method for Column Addition
Scratch Method (Fulkerson, 1963)

```
  ||
  387        3
3 2 3
  936       29
           17  6
+ 245      56  2
------    + 98
 1568     ----
          200
```

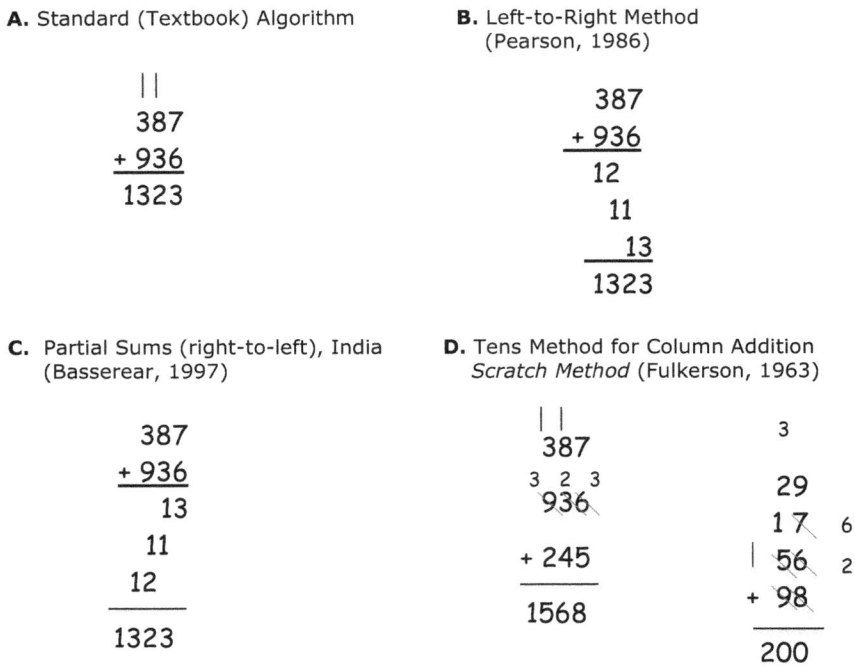

Figure B.7 Alternative Addition Algorithms

alternatives and find the approach that makes the most sense and is most efficient and effective for them. But teachers should not require students to demonstrate mastery with each type of algorithm; that is not the goal of this instruction. Encourage students to find their best fit, whether standard or alternative, and to master that approach.

The standard addition algorithm is depicted in Figure B.7A, followed by the most common alternatives for three-digit and column addition: a partial sums method that may be 1,000 years old, Pearson's left-to-right addition, and Fulkerson's tens method for column addition. These addition methods place fewer demands on students' working memories because they separate rapid recall from renaming. In a study of second-graders' use of invented strategies for two-digit addition and subtraction, Sahin et al. (2020) found that only 34% used an invented strategy occasionally (e.g., combining tens and then ones) and 78% could use standard algorithms. However, students who used the standard algorithm primarily had statistically significantly lower mathematics achievement than students who used a combination of invented and standard strategies, indicating a higher level of conceptual understanding of place value and properties of operations for students who understood alternative methods.

Figure B.8 depicts the standard and alternative subtraction algorithms. The conservation property of number is applied in the equal additions, adding constants, and adding the complement methods, because when equal numbers are added or taken away from both the minuend (top) and subtrahend (bottom) numbers, the difference remains the same. Another alternative subtraction method, termed the *indirect addition strategy*, was studied by Torbeyns et al. (2018) with sixth graders. This number-based strategy involves adding on from the

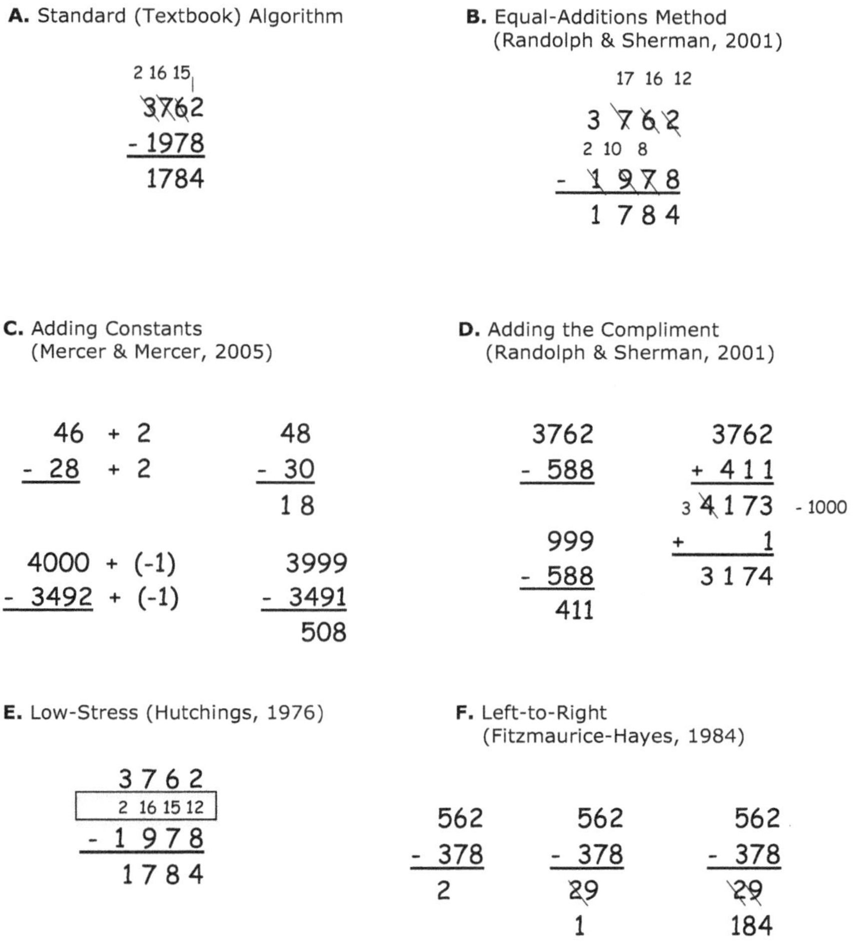

A. Standard (Textbook) Algorithm

```
    2 16 15
    3 7 6 2
  - 1 9 7 8
    1 7 8 4
```

B. Equal-Additions Method
(Randolph & Sherman, 2001)

```
     17 16 12
   3  7  6  2
     2 10  8
 -  1  9  7  8
    1  7  8  4
```

C. Adding Constants
(Mercer & Mercer, 2005)

```
   46  + 2        48
 - 28  + 2      - 30
                  18

 4000 + (-1)     3999
-3492 + (-1)    - 3491
                  508
```

D. Adding the Compliment
(Randolph & Sherman, 2001)

```
  3762          3762
 - 588         + 411
            3 4173   - 1000
  999          +    1
 - 588          3174
  411
```

E. Low-Stress (Hutchings, 1976)

```
    3 7 6 2
   [2 16 15 12]
  - 1 9 7 8
    1 7 8 4
```

F. Left-to-Right
(Fitzmaurice-Hayes, 1984)

```
  562      562      562
- 378    - 378    - 378
  2        29       29
           1        184
```

Figure B.8 Alternative Subtraction Algorithms

subtrahend to the minuend (e.g., 82 − 67 via 67 + **3** = 70 and 70 + **12** = 82, yielding 3 + 12 = 15). The researchers compared this strategy for small difference and large difference subtractions in two- and three-digit problems, with four achievement levels of students (low, below-average, above-average, and high). They concluded that this method was applied frequently, efficiently, and flexibly by all groups of students. Even low achievers, with instruction and demonstration, mastered the strategy.

Multiplication of multidigit factors is complicated in the standard algorithm by keeping track of place values and overloading working memory. These issues are addressed through the partial products and low-stress methods depicted in Figure B.9. Flores and colleagues (2020) investigated the use of the partial-product algorithm taught through a CRA sequence (place-value mats, base-10 blocks). The fourth and fifth graders in the study developed the meanings of composition of numbers and multiplication, increasing from 16% to 80% on direct assessments of multiplication, in just 25-minute triweekly sessions over four weeks. Lattice multiplication, described in Chapter 4, may focus too much on digits, and not the full

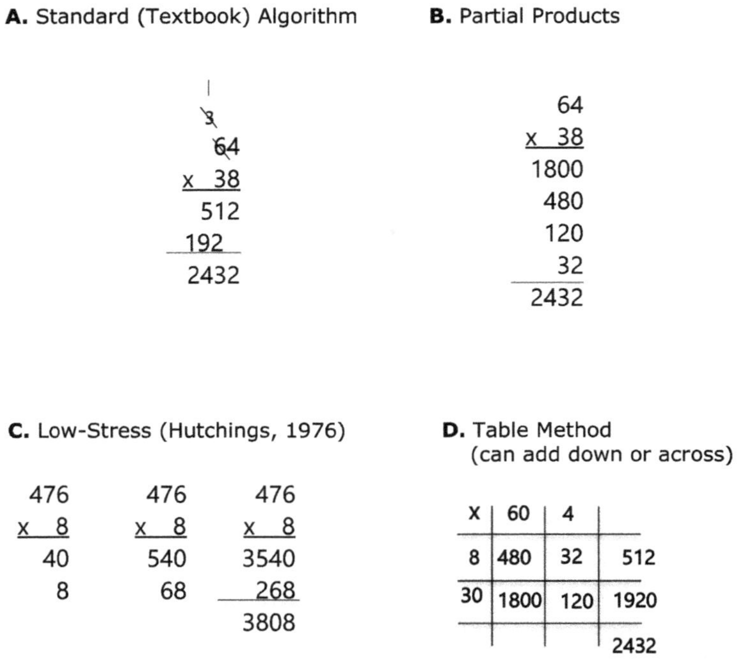

A. Standard (Textbook) Algorithm   **B.** Partial Products

```
          |
          ȝ
         ⁔4
        64
       x  38
        512
        192
       2432
```

```
         64
       x  38
       1800
        480
        120
         32
       2432
```

C. Low-Stress (Hutchings, 1976)   **D.** Table Method
                                             (can add down or across)

```
   476       476       476
  x  8      x  8      x  8
    40       540      3540
     8        68       268
                      3808
```

x	60	4	
8	480	32	512
30	1800	120	1920
			2432

Figure B.9 Alternative Multiplication Algorithms
Note: See Chapter 4 for Lattice Method

numbers, unless their meaning is taught explicitly. A table algorithm, as depicted in Figure B.9D, maintains the number values throughout (Clivaz, 2017).

For multidigit division, numbers can be decomposed by partial quotients, factors, or multiples of 10. Some students may prefer to pyramid the quotient above the problem as depicted in Figure B.10. These alternatives to the standard algorithm help students keep track of the steps applied and are true to multiplicative concepts. Hickendorff et al. (2018) investigated the use of digit-based strategies (standard algorithm), column-based strategies (scaffolding), and number-based strategies (repeated subtraction, addition, or partitioning) that emphasize the meaning of numbers in division computation by fourth to sixth graders. The students used a digit-based (standard) approach 26% of the time, column-based approach 10%, and number-based 56% (of those 25% were repeated subtraction and 17% partitioning). Overall, the students tended to use the strategies emphasized in their classrooms. The researchers concluded that a carefully constructed instructional pathway that focuses on strategy variety and progressively building toward standard procedures with meaning leads to performance levels similar to efficient, standard procedures, but with more conceptual understanding for better problem solving.

Some alternative algorithms are incorporated within curriculum materials, such as the research-based *Everyday Mathematics* (The University of Chicago School Mathematics Project, 2001), in the student reference books and instructor's manuals. These include partial-sums and column methods for addition; trade-first, counting-up, and partial-differences subtraction methods; partial-products and lattice multiplication; and partial-quotient and column division. Teachers should examine the algorithms used within textbooks and curricula to

A. Standard (Textbook) Algorithm

```
        324  r 7
   23 / 7459
      - 69
        55
      - 46
        99
      - 92
         7
```

B. Scaffolding or Partial Quotient
(Reisman, 1977)

```
   23 / 7459
      - 2300    100
        5159
      - 2300    100
        2859
      - 2300    100
         559
      - 230      10
         329
      - 230      10
          99
      - 46        2
          53
      - 46        2
           7

                324
```

C. Pyramid Form of Scaffolding
(Randolph & Sherman, 2001)

```
          131  r 4
            1
           10
           10
           10
          100
     7 /  921
       - 700
         221
       - 70
         151
       - 70
          81
       - 70
          1 1
        - 7
           4
```

Figure B.10 Alternative Division Algorithms

ensure that number and place-value concepts are supported, not simply manipulation of digits.

When teachers try these algorithms for the first time, they may find them confusing. Many parents and educators are skeptical about their usefulness because they are not the "standard" approach. However, for certain students they can be quite effective. These students have had trouble with traditional algorithms, especially with the notation system or memory burdens imposed by those procedures. An often-unexpected benefit is for the student to realize that he knows a math method other students want to learn. Students can also invent their own effective algorithms if they have the requisite understanding of the underlying concepts such as place value, effects of operations, properties, and conceptual connections. Student-developed algorithms should be encouraged but monitored for concept fidelity, effectiveness, and value for the next grade levels.

Finally, it is important to remember to embed the use of these operations with a range of number sizes into different types of problem situations, across all mathematics content areas. Applying operations accurately and efficiently when needed is the goal of computation instruction. The meanings of the four operations are extended with "each new number system—integers, rational numbers, real numbers, and complex numbers. The four operations stay the same in two important ways: they have the commutative, associative, and distributive properties and their new meanings are consistent with their previous meanings" (CCSSM, 2010, p. 58).

Multicultural Connection

The ancient Egyptian civilization spanned 31 dynasties and 3,000 years, until 335 BCE, but their contributions to mathematics and daily life are evident today in architecture, astronomy, time measurement, written language, and textiles. Through sources such as the Rhind and Moscow papyri, scholars have determined that the Egyptians could only express fractions as added unit fractions with numerators of 1 (except for $\frac{2}{3}$ with a special symbol). For example, $\frac{23}{24}$ was expressed as $\frac{1}{2}+\frac{1}{4}+\frac{1}{8}+\frac{1}{12}$. They used addition to perform multiplication and division calculations (Corry, 2020). Multiplication involved progressive doubling and adding. In the example in Box B.3, multiply 14 times 24 by placing the number 1 in the first column and one of the factors, 24, in the second column (the process is quicker using the larger factor but will work with either factor). Double the numbers in each column until the next number in the first column would be larger than the second factor (14). In this case the last double is 8 because the next one would be 16, larger than 14. Check off the numbers in the left-hand column that add up exactly to the other factor (2 + 4 + 8 = 14). Add up the corresponding numbers in the right-hand column for the final product (48 + 96 + 192 = 336).

Box B.3   Egyptian Multiplication

1	24
2 √	48 √
4 √	96 √
8 √	192 √
	336

References

Ambrose, R., Baek, J., & Carpenter, T. P. (2004). Children's invention of multigit multiplication and division algorithms. In A. J. Baroody & A. Dowker (Eds.), *The development of arithmetic concepts and skills: Constructing adaptive expertise* (pp. 305–336). Erlbaum. doi: 10.4324/9781410607218

Anthony, G. J., & Walshaw, M. A. (2004). Zero: A "none" number? *Teaching Children Mathematics, 11*(1), 38–42. doi: 10.5951/tcm.11.1.0038

Ashlock, R. B. (2002). *Error patterns in computation: Using error patterns to improve instruction* (8th ed.). Merrill Prentice Hall.

Askew, M., Venkat, H., Mathews, C., Ramsingh, V., Takane, T., & Roberts, N. (2019). Multiplicative reasoning: An intervention's impact on foundation phase learners' understanding. *South African Journal of Childhood Education, 9*(1), a622. doi: 10.4102/sajce.v9i1.622

Baroody, A. J. (1987). *Children's mathematical thinking: A developmental framework for preschool, primary, and special education teachers.* Teacher College, Columbia University.

Baroody, A. J., Bajwa, N. P., & Eiland, M. (2009). Why can't Johnny remember the basic facts? *Developmental Disabilities Research Reviews, 15*(1), 69–79. doi: 10.1002/ddrr.45

Baroody, A. J., & Purpura, D. J. (2018). Early number and operations: Whole numbers. In J. Cai (Ed.), *Compendium for research in mathematics education* (pp. 308–354). NCTM.

Baroody, A. J., & Tiilikainen, S. H. (2003). Two perspectives on addition development. In A. J. Baroody & A. Dowker (Eds.), *The development of arithmetic concepts and skills: Constructing adaptive expertise* (pp. 75–125). Erlbaum. doi: 10.4324/9781410607218

Basserear, T. (1997). *Mathematics for elementary teachers.* Houghton Mifflin.

Barney, L. (1970, April). Your fingers can multiply! *Instructor*, 129–130.

Berg, D. H., & Hutchinson, N. L. (2010). Cognitive processes that account for mental addition fluency differences between children typically achieving in arithmetic and children at-risk for failure in arithmetic. *Learning Disabilities: A Contemporary Journal, 8*, 1–20.

Byrd, C. E., McNeil, N. M., Chesney, D. L., & Matthews, P. G. (2015). A specific misconception of the equal sign acts as a barrier to children's learning of early algebra. *Learning and Individual Differences, 38*, 61–67. doi: 10.1016/j.lindif.2015.01.001

Carpenter, T. P., Franke, M. L., Jacobs, V., & Fennema, E. (1998). A longitudinal study of invention and understanding in children's multidigit addition and subtraction. *Journal for Research in Mathematics, 29*(1), 3–20. doi: 10.2307/749715

Carrier, J. (2014). Student strategies suggesting emergence of mental structures supporting logical and abstract thinking: Multiplicative reasoning. *School Science and Mathematics, 114*(2), 87–96. doi: 10.1111/ssm.12053

Ching, B. H-H., & Wu, X. (2019). Concreteness fading fosters children's understanding of the inversion concept in addition and subtraction. *Learning and Instruction, 61*, 148–159. doi: 10.1016/j.learninstruc.2018.10.006

Cheeseman, J., Downton, A., Roche, A., & Ferguson, S. (2020). Investigating young students' multiplicative thinking: The 12 little ducks problem. *The Journal of Mathematical Behavior, 60*, 100817. doi: 10.1016/j.jmathb.2020.100817

Chow, J. C., & Wehby, J. H. (2019). Effects of symbolic and nonsymbolic equal-sign intervention in second-grade classrooms. *The Elementary School Journal, 119*(4), 677–702. doi: 10.1086/703086

Clivaz, S. (2017). Teaching multidigit multiplication: Combining multiple frameworks to analyse a class episode. *Educational Studies in Mathematics, 96*(3), 305–325. doi: 10.1007/s10649-017-9770-7

Codding, R. S., Burns, M. K., & Lukito, G. (2011). Meta-analysis of mathematics basic-fact fluency interventions: A component analysis. *Learning Disabilities Research & Practice, 26*(1), 36–47. doi: 10.1111/j.1540-5826.2010.00323.x

Corry, L. (2020). Algebra. *Encyclopædia Britannica.* www.britannica.com/science/algebra

de Chambrier, A.-F., & Zesiger, P. (2018). Is a fact retrieval deficit the main characteristic of children with mathematical learning disabilities? *Acta Psychologica, 190*, 95–102. doi: 10.1016/j.actpsy.2018.07.007

Dennis, M. S., Calhoon, M. B., Olson, C. L., & Williams, C. (2014). Using computation curriculum-based measurement probes for error pattern analysis. *Intervention in School and Clinic, 49*(5), 281–289. doi: 10.1177/1053451213513957

Dennis, M. S., & Gratton-Fisher, E. (2020). Use data-based individualization to improve high school students' mathematics computation and mathematics concept, and application performance. *Learning Disabilities Research and Practice, 35*(3), 126–138. doi: 10.1111/ldrp.12227

DeVisscher, A. & Noël, M. P. (2014). Arithmetic facts storage deficit: The hypersensitivity-to-interference in memory hypothesis. *Developmental Science, 17*(3), 434–442. doi: 10.1111/desc.12135

Fitzmaurice-Hayes, A. (1984). Curriculum and instructional activities: Pre-K through Grade 2. In J. Fawley (Ed.), *Developmental teaching of mathematics for the learning disabled* (pp. 95–114). Aspen Systems.

Flores, M. M., Hinton, V. M., & Schweck, K. B. (2014). Teaching multiplication with regrouping to students with learning disabilities. *Learning Disabilities Research & Practice*, *29*(4), 171–183. doi: 10.1111/ldrp.12043

Flores, M. M., Moore, A. J., & Meyer, J. M. (2020). Teaching the partial products algorithm with the concrete representational abstract sequence and the strategic instruction model. *Psychology in the Schools*, *57*(6), 946-958. doi: 10.1002/pits.22335

Fuchs, L. S., Powell, S. R., Seethaler, P. M., Fuchs, D., Hamlett, C. L., Cirino, P. T., & Fletcher, J. (2010). A framework for remediating number combination deficits. *Exceptional Children*, *76*(2), 135–156. doi: 10.1177/001440291007600201

Fulkerson, E. (1963). Adding by tens. *The Arithmetic Teacher*, *10*, 139–140.

Funckhouser, C. (1995). Developing number sense and basic computation skills in students with special needs. *School Science and Mathematics*, *95*(5), 236–239. doi: 10.1111/j.1949-8594.1995.tb15773.x

Fuson, K. C. (1992). Research on whole number addition and subtraction. In D. A. Grouws (Ed.), *Handbook of research on mathematics teaching and learning* (pp. 243–275). Macmillan.

Fuson, K. C., Carroll, W. M., & Landis, J. (1996). Levels in conceptualizing and solving addition and subtraction compare word problems. *Cognition and Instruction*, *14*(3), 345–371. doi: 10.1207/s1532690xci1403_3

Fyfe, E. R., Matthews, P. G., & Amsel, E. (2020). College developmental math students' knowledge of the equal sign. *Educational Studies in Mathematics*, *104*(1), 65–85. doi: 10.1007/s10649-020-09947-2

Geary, D. C. (1993). Mathematical disabilities: Cognitive, neuropsychological, and genetic components. *Psychological Bulletin*, *114*(2), 345–362. doi: 10.1037/0033-2909.114.2.345

Geary, D. C. (2015). The classification and cognitive characteristics of mathematical disabilities in children. In R. C. Kadosh & A. Dowker (Eds.), *The Oxford Handbook of Numerical Cognition* (pp. 751–770). Oxford University Press. doi: 10.1093/oxfordhb/9780199642342.013.017

Geary, D. C., Hoard, M. K., Nugent, L., & Bailey, D. H. (2012). Mathematical cognition deficits in children with learning disabilities and persistent low achievement: A five-year prospective study. *Journal of Educational Psychology*, *104*(1), 206–223. doi: 10.1037/a0025398

Greer, B. (1992). Multiplication and division as models of situations. In D. A. Grouws (Ed.), *Handbook of research on mathematics teaching and learning* (pp. 276–295). Macmillan.

Hackenberg, A. J. (2010). Students' reasoning with reversible multiplicative relationships. *Cognition and Instruction*, *28*(4), 383–432. doi: 10.1080/07370008.2010.511565

Hickendorff, M., Torbeyns, J., & Verschaffel, L. (2018). Grade-related differences in strategy use in multidigit division in two instructional settings. *British Journal of Developmental Psychology*, *36*(2), 169–187. doi: 10.1111/bjdp.12223

Hickendorff, M., Torbeyns, J., & Verschaffel, L. (2019). Multi-digit addition, subtraction, multiplication, and division strategies. In A. Fritz, V. G. Hasse, & P. Räsänen (Eds.), *International handbook of mathematical learning difficulties* (pp. 543–560). Springer. doi: 10.1007/978-3-319-97148-3_32

Hutchings, B. (1976). Low-stress algorithms. In D. Nelson & R. E. Reys (Eds.), *Measurement in school mathematics* (pp. 218–239). NCTM.

Jong, C., & Magruder, R. (2014). Beyond cookies: Understanding various division models. *Teaching Children Mathematics*, *20*(6), 367–373. doi: 10.5951/teacchilmath.20.6.0366

Kilpatrick, J., Swafford, J., & Findell, B. (Eds.). (2001). *Adding it up: Helping children learn mathematics*. National Academy Press. doi: 10.17226/9822

Kinzer, C. J., & Stanford, T. (2014). The distributive property: The core of multiplication. *Teaching Children Mathematics*, *20*(5), 302–309. doi: 10.5951/teacchilmath.20.5.0302

Kling, G., & Bay-Williams, J. M. (2014). Assessing basic fact fluency. *Teaching Children Mathematics*, *20*(8), 488–497. doi: 10.5951/teacchilmath.20.8.0488

Knuth, E. J., Stephens, A. C., McNeil, N. M., & Alibali, M. W. (2006). Does understanding the equal sign matter? Evidence from solving equations. *Journal for Research in Mathematics Education*, *37*, 297–312.

Kosko, K. W. (2020). The multiplicative meaning conveyed by visual representations. *The Journal of Mathematical Behavior*, *60*, 100800–. doi: 10.1016/j.jmathb.2020.100800

Mancl, D. B., Miller, S. P., & Kennedy, M. (2012). Using the concrete-representational-abstract sequence with integrated strategy instruction to teach subtraction with regrouping to students with learning disabilities. *Learning Disabilities Research & Practice*, *27*(4), 152–166. doi: 10.1111/j.1540-5826.2012.00363.x

Matthews, P. G., & Fuchs, L. S. (2020). Keys to the gate? Equal sign knowledge at second grade predicts fourth-grade algebra competence. *Child Development, 91*(1), e14–e28. doi: 10.1111/cdev.13144

McDonald, P. A., & Berg, D. H. (2018). Identifying the nature of impairments in executive functioning and working memory of children with severe difficulties in arithmetic. *Child Neuropsychology, 24*(8), 1047–1062. doi: 10.1080/09297049.2017.1377694

McNeil, N. M., Hornburg, C. B., Devlin, B. L., Carrazza, C., & McKeever, M. O. (2019). Consequences of individual differences in children's formal understanding of mathematical equivalence. *Child Development, 90*(3), 940–956. doi: 10.1111/cdev.12948

McNeil, N. M., Hornburg, C. B., Brletic-Shipley, H., & Matthews, J. M. (2019). Improving children's understanding of mathematical equivalence via an intervention that goes beyond nontraditional arithmetic practice. *Journal of Educational Psychology, 111*(6), 1023–1044. doi: 10.1037/edu0000337

Mercer, C. D., & Mercer, A. R. (2005). *Teaching students with learning problems* (6th ed.) Merrill/Prentice Hall.

Mercer, C. D., & Miller, S. P. (1991–1994). *Strategic math series.* Edge Enterprises.

Mercer, C. D., & Miller, S. P. (1992). Teaching students with learning problems in math to acquire, understand, and apply basic math facts. *Remedial and Special Education, 13*, 19–35, 61.

Mercer, C. D., & Miller, S. P. (1994). *Subtraction facts 10–18.* Edge Enterprises. www.edgeenterprisesinc.com

Moser Opitz, E., Freesemann, O., Prediger, S., Grob, U., Matull, I., & Hußmann, S. (2017). Remediation for students with mathematics difficulties: An intervention study in middle schools. *Journal of Learning Disabilities, 50*(6), 724–736. doi: 10.1177/0022219416668323

Moser Opitz, E., Grob, U., Wittich, C., Häsel-Weide, U., & Nührenbörger, M. (2018). Fostering the computation competence of low achievers through cooperative learning in inclusive classrooms: A longitudinal study. *Learning Disabilities: A Contemporary Journal, 16*(1), 19–35.

National Council of Teachers of Mathematics (2000). *Principles and standards for school mathematics.* www.nctm.org/Standards-and-Positions/Principles-and-Standards/

National Council of Teachers of Mathematics (2006). *Curriculum focal points for prekindergarten through grade 8.* www.nctm.org/curriculumfocalpoints/

National Council of Teachers of Mathematics (n.d.). Pan Balance. *Classroom Resources: Interactives.* www.nctm.org/classroomresources/

National Governors Association Center for Best Practices & Council of Chief State School Officers. (2010). *Common core state standards for mathematics.* www.corestandards.org/Math/

Nelson, G., & Powell, S. R. (2018). Computation error analysis: Students with mathematics difficulty compared to typically achieving students. *Assessment for Effective Intervention, 43*(3), 144–156. doi: 10.1177/1534508417745627

Pearson, E. S. (1986). Summing it all up: Pre-1900 algorithms. *The Arithmetic Teacher, 33*(7), 38–41. doi: 10.5951/at.33.7.0038

Poncy, B. C., Skinner, C. H., & Jaspers, K. E. (2006). Evaluating and comparing interventions designed to enhance math fact accuracy and fluency: Cover, copy, and compare versus taped problems. *Journal of Behavioral Education, 16*(1), 27–37. doi:10.1007/s10864-006-9025-7

Pope, S. (2012). The problem with division. *Mathematics Teaching, 231*, 42–45.

Randolph, T. D., & Sherman, H. J. (2001). Alternative algorithms: Increasing options, reducing errors. *Teaching Children Mathematics, 7*(8), 480–484. doi: 10.5951/tcm.7.8.0480

Reisman, F. K. (1977). *Diagnostic teaching of elementary-school mathematics: Methods and content.* Rand McNally.

Riccomini, P. J., Stocker, J. D., & Morano, S. (2017). Implementing an effective mathematics fact fluency practice activity. *Teaching Exceptional Children, 49*(5), 318–327. doi: 10.1177/0040059916685053

Richardson, K., Pratt, S., & Kurtts, S. (2010). Utilizing an area model as a way of teaching long division: Meeting diverse student needs. *Oklahoma Journal of School Mathematics, 2*(1), 14–24.

Robinson, K. M. (2019). Arithmetic concepts in the early school years. In K. M. Robinson, H. P. Osana, & D. Kotsopoulos (Eds.), *Mathematical learning and cognition in early childhood* (pp. 165–186). Springer. doi: 10.1007/978-3-030-12895-1_10

Rotem, A., & Henik, A. (2020). Multiplication facts and number sense in children with mathematics learning disabilities and typical achievers. *Cognitive Development, 54*, 100866. doi: 10.1016/j.cogdev.2020.100866

Russell, S. J. & Economopoulos, K. (2008). *Investigations in number, data, and space* (2nd ed.). Pearson Scott Foresman.

Sahin, N., Dixon, J. K., & Schoen, R. C. (2020). Investigating the association between students' strategy use and mathematics achievement. *School Science and Mathematics, 120*(6), 325–332. doi: 10.1111/ssm.12424

Schneider, M., Rittle-Johnson, B., & Star, J. R. (2011). Relations among conceptual knowledge, procedural knowledge, and procedural flexibility in two samples differing in prior knowledge. *Development Psychology, 47*(6), 1525–1518. doi: 10.1037/a0024997

Seo, K. & Ginsburg, H. P. (2003). "You've got to carefully read the math sentence ...": Classroom context and children's interpretations of the equals sign. In A. J. Baroody & A. Dowker (Eds.), *The development of arithmetic concepts and skills: Constructing adaptive expertise.* (pp. 161–188). Lawrence Erlbaum.

Sidebotham, T. H. (2002). *The A to Z of mathematics: A basic guide.* John Wiley & Sons.

Sievert, H., van den Ham, A.-K., Niedermeyer, I., & Heinze, A. (2019). Effects of mathematics textbooks on the development of primary school children's adaptive expertise in arithmetic. *Learning and Individual Differences, 74*, 101716–. doi: 10.1016/j.lindif.2019.02.006

Speiser, R., Schneps, M. H., Heffner-Wong, A., Miller, J. L., & Sonnert, G. (2012). Why is paper-and-pencil multiplication difficult for many people? *The Journal of Mathematical Behavior, 31*(4), 463–475. doi: 10.1016/j.jmathb.2012.08.001

Strother, S. (2010). Developing fact fluency in mathematics. *The Educator, 5* (Winter), www.lplearningcenter.org/wp-content/uploads/2012/01/Mathematics-Winter2010.pdf

Swetz, F. J. (1987). *Capitalism and arithmetic: The new math of the 15th century, including the full text of the Treviso arithmetic of 1478* (D. E. Smith, Trans.). Open Court.

Tillema, E. S. (2018). An investigation of 6th graders' solutions of Cartesian product problems and representation of these problems using arrays. *The Journal of Mathematical Behavior, 52*(1), 1–20. doi: 10.1016/j.jmathb.2018.03.009

Thanheiser, E. (2012). Understanding multidigit whole numbers: The role of knowledge components, connections, and context in understanding regrouping 3+ -digit numbers. *The Journal of Mathematical Behavior, 31*(2), 220–234. doi: 10.1016/j.jmathb.2011.12.007

Torbeyns, J., Peters, G., De Smedt, B., Ghesquière, P., & Verschaffel, L. (2018). Subtraction by addition strategy use in children of varying mathematical achievement level: A choice/no-choice study. *Journal of Numerical Cognition, 4*(1), 215–234. doi: 10.5964/jnc.v4i1.77

The University of Chicago School Mathematics Project (2001). *Everyday Mathematics.* Everyday Learning Corporation. http://everydaymath.uchicago.edu

Usiskin, Z. (1998). Paper-and-pencil algorithms in a calculator-and-computer age. In L. J. Morrow & M. J. Kenney (Eds.), *The teaching and learning of algorithms in school mathematics* (pp. 7–20). NCTM.

Venkat, H., & Mathews, C. (2019). Improving multiplicative reasoning in a context of low performance. *ZDM, 51*(1), 95–108. doi: 10.1007/s11858-018-0969-6

Vergnaud, G. (1988). Multiplicative structures. In J. Hiebert & M. Behr (Eds.), *Number concepts and operations in the middle grades* (pp. 141–161). NCTM, Lawrence Erlbaum.

Strand C
Rational Numbers

After studying this strand, the reader should be able to:

1. describe the relationships among rational numbers—fractions, decimals, percents, and whole numbers—and the five types of rational number representations;
2. provide examples of proportional reasoning;
3. describe the most common difficulties with rational number concepts;
4. implement strategies for teaching about fractions, decimals, ratios, percents, and proportions.

Henry is an 11-year-old student with learning disabilities in reading and mathematics. His sixth-grade teacher at Balsam Middle School is attempting to determine Henry's fraction concept understanding.

MRS. QUAID: I want to show you a picture of something and see if you know what it is. What is in that picture? [Shows a picture of a six pack of soft drinks.]

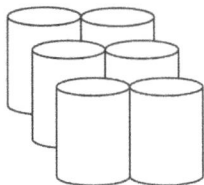

HENRY: I guess just soda cans or whatever.
MRS. QUAID: Yes, soda cans, they are connected so you can carry them. They are called six-packs. If you and your sister drank two of those cans, could you write a fraction to show me how much you drank? Do you know how to write fractions?
HENRY: [Draws six lines on a piece of paper and marks through two of them.] Two-fourths?

MRS. QUAID: Okay, can you write that as a fraction and explain it pointing to the cans?
HENRY: [Writes $\frac{2}{4}$.] There are two here that I drank and four over there.

Rational number extensions beyond whole numbers, such as those of fractions and decimals, are often the first major conceptual challenges for students who have been working primarily

DOI: 10.4324/9781003096733-11

with whole-number concepts. Rational numbers are important for real-world applications and are essential for algebra and other higher-level mathematics understanding. The concepts embedded within these constructs are complex and interconnected: proportion, ratio, equivalence, multiplicative reasoning, measure, decimal, percentage, unit, and rate. This strand will describe and illustrate the critical concepts related to rational numbers, offer common student misconceptions, and provide instructional strategies across the grade levels. The reader should be aware that professional mathematicians still struggle to define terms and explain concepts related to nonintegral rational numbers. Educators often were not provided adequate instruction in these concepts in their own elementary- and secondary-school experiences. With more personal experience with fractions and related concepts and observing the explorations of students and listening to their reasoning, educators can develop a stronger understanding that should continue to evolve and deepen.

Rational Numbers

The diagram of the real number system in Strand A (Figure A.1) illustrates the domain of rational numbers under which numbers are classified as integers (negative and positive whole numbers and zero) or nonintegral numbers. Rational numbers are all numbers that can be expressed as $\frac{a}{b}$ where $b \neq 0$. For example, 8 is a rational number because it can be expressed as $\frac{8}{1}$. The fractions $\frac{10}{15}, \frac{8}{12}$, and $\frac{4}{6}$ all represent the rational number $\frac{2}{3}$. These fractions are not separate rational numbers, they are all equivalent expressions of the same rational number. The expression $\frac{2}{3}$ is one number (not two), the numerals 2 and 3 represent parts (or terms) of the fraction symbol for that number. The rational number $\frac{1}{2}$ can be expressed in decimal form (*decimal fraction*) \, such as .5 or .50 or .500. A mixed decimal includes an integer and a decimal fraction as in 3.94. A mixed number is composed of an integer and fraction, such as $2\frac{2}{3}$ (which means $2 + \frac{2}{3}$). The rational number $\frac{1}{3}$ can be expressed as a repeating decimal fraction: $0.3\overline{3}$. But decimal fractions that are not repeating or terminating (finite), such as π or $\sqrt{2}$, are irrational (not rational).

Rational numbers can also be expressed in percent form: 50%, 25.16%, or $33.3\overline{3}$%. The word *percent* means *for each hundred*, so two decimal places are implied. The expression 50% means 50 per one hundred or $\frac{50}{100}$ or .50 or .5. These equivalent forms also refer to the fraction $\frac{1}{2}$. A more precise definition proposed by Behr et al. (1992, p. 296) characterized rational numbers as "elements of an infinite quotient field consisting of infinite equivalence classes, and the elements of the equivalence classes are fractions." One conceptual problem for students (and many adults) is that rational numbers cannot be counted sequentially. There are an infinite number of rational numbers between each integer (density).

There is some agreement in the literature on five types of rational numbers (Behr et al., 1992; Kieren, 1980): part-whole comparisons, quotient, measure, ratio, and operator, although different terms for these types may be used. Students often have trouble with rational-number concepts because they have not had experience with applications of all five types or their experience was too abstract, not connected to concrete models, real-world applications, or their informal mathematics understanding.

Part-Whole Comparison

Part-whole comparison is the comparison of one or more equal parts of a unit to the total number of equal parts (fraction of a whole). Part-whole comparisons use the familiar fraction notation with the part as numerator and whole as denominator. This is typically the first type of fraction introduced to students, because it is more common in their informal understanding. We can share this sandwich by cutting it in half. If we cut this pizza into four parts, we each have a fourth. A quarter is a fourth of a dollar; it takes four quarters to make a dollar.

But part-whole comparisons are not always as simple as slices of pizza. The unit may be one object (continuous) such as one rectangle or one circle. In that case:

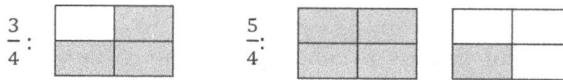

$$\frac{3}{4}: \qquad \frac{5}{4}:$$

If the last diagram above represented one unit rather than two, then the fraction for the part shaded would be $\frac{5}{8}$. The same diagram could be interpreted as $\frac{5}{4}$ or $\frac{5}{8}$, depending on the unit. When the unit is comprised of more than one object as in sets (discrete items), it is often easier to consider familiar units such as one group of children or a bag of marbles. Discrete objects comprising one unit that come in familiar groups are called *composite units*, such as six-packs of soda or a dozen eggs. Students need practice with this *unitizing* or thinking about and visualizing a unit in different-sized chunks (Lamon, 1999). The guiding question for part-whole comparison is, "What is the unit?"

It is typically while working with part-whole comparisons that students are introduced to the concepts of numerator and denominator in the symbolic construction of a fraction (Box C.1). These are not the top number and bottom number; these are the terms of *one* number. Later students will need to consider these parts when comparing fractions, finding factors, performing operations with fractions, and working with expressions and equations. This new syntax or form for representing a number can be confusing and misleading for students comfortable with whole-number representations. The symbolic representation $\frac{1}{2}$ refers to one value, not two. Likewise, $4, \frac{4}{1}$, and $\frac{8}{2}$ all refer to the same, singular value. This is why it is *essential* that students working with whole numbers understand the concepts of unit, value, and equivalence.

Box C.1   Parts of a Fraction

$$\frac{numerator}{denominator} = \frac{portion\ of\ whole}{total\ parts\ of\ whole}$$

$\frac{3}{4}$ is one value that represents 3 of 4 parts of a unit

---

Quotient

The quotient view of rational numbers includes part-whole comparisons (partitioning) and a representation of one share or part (sometimes termed *equal shares*). The quotient is the division of the number of objects by the number of shares. The guiding question is, "How much is one share?" For example, eight children have four sticks of licorice. What part will each child receive? The answer is expressed as division: $\dfrac{4 \, sticks}{8 \, children} = \dfrac{1}{2}$ stick for each child. Many textbooks combine part-whole and quotient views into one, equal shares, representation. What is important is that students see the connection between fractions and division. When they encounter division involving fractions, as in $\dfrac{1}{2} \div \dfrac{1}{3}$, they will be more likely to consider partitioning by division than some mindless algorithm that has no conceptual meaning. The quotient $\dfrac{3}{2}$ will make sense.

Measure

The measure concept of rational numbers is best visualized using a number line, but students should be exposed to other situations such as paper strips, ribbon, and other linear views. Between any two integers along a number line are an infinite number of rational numbers. This is a difficult concept for students comfortable with whole numbers that can be counted, with each followed by another integer. With fractions and decimals, we can compare magnitudes of numbers between or across integers (e.g., $\dfrac{1}{3}$ and $\dfrac{1}{4}$, 0.057 and 0.07, $\dfrac{4}{5}$ and $\dfrac{5}{3}$), but we cannot name the next number as with counting.

Figure C.1 illustrates the division of the interval between whole numbers 0 and 1 into two, four, and eight equal parts (and between 1 and 2). Students can compare the fractions $\dfrac{1}{4}$ and

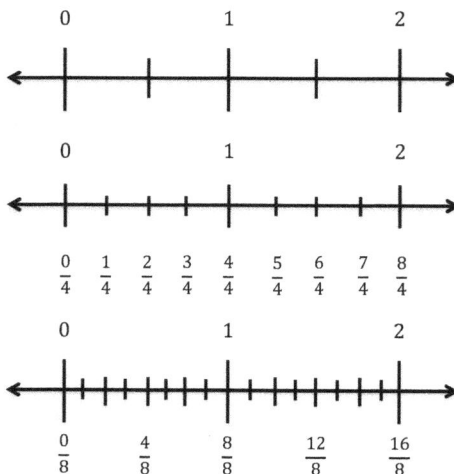

Figure C.1 Measure View of Fractions

$\frac{2}{8}$ on parallel number lines. They can compare the measures of $\frac{3}{4}$ and $\frac{7}{8}$, often very difficult to sketch by hand using rectangles or circles. This measurement model of rational numbers is excellent for recognizing equivalent fractions, developing a sense of magnitude, comparing fractions, and understanding the effects of addition and subtraction on fractions. This model also assists students with concepts of exactness and estimates. A guiding question for using measures is, "In this situation, can a number line be used to illustrate the relationship?"

The measure representation for rational numbers requires a strong understanding of unit. For example, a number line may indicate the positions of 0, 1, and 2 as in Figure C.1. A number line the same length could be labeled with 0, 5, and 10; or 0, 0.5, and 1. A unit on a number line can be very long or very short, depending on the labeled reference points. Typically a unit is the distance between 0 and 1 with subunit intervals divided into segments of the same length. Some students may confuse this coordination of different levels of units and varying lengths of units with the constant units on actual rulers (Zhang et al., 2017). Rulers have precise unit lengths and are often divided into 8ths or 16ths (or ten parts for centimeter rulers), while number lines can be drawn to show longer or shorter units, with subunits based on the denominator of a fraction. The ruler example should not be overemphasized in fraction instruction. Most researchers recommend that educators in Grades 3 to 8 emphasize the measurement model and number-line representation for assessment and instructional tasks, while also providing examples using the other forms (e.g., Fuchs et al., 2017; Obersteiner, Dresler, et al., 2019).

Ratio

Ratio is also a comparison, but between two quantities. There is one teacher for each group of 20 children. This state has two registered cars for every three people. Some ratios compare part to part (5 boys to 6 girls) and some compare part to whole (5 boys in the group of 11 children). A special and powerful type of ratio is *rate*, comparing measures of different types such as $1.90 per gallon, 60 miles per hour, and 80 heartbeats per minute. In science, as depicted in Strand E, some measures are actually ratios: speed, density, ampere. To distinguish ratios, sometimes colon notation is used (3:2) and sometimes the compared measure is hidden in the notation (45 mph). It should be noted that ratios are not always rational (1:$\sqrt{2}$ and 5:0). Ratios may not be operated on in the same way as fractions. Some can be scaled down or up, but one must keep in mind what the parts represent. Students should consider the context of a problem before applying operations arbitrarily. Adding further to the confusion are the careless uses of the terms ratio and rate in everyday language. A guiding question for ratios should be, "What type of measure does each term represent?"

Operator

The operator view of rational numbers is considering the rational number as a function or actively applied to or transforming something. Operators can enlarge, reduce, multiply, divide, increase, and decrease other values. For example, applying $\frac{3}{4}$ to a garment's original price of $25 involves the process of taking $\frac{3}{4}$ of $25 or $18.75. (In this case, *of* means to multiply.) How would we return the garment to its original price? By using the operator $\frac{4}{3}$, or the inverse, the operation becomes $\frac{4}{3}$ of $18.75, resulting in $25.00. Consumers are often confused by percent operators. If the item priced $100 was marked down twice, 25% each time, it would not be on sale for $50—the new price would be $56.25.

Students who were taught that multiplication results in a larger number will be confused by the results of some rational numbers as operators. Multiplying a whole number by a *proper fraction* (e.g., $\frac{1}{4}$) will result in a smaller quantity; by an *improper fraction* (e.g., $\frac{5}{4}$), a larger quantity. Lamon (1999) proposed an exchange model for operators. The operator $\frac{2}{3}$ means that for every three items you give or put in, two will come back, or "two for three." The function-machine concept can help students visualize the operator exchanging amounts. The guiding question for operator views of rational numbers is "Am I being asked to apply a fraction (or decimal, percent) OF another quantity?" Students should have experiences with different types of fractions—proper, improper, and mixed forms—as well as their decimal and percent forms and recognize the results of multiplicative operators on these forms. Some researchers recommend that teachers phrase their discussions of whole number multiplication in a way that can also be used with fractions (Siegler & Lortie-Forgues, 2015). For example, 5×3 can be described as "How much is 5 of the 3s?" Likewise, the problem $\frac{1}{3} \times \frac{1}{5}$ can be described as "How much is $\frac{1}{3}$ of the $\frac{1}{5}$ s?"

Related Concepts

Concepts closely related to fractions, decimals, percents, and ratios are those of proportion and probability. These concepts are critical foundations for algebra, data analysis, and higher-level problem solving.

Proportion

Proportion is one of the most useful concepts for mathematics problem solving and is based on the concept of ratio. As discussed in a previous section, a ratio is the comparison of one quantity with another, the relative sizes of two or more values. A proportion is the relationship of two ratios. A direct proportion is when two ratios are equivalent; inverse proportion involves one quantity increasing while the other decreases (Sidebotham, 2002). For example, 1 is to 2 as 3 is to 6, resulting in the proportion $\frac{1}{2} = \frac{3}{6}$. This direct relationship between elements in a proportion becomes particularly useful when one element is unknown. By applying the rules of mathematics, or properties and *axioms* \, a student can solve for the unknown, regardless of its position within the proportion.

> *We have 6 pizzas for our class of 24 students. How many pizzas would be needed for 30 students?*
>
> 6 is (related) to 24 as the unknown (x) is (related) to 30: $\frac{6}{24} = \frac{x}{30}$.

This problem can be set up with a different proportion:

> 6 is to the unknown as 24 is to 30: $\frac{6}{x} = \frac{24}{30}$.

Proportional reasoning is more than setting up and solving a proportional equation. It is thinking about and comparing the relationships depicted in a variety of ways, through graphs, charts, and equations. It is a way to study and predict changes in the relationship. Some proportional relationships are simple and direct such as using one cup of water for every one-half cup rice. You can reason how much water you would need with two cups of rice. Both quantities change in the same direction. An example of an inverse relationship: we have more workers today so the job will take less time. Proportional relationships also can be complex, with multiple variables changing in different ways.

Probability

Probability is closely related conceptually to proportional reasoning and ratio yet has been treated as a separate domain in most textbooks. Probability refers to the chance of something happening and is useful for a wide range of applications, including many in science, social science, sports, and, of course, gambling. It is often paired with statistics because data sets are used with both domains and within statistics the characteristics of data sets intended to predict or generalize include confidence intervals (errors of measure) that represent the probability of a given event, view, or other situation.

Working with probability problems requires knowledge of the sample space (the set of all possible outcomes) or at least a good estimate of that value. The probability of an event is equal to the number of times something can occur divided by the number of events that could possibly occur. For example, the probability of selecting a red playing card is $\frac{26}{52}$ or $\frac{1}{2}$.

Odds, on the other hand, refer to the probability of one event divided by the probability of the opposite event. For example, the odds in favor of selecting a diamond from a deck of cards is: $\frac{1}{4} \div \frac{3}{4} = \frac{1}{4}\left(\frac{4}{3}\right) = \frac{4}{12} = \frac{1}{3}$. Concepts of probability and odds are best explored with hands-on experiments where students record findings in data charts. Dice, coins, spinners, and random data tools are often used for these explorations. Find more discussion about probability in Strand E.

Curriculum Sequence

A great deal of research in recent years on learning progressions for fractions and related concepts has informed current curriculum sequences in schools. Both the *Common Core State Standards for Mathematics* (CCSSM, NGA, & CCSSO, 2010) and the National Council of Teachers of Mathematics' *Focal Points* (NCTM, 2006) encourage early fraction-concept development primarily through language in the domains of geometry and measurement. Students in kindergarten, first, and second grades use fraction language—half, fourth, third—to describe partitions of equal shares of shapes. Students are actively comparing shapes and portions of those shapes. NCTM standards recommended that students use standard notation for common fractions (e.g., $\frac{1}{2}, \frac{1}{3}, \frac{1}{4}$) by the end of second grade. Most standards recommend that students build on their informal knowledge through involvement with equal-sharing activities and delay using symbolic notation until students have firm fraction concepts (third grade for some students). The CCSSM recommended first graders partition circles and rectangles into equal shares and use language to describe those shares (primarily unit fractions, fractions with 1 in the numerator). Second graders extend this partitioning to more complex portions such as two-thirds and four-fifths and describe equal shares of identical wholes with different shapes (comparing parts of rectangle to those of a diamond shape).

In third grade, students should shift to a measurement model of fractions, typically using a number line. They begin with fractions with denominators 2, 3, 4, 6, and 8 and locate fractions on the number line, make magnitude comparisons, and generate simple equivalent fractions. The measurement model is more powerful for assisting students in understanding that fractions are an extension of their whole-number concepts and for developing deeper understandings of fraction magnitudes, operations with fractions, and for connections with decimals (Siegler et al., 2011). Third-grade students' experiences in other domains such as geometry, measurement, and data analysis also integrate fraction and decimal understanding.

Fourth graders work with addition and subtraction of fractions with like denominators and begin expressing fractional parts using decimal notation (have denominators of 10 or 100). They compare decimal magnitudes and should find decimals on the number line. Fourth graders also begin working with multiplication of a whole number and fraction, which is fundamental for algebraic concepts (e.g., $5 \times \frac{2}{3} = \frac{5}{1} \times \frac{2}{3}$ which equals $\frac{10}{3}$). It is in fourth grade when many students are identified with mathematics learning disabilities (MLD), as their whole-number system is still faulty, lacking conceptual understanding and fluency, and the movement into other rational-number systems isn't connected or well developed. Screening in earlier grades should have predicted these difficulties so that targeted interventions could be provided earlier.

Fifth graders add and subtract the range of fractions and begin multiplication and division. Students are expected to use all four operations using fractions and decimals (to hundredths) in problem solving and with geometry and measurement. Students are expected to be completely fluent with fraction operations and conversions to other equivalent forms by the end of fifth grade. Development of fraction and related decimal concepts over essentially two and one-half years will not be sufficient for some students.

Sixth and seventh graders work primarily with ratio, rate, percents, and proportional relationships. They convert fluently between various forms of rational numbers and apply properties and operations to algebraic expressions and equations. By the end of seventh grade, students' understanding of the full system of rational numbers should be solid, in preparation for eighth-grade work with equations, functions, and fractional exponents $(a^{\frac{1}{2}})$, as well as applications in high-school coursework involving algebra, geometry, measurement, and statistics. Many students who struggle with mathematics will not have mastered rational number concepts by eighth grade (Mazzocco et al., 2013; Siegler & Lortie-Forgues, 2017). Even with excellent instruction, students with MLD may need more time and experiences to develop rational-number concepts that are robust enough for algebraic and other higher-level concepts. Mazzocco et al. (2013) recommended that students with MLD receive stronger grounding in fraction concepts that underly rules, procedures, and strategies, that teachers not just move on to the next topic. They also recommended continued instructional supports, such as visual representations, while concepts and abstract symbol notations are developed, even through eighth grade.

CCSSM standards introduce fractional exponents (e.g., $a^{\frac{1}{2}}$) in the eighth grade and continue their use through high-school coursework, while NCTM does not include those types of exponents until high school. Concepts involving fractional exponents and radicals depend on prior development of exponent, root, negative number, and fraction concepts, so they are quite challenging. The expression $a^{\frac{1}{2}}$ can be translated into the square root of a: $\sqrt{a}$. The expression $b^{\frac{1}{3}}$ can be translated into the cubed root of b: $\sqrt[3]{b}$. And $9^{\frac{2}{3}} = \sqrt[3]{9^2}$. The denominator of the exponent becomes the (index of) root and the numerator the power. High-school students work with the properties and operations of fractional exponents. Radical expressions and fractional exponents have real-world uses in finance, electrical engineering, the sciences, and construction.

Research on Students with Learning Difficulties

Research comparing the knowledge and performance of students with mathematics learning disabilities (MLD) and other low-achieving students (mathematics difficulties, MD) with that of typically achieving students (TA) indicates that rational-number achievement is a significant problem for most students with MLD and MD, preventing many of these students from accessing algebra and other higher-level topics. In a study of fourth- through eighth-grade students, Mazzocco et al. (2013) found that on recognizing representations of one-half (symbolic and representational formats), TA and MD groups tended to reach ceiling performances (mastery) by fourth and fifth grade, respectively, but students with MLD did not begin to approach the criterion until seventh grade and demonstrated only 75% accuracy by eighth grade. The authors concluded that students with MLD have a persistent problem with symbolic notations of fractions through middle school and may need visual representations longer, if those promote concept understanding.

The Delaware Longitudinal Study of Fraction Learning (Jordan et al., 2017) followed the fraction instruction of 536 students from Grades 3 to 6. The researchers focused on knowledge of relative magnitudes of whole numbers, other mathematics skills (math calculation, nonsymbolic-proportional reasoning), attention, and working memory as possible predictors of fraction achievement and growth. The researchers concluded that fraction learning involves a range of cognitive processes and math-specific skills. The ability to estimate placement of whole numbers on a number line in Grade 3 was the strongest predictor of fraction knowledge (concepts and procedures) at the end of Grade 4. Attentive behavior, calculation fluency, verbal (language), and nonverbal (visual-spatial) abilities also made significant contributions. Whole number line estimation and attention were also predictors in Grade 5 for Grade 6 fraction outcomes, in addition to nonsymbolic-proportional reasoning, long division, and working-memory abilities. The researchers concluded that a significant number of students (42%) had persistent difficulties with fractions and demonstrated low growth in fraction knowledge, predicting later mathematics failure. The lowest-performing students with minimal growth had poor calculation fluency, weak classroom attention, and inaccurate whole number line estimation skill in Grade 3. About half the students with low growth in fraction concepts had average growth in procedures, indicating they performed procedures by rules without understanding the concepts involved.

Siegler and Pyke (2013) examined the fraction knowledge of students with MD and their TA peers in Grades 6 to 8, using fraction magnitude tasks on number lines (estimating fraction locations on 0–1 and 0–5 number lines and comparing two fractions between 0 and 1) as well as fraction arithmetic problems. Students with MD were less accurate and used less sophisticated strategies than their peers. A persistent problem was basing number-line estimates on a single factor (numerator or denominator). Students with MD also tended to add and subtract numerators and denominators separately, use the wrong fraction operations (e.g., mixing rules for addition and multiplication in parts of an operation), and demonstrate correct responses inconsistently on the same types of problems. The researchers concluded that students with MD started middle school with less knowledge about fraction arithmetic and they made slower progress from Grade 6 to 8, falling even further behind.

Issues with fraction understanding can persist into adulthood. Lewis (2016) conducted clinical interviews with two adult students (18 and 19) with MLD. Both students demonstrated difficulties on comparison problems with the same denominator and comparisons with the fraction $\frac{1}{2}$. These errors were associated with three persistent misconceptions: fractional-complement understanding (tendency to attend to the fractional complement rather than quantity), single-factor understanding (comparisons based on single term, most often denominator), and viewing the fraction $\frac{1}{2}$ as halving something (operator) rather than as a quantity. For example, a fractional-complement interpretation would lead to misinterpreting $\frac{1}{3}$ as $\frac{2}{3}$ with area models of the fraction. Single-factor understanding in comparing two fractions, such as $\frac{3}{4}$ to $\frac{2}{6}$, might lead to comparing only denominators, considering $\frac{2}{6}$ larger. Halving a diagram would involve cutting it in half, without considering the internal portions. Powell and Nelson (2021) assessed the rational-number knowledge of 331 undergraduate students. Although these students averaged 76% on the measure, many error types were noted: procedural errors (e.g., finding least common denominator and greatest common multiple when required, finding a reciprocal), miscalculations (especially with numerators, multiplication, and division with fractions and decimals), and word-problem solving (procedural errors, miscalculation, and unreasonable answers). These are concept issues also prevalent across the grade levels.

These studies and others suggest that the development of rational-number understanding for students with MLD and other low-achieving students will take longer than that of other students. These struggling students need time to develop concepts and not jump too quickly into purely symbolic-level work. Some students may need supports—such as visual representations and tools that assist working memory and attention—in order to achieve at expected levels. They need more explicit connections between representations and more examples of rational-number comparisons and operations.

Common Difficulties and Instructional Strategies

Rational numbers are among the most difficult concepts for students to understand and apply. For example, on the *National Assessment of Educational Progress* (NCES, 2019), only 32% of fourth graders could compare fractions such as $\frac{1}{3}$, $\frac{2}{3}$, and $\frac{4}{6}$ to $\frac{1}{2}$, applying less than, equal to, and greater than. Only 58% could match a decimal number (e.g., .02, .20, .25, 2.0, 2.5) to shaded 100s grids and only 70% could place a decimal number (2.6) on a number line. For eighth graders, only 27% could locate decimal fractions on the number line and a decimal fraction in the middle of those (e.g., .8 and 1.4). Only 47% could calculate percent off a purchase price. In the past 30 years, little progress has been made in students' knowledge of rational numbers (Tian & Siegler, 2018). Competence in fractions in Grade 5 was demonstrated to be a unique predictor of gains in mathematics knowledge by Grade 10 (Siegler et al., 2012). Rational-number knowledge is critical for success in high school,

a) Place <, +, or > between each pair of fractions (e.g., Schumacher et al., 2015).

$$\frac{3}{4} \quad \frac{3}{6} \qquad\qquad \frac{3}{4} \quad \frac{6}{8} \qquad\qquad \frac{3}{4} \quad \frac{3}{2}$$

b) Order these numbers from smallest to largest (e.g., Van Hoof et al., 2015).

$$3.682 \qquad\qquad 3.2 \qquad\qquad 3.35$$

c) Indicate where each fraction should be placed on the number line (e.g., Siegler et al., 2011).

$$\frac{3}{8}$$

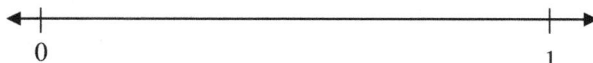

$$\frac{11}{4}$$

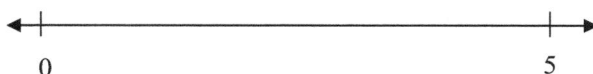

d) Circle the correct operation (e.g., Van Hoof et al., 2015).

I paid 4 dollars for ¾ pounds of candy. What is the price of 1 pound of candy?

A. $4 \div \frac{3}{4}$ B. $4 \times \frac{3}{4}$ C. $\frac{3}{4} \div 4$

Figure C.2 Example Items for Screening Rational Number Understanding

college, and in many careers, such as medical, construction, social sciences, business, and the arts. Therefore, the emphasis on developing rational-number understanding should be increased within each grade level, connected with other domains (e.g., measurement, geometry, statistics), and with targeted interventions as soon as screenings identify risks for low achievement. Example items for screening for rational-number concepts are depicted in Figure C.2.

Even average-achieving students struggle with rational-number concepts, but for a range of reasons. Some researchers pointed to the complexity of the domain—rational numbers are not absolute; they represent relationships that are inherently complex (Baroody & Hume, 1991; Siegler & Lortie-Forgues, 2017). Students must understand the meaning of individual rational numbers, their notations, and their relationship to other rational numbers (including whole numbers) and their properties and operations. Other researchers explored the effects of the many underlying structures (number sense, spatial sense, multiplicative reasoning, proportional reasoning) and subconstructs (measure, quotient, ratio, operator, part-whole) that challenge students' understanding of rational numbers (Grobecker, 2000; Kieren, 1988). Some researchers identified *whole-number bias* as students move from working with whole-number concepts to the full range of rational numbers (Behr et al., 1992; Vamvakoussi, 2015). This bias is most likely due to incomplete models of rational numbers (too much emphasis on part-whole models) and lack of instructional attention to similarities and differences in natural and rational-number properties. Some students with MLD are affected more severely by this bias than their peers, equally strong for fractions and decimals, perhaps due to problems with inhibition of whole-number concepts (Van Hoof et al., 2017). Teachers should attend to the significant conceptual change involved when beginning instruction with rational numbers, using aspects of prior knowledge, analogies, representations, concept connections, and language to scaffold this change.

Another often-identified factor for low rational-number understanding is students' lack of access to quality instruction by teachers knowledgeable in rational-number concepts and effective pedagogy specific to the domain (Obersteiner, Dresler, et al., 2019). Study after study with preservice and practicing teachers identified major conceptual and instructional issues for many educators (Depaepe et al., 2018; Siegler & Lortie-Forgues, 2015; Utley & Reeder, 2012). Teachers had difficulty explaining the concepts supporting procedures, in reasoning about multiplication and division of fractions, in using clear and accurate language to describe fraction concepts, in presenting and interpreting fraction and decimal representations, and in developing problem situations that represented computations. Teachers tended to have some of the same types of problems with rational-number concepts as their K-8 students and, in some cases, had deeper misconceptions (Park et al., 2013; Rathouz, 2011; Siegler & Lortie-Forgues, 2017). Woodward (2017) emphasized that it is particularly important that teachers providing Tiers 2 and 3 interventions be knowledgeable in both content and specific pedagogy for teaching about rational numbers. Generic effective-instructional approaches and small group size will not be enough to effect improvements. The following sections discuss the most common errors and misconceptions related to fraction, decimal, percent, ratio, and proportion concepts. After each discussion of difficulties is a list of recommended strategies for instruction and intervention.

Fractions: Common Difficulties and Instructional Strategies

Fractions are the earliest nonintegral rational numbers in the mathematics curriculum. When students cannot connect fraction concepts with those of whole numbers or have not had the time and supports to develop accurate fraction concepts, they may fall further behind their peers. The most common difficulties involve the conceptualization of fractions—making the shift from whole-number reasoning and representations. Some students still lack a firm understanding of whole-number concepts that do extend to fractions: commutative, associative, identity properties; zero and units on the number line; factors and multiples, and equivalence concepts. Many students overgeneralize whole-number operations across rational numbers such as thinking multiplication will result in a larger number, division a smaller one. Both similarities and differences between the whole-number system and the full rational-number system should be emphasized during instruction (Obersteiner, Dresler, et al., 2019; Van Hoof et al., 2015).

One of the most significant predictors of fraction achievement is strong understanding of fraction magnitude (Jordan et al., 2017; Obersteiner, Reiss, et al., 2019; Siegler et al., 2012). If students lack this basic understanding, it will impact their ability to compare and order fractions mentally or on the number line. Malone and Fuchs (2017) found that 81% of at-risk fourth graders (< 35% percentile in mathematics) made fraction-ordering errors, with whole-number bias and ordering by size of terms the greatest obstacles. These students also demonstrated poor part-whole understanding. A related concept—understanding fraction equivalents—impacts students' abilities in working through algorithms, solving problems, and general reasoning about fractions (Jordan et al., 2017). Knowing equivalent fractions for

$\frac{1}{4}, \frac{1}{2}, \frac{1}{3}, \frac{2}{3},$ and $\frac{3}{4}$ at a minimum will reduce the burden on students' working memory and provide benchmarks for using number lines.

Some students lack understanding of fraction symbols and their concepts. They confuse the terms within a fraction (e.g., 3 and 5 in the number $\frac{3}{5}$) and the meaning of mixed numbers (e.g., $2\frac{5}{8}$). What is the unit of the fraction $\frac{5}{4}$? These misunderstandings lead to procedural errors such as applying the wrong algorithm, not being able to set up an algorithm to represent a problem, not completing the steps, and not understanding the relationships among fraction algorithms and those of whole numbers. Some students erroneously add the numerators and denominators in a multiplication problem ($\frac{1}{2} \times \frac{1}{8} \neq \frac{2}{10}$). Others will try to cross multiply or invert and multiply! These procedural errors are due to incomplete concept knowledge and connections. Braithwaite and Siegler (2018) found that students tend to choose addition and subtraction procedures when denominators are the same and multiplication/division procedures when they are different. The researchers posited that these spurious associations are due to how textbooks present practice problems.

Students with MD often struggle with reasoning about fractions and problem solving. This situation is due to several factors: lack of practice with sufficient problem types during instruction, algorithms taught in isolation, not enough time with concrete models and diagrams during concept development, and lack of a toolbox of generalizable strategies (e.g., partitioning, simplifying). Further, when faced with word problems involving multiplication and division, many students fall back on additive reasoning instead (Fuchs et al., 2017). Reasoning skills, including adaptive expertise, are the most valuable skills for solving novel and higher-order problems and determining the soundness of answers. McMullen and colleagues (2020) described adaptive expertise as "richly connected knowledge that can be flexibly applied in novel contexts," rather than isolated skills (p. 1). In a study of seventh and eighth graders, the researchers found that adaptive expertise for rational numbers, over and above strong procedural and conceptual knowledge, predicted later algebra knowledge and that only 10% demonstrated adaptive expertise.

Many difficulties with fractions are systematic, persistent over time, and are found in different contexts (Obersteiner, Dresler, et al., 2019). Some difficulties are related to the inherent nature of fractions as a complex number system. Others are related to teacher knowledge, both content and pedagogical, and curriculum materials. Many students with MD have deficits in working memory, language skills, prior knowledge, and reasoning abilities that impact their fraction learning and application. There is no doubt that the conceptual change needed to understand fractions requires competent teachers and more instructional time to develop concepts and their connections.

Instructional Strategies for Fraction Concepts

Many students, especially those in third through fifth grades, will need intensive fraction interventions identified through screening or observation. An example intervention is depicted in Box C.2. Specific strategies for teaching about fraction concepts, procedures, and problem solving include:

Box C.2   Intensive Intervention for Fraction Magnitudes

Target: Three fourth-grade students demonstrate difficulties with fraction addition and subtraction. Upon further screening, teachers found these students have little conceptual understanding of fraction meaning or magnitudes.

Intervention Schedule: 20-minute sessions, three times a week until mastery (estimate 10–12 weeks).

Nature of Intervention: Each session begins with quick multiplication or fraction naming practice and ends with review of major concept with one practice problem. All sessions include student explanations and immediate feedback. Sequence of topics:

1. Using manipulative objects (connecting cubes, strips of paper), students name the unit and fraction terms for examples. Units include sets, connecting cubes (discrete), and folded paper (continuous). Fraction symbolic notation is introduced and written on portions of paper strips.
2. Working with number lines (0–1 and then 0–2), students practice placing benchmark fractions. They discuss their reasoning and compare fractions with denominators 2, 3, and 4, later adding 5, 6, and 8.
3. Students practice comparing simple fractions using symbols, drawing number lines as needed. At least one equivalent fraction for each fraction is explored using parallel lines.
4. Students practice comparing fractions between 0–2, including improper and equivalent fractions. They convert between improper and mixed fractions in their notations. The number lines vary in length.
5. Practice with comparing fractions is interspersed with simple word problems involving comparing. Practice with 0–5 number lines. Adding and subtracting fractions with like denominators, with focus on why the denominators must be the same to add or subtract.

Data Collection (Weekly): CBM of fraction magnitudes, later adding addition and subtraction with like denominators and related word problems.

---

*   Develop fraction concepts before introducing symbols. Draw on students' informal knowledge of halves, fourths, and thirds, and use everyday objects and activities to apply those labels.
*   Introduce part–whole fraction concepts using concrete and representational devices that will emphasize important aspects of fractions accurately: the parts must be equal sizes, the unit must be named for the denominator, the numerator indicates the selected portion of equal parts. The part-whole representation of fractions is concrete but has many limitations including poor modeling for negative numbers, fractions with large terms, improper fractions, and equal parts (Siegler et al., 2011). Some students may focus on the inverse part unless portions are explicitly discussed and named. Focus on varying composite units to help students think about the unit (a unit of one egg carton—one portion is $\frac{1}{12}$, a unit of two egg cartons—one portion is $\frac{1}{24}$). What is the unit?
*   Use linear representations (number lines, paper strips, etc.) to connect rational numbers with whole-number concepts and to develop fraction magnitude and comparison

concepts as well as equivalencies and operations with fractions. In fact, number lines should be the primary vehicle for fraction instruction beyond the initial part–whole concepts in first and second grades.

- Encourage students to use familiar fraction benchmarks ($\frac{1}{2},\frac{1}{4},\frac{3}{4}$) for magnitude comparison, estimation, and problem solving. Fluency with these benchmarks will relieve working memory while computing and problem solving.

- When the situation arises within concept development for rule teaching, teach rules explicitly. For example, in a common fraction (e.g., $\frac{3}{5}$, when the numerator increases so does the magnitude (or value); when the denominator increases the magnitude decreases. These rules may cause overgeneralization, however, and need to be closely connected with reasoning skills—how is this fraction's value related to 1 and other benchmarks?

- For fraction equivalents, begin with $\frac{1}{2}$, a concept that is often in informal knowledge. Draw on students' multiplication knowledge to recognize $\frac{1}{2}:\frac{2}{4},\frac{3}{6},\frac{4}{8},\frac{5}{10}$ as equivalents. Have students develop diagrams or tables of other common equivalent fractions and use bar diagrams, number lines, or Cuisenaire® rods as representations (see Appendix II).

- Ask students to demonstrate fraction problems using manipulatives or drawings. In studying the visual representations created by middle-school students with MD, Morano and Riccomini (2020) identified bar models (rectangles) as the most efficient and effective for fraction instruction. These drawings were useful for word-problem-solving, number-line estimation, and fraction magnitude comparisons. Although most students attempted circular models, they struggled with making those drawings and using them for reasoning about fractions. Some students need explicit instruction in making and interpreting their drawings.

- Encourage estimation. For $\frac{3}{4}+\frac{1}{2}$, my answer will be more than 1 but less than 2. For $\frac{1}{2}\times\frac{5}{7}$ my answer will be only half of $\frac{5}{7}$ between $\frac{2}{7}$ and $\frac{3}{7}$ on my number line.

- Use fraction notations consistently. For example, avoid using ¾ when $\frac{3}{4}$ is a clearer symbolic representation that leads more directly to algebraic language.

- Spend a lot of time (and frequent review) with magnitude comparisons and related reasoning. These are difficult tasks, even for adults, but measures of fraction-magnitude comparisons correlate highly with fraction skill and overall mathematics achievement from fifth through eighth grades (Bailey et al., 2012; L. S. Fuchs et al., 2017; Obersteiner, Dresler, et al., 2019). Some researchers recommend delaying operations until students have a strong understanding of order and equivalence (Brown & Quinn, 2006; Siegler & Pyke, 2013). Examples of magnitude-comparison tasks:

  - Estimate the location of these fractions on the number line in Figure C.1: $\frac{1}{3},\frac{2}{5},\frac{7}{8}$, and $\frac{5}{3}$.

  - Place a <, =, or > between each set of fractions: $\frac{2}{5}$ $\frac{3}{7}$ and $\frac{4}{5}$ $\frac{2}{3}$ and $\frac{6}{12}$ $\frac{4}{8}$.

  - Order the following fractions from smallest to largest: $\frac{3}{2},\frac{1}{5},\frac{2}{7},\frac{3}{3},\frac{4}{5},\frac{5}{7}$.

a) Rewrite as multiplication problem.

$$\frac{9}{12} = \frac{\cancel{3} \times 3}{\cancel{3} \times 4} = \frac{3}{4}$$

b) Divide by a factor.

$$\frac{9}{12} \div \frac{3}{3} = \frac{3}{4}$$

c) Finding common denominators for $\frac{3}{4}$ and $\frac{1}{5}$.

Divide two rectangles—one into 4ths in one direction and the other into 5ths in the other. Twenty is a common denominator (4×5). Using the diagrams, we see that $\frac{3}{4}$ has the same value as $\frac{15}{20}$, and $\frac{1}{5}$ has the same value as $\frac{4}{20}$.

$\frac{3}{4}$

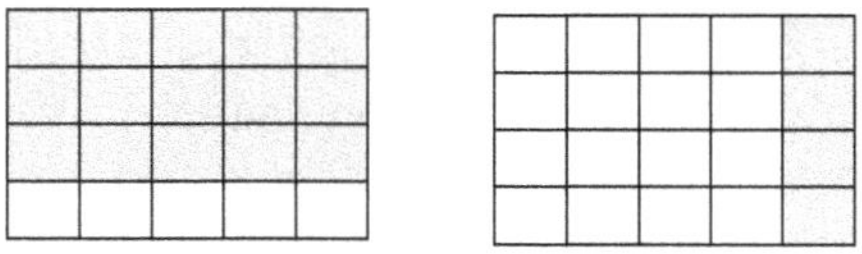

$\frac{1}{5}$

Adding results in $\frac{19}{20}$. Subtraction yields $\frac{11}{20}$.

d) 1897 (Wentworth) method for finding LCM and GCF* of the numbers 24 and 30:

2	**24**	**30**
3	12	15
2	4	5
	2	

LCM = 2 x 3 x 2 x 2 x 5 = 120
GCF = 2 x 3 = 6

*LCM was least common multiple as it is today; GCF (greatest common factor) was GCM or greatest common measure in 1897.

Figure C.3 Fraction Simplification Methods

- Teach effective strategies for fraction simplification to equivalent lower or lowest terms. Avoid calling this process reduction because that implies a lesser value. Lewis (2010) introduced a method of rewriting the fraction as either a multiplication or division problem. This and other methods for fraction simplification are depicted in Figure C.3. Fraction simplification and finding equivalent fractions are essential foundations for operations, ratios, proportional reasoning, and algebra.
- To prevent fraction learning from becoming too compartmentalized (and not connected with other mathematics), expose students to diverse ways of working with fractions and their representations. Moseley (2005) found that the types of perspectives that are embedded in the curriculum can have significant effects on students' understanding

of rational-number concepts as well as their connections with related concepts in mathematics.

- To encourage reasoning about fractions and inhibit whole-number intrusions, teach metacognitive prompts, such as "stop and think" (Obersteiner, Dresler, et al., 2019). Does this answer make sense? Can you explain your reasoning?
- Have students explain their thinking during initial learning, practice, and problem solving. Not only does this practice assist the teacher in identifying misconceptions, self-explaining and think-alouds can assist students' concept understanding if their explanations are scaffolded to include accurate vocabulary and concepts (Foreman-Murray & Fuchs, 2019).
- When teaching fraction operations, begin at the concrete level with valid materials such as strips of paper. Students can fold, label, compare, and manipulate these strips with careful teacher direction. Then move on to number-line representations of operations to make sure the concepts for procedures and their properties are firm before introducing operations at the symbolic level. Even at the symbolic level, some students need simultaneous visual representations (e.g., number line, bar model) to support their understanding.
- When teaching about adding and subtracting fractions, emphasize that the concepts are the same as for whole numbers (operations will have the same effects). Addition is composing and subtraction is decomposing. Begin with fractions of the same denominator and the number line as a representation for adding or subtracting equal sections. Next move to adding and subtracting with fractions and whole numbers, converting whole numbers to fractions with the same denominator. Using fraction strips and number lines, demonstrate that the units (denominators) must be the same in order to add or subtract. Finally, develop the concept of adding or subtracting with unlike denominators, using a CRA sequence to build understanding.
- Educators have used other valid models to emphasize the concepts of addition and subtraction of fractions. For example, one study used musical notation and temporal values of musical symbols for whole, half, quarter, and eighth notes to develop concepts of equivalence, adding and subtracting, and connecting number-line representations (Courey et al., 2012). Others have used carefully selected children's literature (e.g., *Fraction Fun*, Adler, 1996) and accurate schemas (diagrams of problems).
- Finding a common denominator for two fractions can be challenging. Begin with those benchmark fractions that have familiar equivalent fractions. Phelps (2012) used a rectangle, with the divisions in one direction for the first fraction and divisions in the other direction the second fraction (see Figure C.3). With larger terms, students can make lists of multiples until a common multiple appears (least common multiple).
- Multiplication of fractions should be connected with multiplication of whole numbers using similar linguistic phrasing (Siegler & Lortie-Forgues, 2015). "How much is $\frac{1}{3}$ of the $\frac{1}{5}$?" is comparable to "How much is $\frac{1}{2}$ of the 5?" Using fraction strips and number lines to illustrate multiplication further connects these operations. Emphasize and demonstrate the rules that multiplying two positive numbers less than one always results in a number less than one and multiplying two numbers greater than one always results in a number greater than one.
- Begin division of fractions with 1 divided by a unit fraction, such as $1 \div \frac{1}{3}$ (Cavey & Kinzel, 2014). Begin with concrete (fraction strips or towers) and representational models (number lines or fraction bars), then label the parts using symbols. This problem leads

to writing the corresponding multiplication statement: $3 \times \dfrac{1}{3} = 1$ (the inverse property).

Next, have students make a list of similar statements. Then move on to a problem such as $1 \div \dfrac{2}{3}$, and then to $15 \div \dfrac{2}{3}$, again with a CRA sequence. An instructional sequence through progressively more advanced forms of division with fractions will lead to stronger concept understanding. Develop division concept understanding so that when the "invert and multiply" algorithm is introduced, it makes conceptual sense.

- When faced with division problems, students should first consider whether a larger number is divided by a smaller number or the opposite (Siegler & Lortie-Forgues, 2015). If the larger number is divided by a smaller number $(\dfrac{5}{6} \div \dfrac{1}{6})$ say, "How many times can $\dfrac{1}{6}$ go into $\dfrac{5}{6}$?" If it is division of a smaller by a larger number $(\dfrac{1}{6} \div \dfrac{5}{6})$ ask, "How much of $\dfrac{5}{6}$ can go into $\dfrac{1}{6}$?" These linguistic phrasings are the same for whole numbers, so include whole numbers with practice.

Studies involving students with MLD or other low-achieving students have confirmed that effective interventions must include both understanding the concepts of rational numbers and their procedures. Hecht and Vagi (2010) compared students with MLD and typically achieving students from fourth through fifth grades (over two years). The researchers found significant group differences, with conceptual knowledge a significant mediator of group differences and an important predictor of achievement in computation, estimation, and word-problem solving. They also found bidirectional relationships between conceptual knowledge and fraction skills—each influenced the development of the other. Instruction that connected conceptual and procedural knowledge, especially if the links were explicitly taught, was especially beneficial for students with MLD.

The Fuchs research group at Vanderbilt University investigated the effectiveness of various instructional components for Grades 3 to 5 fraction interventions titled *Super Solvers* (Fuchs, 2017; Fuchs et al., 2016, 2021; Wang et al., 2019). Findings from a series of studies supports the importance of understanding the magnitude of fractions and using number lines for instruction, emphasizing concept understanding. The interventions also integrated multiplication practice, word-problem solving via schemas, attention to students' skill and cognitive deficits (e.g., working memory, attention), and systematic, explicit instruction. The students in the at-risk range (< 35th percentile) benefited the most from these Tier 2 interventions, while those with lower pre-intervention scores (< 13th percentile) improved but did not reach performance goals after 12 weeks. These studies underscore the importance of fraction-magnitude understanding for mathematics achievement and the need to address grade-level content while attending to foundational skills for elementary students with MD.

In a meta-analysis of studies with students with MLD (Grades 3–12), Shin and Bryant (2015) found that fraction interventions that included the CRA sequence, explicit and systematic instruction, a range of well-chosen examples, real-world problems, and strategies such as think-alouds resulted in improvement. However, few studies employed number-line instruction and other content-specific interventions that are essential for conceptual change. The authors emphasized the need for teachers to be content competent, to focus on conceptual and procedural understanding (not tricks), and to understand that students learn at different rates and some will need more time and intensive interventions.

Decimals and Percents: Common Difficulties and Instructional Strategies

Typical problems for students working with decimals and percents include many of the issues described with fractions. Whole-number bias also affects understanding these numbers, their magnitude, density, representations, and operations. Not having strong conceptual understanding leads to difficulties setting up problems, applying algorithms, recognizing equivalent values, and determining the reasonableness of results. In an early study of rational number judgments by middle-school students, Mazzocco and Devlin (2008) found that students with MLD had difficulty rank ordering both fractions and decimals and 83% could not read decimal numbers accurately, significantly lower than low-achievers (39%).

Many students misapply place-value concepts from whole-number understanding to decimal systems (Resnick et al., 2019; Siegler & Lortie-Forgues, 2017). You cannot count using decimal numbers; there are an infinite number of decimal numbers between any two numbers. If you add a zero to the end of a decimal number (e.g., 0.345 to 0.3450) it does not change the value of the number, but if you add a zero to the right of the decimal point (e.g., 0.0345) the number is smaller. Longer numbers (by digits) are not necessarily larger (e.g., $34.1078964 \not> 259$). Multiplying by 10 doesn't mean you simply add a zero (e.g., $4.5 \times 10 \neq 4.50$).

Magnitude understanding with decimals is a predictor of achievement but is less difficult than magnitude reasoning with fractions (Hurst & Cordes, 2015; Resnick et al., 2019). Students with MLD may take longer than their peers in developing magnitude comparisons (e.g., comparing and understanding decimal fractions $\frac{1}{10}, \frac{1}{100}, \frac{1}{1000}, \frac{4}{10}, \frac{25}{100}$) and making the transition to their decimal equivalents (0.1, 0.01, 0.001, 0.4, 0.25). Some students have difficulty comparing decimal numbers (e.g., 23.045 and 23.4), even with number-line supports. They lack concepts of the value of each digit in the decimal number and how to compare digits when comparing numbers.

Related to place-value and magnitude concepts are misconceptions associated with arithmetic operations with decimals. There is some evidence that these misconceptions are more persistent than those of fractions, possibly because of the time and emphasis devoted to decimal-number concepts (Hurst & Cordes, 2015). Students often incorrectly generalize decimal-point placements (Siegler & Lortie-Forgues, 2017). For example, in addition and subtraction the decimal points are lined up and brought directly down, while in multiplication they are placed in the product based on reasoning about the decimal placement of the factors. How are zeros used when multiplying 4.05 by 500? Even in addition and subtraction, students have difficulty with problems where decimal points and digits aren't neat as in 34.08 + 703.2402. Students have difficulty reasoning about their answers involving decimals, for example, in the problem above, should the sum be close to 1000, 740, or 100?

Difficulties with percents are related to those with fractions and decimals. Percents can represent a range of values; they are not just numbers between 0 and 100. How much is 125% of $10? Would it be more or less than $10? What does 0.005% mean? How do you interpret .03%? What about −10.2%? Percents also represent a portion of a unit, but the portion is implied with the wording (per hundred) and location of the decimal point. Many students have difficulties interpreting percent problems. *Fifty percent of 120 fifth graders went on a field trip.* Fifty percent (50%) implies multiplication by 0.50. A newspaper article included the phrase: *the budget portion in question was only one-fifth of one percent.* How can that amount be determined? Converting among fractions, decimals, and percents is difficult if concepts are not strong.

Instructional Strategies for Decimal and Percent Concepts

Decimal numbers seem similar to whole numbers and have a place-value structure as with whole numbers. However, there are an infinite number of decimal numbers between each integer. One conceptual problem many students have moving from fractions to decimal forms, including percents, is the repeating decimal, as in $0.3\overline{3}$ for $\frac{1}{3}$. Instructional strategies for decimal concepts include:

- Make a connection between fractions and decimal numbers by using decimal fractions: $\frac{1}{10}$ for 0.10, and the corresponding point on the number line for both symbols for the same value. The fraction $\frac{4}{100}$ translates as 0.04, $\frac{4}{5}$ is equivalent to $\frac{8}{10}$ and 0.8, and $\frac{15}{10}$ has the same value as 1.5 (and the mixed number $1\frac{1}{2}$). Students need a lot of practice moving back and forth among these equivalent values. Use a range of practice items beginning with familiar numbers.
- Another connection can be made by drawing on students' division concepts. The fraction $\frac{2}{5}$ can be reinterpreted as 2 divided by 5, or 0.4. This connection is important for understanding algebraic expressions, such as: $\frac{2x+30}{10}$ simplified to $\frac{x+15}{5}$ or $\frac{1}{5}x+3$.
- Use consistent decimal notation, with a zero in the ones position for decimal numbers less than 1. For example, 0.456, 0.05, and 0.3. Students should look first to the tenths' position (numeral to the right of the decimal point) for decimal interpretation and comparison. Cue the tenth's position by color coding. Ask, "Where do you look first when comparing the values of these numbers: 0.047 and 0.1695?"
- Practice decimal-number magnitude comparisons and locating decimals on the number line. These tasks are generally easier for students than those with fractions, because decimals have place value and a similar distance effect as with whole numbers (e.g., it is easier to compare 0.05 and 0.008 than it is to compare 0.1556 and 0.156). However, some students may be misled by the number of digits or positions of digits and zeros within decimal numbers. Use a number line with 10 sections between the 0 and 1 so that both fractions and decimals can be labeled (Shaughnessy, 2011).
- Use decimal grids to supplement number lines for representing decimal numbers. Start with the grid of tenths, but ask students to show two, three, and even four decimal places (D'Ambrosio & Kastberg, 2012). As shown in Figure C.4, decimal grids are flexible, but teachers need to emphasize meanings of various units represented. Ask, "What is the unit?"
- When using decimal grids to model addition and subtraction, color code each place so students can see they are adding tenths to tenths, ones to ones, and so forth (Cramer et al., 2009).
- Operations with decimals can be easier than those with fractions because of whole-number concepts. With addition and subtraction, teachers should focus on making sense, not just the rule to line up the decimals. When we add 0.537 + 0.03, will the sum be greater or less than one?
- Multiplication with decimals can be modeled with rectangular arrays and decimal grid paper (Rathouz, 2011). Students should understand why we move the decimal point in

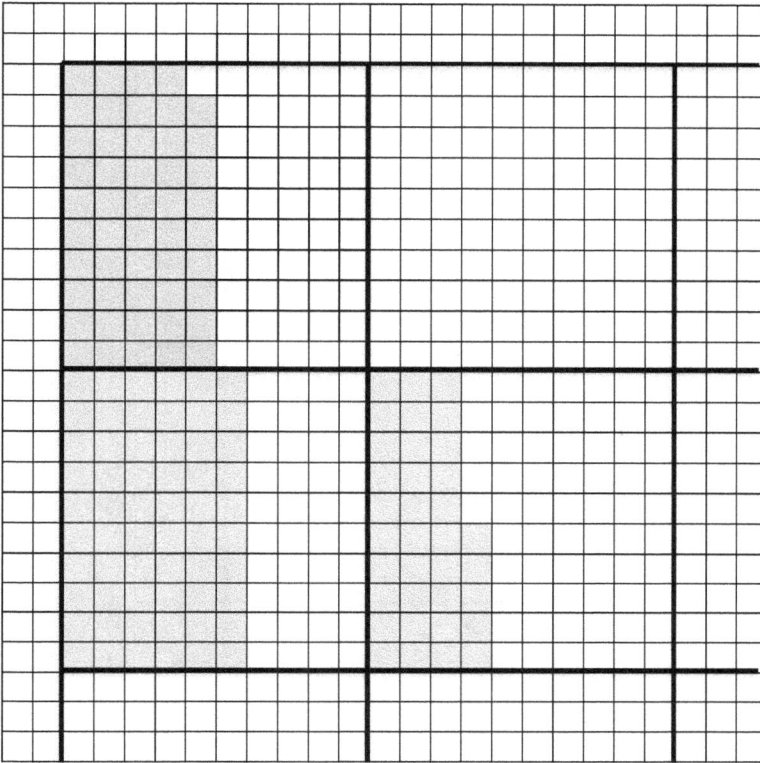

The unit of the first large square is one with rows and columns tenths and small squares within the grid hundredths. The first square shows 0.49.

The large squares on the bottom row are tenths and each row or column hundredths, and small squares thousandths. The two squares represent 0.06 + 0.035 = 0.095.

Figure C.4 Decimal Grid with Different Unit Values

problems such as those depicted in Figure C.5. Connecting with whole-number multiplication can assist understanding, as with the problem 143 × 2000. What was the effect of multiplying by 2000? What would be the effect of multiplying by 0.2 or by 0.02?

- Encourage students to use estimation before division of decimals. In the problem 100.5 ÷ 0.5, would the quotient be closer to 20, 200, or 2000? Practice with estimation problems using easier values to develop place-value concepts before introducing the "move the decimal" rule. Use number lines and decimal grids to support concepts.
- For the concept of repeating or recurring decimals, practice division problems with calculators: $\frac{1}{3}, \frac{5}{9}, \frac{12}{99}$. Demonstrate that the representation eventually becomes periodic (the same sequence of digits repeats indefinitely) and ways to record that decimal: $\frac{3}{11} = .\overline{27}$, with the line over the sequence of digits that repeat. Denominators that can have repeating decimals include 3, 6, 7, 9, 11, 12, 13, 14, and 15.
- Students should have exposure to decimal numbers before Grade 4 because decimal knowledge supports fraction understanding and reinforces place-value knowledge (Resnick et al., 2019). Students should practice reading, writing, and comparing decimal

2.1

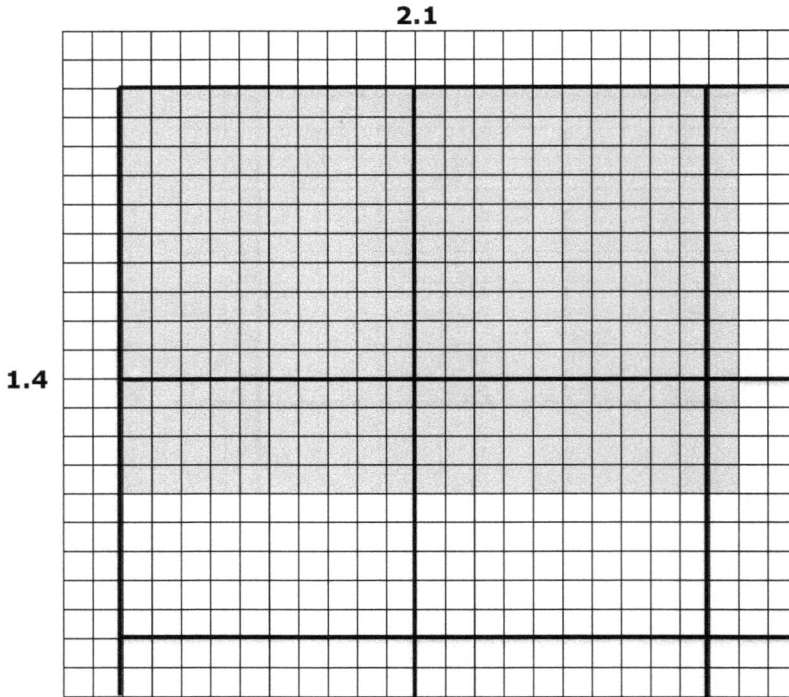

1.4

Area Model for 2.1 × 1.4. The horizontal depicts 2.1, while the vertical depicts 1.4, with each small square representing a hundredth, a row or column a tenth, and a decimal grid a whole.

Counting up for the product: 2 wholes, 9 rows/columns, and 4 small squares = 2.94

Division is the inverse: the area in the middle divided by one of the sides would equal the other side.

Figure C.5 Decimal Grid for Multiplication and Division

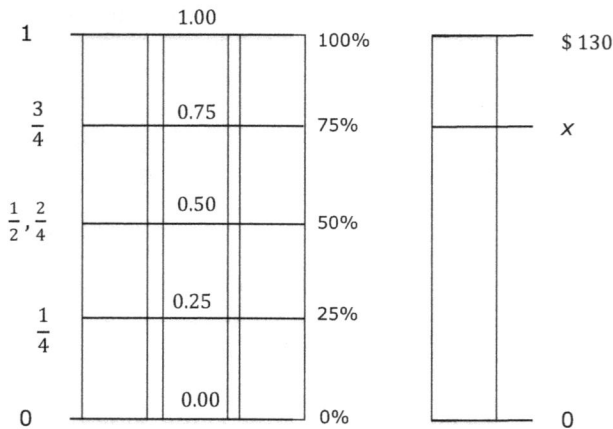

Compare equivalent numbers in fraction, decimal, and percent form.

Find percent of an amount: 75% of $130 = x.

Figure C.6 Visual Model for Fraction-Decimal-Percent Equivalents

numbers focusing on place-value structures, at least to the hundredths place, such as reading 0.25 as two tenths and five hundredths or twenty-five hundredths (not "point two five"). However, fourth-grade interventions designed to integrate fraction- and decimal-magnitude understanding may not offer an advantage over fraction-magnitude interventions alone (Malone et al., 2019). It may be that transcribing across two notation systems at once creates cognitive overload and students may need a more staged sequence of emphasis before integrating these concepts.

- Connect percent concepts with those of fractions and decimals by using the most common benchmarks: 25%, 50%, 75%, and 100%. Use visual models, as in Figure C.6, to show these relationships as well as concrete materials such as labeled fraction/decimal/percent towers. Teach the meaning of percent (per hundred) explicitly. Be sure to extend instruction to examples with percents greater than 100 (125%, 200%) and less than zero (−25%).

- Use pie and bar charts, such as those created with data sets on spreadsheets, to model the relationship between fractions and percents (Whitin & Whitin, 2012). Have students collect data and create charts with both forms of rational numbers.
- Plan activities where students read and compare numbers in decimal and percent forms, convert numbers in both directions, and create word problems involving decimals and percents.

Ratios and Proportions: Common Difficulties and Instructional Strategies

Ratios and proportions are powerful concepts and foundational to solving everyday problems and understanding algebraic, geometric, and scientific concepts. These concepts are taught in Grades 6 and 7 and critical for topics in later grades. Weak foundational concepts of whole-number and rational-number meanings, magnitudes, properties, and operations can lead to weak ratio and proportion understanding, perhaps due to too many part–whole representations, misleading instructional language, and the lack of experiences making conceptual connections. Many students continue to rely on additive rather than multiplicative reasoning well into middle school. Additive thinkers tend to view problems in terms of differences in quantities, while multiplicative thinkers see the relationship as a rate of change, how one value's change affects the other value's change (Lamon, 1993). Additive reasoning is reinforced by referring to multiplication as repeated addition long beyond third grade and using faulty key words for problem solving.

Working with ratios and proportions requires reasoning about the relationships of multiple quantities at once, difficult for students who struggle even with the two terms in a simple fraction. When two ratios are compared, the student must reason with four quantities at once and use strong equivalence concepts. For example, *if it takes three bottles of ginger ale to make two batches of punch, how many bottles are needed for eight batches?* The ratio 3 to 2 is expanded until it reaches an equivalent ratio of 12 to 8. Ratios can be scaled up or down or operate in different directions using proportional reasoning and related strategies. Ratios can be scaled by other rational numbers, not just whole-number equivalents (e.g., 15:12 scaled down to 12:9.6).

Like working with fractions and decimals, ratios have equivalent values. Many students fail to recognize equivalent ratios that are not necessarily integral multiples (Dougherty et al., 2017). For example, 6:9 is equivalent to 4:6; 15:20 is equivalent to 6:8. But ratios are not operated on in the same way as fractions. Other students use single-unit strategies (*building-up* strategies) rather than more sophisticated *composite-unit* strategies. A building-up strategy would expand each term in the ratio using multiples or count-bys. In the composite-unit strategy the student would divide 2 into 8 for 4 groups of batches. Then 4 times 3 equals 12 bottles needed. But the student needs to be able to reason with the units of bottles and batches (unitize) simultaneously to accomplish this method (Lamon, 1993).

Many students lack effective strategies for solving problems involving ratios and proportions (Beckmann & Izsák, 2015; Hunt & Vasquez, 2014). Students can certainly be taught the cross-product algorithm for finding the unknown in a direct proportion but may not be able to reason through the process or know when to apply the algorithm. Other strategies that can extend a student's toolbox include creating tables, double number lines, and graphs. Some students set up an incorrect equation, one that does not reflect the relationship of the quantities. For example, one student attempted to solve the problem: *You bought 3 notebooks for $1.69. What is the cost of 10 notebooks?* She attempted to set it up as 10 over $1.69 = the unknown over $1.69. However, there should be consistencies in rows and columns in a proportional equation. If the first ratio represents what you bought (3 notebooks over their cost of $1.69), then the equivalent ratio will be what the question asks, parallel to the first ratio: 10 notebooks over the unknown cost: $\dfrac{3\,notebooks}{\$1.69} = \dfrac{10\,notebooks}{x}$. Labeling the parts of ratios can support reasoning. When working with tables of equivalent ratios, some students only look at patterns from row to row rather than how two quantities vary together (Dougherty et al., 2017). For example, a table with columns for number of fruit baskets sold and the total earned reflects a linear relationship (1:$3.50, 2:$7.00, 3:$10.50, etc.). Students may focus on the 1 to $3.50 additively and not understand the proportional growth of this model, not able to find the total earned for 10 baskets.

In a study of seventh-grade students, Im and Jitendra (2020) examined misconceptions and errors related to ratios and proportional reasoning. Even after a successful intervention (schema-based instruction, $g = 0.50$ on direct measures), students with MLD still demonstrated "limited understanding of fractional and ratio representations and persistent numerical and additive errors" (p. 15). Students often did not know how to express fractions and ratios, to identify the type of ratio (part-to-part or part-to-whole), and to convert ratios to percents using multiple steps. Their reasoning was better than that of a matched control group, but still slow and limited, reinforcing the need for targeted interventions and increased instructional time to address fundamental concepts and their connections.

Instructional Strategies for Ratio and Proportion Concepts

The National Research Council (Kilpatrick et al., 2001) called proportional reasoning the capstone of elementary mathematics. Proportional reasoning develops over time and one level forms a foundation for higher levels (Lamon, 2007). Engaging in proportional tasks requires the transition from additive to multiplicative reasoning, focusing on the multiplicative relationship between two or more rational quantities (Nelson et al., 2020). This reasoning involves the covariance of quantities and invariance of ratios, requiring multiple comparisons in relative terms (Fielding-Wells et al., 2014). The learning progression involves linking two composite units (a composite unit of 15 might be a unit of 5 iterated 3 times, then iterating linked composite units (3×5 can be iterated up to 15×25 either additively or multiplicatively). Effective instruction in ratios and firm understanding of equivalent expressions supports proportional reasoning.

Like work with fractions and decimals, instruction about proportions, ratios, and related concepts require tasks in a wide range of contexts and conceptual instruction before symbolic strategies (Jitendra et al., 2011). Proportional understanding is critical for algebra (functions), geometry, and data analysis in high-school mathematics. Specific strategies for instruction in ratios and proportions include:

- Ratios are situations in which two or more quantities are compared, sometimes with the same unit and sometimes with different units. Some textbook authors distinguish these by using ratio for same unit comparison and rate with different units. The CCSSM use the term ratio to indicate both types of situations. Ratios can have associated rates: the ratio 3 mistakes for every 10 pages is the ratio 3:10 or the rate $\frac{3}{10}$ or 0.3 mistakes per page. Percent (per hundred), therefore, is a ratio with 100 as the value of the unit of comparison (45% on a test represents 45 out of 100 units correct). Connecting these concepts under the concept of ratio will support problem solving and reasoning. The concept of percent can serve as a bridge to ratios and proportional concepts.

- Use carefully selected graphic representations to model ratios and proportional relationships. Tape (strip) diagrams may be best when two quantities have the same unit and double number lines when quantities have different units (Institute for Mathematics and Education, 2011).

- Use interventions based on instructional trajectories of the domain. For example, Hunt and Vasquez (2014) used a sequence of build-up, emergent unit, and unit strategies to build ratio-equivalence concepts of students with MLD, over 15 lessons using a CRA sequence. This sequence moved students from additive to multiplicative thinking. Build-up strategies included iterating composite units using addition and multiplication with concrete and tabular representations. Emergent–unit strategies shortened the process to the use of multiplication and division to scale the units up or down. Finally, students extended their multiplicative thinking to rational numbers other than whole numbers with the unit-rate strategy, such as scaling 4:6 to other equivalent ratios. See the progression in Figure C.7.

- Delay introducing the cross-multiplication algorithm until students have acquired requisite algebraic-reasoning skills (Cohen, 2013; Siegler et al., 2010). Instead, develop proportional concepts using parallel paper strips, bar diagrams, tables, and double number lines. Begin with problems without numbers to develop multiplicative reasoning, such as: *Mary Jo and Caitlin decided to roller skate along the park path. One skates faster than the other, but both started at the same time and maintained a constant speed. Mary Jo took longer to reach the end of the path. Who was skating faster? Explain why.* Then use parallel representations to introduce covariance: *There are 3 red marbles for every 4 blue marbles in the bag. Each segment on this strip represents 3 red marbles and on the strip beside that each segment is 4 blue marbles. When we reach 9 red marbles, how many blue marbles will there be?*

- Use arrows with the symbolic forms of proportion equations to show scalar and functional relationships (see Figure C.8).

- Use schema-based instruction (SBI) for word-problem solving with ratios and proportions. This approach assists students in identifying problem types, representing problem information in an appropriate visual representation, estimating answers to support strategy selection, and solving problems. Jitendra and colleagues (2011) used an emphasis on underlying problem structure and schemas (diagrams) depicting proportional relationships to teach seventh graders how to set up the important information in a problem in a way that highlights relationships in ratios, percents, scales, and proportions. The visual diagrams reflect the relationships. Jitendra et al. (2019) replicated previous research on the performance of 7th-grade students in ratio, proportion, and percent word problems in 36 schools. The intervention groups with schema-based instruction (SBI, visual diagrams) taught by classroom teachers outperformed control classrooms studying the same topics on proximal ($d = 0.45$) and distal ($d = 0.31$) measures. This research demonstrated that SBI is an effective approach for teaching a diverse group of students on these topics. See schemas for word problems in Chapter 5 and Jitendra and Star (2017).

Sam can run around the track at school 3 times in the same amount of
time it takes Bebe to run two times around. How many times will Bebe
have run if Sam runs 15 times around the track?

Additive reasoning.

$$+3 \quad +3$$

Sam	3	6	9	12	15
Bebe	2	4	6	8	10

$$+2 \quad +2$$

Multiplicative reasoning.

$$\times 5$$

Sam	3	15
Bebe	2	10

$$\times 5$$

Proportional reasoning.

3 is to 2 as 15 is to x:

$$\frac{3}{2} = \frac{15}{x}$$

$$3x = 30$$

$$x = 10$$

Figure C.7 Trajectory of Ratio Equivalence Concepts

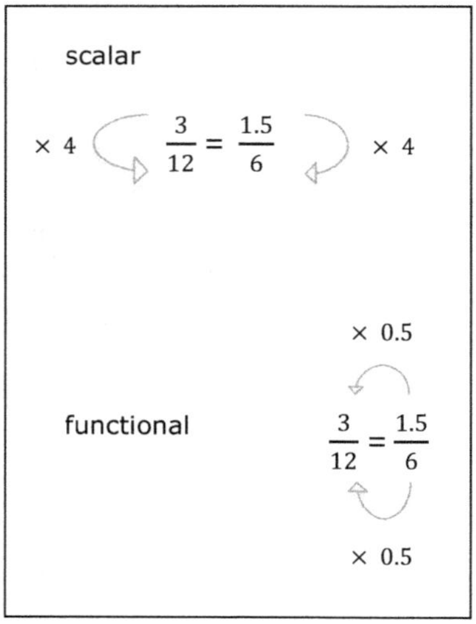

scalar

$$\times 4 \quad \frac{3}{12} = \frac{1.5}{6} \quad \times 4$$

$$\times 0.5$$

functional

$$\frac{3}{12} = \frac{1.5}{6}$$

$$\times 0.5$$

Figure C.8 Diagramming Proportional Relationships

- Graphing a set of equivalent ratios on a coordinate plane, and the corresponding equation, makes the connection with functions (Beckmann & Izsák, 2015). For the marbles problem, graphing the coordinate points 3, 4; 6, 8; and 9, 12 results in the linear representation (unit rate) for the relationship $y = \frac{4}{3}x$. Setting up the proportion using algebraic reasoning results in $\frac{3}{4} = \frac{9}{x}$ and with cross-multiplication $3x = 36$, $x = 12$.

Effective Elements of Rational-Number Instruction

Instruction about rational numbers, whether for initial learning or intervention and remediation, should be designed around several effective elements drawn from research with students with mathematics difficulties. First, build on students' informal knowledge. Even young children have informal understanding of sharing, portions, and proportions. Use that knowledge with familiar activities and contexts to develop more advanced concepts (Mazzocco et al., 2013; Siegler et al., 2010). For example, Cwikla (2014) described asking 36 preschool and kindergarten children snack-sharing questions such as, *Jade wanted to share six carrot sticks with her four friends. How could she do this fairly?* The researcher found that these children could make illustrations (except the two youngest) and describe how to give each child the same number of pieces, and some could use terms such as half and whole. An example for middle-school students beginning a unit on proportions would be to use their informal knowledge about scales, how a map represents a place but is to scale and not an exact measure of the place.

Second, make connections with students' whole-number knowledge and emphasize the similarities and differences with new concepts. Fractions and other rational numbers are not separate topics. They are part of the same, expanded number system. In the snack-sharing example previously, we see the connection between division and part-whole concepts of fractions. Siegler et al. (2011) challenged previous thinking about how fraction concepts are developed. Previous theorists posited that whole-number learning was fundamentally different and not connected with rational-number learning. They thought that children understand whole numbers naturally but have difficulties with fractions because those are not so naturally understood. The integrated theory of numerical development, however, asserts that learning about numerical magnitudes is a basic process uniting the development of understanding of all real numbers (Siegler et al., 2011). The researchers suggested that difficulties in learning about fractions stems not from whole-number bias but from drawing inaccurate analogies to whole numbers. Drawing correct analogies can enhance fraction understanding. For example, whole numbers, fractions, and decimals have magnitudes that can be ordered and represented on number lines. A good representational analogy is an elastic band that never breaks to represent a number line with infinite density (Vamvakoussi, 2017). This bridging analogy can assist students to move beyond discrete number representations. Siegler and Lortie-Forgues (2015) expanded the integrated theory of numerical development to include the role of understanding magnitudes produced by arithmetic with rational numbers, specifically understanding the direction of effects of arithmetic operations.

Third, plan instruction in a graduated sequence—concrete, representational, abstract—with needed supports at the abstract (symbolic) level. The concrete level offers students objects that can be manipulated—such as fraction circles, interconnecting cubes, and fraction bars. The representational level includes diagrams and drawings. Students with MLD may need visual representations longer than other students, to support concept understanding (Mazzocco et al., 2013). Number lines should be the *primary* representational tool for teaching about rational number concepts across the grade levels (Gersten et al., 2017; Jordan et al., 2017; NMAP, 2008). Representing fractions and other rational numbers on number lines is a critical method for linking conceptual and procedural knowledge. Number-line representations also connect whole-number systems with those of rational numbers, promote magnitude understanding, and assist students' understanding of the relationships among rational-number types. Other powerful representations include decimal grids (graph paper with tenths, hundredths), pie charts (for connecting with data-analysis concepts), and students' own drawings. The symbolic level involves all the symbols of rational numbers and the syntax for using those symbols. Moving to the symbolic level too quickly limits essential concept understanding. Some students will need simultaneous representations to support symbolic notations.

Fourth, use explicit and accurate language for instruction, feedback, and assessment. Teachers need to select their instructional language carefully. For example, Muzheve and Capraro (2012) found, in a study of sixth- through eighth-grade teachers on a range of rational numbers, that there was widespread use of natural language by teachers, which could be helpful in some cases, but was often ambiguous, erroneous, and confusing. For example, using phrases such as "6 out of 12" for the fraction $\frac{6}{12}$ implied the part-whole view only; saying the phrase "smaller equivalent fraction" was unclear when they meant, "write the same value in simpler terms;" and not drawing attention to the equal sign to show equivalent relationships lessens that understanding. Foreman-Murray and Fuchs (2019) found that the quality of fourth-grade student explanations of fraction-magnitude understanding was moderately correlated with accuracy in comparing magnitudes, especially when student self-explaining was supported by teachers for higher-quality explanations with accurate vocabulary.

Finally, focus on concept understanding and connections, with procedures developed iteratively. Many students who struggle with learning procedures for fractions, decimals, and other rational numbers lack concept understanding. Students may have memorized an algorithm, such as "invert and multiply" for fraction division, but not understand why it works. They may know to find the common denominator for some operations, but unless they also have concept understanding, students will not know when to find a common denominator or more efficient ways for finding one. Learning by rote limits students' progress in higher-level topics, problem-solving abilities, and facility with reasoning. Rittle-Johnson et al. (2001) proposed the iterative model of learning, with conceptual understanding and procedural skill developing in tandem. Increases in one type of knowledge lead to increases in the other, which triggers increases in the first, and so forth. In studies with fifth- and sixth-grade students on decimal fractions, the researchers found that students' initial conceptual knowledge predicted gains in procedural knowledge, and gains in procedural predicted improvements in conceptual.

As students progress from whole numbers to the full range of rational-number concepts, they often face difficulties with representations of these concepts, expressing numbers in different forms, and using the range of rational numbers to solve problems. Rational-number concepts are critical for success with algebra and other higher-level mathematics concepts. Some students will require more time and special interventions to be proficient with rational numbers and their properties, operations, and applications.

Multicultural Connection

A Chinese text, *Chiu Chang* (about first century CE), discussed a method for simplifying fractions (Joseph, 1991). Begin with a common fraction and follow these steps:

- If both numbers can be halved, then halve them.
- Set the denominator below the numerator and subtract the smaller number from the larger number.
- Continue this process until the lowest common divisor (teng) is obtained.
- Simplify the original fraction by dividing both numbers by teng. (See the examples in Box C.3.)

Box C.3   Multicultural Connection: Chinese Fraction Simplification

For the fraction $\dfrac{78}{130}$ the terms can first be halved: $\dfrac{39}{65}$.

Next, carry out the subtraction steps until you reach a common divisor:

39	39		
65	26	13	13 is *teng*

$39 \div 13 = 3$

$65 \div 13 = 5$ $\qquad$ The equivalent fraction is $\dfrac{3}{5}$.

For the fraction $\dfrac{49}{91}$, the numbers cannot be halved. Move on to the second step, subtraction:

49	49		
91	42	7	7 is *teng*

$49 \div 7 = 7$

$91 \div 7 = 13$ $\qquad$ The equivalent fraction is $\dfrac{7}{13}$.

References

Adler, D. A. (1996). *Fraction fun*. Holiday House/Penguin Random House.

Bailey, D. H., Hoard, M. K., Nugent, L., & Geary, D. C. (2012). Competence with fractions predicts gains in mathematics achievement. *Journal of Experimental Child Psychology, 113*(3), 447–455. doi: 10.1016/j.jecp.2012.06.004

Baroody, A. J., & Hume, J. (1991). Meaningful mathematics instruction: The case of fractions. *Remedial and Special Education, 12*(3), 54–68. doi: 10.1016/j.jecp.2012.06.004

Beckmann, S., & Izsák, A. (2015). Two perspectives on proportional relationships: Extending complementary origins of multiplication in terms of quantities. *Journal for Research in Mathematics Education, 46*(1), 17–38. doi: 10.5951/jresematheduc.46.1.0017

Behr, M. J., Harel, G., Post, T., & Lesh, R. (1992). Rational number, ratio, and proportion. In D. A. Grouws (Ed.), *Handbook of research on mathematics teaching and learning* (pp. 296–333). Macmillan.

Braithwaite, D. W., & Siegler, R. S. (2018). Children learn spurious associations in their matl textbooks: Examples from fraction arithmetic. *Journal of Experimental Psychology, 44*(11), 1765–177' doi: 10.1037/xlm0000546

Brown, G., & Quinn, R. J. (2006). Algebra students' difficulty with fractions: An error analysis. *Austral Mathematics Teacher, 62*(4), 28–40.

Cavey, L. O., & Kinzel, M. T. (2014). From whole numbers to invert and multiply. *Teaching Chil Mathematics, 20*(6), 374–383. doi: 10.5951/teachchilmath.20.6.0374

Cohen, J. S. (2013). Strip diagrams: Illuminating proportions. *Mathematics Teaching in the Middle S 18*(9), 536–542. doi: 10.5951/mathteachmiddscho.18.9.0536

Courey, S. J., Balogh, E., Siker, J. R., & Paik, J. (2012). Academic music: Music instruction to e third-grade students in learning basic fraction concepts. *Educational Studies in Mathematics 251–278. doi: 10.1007/s10649-012-9395-9

Cramer, K. A., Monson, D. S., Wyberg, T., Leavitt, S., & Whitney, S. B. (2009). Models for initial ideas. *Teaching Children Mathematics, 16*(2), 106–117. doi: 10.5951/tcm.16.2.0106

Cwikla, J. (2014). Can kindergarteners do fractions? *Teaching Children Mathematics*, *20*(6), 354–364. doi: 10.5951/teachchilmath.20.6.0354

D'Ambrosio, B. S., & Kastberg, S. E. (2012). Building understanding of decimal fractions. *Teaching Children Mathematics*, *18*(9), 558–564. doi: 10.5951/teachchilmath.18.9.0558

Depaepe, F., Van Roy, P., Torbeyns, J., Kleickmann, T., Van Dooren, W., & Verschaffel, L. (2018). Stimulating pre-service teachers' content and pedagogical content knowledge on rational numbers. *Educational Studies in Mathematics*, *99*(2), 197–216. doi: 10.1007/s10649-018-9822-7

Dougherty, B., Bryant, D. P., Bryant, B. R., & Shin, M. (2017). Helping students with mathematics difficulties understand ratios and proportions. *TEACHING Exceptional Children*, *49*(2), 96–105. doi: 10.1177/0040059916674897

Fielding-Wells, J., Dole, S., & Makar, K. (2014). Inquiry pedagogy to promote emerging proportional reasoning in primary students. *Mathematics Education Research Journal*, *26*, 47–77. doi: 10.1007/s13394-013-0111-6

Foreman-Murray, L., & Fuchs, L. S. (2019). Quality of explanation as an indicator of fraction magnitude understanding. *Journal of Learning Disabilities*, *52*(2), 181–191. doi: 10.1177/0022219418775120

Fuchs, L. S., Malone, A. S., Schumacher, R. F., Namkung, J., & Wang, A. (2017). Fraction intervention for students with mathematics difficulties: Lessons learned from five randomized controlled trials. *Journal of Learning Disabilities*, *50*(6), 631–639. doi: 10.1177/0022219416677249

Fuchs, L. S., Schumacher, R. F., Long, J., Namkung, J., Malone, A. S., Wang, A., Hamlett, C. L., Jordan, N. C., Siegler, R. S., & Changas, P. (2016). Effects of intervention to improve at-risk fourth graders' understanding, calculations, and word problems with fractions. *The Elementary School Journal*, *116*(4), 625–651. doi: 10.1086/686303

Fuchs, L. S., Wang, A. Y., Preacher, K. J., Malone, A. S., Fuchs, D., & Pachmayr, R. (2021). Addressing challenging mathematics standards with at-risk learners: A randomized controlled trial on the effects of fractions intervention at third grade. *Exceptional Children*, *87*(2), 163–182. doi: 10.1177/0014402920924846

Gersten, R., Schumacher, R. F., & Jordan, N. C. (2017). Life on the number line: Routes to understanding fraction magnitude for students with difficulties learning mathematics. *Journal of Learning Disabilities*, *50*(6), 655–657. doi: 10.1177/0022219416662625

...ecker, B. (2000). Imagery and fractions in students classified as learning disabled. *Learning Disability
      ...uarterly*, *23*(2), 157–168. doi: 10.2307/1511143

...S. A., & Vagi, K. J. (2010). Sources of group and individual differences in emerging fraction skills.
      ...al of Educational Psychology*, *102*(4), 212–229. doi: 10.1037/a0019824

..., & Vasquez, E. (2014). Effects of ratio strategies intervention on knowledge of ratio equivalence
      ...ents with learning disability. *The Journal of Special Education*, *48*(3), 180–190. doi: 10.1177/
      ...912474102

...Cordes, S. (2015). Rational-number comparison across notation: Fractions, decimals and
      ...bers. *Journal of Experimental Psychology*, *42*(2), 281–293. doi: 10.1037/xhp0000140

...endra, A. K. (2020). Analysis of proportional reasoning and misconceptions among
      ...mathematical learning disabilities. *The Journal of Mathematical Behavior*, *57*, 100753–.
      ...jmathb.2019.100753

...matics and Education (2011). Progression on ratios and proportional relationships
      ...documents for the common core math standards* (draft). http://ime.math.arizona.edu/

...M. R., Im, S.-h., Karl, S. R., & Slater, S. C. (2019). Improving student learning of
      ...percent: A replication study of schema-based instruction. *Journal of Educational
      ...5–1062. doi: 10.1037/edu0000335

...2017). *Solving ratio, proportion, and percent problems using schema-based instruction.*
      ...Learning, University of Oregon. https://dibels.uoregon.edu/market/sbi

...driguez, M., Lindell, M., & Someki, F. (2011). Improving students' propor-
      ...a-based instruction. *Learning and Instruction*, *21*(6), 731–745. doi: 10.1016/

...ues, J., Hansen, N., & Dyson, N. (2017). Delaware longitudinal study of
      ...for helping children with mathematics difficulties. *Journal of Learning
      ...0.1177/0022219416662033

Joseph, G. G. (1991). *The crest of the peacock: Non-European roots of mathematics*. Penguin Books.

Kieren, T. E. (1980). The rational number construct—its elements and mechanism. In T. E. Kieren (Ed.), *Recent research on number learning* (pp. 125–150). (ED212463). ERIChttp://files.eric.ed.gov/fulltext/ED212463.pdf

Kieren, T. E. (1988). Personal knowledge of rational numbers: Its intuitive and formal development. In J. Hiebert & M. Behr (Eds.), *Number concepts and operations in the middle grades* (pp. 162–181). NCTM, Lawrence Erlbaum.

Kilpatrick, J., Swafford, J., & Findell, B. (Eds.). (2001). *Adding it up: Helping children learn mathematics*. National Research Council, The National Academies Press. www.nap.edu/catalog/9822/adding-it-up-helping-children-learn-mathematics

Lamon, S. J. (1993). Ratio and proportion: Children's cognitive and metacognitive processes. In T. P. Carpenter, E. Fennema, & T. A. Romberg (Eds.), *Rational numbers: An integration of research* (pp. 131–156). Routledge.

Lamon, S. J. (1999). *Teaching fractions and ratios for understanding: Essential content knowledge and instructional strategies for teachers*. Erlbaum/Routledge.

Lamon, S. J. (2007). Rational numbers and proportional reasoning: Toward a theoretical framework for research. In F. K. Lester, Jr. (Ed.), *Second handbook of research on mathematics teaching and learning* (pp. 629–668). NCTM/Information Age Publishing.

Lewis, K. E. (2010). Understanding mathematical learning disabilities: A case study of errors and explanations. *Learning Disabilities: A Contemporary Journal, 8*(2), 9–18.

Lewis, K. E. (2016). Beyond error patterns: A sociocultural view of fraction comparison errors in students with mathematical learning disabilities. *Learning Disability Quarterly, 39*(4), 199–212. doi: 10.1177/0731948716658063

Malone, A. S., & Fuchs, L. S. (2017). Error patterns in ordering fractions among at-risk fourth-grade students. *Journal of Learning Disabilities, 50*(3), 337–352. doi: 10.1177/0022219416629647

Malone, A. S., Fuchs, L. S., Sterba, S. K., Fuchs, D., Foreman-Murray, L. (2019). Does an integrated focus on fractions and decimals improve at-risk students' rational number magnitude performance? *Contemporary Educational Psychology, 59*, 101782. doi: 10.1016/j.cedpsych.2019.101782

Mazzocco, M. M. M., & Devlin, K. T. (2008). Parts and "holes": Gaps in rational number sense among children with vs. without mathematical learning disabilities. *Developmental Science, 11*(5), 681–691. doi: 10.1111/j.1467-7687.2008.00717.x

Mazzocco, M. M. M., Myers, G. F., Lewis, K. E., Hanich, L. B., & Murphy, M. M. (2013). Limited knowledge of fraction representations differentiates middle-school students with mathematics learning disability (dyscalculia) vs. low mathematics achievement. *Journal of Experimental Child Psychology, 115*(2), 371–387. doi: 10.1016/j.jecp.2013.01.005

McMullen, J., Hannula-Sormunen, M. M., Lehtinen, E., & Siegler, R. S. (2020). Distinguishing adaptive from routine expertise with rational number arithmetic. *Learning and Instruction, 68*, 101347–. doi: 10.1016/j.learninstruc.2020.101347

Morano, S., & Riccomini, P. J. (2020). Is a picture worth 1,000 words? Investigating fraction magnitude knowledge through analysis of student representations. *Assessment for Effective Intervention, 46*(1), 27–38. doi: 10.1177/1534508418820697

Moseley, B. (2005). Students' early mathematical representation knowledge: The effects of emphasizing single or multiple representations of the rational number domain in problem solving. *Educational Studies in Mathematics, 60*(1), 37–69. doi: 10.1007/s10649-005-5031-2

Muzheve, M. T., & Capraro, R. M. (2012). An exploration of the role natural language and idiosyncratic representations in teaching how to convert among fractions, decimals, and percents. *The Journal of Mathematical Behavior, 31*(1), 1–14. doi: 10.1016/j.jmathb.2011.08.002

National Center for Education Statistics (2019). *2019 NAEP Mathematics Assessments*. US Department of Education. www.nationsreportcard.gov

National Council of Teachers of Mathematics (2006). *Curriculum focal points for prekindergarten through grade 8*. www.nctm.org/curriculumfocalpoints/

National Governors Association Center for Best Practices & Council of Chief State School Officers (2010). *Common core state standards for mathematics*. www.corestandards.org/Math/

National Mathematics Advisory Panel (2008). *Foundations for success: The final report of the National Mathematics Advisory Panel*. US Department of Education. www2.ed.gov/about/bdscomm/list/mathpanel/index.html

Nelson, F., Hunt, J. H., Martin, K., Patterson, B., & Khounmeuang, A. (2020). Current knowledge and future directions: Proportional reasoning interventions for students with learning disabilities and mathematics difficulties. *Learning Disability Quarterly*, 73194872093285–. doi: 10.1177/0731948720932850

Obersteiner, A., Dresler, T., Bieck, S. M., & Moeller, K. (2019). Understanding fractions: Integrating results from mathematics education, cognitive psychology, and neuroscience. In A. Norton & M. W. Alibali (Eds.), *Constructing number: Merging perspectives from psychology and mathematics education* (pp. 135–162). Springer. doi: 10.1007/978-3-030-00491-0

Obersteiner, A., Reiss, K., Van Dooren, W., & Van Hoof, J. (2019). Understanding rational numbers: Obstacles for learners with and without mathematical learning disabilities. In A. Fritz, V. G. Haase, & P. Räsänen (Eds.), *International handbook of mathematical learning difficulties* (pp. 581–594). Springer. doi: 10.1007/978-3-319-97148-3

Park, J., Güçler, B., & McCrory, R. (2013). Teaching prospective teachers about fractions: Historical and pedagogical perspectives. *Educational Studies in Mathematics*, *82*(3), 455–479. doi: 10.1007/s10649-012-9440-8

Phelps, K. A. G. (2012). The power of problem choice. *Teaching Children Mathematics*, *19*(3), 152–157. doi: 10.5951/teachchilmath.19.3.0152

Powell, S. R., & Nelson, G. (2021). University students' misconceptions about rational numbers: Implications for developmental mathematics and instruction of younger students. *Psychology in the Schools*, *58*(2), 307–331. doi: 10.1002/pits.22448

Rathouz, M. (2011). Visualizing decimal multiplication with area models: Opportunities and challenges. *Issues in the Undergraduate Mathematics Preparation of School Teachers*, *2*, 1–12. https://scinapse.io/papers/1590361

Resnick, I., Rinne, L., Barbieri, C., & Jordan, N. C. (2019). Children's reasoning about decimals and its relation to fraction learning and mathematics achievement. *Journal of Educational Psychology*, *111*(4), 604–618. doi: 10.1037/edu0000309

Schumacher, R. F., Namkung, J., Malone, A., Wang, A., Abramson, R., & Fuchs, L. (2015). *Fraction Battery–Revised*. Fuchs Research Group, Vanderbilt University.

Shaughnessy, M. M. (2011). Identify fractions and decimals on a number line. *Teaching Children Mathematics*, *17*(7), 428–434. doi: 10.5951/teachchilmath.17.7.0428

Shin, M., & Bryant, D. P. (2015). Fraction interventions for students struggling to learn mathematics: a research synthesis. *Remedial and Special Education*, *36*(6), 374–387. doi: 10.1177/0741932515572910

Sidebotham, T. H. (2002). *The A to Z of mathematics: A basic guide.* Wiley. www.wiley.com

Siegler, R. S., Carpenter, T., Fennell, F., Geary, D., Lewis, J., Okamoto, Y., Thompson, L., & Wray, J. (2010). *Developing effective fractions instruction for kindergarten through 8th grade: A practice guide* (NCEE #2010–4039). National Center for Education Evaluation and Regional Assistance, Institute of Education Sciences, US Department of Education. https://ies.ed.gov/ncee/wwc/Docs/PracticeGuide/fractions_pg_093010.pdf

Siegler, R. S., Duncan, G. J., Davis-Kean, P. E., Duckworth, K., Claessens, A., Engel, M., Susperreguy, M. I., & Chen, M. (2012). Early predictors of high school mathematics achievement. *Psychological Science*, *23*(7), 691–697. doi: 10.1177/0956797612440101

Siegler, R. S., & Lortie-Forgues, H. (2015). Conceptual knowledge of fraction arithmetic. *Journal of Educational Psychology*, *107*(3), 1–10. doi: 10.1037/edu0000025

Siegler, R. S., & Lortie-Forgues, H. (2017). Hard lessons: Why rational number arithmetic is so difficult for so many people. *Current Directions in Psychological Science: A Journal of the American Psychological Society*, *26*(4), 346–351. doi: 10.1177/0963721417700129

Siegler, R. S., & Pyke, A. A. (2013). Developmental and individual differences in understanding of fractions. *Developmental Psychology*, *49*(10), 1994–2004. doi: 10.1037/a0031200

Siegler, R. S., Thompson, C. A., & Schneider, M. (2011). An integrated theory of whole number and fractions development. *Cognitive Psychology*, *62*(4), 273–296. doi: 10.1016/j.cogpsych.2011.03.001

Tian, J., & Siegler, R. S. (2018). Which type of rational numbers should students learn first? *Educational Psychology Review*, *30*(2), 351–372. doi: 10.1007/s10648-017-9417-3

Utley, J., & Reeder, S. (2012). Prospective elementary teachers' development of fraction number sense. *Investigations in Mathematics Learning*, *5*(2), 1–13. doi: 10.1080/247274466.2012.11790320

Vamvakoussi, X. (2015). The development of rational number knowledge: Old topic, new insights. *Learning and Instruction, 37*, 50–55. doi: 10.1016/j.learninstruc.2015.01.002

Vamvakoussi, X. (2017). Using analogies to facilitate conceptual change in mathematics learning. *ZDM, 49*(4), 497–507. doi: 10.1007/s11858-017-0857-5

Van Hoof, J., Verschaffel, L., Ghesquière, P., & Van Dooren, W. (2017). The natural number bias and its role in rational number understanding in children with dyscalculia. Delay or deficit? *Research in Developmental Disabilities, 71*, 181–190. doi: 10.1016/j.ridd.2017.10.006

Van Hoof, J., Verschaffel, L., & Van Dooren, W. (2015). Inappropriately applying natural number properties in rational number tasks: Characterizing the development of the natural number bias through primary and secondary education. *Educational Studies in Mathematics, 90*(1), 39–56. doi: 10.1007/s10649-015-9613-3

Wang, A. Y. Fuchs, L. S., Fuchs, D., Gilbert, J. K., Krowka, S., & Abramson, R. (2019). Embedding self-regulation instruction within fractions intervention for third graders with mathematics difficulties. *Journal of Learning Disabilities, 52*(4), 337–348. doi: 10.1177/0022219419851750

Wentworth, G. A. (1897). *A practical arithmetic*. Ginn and Company.

Whitin, D. J., & Whitin, P. (2012). Making sense of fractions and percentages. *Teaching Children Mathematics, 18*(8), 490–496. doi: 10.5951/teachchilmath.18.8.0490

Woodward, J. (2017). The concept of magnitude and what it tells us about how struggling students learn fractions. *Journal of Learning Disabilities, 50*(6), 640–643. doi: 10.1177/0022219416659445

Zhang, D., Stecker, P., & Beqiri, K. (2017). Strategies students with and without mathematics disabilities use when estimating fractions on number lines. *Learning Disability Quarterly, 40*(4), 225–236. doi: 10.1177/0731948717704966

Strand D

Spatial and Geometric Reasoning

After studying this strand, the reader should be able to:

1. describe the applications of spatial reasoning across the mathematics curriculum;
2. provide examples of student abilities for the van Hiele levels of geometric thinking;
3. discuss issues related to the assessment of spatial abilities;
4. describe typical difficulties with learning spatial and geometric concepts;
5. describe key instructional elements for promoting spatial and geometric reasoning.

Steven is a tenth-grade student in Mr. Lopez' geometry class at Pine Road High School. Steven has received special education services for reading and language disabilities since the third grade and now has sufficient strategies to be enrolled in general education classes for all his core subjects. His mathematics skills had always been average, with strong computation skills and concepts, average problem-solving ability, and some difficulty with rational-number representations. But high-school geometry has been very difficult for Steven. Consider the following conversation between Steven and Mr. Lopez:

MR. LOPEZ: Steven, what three-dimensional object is depicted on page 137 of your text?

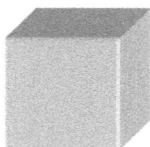

STEVEN: It looks like a square with other pieces.
MR. LOPEZ: There are squares in that diagram, but a square is two-dimensional. What is the three-dimensional form?
STEVEN: What do you mean, two-dimensional?
MR. LOPEZ: It has length and width, the figure is only on a plane, it is flat.
STEVEN: This page is flat so that diagram is two-dimensional?
MR. LOPEZ: The drawing has been done with perspective so that we can see the figure as three-dimensional. It is a cube with six squares as sides.
STEVEN: I see only one square and two other smaller parts.

Spatial reasoning and related concepts in geometry are among the most neglected topics in mathematics, yet the development of spatial abilities is critical for overall mathematics achievement and problem-solving abilities. Textbooks often relegate work with spatial concepts and geometry to one of the last chapters and fail to integrate the development

DOI: 10.4324/9781003096733-12

of spatial ability and applications across topics. However, spatial reasoning is required for all mathematics tasks (Bronowski, 1947) and is a predictor for general mathematics achievement (Gilligan et al., 2019; Uttal et al., 2013; Xie et al., 2020). Spatial ability has been connected with numeric understanding, calculation skill, problem-solving ability, proportional reasoning, algebraic reasoning, achievement in upper-level mathematics, and the development of concepts in science, geography, technology, engineering, medicine, art, sports and other fields (Boonen et al., 2014; Cheng & Mix, 2014; Gagnier et al., 2017; Matthews & Hubbard, 2017). Strong spatial skills also predict participation in STEM (science, technology, engineering, and mathematics) careers (Atit et al., 2020; Ontario Ministry of Education, 2014; Ramey et al., 2020; Sorby et al., 2018; Uttal et al., 2013). In a meta-analysis of 73 studies, Xie et al. (2020) found an overall positive relationship between spatial and mathematical abilities ($r = 0.25$) and that each domain of spatial ability that was sampled sufficiently showed a significant relationship with mathematics ability. Further, each domain of mathematical ability was significantly correlated with spatial ability, especially logical reasoning.

The National Research Council's 2006 report *Learning to Think Spatially* called for spatially literate students, in order to meet the needs of today's higher education, careers, and everyday life. "Spatial thinking can be learned, *and* it can and should be taught at all levels in the education system" (p. 3). The report continued with a description of spatially literate students as those "who have developed appropriate levels of spatial knowledge and skills in spatial ways of thinking and acting, together with sets of spatial capabilities" (p. 4). These students: 1) have the habit of mind of thinking spatially; 2) practice spatial thinking in an informed way; and 3) adopt a critical or evaluative stance to spatial thinking. Unfortunately, both the National Council of Teachers of Mathematics curriculum standards (NCTM, 2000) and the *Common Core State Standards for Mathematics* (CCSSM, NGA, & CCSSO, 2010) failed to emphasize the importance of spatial reasoning across the curriculum. The NCTM standards focused discussion of spatial skills within geometry and measurement standards. The CCSSM standards emphasized spatial skills for kindergarten, but only implied their importance in other levels and domains (e.g., recognize attributes, decompose shapes, draw angles, represent data). In fact, spatial skills, with their range of subskills, support learning in number, operations, fractions, proportions, algebra, statistics, and most other mathematics topics. Because of this neglect in national standards, local curricula rarely emphasize spatial skills or develop those intentionally. Lowrie and Logan (2018) noted the importance of spatial reasoning for all problem solving, not just geometry-based topics, and that spatial reasoning is not taught explicitly or in depth in most curricula. They emphasized that spatial reasoning can be improved and will be increasingly important in our technological society.

Examples of spatial reasoning are everywhere: a florist preparing a flower arrangement, a quarterback throwing a football to a specific point on the field, a friend giving driving directions over the phone, a surgeon using a laparoscope, and a motorist pulling out to pass a slower car on a two-lane road. Spatial reasoning is fundamental for any task that requires comparison, problem solving, visualization, making representations, measurement, orientation, or navigation.

Teachers with lower spatial abilities tend to have lower geometric understanding (Unal et al., 2009). Teachers with inadequate geometry knowledge or poor understanding of using visual representations for instruction have an impact on their students' achievement across mathematics topics (Clements & Sarama, 2011). Some studies have found that teachers' spatial abilities were lower than those of their students and the requirements of the curriculum (de Villiers, 2010). Yi et al. (2020) found that preservice teachers' geometry content knowledge about two-dimensional (2D) shapes and relationships was low (e.g., 58% knew properties of rectangles, 17% for rhombi) and their understanding of student knowledge was also low (e.g., 38% could interpret student knowledge of triangles). The researchers implemented a twelve-hour geometry module within a mathematics-methods course that focused on understanding 2D shapes, identifying students' levels of understanding, and selecting instructional activities to develop these concepts. This training improved teachers' geometry content knowledge, knowledge of students' levels of understanding, and pedagogical knowledge. Each type of knowledge growth contributed to the growth of the others.

There is considerable evidence that students in the United States lag behind their peers in other countries on spatial and geometric concepts. On the *Trends in Mathematics and Science Study* (Mullis et al., 2020), although students in the US averaged at or just above the international mean for geometry and measurement, fourth-grade students in the US scored below 24 other countries and equal to Portugal, decreasing since 2011. Eighth graders in the US averaged at the mean for geometry, below 16 other countries. Increasingly, national and international assessments include items that require spatial reasoning beyond geometry, such as graphic displays and their conventions—maps, graphs, diagrams with keys, legends, and scales (Lowrie et al., 2019).

On the *National Assessment of Educational Progress* mathematics assessments of 2017 (National Center for Educational Statistics), most spatially connected items were in the geometry and measurement domain, although other domains required spatial reasoning to interpret graphs, charts, and other visual information. Only 23% of fourth graders could identify pairs of congruent figures, only 35% could select sentences describing how area and perimeter change by rearranging rectangular figures, and only 3% could select *nets* ᴛ that would not fold into a cube. For eighth graders, only 13% could answer questions about lines and angles (parallel, right, perpendicular, degrees), only 35% could identify a solid prism with fewer than five rectangular faces, and only 7% could mentally join two shapes to form a new shape. These students were not performing at expected levels on tasks involving spatial abilities and reasoning such as understanding visual representations, mentally visualizing and rotating images, and working with geometric concepts in real-world applications. Spatial abilities and geometric concepts require more emphasis across the mathematics curriculum if students are to succeed in school and post-school endeavors.

This strand addresses the development of spatial abilities, with attention to assessment and the spatial abilities of students with learning difficulties. It continues with an overview spatial reasoning and geometry across the curriculum and recommendations for promoting spatial reasoning. The strand's final section addresses challenging concepts in geometry with intervention recommendations.

Spatial Ability

The terminology and typology used in psychological and educational research related to spatial abilities can be confusing and is not always consistent. Psychologists in cognitive, developmental, educational, social, cultural, and biological branches have different lines of research in spatial abilities, and very little has been applied to individual differences in learning. In fact, when considering multidisciplinary applications of spatial abilities, researchers have identified 129 specific concepts incorporated within spatial studies (Grossner & Janell, 2014). There is agreement that spatial abilities are an important component of general intelligence (Johnson & Bouchard, 2003), and, more specifically, of the cognitive processes involved in working memory (Baddeley, 1986). As discussed in Chapter 2, working memory is the cognitive-processing center for both verbal and visual-spatial tasks, where information is held consciously while it is being used.

Some researchers have proposed two primary constructs related to spatial ability: object-based mental representations and manipulations (of two- and three-dimensional objects in space; 2D, 3D), and environmental-based abilities such as mapping, orientation, and perspective taking (Casey, 2013). Most psychoeducational tests of spatial ability are limited to small-scale tasks using visual images and paper folding rather than large-scale tasks such as spatial orientation and navigation (Atit et al., 2020). The interdisciplinary Spatial Reasoning Study Group developed a definition of spatial reasoning through research and consensus. Spatial reasoning:

> refers to the ability to recognize and (mentally) manipulate the spatial properties of objects and the spatial relations among objects. Examples … include: locating, orienting, decomposing/recomposing, balancing, diagramming, symmetry, navigating, comparing, scaling, and visualizing (Davis et al., 2015, p. 48).

This group found, through extensive research reviews, that spatial reasoning is important for many disciplines, but much research is often conducted and reported in isolation (Woolcott et al., 2020). Critically, educational research appears to be in a silo, not contributing to or integrating with the research of fields also studying spatial concepts and their contributions.

Clements and Battista described two major components of spatial ability: *spatial orientation* as "understanding and operating on the relationships between positions of objects in space with respect to one's own position" and *spatial visualization*, the "comprehension and performance of imagined movements of objects in two- and three-dimensional space" (1992, p. 444). Maier (1994) proposed a five-factor model for spatial abilities: *spatial perception*, determine vertical and horizontal position of object despite distracting information; *mental rotation*, rotate visual images of 2D or 3D objects; *spatial relations*, comprehend spatial configurations of objects or their parts, as well as *spatial orientation* and *spatial visualization*. Some researchers identify *disembedding*, or figure-ground perception tasks, as a distinct category (Mix & Cheng, 2012), but it can be considered a perception task.

While there is a lack of agreement on defining and classifying spatial abilities, a useful framework is offered by Uttal et al. (2013). The researchers proposed a two-by-two model, with one dimension considering whether information is intrinsic or extrinsic and the other dimension making a distinction between static and dynamic tasks. Intrinsic information is the specification of parts of an object and the relation among the parts while extrinsic information involves the relation of an object to others. Static information is stable while dynamic involves movement such as folding, cutting, rotating, and flipping. For example, if a student is considering the attributes of a triangle and imagining the same triangle rotated 180 degrees, she is engaged in an intrinsic, dynamic task. These classifications can be used to describe assessment or instructional tasks within geometry and other domains of mathematics (see Figure D.1).

In research with fifth and sixth graders engaged in *making* challenges (using 3D printers, CAD software, circuit boards, and robots), Ramey et al. (2020) identified 13 different types of spatial reasoning in which students were engaged, spanning the two-by-two model. The most commonly used were ES skills (57%; visualizing or describing spatial relations, describing relative size, map reading static), followed by IS (24%; disembedding, categorizing space, quantifying space), ED (11%; perspective taking, dynamic spatial relations), and ID (8%; 2D to 3D translation, mental rotation, mental simulation, mental folding, scaling). The researchers also concluded that students' spatial reasoning developed over time through the use of maker activities plus other tools such as videos and diagrams, was diverse, and contributed to STEAM (STEM + the arts) concept learning and problem solving. In studying the relationship between spatial skills (two-by-two model) and mathematics achievement from ages 6 to 10, Gilligan and colleagues (2019) found that ES (spatial mapping) was a predictor of all math outcomes

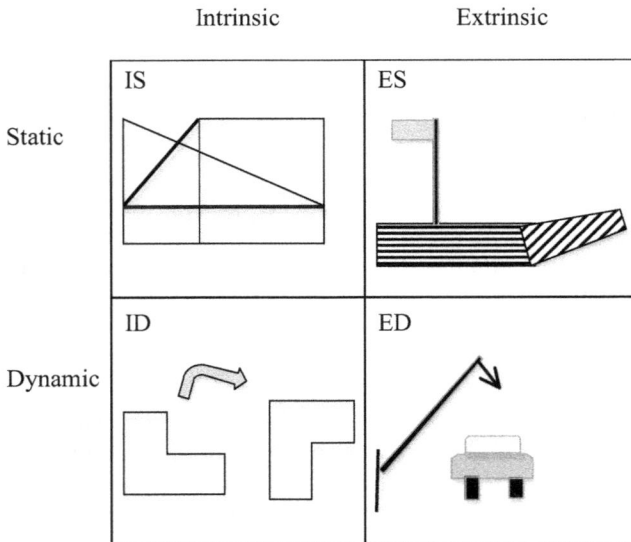

Intrinsic   Extrinsic

Static

IS   ES

Dynamic

ID   ED

Visualization: Can you mentally extract a square from the other parts of the figure in IS? Can you mentally piece together the objects in ID to form a rectangle?

Mental rotation: Can you mentally rotate the figures in ID to new positions?

Orientation: How is the flagpole in ES oriented with relation to the fence?

Spatial relations: How is the larger right triangle in IS related to the rectangle in the lower portion of the figure?

Spatial perception: If you were standing at the flagpole in ES, how would the fence appear? Will the driver of the car in ED pass the crossing gate before it decends?

Figure D.1 Components of Spatial Ability

Adapted from D. H. Uttal et al. (2013). The malleability of spatial skills: A meta-analysis of training studies. *Psychological Bulletin, 139*, p. 354. American Psychological Association. Adapted with permission.

for all age groups and ID (mental rotation) was a predictor at ages 6 and 7. Spatial skills were significant predictors of overall mathematics performance, explaining 5%–14% of individual variation. Resnick et al. (2019) emphasized that, although spatial skills can be disassociated for study, they are necessarily interconnected cognitively and the strength of the connection between specific spatial skills and mathematics content depends on the task. Tasks can vary from close, "desktop" tasks such as interpreting diagrams, to broader and more complex tasks such as drawing a cross section of the geology of a hill given a scale model and point of perspective.

The NRC illuminated the differences in forms of thinking (e.g., verbal, logical, statistical, spatial) by differences in their representation systems (NRC, 2006). Spatial thinking is "a constructive amalgam of three elements: concepts of space, tools of representation, and processes of reasoning" (p. 12). The concept of space, asserted the report, makes spatial thinking distinctive. We use properties of space (e.g., dimensionality, proximity, separation, continuity) for the problem-solving process. We use representations in various modes and media to describe, explain, and communicate about objects and their relationships.

The concept of *spatial sense* was introduced to most mathematics educators by the National Council of Teachers of Mathematics' 1989 curriculum standards (NCTM). Spatial sense was described as "an intuitive feel for one's surroundings and the objects in them" (p. 49). Spatial understandings, the document clarified, "are necessary for interpreting, understanding, and appreciating our inherently geometric world" (p. 48). Spatial sense was used to describe the abilities underpinning geometry as number sense was used as a basis for numerical understanding, to emphasize the inherent nature of the ability; however, the document did not extend the impact of spatial abilities to other curriculum domains. Although spatial abilities are biologically situated and activated (e.g., angular gyrus, middle temporal gyrus, and fusiform), they are not rigid (Pyke et al., 2015). Spatial abilities are, in fact, malleable and a wide variety of informal activities, training procedures, and classroom experiences can lead to lasting improvements (Cheng & Mix, 2014; Owens, 2015; Uttal et al., 2013).

Development and Assessment of Spatial Abilities

Everyone develops some sense of space by interacting with objects in the environment. As discussed in Chapter 2, young children come to school with informal mathematics knowledge, but older students and adults also continue to learn informally through everyday experiences. This informal knowledge of students can be tapped to enhance formal and guided-learning activities. Often, individuals are not even aware of their own tacit knowledge (Gravemeijer, 1998). For example, you may have never studied art or architecture but when you view a drawing such as the one in Figure D.2, you know that the tree is really larger than the window. You are able to interpret, without formal training, the relative sizes and distances depicted in the drawing.

Teachers can determine students' informal knowledge (or misconceptions) about spatial concepts through informal assessments, listening to students describe objects and their relationships. Before teaching a unit on quadrilaterals, ask students to name and describe everyday shapes using their own language. Before teaching a unit on volume, ask students to explain, in their own words, how people measure the contents of a car's gasoline tank. Before you teach about number magnitude on a 0 to 20 number line, ask students to show where benchmark numbers (5, 10, 15) should be located. Pay attention to spatial language used and features of objects that were of focus. This knowledge of students' prior formal and informal

Figure D.2 Perspective at Work

understanding is invaluable for teachers in planning meaningful instruction and assessing new learning. However, mathematical reasoning about space based on informal knowledge develops slowly and does not evolve spontaneously (Lehrer et al., 1999). Spatial ability training should begin early, with preschoolers, and continue throughout the grade levels (Ontario Ministry of Education, 2014; Xie et al., 2020). Teachers must provide high-quality and explicit guided-learning experiences.

Development of Spatial Reasoning

Children develop initial spatial abilities before their first birthday: one-to-two-month-old infants respond to expanded distances of objects; two-month-olds notice depth differences, three-month-olds discriminate between shapes such as rectangle and trapezoid; four-month-olds are sensitive to rotations of two-dimensional shapes; and five-month-olds recognize that an object has been moved just a few inches (Liben & Christensen, 2010). These young children tend to use egocentric frames of reference for object location, using their own body positions as references. Children between 12 and 24 months begin attending to external cues to locate objects, relying on landmark and geometric cues (e.g., location of door, position of a dot). By three years old, toddlers have a very basic sense of abstract representations, that a block on a table can represent the chair in the room. By kindergarten, children are increasingly accurate in reconstructing a model town based on their own experiences walking around in the town. Preschoolers often use inconsistent scales and perspectives but are able to locate isolated landmarks or clusters of places. Young children do better on spatial tasks when they have been actively involved with the environments and objects and when their attention is focused through guided questions to portions of the environment or object. Children's strategies for interpreting these aspects of the environment become more diverse and flexible with experience.

Researchers have also noted gender differences in spatial abilities, from both biological and environmental sources. Biological factors may include spatial abilities controlled by the X-linked recessive gene (phenotype would be seen in males) and exposure to sex hormones (e.g., prenatal, adolescent, stress). Experiential effects, such as encouraging boys to play with construction objects while encouraging girls to play with dolls, have been associated with differences in spatial abilities (Halpern, 2000). Further, the effects strengthen when children with higher spatial skills gravitate toward spatially challenging activities (NRC, 2006). Peterson and colleagues (2020) found that adolescents who participated in more spatial activities during childhood participated in more spatial activities and demonstrated stronger spatial-thinking skills and strategies in high school, a relationship that was firmer for males than females. Females, because of fewer early spatial experiences, may have lower confidence in their abilities and avoid spatial reasoning as a tool (Nardi et al., 2013). Training in spatial tasks results in improvements for both genders, but the gender gap will not necessarily be eliminated, except for individuals who have increased interest in spatial challenges and are more motivated to engage in spatial activities (Geer et al., 2019; Uttal et al., 2013). In a study of fifth and eighth graders, Harris et al. (2020) found that mental rotation and spatial visualization varied by gender, and that perspective-taking did not. The researchers found that verbal reasoning could support females' spatial reasoning on more complex tasks in eighth grade. Content knowledge and spatial reasoning supported more flexible strategies for male students on more complex tasks. Overall, the research on spatial skills and gender has been mixed (Peterson et al., 2020). Important contributors for an individual student may be early childhood spatial experiences, cognitive development (e.g., working memory), types of mathematical tasks (spatial demands, familiar or new concepts), and experience with spatial activities including technologies. Teachers and caregivers should be aware of how students view their spatial abilities, not reinforce negative self-assessments, and build in a range of experiences that engage all students in spatial activities.

The psychologists Piaget, Inhelder, and Szeminska studied spatial concepts in children in the 1940s (1948/1960). Major tenets of their work included:

- Representations of space are constructed through a developmental process involving the organization of motor and cognitive (internalized) skills, therefore spatial reasoning is not just perceptual but an accumulation of prior experience with the environment and cognitive activity.
- The organization of geometric ideas follows a specific order—topological relationships, projective space, and Euclidian relations. Topological discrimination is the ability to recognize features of figures such as closed or open and curved or linear (ages three to four for simple shapes; ages six to seven for more difficult shapes). Projective space begins when the figure is considered in relation to a point of view or perspective (about age seven). Euclidian space involves the development of a two-dimensional framework and metric relations for making spatial connections, about nine years of age.

The Piagetian theories have been criticized for inaccurate uses of mathematical terms such as *topological* and *Euclidian* , and children's drawings to accurately represent their thinking. Others have criticized the underestimation of young children's spatial abilities. Further, there is some evidence that all three relational abilities are developing somewhat simultaneously and at some point become integrated (Clements & Battista, 1992). Piaget also incorrectly implied that adults' spatial abilities are developed and accurate, when, in fact, some adults' spatial abilities are at the level of elementary children (Newcombe & Huttenlocher, 2000).

One theoretical model of the development of spatial and geometric reasoning is a modification of the van Hieles' levels of geometric thinking (Clements & Battista, 1992; Clements et al., 1999; van Hiele, 1959/1984):

- *Level 0: Pre-recognition (proposed by* Clements & Battista, 1992*)*. Children perceive geometric shapes but attend to only part of the shapes' characteristics. For example, a four-year-old sorts attribute blocks by color. When asked to sort another way, she is unable to view other attributes.
- *Level 1: Recognition (Visual)*. Students identify and operate on shapes according to their appearance and can mentally represent figures as visual images. This was the first level in the original van Hiele model (Figure D.3). Clements et al. (1999) termed this the *syncretic* level because of their findings that the synthesis of verbal declarative and imagistic knowledge is at work, not just visual matching. For example, young students worked with cut-out rectangles and triangles one day. The next day the teacher asked them to draw a rectangle and a triangle from memory. They could visualize a three-sided shape and a four-sided shape.
- *Level 2: Analysis*. Students can characterize shapes by their properties but do not see the relationships between classes of figures. For example, a student can sort two-dimensional objects by shape, size, and texture but do not see the similarities between quadrilaterals, rectangles, and squares.
- *Level 3: Logical ordering (Informal deduction, abstraction)*. Students can classify figures and give informal arguments. One student justified his argument that a square is also a rectangle by comparing their defining attributes. He could relate previously discovered properties and argue informally.

0 - Prerecognition (Clements & Battista, 1992)	Children begin to visualize shapes, such as triangles and circles, but are not consistently discriminating among shapes. They are beginning to form concepts of shapes. 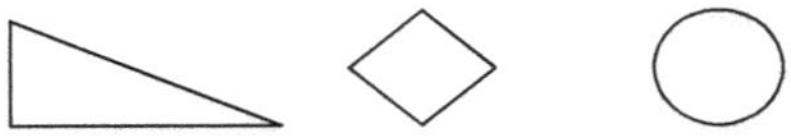
1 – Visualization/ Recognition	Children at this level appear to be matching prototypes of shapes and are beginning to recognize components and properties of shapes. 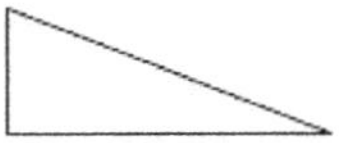 *Three lines form this shape, and it has three corners.*
2 – Analysis	Students begin to understand the properties of shapes and the names for those characteristics but do not make connections yet. *This is a triangle because it has three straight sides and three angles.* 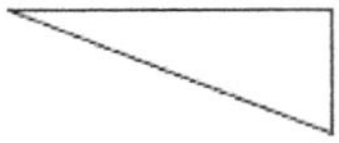 *This is a triangle, too.* 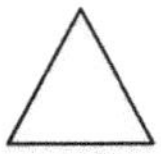
3 – Abstraction/ Logical Ordering	Students can classify shapes and make informal arguments. *This is a rectangle because it has four sides that are parallel and four angles that are 90°.* 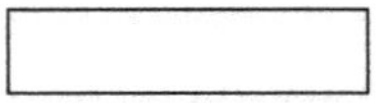 *This is a special rectangle called a square because all four sides are equal.*
4 – Formal Deduction	Students can construct proofs. *Through a series of arguments, I can prove that the sum of the three angles of this triangle equal 180°.* 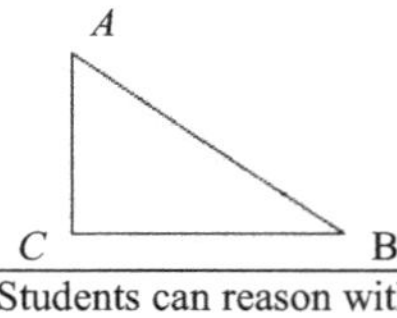 $A + B + C = 180°$
5 – Rigor/ Meta-Mathematical	Students can reason without reference models. If the area of a rectangle is L x W, then the area of a right triangle will be (L x W)/2, or one-half the base times the height.

Figure D.3 Van Hiele Levels Illustrated

- *Level 4: Formal deduction.* Students can use an axiomatic system to establish theorems and construct original proofs. A middle-school student can use Euclid's axioms about lines, circles, and angles to construct a proof of the length of a line segment in relation to a circle's radius.
- *Level 5: Rigor/meta-mathematical.* Students can reason formally about mathematical systems, thinking about figures and relationships without reference models. A high-school student views the formula $a^2 + b^2 = c^2$ where c is the hypotenuse of a right triangle and solve for c by taking the square root of the sum of squares.

Many studies have confirmed these levels as useful in describing students' geometric concept development. Initially the levels were interpreted as discrete, however research since the 1980s suggests that students may display different levels of thinking for different tasks or contexts. Clements and Battista (1992) asserted that the levels may develop simultaneously in some situations, and that there may be transition periods between levels. It is important to note that moving up in levels is a process of achieving a higher level of abstraction and generalization and that all students will work at lower levels when they choose. Further, communication (discourse) between teacher and student and the use of tools during learning activities, the interaction between verbal and visual, will have an impact on students' spatial development (Sinclair et al., 2018). There is some evidence that students do not enter Level 2 until sixth or seventh grade and some students, especially low achievers, will still be performing at Level 0 or 1 in middle school (Fuys et al., 1988). Movement to Level 2 requires guided instruction because it involves technical language and recognizing new concept relationships. Some researchers have estimated that 40% of students in the US finish high school at or below Level 2, clearly not prepared for college mathematics (Burger & Shaughnessy, 1986). High-school teachers of mathematics may be thinking, teaching, and using spatial language at a Level 4 or 5, while most students beginning high-school geometry are at Level 1 or 2, causing a lack of concept understanding and students resorting to memorization, not being equipped to apply the concepts (Mason, 1998).

The van Hieles asserted, and others confirmed, that progress through these levels is not due to biological maturation but by experience, both informal and formal learning. Teachers should facilitate student growth through active and constructive but guided-learning opportunities. Clements and Battista (1992) proposed setting earlier geometric goals for children, achieving Levels 2 or 3 in the presecondary curriculum; employing more precise language in geometry instruction; and using manipulatives and real-world objects to support concept understanding. Additionally, teachers' levels of spatial reasoning should be more advanced than that of their students, to "do no harm" and to be able to extend and expand students' spatial-reasoning abilities (Clements & Sarama, 2011, p. 136). Some studies have found that only 30%–45% of preservice teachers had achieved Level 3, and as many as 67% were at Level 1 or below (Clements & Sarama, 2011; de Villiers, 2010).

Other researchers have extended the van Hiele levels to three-dimensional shapes (Gutiérrez et al., 1991). At the first level, students view solids based on the whole, without considering components. At the second level, attributes such as shapes of objects' faces are recognized but not related. The third level brings meaningful definitions and logical classification based on attribute relationships and the fourth level involves proving theorems related to three-dimensional geometry.

Assessment of Spatial Abilities

Research on the development of spatial abilities relies on valid and reliable assessment of cognitive abilities, elusive and uneven abilities measured indirectly using overt tasks. Assessment, in turn, often depends on language or drawing skills that can skew results in favor of some children and to the detriment of others. Educators should consider appropriate assessment strategies when evaluating research on spatial abilities and screening students for potential spatial deficits.

Yakimanskaya (1980/1991) characterized the spatial aptitude tests of Western psychologists as quick measures of "visual acuity, keenness of observation, and quick-wittedness" because subjects were asked to view artificial symbols or diagrams and evaluate interrelationships and patterns (p. 141). These tests assess visual acuity, the ability to evaluate lines for size and position, and the ability to modify figures visually, and are only part of spatial reasoning and not connected with mathematical or other concept development or instruction. Yakimanskaya and her fellow Soviet researchers theorized levels of spatial thinking that, unlike the van Hieles' and Piagetian theories, were sensitive to individual differences and assumed that developmental differences were due to instructional differences (social learning theory). The Soviets described spatial thinking as "a multi-faceted, hierarchical whole, and essentially multi-functional" (1980/1991, p. 102).

Yakimanskaya's research team developed diagnostic problems that included a range of complexity levels, offered a view into problem-solving processes (not just solutions), were not dependent on specific curricular backgrounds, varied in graphic type, and were relatively brief in form. For example, one problem asked subjects to view a drawing of an overlapping circle, triangle, and rectangle with instructions to shade the intersection of the three figures. Another problem depicted four congruent right triangles and asked subjects to mentally create and then draw other figures that use all four triangles (triangle, rectangle, rhombus, trapezoid, hexagon, and non-rhombus, non-rectangular parallelogram).

These diagnostic problems were used in research with middle-school students to determine individual differences in spatial reasoning (Yakimanskaya, 1980/1991). Three groups of students emerged: a highly creative group that solved most of the problems (80%–90%) quickly and with flexibility (3–4 min each); a second group that could handle many of the problems (60%–70%) but less efficiently (5–6 min) and with less creativity; and a third group that found most problems difficult and solved only 30%–40% in 6 to 9 minutes each and were described as stubborn problem solvers with low independence. In this and subsequent studies, the researchers concluded that individual student differences can be observed in ability to create static and dynamic images and flexibility in creating and manipulating images.

Two general types of visual-spatial assessments that inform mathematics instruction can be described as static and dynamic. *Static* assessments are those that are snapshots of a student's ability, typically by viewing drawings of two-and three-dimensional objects and finding the same object flipped or turned. *Dynamic* assessments attempt to capture processes, movement, and change through student interaction with objects, drawings, and concepts. For example, a student is shown a rectangle and asked to draw as many lines of symmetry for the rectangle as possible. In this task the researcher is able to detect the student's approach to the problem (such as paper folding), self-corrections, comments about the task, and duration of the activity. Mathematics skills are related to both static and dynamic tasks (Reuhkala, 2001), so both types of assessments may be needed. However, many studies and screenings incorporate only static assessments.

Some educators are concerned that students with spatial deficits (and related MD) are not identified as early as those with verbal deficits, sometimes not until the geometric concept demands of middle school. Spatial abilities should be screened early, upon school entry, and in elementary and secondary grades within multi-tiered systems of support (Tier 1). Concerned with the paucity of valid means for early identification, Cornoldi and colleagues (2003) developed an 18-item screening measure for visual-spatial learning disabilities. The *Shortened Visuospatial Questionnaire* (SVS) was developed for teachers to rate items such as comprehension of visual-spatial relations, spatial orientation, skills in observing the environment, and dealing with novel objects. The purpose of this simple screening device was for earlier identification of children who may need further assessment for visual-spatial deficits. Research with Italian and British children demonstrated validity for this purpose (0.90) and appropriate reliability (0.95) for teacher use.

Hawes et al. (2015) found that their measure of mental rotation for four-to-eight-year-olds was valid and reliable for differentiating one aspect of spatial ability. The researchers used three-dimensional blocks, rather than paper-and-pencil tasks, and found that children progressed from guessers to mirror confused (confused a block group with its mirror image), to successful rotators, indicating steady growth in mental rotation between ages four to eight, with no gender differences in their low SES sample of children. The authors of the study suggested that the sequence of mental rotation development they identified be the instructional sequence for young children. Early screening measures, such as this mental-rotation task, can target students for interventions. Early interventions have the largest and longest effects (Heckman, 2006).

Ramful et al. (2017) developed and validated a spatial ability measure for middle-school students, the *Spatial Reasoning Instrument*. The three constructs measured are mental rotation, spatial orientation, and spatial visualization, tasks developed from the examination of curricula, textbooks, and a range of spatial reasoning instruments that are primarily related to geometry and measurement. The 30-item instrument exhibited sound psychometric properties, included normative tables, and employed a multiple-choice format that can be administered to groups of students.

The best assessment tasks for informing instruction are informal, diagnostic tasks where the evaluator can watch and listen to the student's reasoning during the task (Gillum, 2014; Mason, 1998). Owens (2015) proposed a framework for classifying these informal visual-spatial reasoning tasks and subsequent instructional activities by the strategies they employ. Children move from emerging strategies to perceptual, pictorial, pattern and dynamic, and finally efficient strategies. Figure D.4 illustrates how these levels might be applied to an assessment task for triangles.

Example Assessment Task:

"Look at these shapes and see if you can pick out the triangles."

If student cannot pick out a triangle, pick one out and ask, "This is a shape, what do you know about it?"

If student picks out several triangles, ask, "What can you tell me about them?"

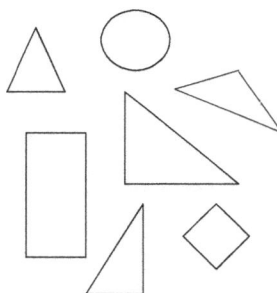

If the student uses strategies that are:	Example responses:
Emerging	…she may say "this one looks like a street sign" or "this one is round."
Perceptual	…she puts the triangles and rectangles together to make a new shape. When asked about a specific triangle, she says "triangle" and matches it to another triangle.
Pictoral Imagery	…she points to the triangle and says, "The triangle has three corners and three sides." When shown a tangram with triangles embedded, she can point to the triangles.
Pattern and Dynamic Imagery	..she looks at the triangle shapes and says, "I can make a new, bigger shape using those triangles. Let me show you." She plans and can mentally picture the new shape before rotating and flipping the triangles to make the shape.
Efficient	…she says, "Well, I know that these two are right triangles and that one is isoceles. Two right triangles placed together will have two angles with 90 degrees each for 180 degrees, or a straight line on their edge."

Figure D.4 Assessment Task for Visual-Spatial Reasoning

Research with Students with Mathematics Disabilities and Difficulties

Visual-spatial deficits have been hypothesized as a subtype of mathematics learning disabilities (MLD) by a number of researchers (Cornoldi et al., 2003; Geary, 2003; Gross-Tsur et al., 1995). However, the nature and impact of these deficits has not been adequately developed by research. There is evidence that elementary and secondary students with mathematics learning difficulties (MD) differ from their typically achieving (TA) peers on measures of visual-spatial processing and visual-spatial working memory (VSWM), especially for the lower-achieving students with MLD (Li & Geary, 2017; Passolunghi & Mammarella, 2012; Swanson, 2012). In a five-year longitudinal study from kindergarten through fourth grade, Zhang et al. (2020) found significant differences in spatial visualization among subgroups of students. Students with MLD lagged increasingly behind low-achieving (MD) and TA students in mathematics. Spatial visualization measured before school entry, as well as counting sequence knowledge and rapid automatized naming, differentiated student groups. The authors concluded that spatial-visualization deficits should be considered a risk factor for MLD. These visual-spatial deficits have an impact across mathematical domains, including number representation and problem solving.

Brain-imaging studies have found that students with MLD use VSWM differently than their TA peers (e.g., Ashkenazi et al., 2013). Differences in performing numerical operations, acquiring arithmetic concepts, and reasoning during problem solving can be explained by VSWM abilities. Students with MLD tended to not engage their VSWM resources appropriately during problem solving. Additionally, other areas of mathematics that require nonverbal skills, such as representing magnitudes on a number line, estimation of quantity, counting, and subitizing, were impacted by VSWM deficits. The researchers concluded that, "a common locus of deficits in visual-spatial representations and memory contributes to both numerical and arithmetic deficits in children with MLD" (Ashkenazi et al., 2013, p. 2315).

Other researchers have studied problem-solving skills of students with MLD and found spatial abilities key in identifying differences in student performance. Passolunghi and Mammarella (2012) studied the problem-solving skills of students in Grades 3 through 5 with MLD and their TA peers. Students with MLD failed simple and complex spatial working-memory (WM) tasks but were comparable to their peers on visual WM tasks. Students with MLD had problems with controlled attention and a higher number of intrusion errors (inability to ignore irrelevant information). The authors recommended that students with MLD be trained in using spatial (not pictorial) representations as a tool for problem solving, beginning with low-level, attention-demanding problems. Van Garderen (2006) studied sixth-grader problem solving and found that gifted students used significantly more visual images than students with MLD. Students who used schematic imagery performed better on problem solving than those who used pictorial imagery, and students with low spatial-visualization ability tended to use simplistic pictorial imagery.

Mix and Cheng (2012), in a literature review on spatial abilities and mathematics, noted that spatial abilities are related to mathematics very early and that the relationship is bidirectional: spatial ability supports mathematics learning and mathematics learning boosts spatial ability. Spatial visualization in kindergarten predicted mathematics achievement in first grade and beyond. Spatial visualization for ninth graders was predictive of upper-level mathematics and career achievement. Students with visual-spatial deficits lose ground as they age, therefore early identification and intervention is vital. Across two studies of visuospatial memory gains, Li and Geary (2013, 2017) concluded that students with the largest gains from first to fifth grade, and on to the end of ninth grade had higher mathematics achievement by the end of fifth and ninth grades. Visuospatial memory emerged as a unique contributor to mathematics achievement after elementary school. A one-time assessment of visuospatial memory in young children will not be enough; it would underestimate the importance of these competencies for later achievement.

Research on content-specific geometry interventions for students with MLD and MD has increased in recent years, perhaps with the growing knowledge of the impact of spatial abilities on mathematics learning. Liu and colleagues (2021) located nine geometry intervention studies with students with MLD (since 1975) and synthesized their results. Eight of the studies were conducted in middle schools and one with fourth and fifth graders; however, the topics ranged from third to fifth grade geometry standards (e.g., perimeter, area, angle, volume) except for one study involving seventh-grade geometry word problems (Satsangi et al., 2019). The studies investigated instructional strategies (e.g., CRA instructional sequence) and/or technologies (e.g., video modeling, virtual manipulatives). Students receiving one-on-one instruction improved their geometry performance with large effect sizes (Tau-U > 0.80) and students in small-group instruction showed medium to large effect sizes for these interventions. The researchers concluded that students with MLD can improve their geometry concepts and skills through interventions that employ explicit instruction, sequential practice, multiple representations, selected technology, and real-life examples for applications. They emphasized the importance of effective elementary geometry instruction as a foundation for more advanced topics.

Developing Spatial Abilities across the Mathematics Curriculum

The influential NCTM curriculum standards (2000) emphasized geometry throughout the PreK-12 mathematics curriculum, with important applications for the process standards. The CCSSM (NGA & CCSSO, 2010) also emphasized geometric concepts and related skills across the grade levels, often integrated with measurement and practice standards. However, while both sets of standards emphasized the use of patterns, models, and representations, neither explicitly describes the role of spatial reasoning—including visualization, orientation, mental manipulation—for topics across the mathematics curriculum, beyond geometry and measurement.

Spatial concepts and skills should be incorporated across curricular topics and grade levels. For example, young children at play should be encouraged to use spatial words such as near and far, beside and behind, bigger and smaller, sides and middle, and so forth. Third-grade science investigations into speed and motion should include data collection and graphing for students to study patterns. A sixth grader studying world geography will examine maps to understand the movement of civilizations with relation to natural features such as rivers and mountains. A class of third graders, reading *Scarecrow* (Rylant, 1998), discusses the pieces that form the scarecrow's body and the shapes of various flowers and vegetables in the field, how the author's language describes those objects. The teacher explains that they see the story from the scarecrow's perspective and then challenges his students to consider it from the perspectives of the birds and the girl planning her garden. High-school students in an integrated algebra and geometry course are solving systems of linear and quadratic equations in two variables and finding the points of intersection of a line ($y = -3x$) and circle ($x^2 + y^2 = 3$) algebraically and graphically.

Spatial abilities have been connected with skills in relative magnitude of number (Gilligan et al., 2020; Tian & Siegler, 2017), arithmetic operations (especially subtraction; Cheng & Mix, 2014), fractions (Jordan et al., 2017), proportional reasoning (Möhring et al., 2016), algebra (Lowrie et al., 2019), logical reasoning (Xie et al., 2020), and word-problem solving (Lowrie et al., 2019). Connections have been found across the grade levels, however there is some evidence that specific spatial subdomains are more closely related to specific mathematics tasks at different levels (Gilligan et al., 2019; Mix et al., 2016). For example, mental rotation for six- and seven-year-olds on number-line estimation or spatial scaling for seventh-grade proportional reasoning. Mix and colleagues (2016), in studying K, 3rd, and 6th graders, found spatial skills were most connected with place value across these grades, word problems in K and 6th, calculation in K, fraction concepts in 3rd, and algebra in 6th grade, indicating the significant cross-domain relations for specific tasks. Spatial reasoning is integral for all forms of learning and cognition (Kell et al., 2013). Mathematics curricula today focus too narrowly on number concepts and should be refocused to include more explicit spatial instructional activities. Most importantly, spatial-skill training should begin early, in preschool, so that children entering school can begin formal mathematics with additional tools.

In addition to integrating spatial concepts and skills across the curriculum, there are other universal recommendations, regardless of grade level.

Students should be exposed to clear and specific language for describing spatial concepts. After children use their informal language in preschool and kindergarten (e.g., the pointy part of the shape), teachers should introduce more exact vocabulary so that students can explain their thinking without grasping for words. For example, the terms line, point, and plane have very specific applications in Euclidian geometry. A line is not just any marking but "the set of points that move directly between two points" and continue infinitely in both directions. A line segment has two ending points. A ray has one ending point (Sidebotham, 2002). But the side of a three-dimensional shape is a special form of line called an *edge* and edges and planes meet at special points called *vertices* (*vertex* is the singular). A line graph uses lines segments between points identified by coordinates. Special Greek- and Latin-based morphemes, such as quad-, poly-, and tri-, should be taught as units of meaning that can be connected to form mathematics terms.

Teachers should explicitly teach students how to interpret spatial representations, such as diagrams, graphs, and objects, drawing attention to critical attributes and relationships among objects and their parts and positions. Most textbooks include two-dimensional representations (diagrams, graphs, charts) and some include drawings that represent three-dimensional objects. However, few textbooks provide assistance to students or teachers in using the representations provided (van Garderen et al., 2012). Students need to use representations effectively in order to understand concepts and solve problems. Students with MD may need explicit and focused instruction on what a representation is, how it can be interpreted and used, and how to generate one's own representations. Teachers should select spatial representations carefully and judiciously, providing more opportunities for students to generate their own.

Teachers at all grade levels should model cognitive-imaging processes by employing think-aloud techniques. For example, if the teacher is demonstrating how to create and manipulate a mental image of a transformed triangle, he could describe aloud his own thoughts:

> I am now holding the image of this triangle in my mind, like a picture inside a camera. I see the three sides of the triangle and one angle is a 90-degree corner, called a right angle. Can you draw that? Your image might look different than mine, but both have a right angle and three sides. Now I am changing the triangle in my head so that none of the angles is a right angle. Can you imagine that triangle also? I'm going to draw mine and I want you to draw what you think I am seeing in my head.

Teachers should plan and use good models, especially for novel content. Hands-on objects, drawings, and computer simulations are technologies that enhance spatial abilities and concept understanding. Students should be encouraged to create their own models, at an age-appropriate level, and to describe and manipulate them. Manipulative use that is guided and connected to concepts has been demonstrated to benefit spatial concept learning at all grade and ability levels (Sowell, 1989). Younger students can trace and cut out two-dimensional shapes that can also be turned and folded. They can use common containers to represent three-dimensional shapes such as cubes and cylinders. Older students can construct two- and three-dimensional objects with increasingly sophisticated tools and precision. Students with fine-motor problems may need to use computer models to represent shapes at their level of understanding. Computer modeling is excellent for demonstrating a range of spatial concepts once children are able to understand what a computer image represents.

Spatial reasoning activities should be embedded within mathematics topics across the school year, not isolated in a geometry unit, in appropriate contexts. For example, when teaching about fractions, use number lines and fraction bars to demonstrate units and portions of units. When providing instruction in problem solving, teach students to diagram the problem type. For working with linear or quadratic equations, have students visualize the effects of changing the equation graphed. When working with 3D objects, have students sketch the object from a different perspective or a cross section. In a measurement unit, have students mentally estimate the number of units for a linear measurement or visualize an array for area.

Lowrie and colleagues (2019) implemented a spatial-intervention program within 32 eighth-grade classrooms taught by their regular teachers in 12 hours over ten weeks. The framework for the intervention was an experience (prior knowledge)-language-pictorial-symbolic-application (ELPSA) approach, involving a structured way to embed spatial lessons that included many elements of explicit instruction with a CRA sequence. Lessons involved 2D and 3D representations and the spatial constructs of mental rotation, spatial orientation, and spatial visualization. The intervention significantly improved students' mental rotation and perspective taking performance, but not spatial visualization, as compared with control classrooms studying geometry and measurement. Improved performance by the intervention students extended beyond geometry outcomes to number and algebra content as well as graphic and nongraphic word-problem solving.

Others have concluded that, given the complexity of spatial and mathematical domains and their interconnections, it is difficult to target spatial-skill training (Mix et al., 2016; Young et al., 2018). It may be more effective to focus on general spatial skills development for younger students, in preschool and the early elementary grades, and to leverage spatial skills when introducing novel content across the grade levels. For example, Mix and colleagues (2016), in studying kindergarten, third-, and sixth-grade students, analyzed the roles of novel versus familiar mathematics content for their relationship to spatial skills. The researchers found some support for this theory, such as for block-design skills at each grade level related to success in new mathematics content. Block design requires mental visualization, maintaining multiple orientations, and analyzing parts. They also noted a developmental shift from strong relations with spatial visualization when tasks are new or need grounding to strong relations between spatial perception (quickly recognizing shapes and their parts) and mathematics when tasks are more automatic and procedural. Mathematics may be an inherently spatial domain, with a range of interrelated skills required at different levels of development and for varying content.

Other research-supported materials and methods for developing spatial skills include:

- Dot paper, graph paper, and other specially designed paper is easily accessible on the internet and can be used across the grade levels to support two-dimensional concepts such as area, perimeter, congruence, symmetry, angle, patterns, shapes, and *tessellations*. Teachers should guide activities such as representing area in different ways or demonstrating visual patterns.
- Paper-folding activities can be adapted to any grade level. For example, Arici and Aslan-Tutak (2015) demonstrated that *origami*, the Japanese art of paper folding, used with tenth-grade geometry over a four-week period, had a significant impact on geometry achievement and reasoning in the experimental group as compared with a traditional-instruction

control group. The lessons addressed the properties of triangles, such as classification, angle bisectors, and relationships among bisectors and other measures of triangles.

- When using visual representations for concepts such as fractions, geometric shapes, or graphing, many students have difficulties perceiving parts from others or recognizing familiar elements. Encourage students to use shading and color coding to bring out important aspects of the objects or their drawings.

- Geometry games can link verbal or written descriptions with visual representations. For example, *The Barrier Game* (Moss et al., 2016) involves two students—the designer and the builder, with a barrier in between. Students are involved in visualizing, verbalizing, and verifying as one builds a shape (e.g., with square tiles) and describes the shape for the other student to draw and ask questions about as they attempt to draw. Students build their spatial vocabulary and verify their results.

- Students can create nets to represent three-dimensional figures, using paper or two-dimensional net pieces. For example, there are 11 possible nets for creating a cube. Using nets can reinforce concepts such as face, vertices, and edges. Figure D.5 provides several examples of nets.

- The Quick Draw strategy (Wheatley, 1996) promotes imaging through training students to hold specific mental images in their working memory. The teacher projects a geometric drawing (visual image) for two to three seconds, then asks students to draw what they saw. Sometimes a quick second look is provided. Finally, the original image is shown for checking and discussion.

- Having students make geometric drawings, such as area arrays and shapes within shapes can promote spatial reasoning. Sinclair et al. (2018) used 12 geometric drawings in two lessons with kindergarteners. The act of drawing, along with language and gestures used during teacher interactions, enabled the children to develop more geometric language, new concepts (congruence, symmetry, structure), and more flexible spatial-reasoning abilities.

- Upper-elementary and middle-school students have benefited from construction activities creating 3D objects such as pyramids and prisms to explore and understand concepts of surface area, volume, base, and height (Rowan, 1990). Constructions can be made with paper, modeling clay, or plasticine (Casey & Fell, 2018). Students can translate "bird's eye," side views, or cross-sections of their objects (and even contour maps) onto graph paper. Younger students, who may not be as adept with accurate cutting or molding, can use

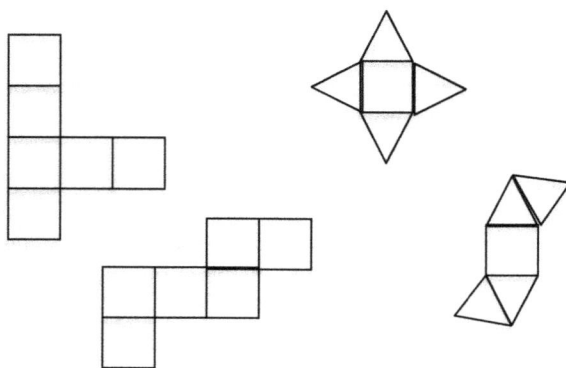

Figure D.5 Nets for Cube and Pyramid

interlocking blocks, wooden proportional blocks, pattern blocks, and common objects to develop concepts about shape, area, and three-dimensional objects.

- Graphing activities can be transformed into orienting and mapping using a grid, scale, and compass. Students can draw a map using a coordinate grid, follow a map given directions, or create a map with a scale during an outdoors adventure.
- Computer-based technology provides endless tools and applications for exploring spatial concepts. Many teachers use *dynamic geometric software* for student constructions. Others use GIS technology to locate landmarks and create maps. Graphing calculators, virtual manipulatives, dynamic and interactive websites, and other tools are available for teachers whose schools have reliable internet access, projection capabilities, and computers for students. Even young students can create 3D representations of 2D objects using technology such as SketchUp Make™, if carefully developed activities are guided by the teacher (Singer & Shafer, 2018). Using dynamic and interactive technology requires more planning by teachers and more initial instructional time for students, but these skills are necessary for upper-level mathematics and many careers.

Teachers who are not confident of their own spatial abilities and skills for geometry tend to rely on textbooks for instruction (van Garderen et al., 2012). Most textbooks promote rule, procedure, and definition instruction in geometry and other mathematics areas involving spatial abilities such as measurement and graphing. This approach to instruction can cause dislike for these important mathematics topics, little concept understanding, and not much generalization. Perhaps more than any other domain, spatial thinking requires interaction with models for understanding. The preceding activities work best in teacher-guided instructional settings that employ explicit language, scaffolding, corrective feedback, individual accommodations or differentiation, and planned connections with grade-level mathematics content.

Challenging Concepts in Geometry

There are a number of major concepts in geometry that are developed over the grade levels and are often integrated with other mathematics domains, especially measurement and algebra. This section illuminates these concepts and provides suggestions for instruction for students with mathematics learning difficulties (MD). The research on interventions in geometry for students with MLD is still sparse.

Attributes and Properties

Beginning in kindergarten, children are expected to analyze and compare two- and three-dimensional shapes and objects, using informal language to describe the attributes that describe and distinguish objects. Attributes in mathematics are the characteristics of these shapes and objects. Children use attributes to sort, compare, and classify. In first grade, students should distinguish between defining attributes (e.g., triangles are closed figures with three sides) and nondefining ones (e.g., the larger triangle or the red square). By third grade, students use shared attributes in classification systems (e.g., all rectangles share specific attributes) and this classification and sub-classification process continues to develop over the grade levels. Most mathematicians use the term *attribute* for a variant characteristic and *property* for a characteristic that is invariant (Panorkou & Greenstein, 2014), consistent with curriculum standards.

Often, young children form narrow and prototypical images of geometric concepts. For example, a child may view a triangle as a shape that sits on its base to the extent that a triangle that is slanted or "sitting" on its pointed end would not be considered a triangle. By viewing shapes, such as the triangle, holistically, children overlook critical properties that distinguish types of triangles and are unable to work with triangles dynamically and retain their properties within the concept of *triangle* when the shape is flipped, turned, or transformed in other ways. Panorkou and Greenstein (2014) proposed a learning trajectory for transformation-based reasoning that begins with attention to attributes in kindergarten, properties in Grades 1 and 2, and looking for commonalities and classifications in Grades 3 and beyond. This hypothetical trajectory bridges Euclidian and topological geometries.

Relationships within and among Shapes

Much of geometry involves reasoning about shapes and their parts, as well as the transformation and comparison of shapes. Two-dimensional shapes take up space on a plane while three-dimensional objects occupy a third dimension as well. Students need time to explore the properties of shapes and objects to develop deep concept understanding. For example, have second-grade students explore different types of triangles after they have a firm concept of triangle. Provide paper shapes for *right, equilateral, isosceles,* and *scalene* triangles with several examples of each, without their labels. *Acute* and *obtuse* are more general categorizations that can be added. Have students in pairs or small groups analyze, discuss, fold, and even cut apart these triangles to explore their properties. Encourage students to use comparison words such as larger, smaller, longer, shorter, part, and combine, as well as math language such as line, angle, vertex, and side. Remind students that the size and color of the whole shape are nondefining attributes and should not be considered. Encourage students to make records of their findings using graph paper and tracings. One group may find that there is a special type of triangle with equal sizes. Another group may discover that cutting off the corners of any of the triangles and connecting those corners forms a flat line on one side (180 degrees). A third group finds that some of the triangles can be folded perfectly in half to form two new triangles that are reflections of each other. There is research support for children's acquisition of geometric categorization concepts by comparing two or more examples within a single category (e.g., two examples of scalene triangles) and also providing non-examples to assist learners in identifying critical category boundaries (Smith et al., 2014).

Rather than teaching 2D and 3D shapes and objects by rules and definitions, take advantage of the sensual nature of geometry—through hands-on activities and multiple representations. For example, Clements and Sarama (2007) developed a curriculum for young children that included the learning trajectories for shape composition, properties of shapes, transformations, and measurement using manipulatives and computer simulations. For shape composition, children moved from pre-composer (unable to combine) to piece-assembler (placing pieces next to each other) to picture-maker (combine shapes to make pictures) and finally to shape-composer, combining shapes to make new shapes or to fit within other shapes with planning and intention. The experimental-group score was significantly greater than that of the control group (like individual tutoring effects). The researchers concluded that organized experiences based on learning trajectories result in greater mathematics knowledge for children entering kindergarten. In a study of the shape composition learning trajectory (LT) with preschoolers (four-year-olds) who were at least two levels below the target instructional level, Clements and colleagues (2019) contrasted instruction following the trajectory and instruction that skipped levels and focused on the target level. Although both interventions had the same goal (shape-composer abilities), the LT treatment was more effective than skip-levels treatment, after eight 9-minute sessions over five weeks ($g = .55$). There are also research-supported LTs for spatial orientation and spatial visualization/imagery (Clements & Sarama, 2014). Supporting children's spatial development through carefully planned and sequenced interventions that involve individualized scaffolding using facilitative language and hands-on materials (e.g., puzzles) can prepare preschoolers for school-entry spatial expectations.

Other geometric relationships that can be explored before measurement and formulas are introduced include congruence, similarity, symmetry, diagonal, perpendicular, parallel, translations, reflections, rotations (from different points), and dilation. During these activities, students are building vocabulary, deepening concept understanding, and extending skills. They are working with materials and representations that build mental images. They are forming the foundations for formulas that make sense and can be applied and generalized. The teacher's role is to provide carefully selected materials, pose questions that prompt exploration and scaffold understanding, encourage students to explain their thinking using precise terminology, and challenge students to create mental images and use what they have learned for new problems. Students will have difficulties with geometric relationships if they lack experiences with both examples and non-examples, are taught rules and formulas by rote, are not given the time to explore concepts, and their mathematics language doesn't support their learning. Many students have gaps in geometric concepts due to isolated and limited instruction. An intensive intervention that employs a learning trajectory approach for attributes, properties, and shapes is depicted in Box D.1.

Box D.1   Intensive Intervention for Geometric Concepts

Target: Four fifth-grade students are struggling in a geometry unit. On the pretest results, teachers noted gaps in geometric language and concept understanding of 2D shapes and their attributes and classifications from 3rd- and 4th-grade competencies.

Intervention Schedule: The small-group instruction will supplant the students' regular classroom instruction in geometry. Daily for 50 minutes until grade-level unit objectives mastered (5–6 weeks).

Nature of Intervention: Each class begins with review of previous concepts and vocabulary, followed by new geometry/spatial task (with construction paper, tools, diagrams), practice with the task emphasizing student explanations, drawings, scaffolding, and feedback. Next is a problem-solving segment with previous concepts. The class ends with active review. Sequence of topics:

1. Attributes of quadrilaterals (parallelogram, rectangle, square, rhombus, trapezoid, kite) and related vocabulary (edges, vertices, angles, etc.).
2. Attributes of other polygons, closed with straight edges (pentagon, hexagon, triangle) with emphasis on prefixes and creating shapes on paper and on the computer, describing the features in mathematical terms.
3. Classification activities and language with previous shapes, examples and non-examples.
4. Combining and decomposing shapes tasks, flipping and rotating, naming shapes and defining attributes.
5. Construct lines, line segments, rays, angles. Discuss and measure angles in terms of degrees and identify their types by defining attributes.
6. Examine triangles with attention to their angles. Identify and construct right triangles and discuss their properties (perpendicular, right angle).
7. Review 2D shapes and their attributes and properties. Explore concept of symmetry and how to find symmetry for the range of polygons.
8. Review classification of 2D figures and create a hierarchy.
9. Integrate problems with perimeter and area (review of measurement objectives and related concepts, vocabulary).

10. Review and practice line graphing on one axis. Introduce the coordinate plane as a pair of perpendicular number lines (called axes). Practice locating points for coordinate pairs, using math language (on paper and computer).
11. Review the concept of volume with real-world examples. Incorporate 2D language when describing parts of prisms (including cubes). Practice with a range of examples, leading to volume formulas.
12. Culminating project: Cereal box design.

Data Collection (Weekly): CBMs on range of geometric objectives (Grades 3–5). Adjust scope and intensity based on individual responses.

Perimeter, Area, and Volume

The concepts of perimeter, area, and volume build upon students' understanding of the properties of shapes and their conceptions about the measurement of length. Measurement involves the concepts of unit and the iteration of units. Measuring length, as is discussed in Strand E, progresses from comparing lengths to using unnamed units to using recognized units consistently. To measure perimeter, students must coordinate their conceptions of length and consider the sum of the lengths of each side of a polygon, or the total distance around a circle. Activities using everyday problems make perimeter concepts less abstract. For example, "We need to fence in the school garden." The units for perimeters will be linear and one-dimensional: inches, feet, meters.

Area is a more challenging concept, however, and instruction often centers on procedures for measuring rather than area concepts. Students need foundational experiences with partitioning a region (with a two-dimensional unit), iterating a unit to cover a region (without overlaps and spaces), conservation of area (cutting and rearranging), and structuring an array (McDuffie & Eve, 2009). These concepts are generally developed between Grades 2 and 4. Like the trajectory for linear measurement, area measurement begins with nonstandard units, as in the example: *How many of the blue tiles (units) will cover the entire area of the trapezoid?* Students can compare areas of shapes using these nonstandard units: *The area of the trapezoid is 3 units, while the area of the rectangle is 6 units.* Only when the concept of area is firm should standard units and measurements be introduced. Since area is a measure of two lengths (commonly length by width), then the standard unit of measure is a square unit: *The rectangle measures 5 inches on one side and 7 inches on another. The area of the rectangle is 35 square inches.* Students will not adopt the concept of *square units* without first moving through other explorations of surface area of shapes. Further, they will not understand the area formula for a triangle and other shapes without these prior explorations.

Satsangi, Bouck, and colleagues (2015) used virtual manipulatives (VM; polynomials formed by arranging square tiles) and learning sheets with diagrams to teach perimeter and area concepts and procedures as interventions for students with MLD. The VMs allowed students to create and transform shapes for a better understanding of properties than static images. Students mastered the concepts quickly (over 5–9 sessions), but perimeter took longer than area for two of the three students, possibly due to a higher cognitive load. The researchers also used video modeling to teach geometry word-problem solving to three secondary students with MLD (Satsangi et al., 2019). Five-minute videos were created to show four geometry skills: find perimeter (P) of a square using area (A), find A of a square using P, solve for length of a side of rectangle given width and P, solve for P of rectangle given length and width. Each video used voice-over narration, explicit instruction, review of key vocabulary and concepts, and step-by-step demonstrations. The students needed researcher prompts and error corrections during the treatment phases, indicating a teacher is still needed to support student learning for video modeling. All three students improved from baseline (0%–8%) to post-intervention (90%–100%) and maintenance (96%–100%), after five to six sessions, needing less teacher support and prompting after mastery. The research demonstrated that students with MLD can master and maintain grade-level geometry concepts with well-designed interventions.

Volume builds on length and area concepts, as well as the properties of 3D objects, and begins approximately during fifth grade. Nets and 3D containers, such as cereal and tea boxes, provide excellent representations to study the properties of solids such as prisms and cubes (and increasingly more complex objects) and measurements of those properties. The concept of volume is the space within the object. To measure volume, you need linear measurements of three dimensions: length, width, and height. As with area, other properties become key in the computation of volume depending on the shapes of the faces of the object (e.g., if four faces are triangles and one face is a square, then for this pyramid shape, we will start with a cube and remove portions). The unit of measure is now a cubed unit because it represents three dimensions (e.g., the volume of the cereal box is 1539 cm^3).

Students with MLD will struggle with these measurement aspects of geometry if foundations are not developed for each concept, including key vocabulary. They will have difficulty with very abstract formulas and diagrams if concepts are not introduced through concrete explorations. Concepts that can be connected are often taught in isolation, such as the formulas for volume for different types of objects, overloading working memory and not providing sufficient strategies for problem solving. Even younger students can be taught about challenging geometric topics if teachers employ engaging experiences that promote the development of spatial vocabulary and properties of shapes; explore relationships, orientations, and sizes; and involve models and drawing (e.g., prisms with Grade 1 and 2 students; Livy et al., 2019).

Concepts of perimeter, area, and volume are applied to circles (and spheres, cones, and cylinders) in middle school, as they are much more abstract and difficult to measure. High-school students make constructions, explore concepts such as arcs, sectors, and conics, and compose more formal proofs of geometric theorems using algebra and coordinate-plane modeling. Previous concepts should be connected with these more challenging ones.

Angle

Euclidian geometry is built around three undefined terms: point, line, and plane. A point has no dimension (size), it is a place represented by a dot. A line, unless otherwise indicated, is straight and passes through at least two points to infinity; it is one-dimensional. A plane is two-dimensional and extends indefinitely. A ray is a line with one ending point. An angle is the space formed by two intersecting lines or rays. The point of intersection is the vertex; the sides of an angle are termed legs; and angles are measured in degrees. Standard notation for identifying angles is depicted in Figure D.6. Like triangles, angles can be classified as acute, obtuse, and right, but also *straight* (180 degrees), *reflex* (between 180 and 360 degrees), and *full* (360 degrees). In trigonometry, angles have additional properties, including positive and negative aspects, and are measured in radians rather than degrees, based on the radius of a circle (1 radian = 57.296 degrees), because thinking of any angle is like thinking of a piece of a circle. Calculations using radians are simpler for higher-level topics. A good proportion to help recall formulas is $\dfrac{radian}{2\pi} = \dfrac{degrees}{360}$, allowing for substitution of either radians or degrees to find the

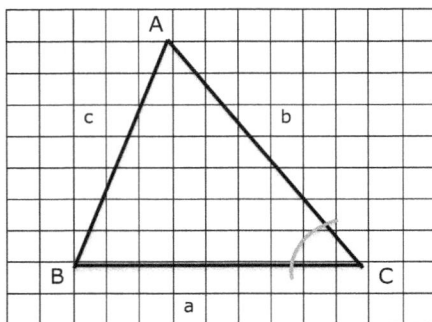

Conventions for labeling triangles:
A, B, and C represent points, the vertices of angles.
The lower-case letters are labels for line segments.
Angles are named using side and vertex letters. ∠ BCA
Other special markings are used on line segments and angles
to indicate measures, congruency, and other properties.

Figure D.6 Conventions for Angle Notation

other. A radian is the angle whose arc length on the perimeter of a circle is equal to the radius of the circle. Using a piece of string the measure of the radius and iterating that around the circle will show how the radius and radian are related.

Angles are difficult concepts for students because they are based on undefined terms, are abstract concepts, and, when depicted in geometric representations, are often a portion of a complex diagram. The nomenclature for describing and recording angles and their relationships to figures can be like a new language for some students. Some angles are static, while others are rotated and even increased and decreased during problem solving. It is more difficult to compare dynamic angles. Some students confuse angle directions (right and left); others confuse the size of an angle with the length of its rays or the size of the arc that identifies an angle. Many students have difficulty extracting the relationships between angles that are *supplementary* (measures add to 180 degrees), *complementary* (measures add to 90 degrees), *adjacent* (share a vertex and one ray), and *opposite* (share vertex but are not adjacent, are equal in measure). Some students have misconceptions such as right angles point to the right and straight angles are horizontal. Later, when students are required to measure or construct angles using tools such as protractors and compasses, they may have difficulty with matching the correct attributes to measure with readings on their tools. These students need explicit and precise instruction in how to use geometric tools. Students with fine-motor issues into upper elementary school may need to use computer tools instead, with similar explicit instruction (see virtual manipulatives and safe construction tools in Chapter 7).

The informal concept of angle is introduced in first grade, with descriptions of triangles. But it is not until fourth grade that point, line, ray, and angle are described and classified, drawn and used for problem solving. The learning trajectory for angles in the CCSSM begins in fourth grade with recognizing, drawing, and labeling different types of angles and understanding that angle measurements are in reference to a circle, in degrees. These concepts and skills are foundational for exploration of two- and three-dimensional figures through eighth grade, and for high-school topics such as geometric transformations, analytic geometry (connecting algebraic equations), and trigonometry. Students should be fluent with interpreting and constructing angles and solving problems that include basic angles by the end of the elementary grades.

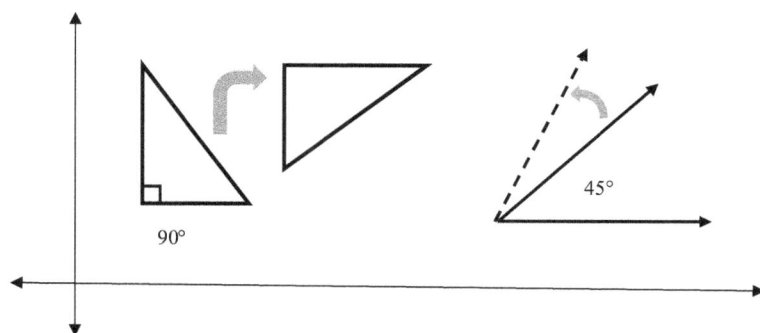

Figure D.7 Visualizing Triangle and Angle Rotations

Battista et al. (2018) challenged educators to consider ±90° rotations of triangles on a coordinate grid. Many students visualize an L-shaped figure attached to the triangle, with one side horizontal and the other vertical. Now if a student visualized a +90° rotation of the triangle, their mental model preserved both static and dynamic properties—the side measurements remained the same during the rotation. Visualizing and predicting angle rotations can support reasoning about angles and triangles. Triangle and angle rotations are depicted in Figure D.7. Spatial visualization supports spatial reasoning. Spatial reasoning can break down from poor visualization (not enough experiences) or lack of connections with underlying geometric properties.

Proof

Of the NCTM standards for practice (2000), reasoning and proof is arguably the most challenging for teachers and their students across the grade levels, and even into college-level mathematics. Children from the youngest ages should be encouraged to support their answers using reasoned responses to questions such as, "Why do you think that is true?" and "Can you explain your reasoning to get that answer?" Students in the early grades are encouraged to make conjectures, informed guesses, and to investigate their conjectures using materials, tools, and technology. They must explain and defend their reasoning for classmates and question the reasoning of others (CCSSM, NGA, & CCSSO, 2010). Reasoning moves from informal arguments to more formal and systematic deductions based on established mathematics truths.

While reasoning and proof are important for all mathematics domains, "geometry has long been regarded as the place in the school mathematics curriculum where students learn to reason and to see the axiomatic structure of mathematics" (NCTM, 2000, p. 41). There is a strong focus within geometry on the development, over the grade levels, of careful reasoning and proof, using specific definitions and facts. Developing understanding and skill with *deductive reasoning* assists students in moving from the descriptive level (2) in the van Hiele stages of development to the abstract and more logical reasoning (3) to formal deduction (4). In mathematics, proof depends on deductive reasoning, involving a logical process that leads to a conclusion, a series of steps of establishing multiple premises that are assumed to be true. For example, if we consider the following premise true, then the conclusion is supported through deductive reasoning.

Premise: Odd numbers, when grouped into twos have one left over.
Prove: odd + odd = even.

Statement: We can express an odd number in the form $2n + 1$. (Where n is an integer.)

Statement: We can add two odd numbers: $(2n + 1) + (2m + 1)$ and applying the commutative property of addition: $(2n + 2m) + (1 + 1) = 2n + 2m + 2$.

Statement: $(2n + 2m + 2)$ is an integer divisible by 2 (distributive property).

Therefore, the sum of two odd numbers is even (an integer divisible by two).

Proofs are "arguments consisting of logically rigorous deductions of conclusions from hypotheses" (NCTM, 2000, p. 55). A more useful definition for teachers may be "a mathematical argument, a connected sequence of assertions for or against a mathematical claim" that uses "statements accepted by the classroom community …, employs forms of reasoning that are valid and known to the community (modes of argumentation)," and is communicated in a way that is understood by the community (modes of representation of arguments; Stylianides, 2007, p. 291). As noted by Bieda (2010), this view of proofs makes the process more accessible for students at their grade level and proofs become vehicles for establishing stores of knowledge developed by the classroom. In a study of teacher and student discourse using proofs in the middle grades, Bieda found that while students were responsive to proof-related tasks, teachers often provided insufficient feedback (permitting non-proof arguments and examples) for students to develop these high-level skills. Teachers' instruction and feedback was quite superficial, and many teachers were uncertain about their students' capabilities in the area of proof even after instruction.

Most students have difficulties with proof, at all grade levels, not just those with MLD. Rather than constructing a mathematical proof using deductive reasoning, students tend to generate empirically based justifications (the data support this conclusion), provide examples or non-examples from their experience, or attempt non-mathematical proofs, such as those used to solve crimes. Other students may assume that the only valid proofs are those published and accepted by mathematicians. As students move to high school and beyond, mathematical arguments become more abstract and depend on formal notations that can be difficult to follow. Students may have insufficient foundational concepts to support reasoning at this level. Finally, working through the proof process requires strategies that some students have not developed: where to begin, how to build and assess each step, how to hold arguments within working memory or represent their thinking graphically, and how to justify their line of reasoning. Some mathematics educators recommend three types of visual representations for working with proofs as instructional scaffolds: the proof tree (an outline where a student can move forward from the given or backward from the statement to be proven), a flow proof (using a logic flow-chart with arrows), and the two-column proof (Cirillo & Herbst, 2012). Two-column proofs allow students to track their reasoning, move back and forward in steps, leave out steps and return to those, and not have to write complete sentences (Weiss et al., 2009). However, upper-level courses typically require proofs in paragraph form, which some students find difficult for keeping track of their arguments. A proof representation could serve as a support for a proof narrative. Figure D.8 illustrates the elements of a two-column proof in geometry.

Textbooks and other curricular materials may provide insufficient and inconsistent support for proof-related tasks across mathematics domains. Reasoning and proof are not separate topics, they are *ways of thinking* in mathematics and ways to enhance concept understanding. Teachers should attempt to embed reasoning and deductive arguments across mathematics topics, teach specific aspects of the deductive process explicitly, provide time for students to develop and critique arguments, and give feedback that is targeted and corrective. Cirillo and Hammer (2019) studied the misconceptions, or conceptual obstacles, of high-school students attempting proofs and recommended that teachers preteach specific proof competencies separately and explicitly rather than simply modeling examples of proofs. Proof competencies

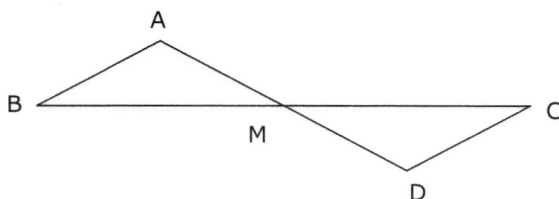

Given: Segment AD bisects segment BC.
              Segment BC bisects segment AD.

Prove: Triangles ABM and DCM are congruent.

Statements	Reasons
1. Segment AD bisects segment BC.	1. Given.
2. Segments AM and MD are congruent.	2. When a segment is bisected, the two resulting segments are congruent.
3. Segment BC bisects segment AD.	3. Given.
4. Segments BM and CM are congruent.	4. Defintion of bisect (see #2).
5. Angles AMB and DMC are congruent.	5. Vertical angles are congruent (definition).
6. Triangles ABM and DCM are congruent.	6. SAS postulate (Side-Angle-Side: if two sides and the included angle are congruent to the corresponding parts of another triangle, then the triangles are congruent).

Figure D.8 Two–Column Proof

include drawing a conclusion, distinguishing what you can or cannot conclude from a diagram, stating an accurate definition without too much information (e.g., a parallelogram is a quadrilateral that has two pairs of opposite parallel sides), practicing with types of bisectors (line segment, angle, perpendicular), and rewriting conjectures as conditional statements. When students demonstrate difficulties with these competencies when attempting to create proofs, teachers should explicitly reteach with guided practice.

High-School Concepts

High-school geometry, algebra, and trigonometry build on the previous foundational concepts. Students who move from elementary to middle school without strong concepts about the properties of shapes and the skills for reasoning about them will have difficulty. Those who move from middle to high school without the foundations for analyzing two-and three-dimensional space and figures, and the integration of topics in algebra and geometry, will have difficulty. Those students who do not have firm understanding of prerequisite concepts will attempt to memorize definitions and procedures by rote and will be unable to apply their learning to new problems. High-school topics within geometry can be extremely interesting but challenging if students lack foundations. Students explore the properties of circles, deepen their understanding of angles through trigonometry, and study special applications such as fractals and transformations. They use dynamic software and other tools with increasing competence. But those students without prerequisite concepts and skills will be limited in future coursework and careers.

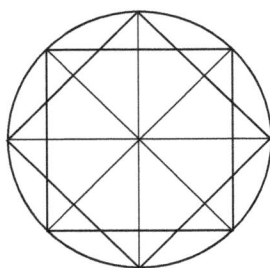

Figure D.9 Islamic Ornamentation with Geometric Properties

Development of spatial abilities must begin early, before children begin formal schooling, and be integrated throughout the PreK-12 mathematics curriculum. Spatial abilities have the power to transform students' mathematics achievement.

Multicultural Connection

Mathematically significant spatial concepts are evident in the art and architecture of many cultures. Native Americans (Yup'ik, Hopi, and Navajo), Africans (Asante and Kuba), and natives of South America (Inca, Mayan), Europe (Celtic, Hungarian, Scandinavian), Asia, Australia, and the Pacific Islands incorporated traditional geometric patterns for weaving, pottery design, house decoration, and jewelry (e.g., Zaslavsky, 1996). Shahbari and Daher (2020) used a sequence of Islamic ornamentations (e.g., star-shaped polygons within circles) to teach about congruent triangles and related theorems with students with MD, resulting in significant gains. Consider the shapes in Figure D.9. How can the concepts of symmetry, rotation, and tessellations be developed? How many triangles are in the figure? How can you reason about the angle measures of the triangles and their sides? Which triangles are congruent?

References

Arici, S., & Aslan-Tutak, F. (2015). The effect of origami-based instruction on spatial visualization, geometry achievement, and geometric reasoning. *International Journal of Science and Mathematics Education*, *13*(1), 179–200. doi: 10.1007/s10763-013-9487-8

Ashkenazi, S., Rosenberg-Lee, M., Metcalfe, A. W. S., Swigart, A. G., & Menona, V. (2013). Visuo-spatial working memory is an important source of domain-general vulnerability in the development of arithmetic cognition. *Neuropsychologia*, *51*, 2305–2317. doi: 10.1016/j.neuropsychologia.2013.06.031

Atit, K., Uttal, D. H., & Stieff, M. (2020). Situating space: Using a discipline-focused lens to examine spatial thinking skills. *Cognitive Research: Principles and Implications*, *5*(1), 19. doi: 10.1186/s41235-020-00210-z

Baddeley, A. D. (1986). *Working memory*. Oxford University Press.

Battista, M. T., Frazee, L. M., & Winer, M. L. (2018). Analyzing the relation between spatial and geometric reasoning for elementary and middle school students. In K. S. Mix & M. T. Battista (Eds.), *Visualizing mathematics: The role of spatial reasoning in mathematical thought* (pp. 195–228). Springer. doi: 10.1007/978-3-319-98767-5

Bieda, K. N. (2010). Enacting proof-related tasks in middle-school mathematics: Challenges and opportunities. *Journal for Research in Mathematics Education*, *41*(4), 351–382. doi: 10.5951/jresematheduc.41.4.0351

Boonen, A. J. H., van Wesel, F., Jolles, J., & van der Schoot, M. (2014). The role of visual representation type, spatial ability, and reading comprehension in word problem solving: An item-level analysis in

elementary school children. *International Journal of Educational Research, 68,* 15–26. doi: 10.1016/
j.ijer.2014.08.001

Bronowski, J. (1947). Mathematics. In D. Thompson & J. Reevers (Eds.), *The quality of education.* Muller.

Burger, W., & Shaughnessey, J. M. (1986). Characterizing the van Hiele levels of development in geometry. *Journal for Research in Mathematics, 17*(1), 31–48. doi: 10.2307/749317

Casey, B. M. (2013). Individual and group differences in spatial ability. In D. Walker & L. Nadel (Eds.),
Handbook of spatial cognition (pp. 117–134). American Psychological Association. doi: 10.1037/
13936-007

Casey, B. M., & Fell, H. (2018). Spatial reasoning: A critical problem-solving tool in children's mathematics strategy tool-kit. In K. S. Mix & M. T. Battista (Eds.), *Visualizing mathematics: The role of spatial
reasoning in mathematical thought* (pp. 47–76). Springer. doi: 10.1007/978-3-319-98767-5_3

Cheng, Y.-L., & Mix, K. S. (2014). Spatial training improves children's mathematics ability. *Journal of
Cognition and Development, 15*(1), 2–11. doi: 10.1080/15248372.2012.725186

Cirillo, M., & Herbst, P. G. (2012). Moving toward more authentic proof practices in geometry.
Mathematics Educator, 21(2), 11–33. ERIChttps://files.eric.ed.gov/fulltext/EJ961514.pdf

Cirillo, M., & Hummer, J. (2019). Addressing misconceptions in secondary geometry proof. *The
Mathematics Teacher, 112*(6), 410–417. doi: 10.5951/mathteacher.112.6.0410

Clements, D. H., & Battista, M. T. (1992). Geometry and spatial reasoning. In D. A. Grouws (Ed.),
Handbook of research on mathematics teaching and learning (pp. 420–464). Macmillan.

Clements, D. H., & Sarama, J. (2007). Effects of a preschool mathematics curriculum: Summative research
on the *Building Blocks* project. *Journal for Research in Mathematics Education, 38,* 136–163. doi: 10.2307/
30034954

Clements, D. H., & Sarama, J. (2011). Early childhood teacher education: The case of geometry. *Journal
of Mathematics Teacher Education, 14,* 133–148. doi: 10.1007/s10857-011-9173-0

Clements, D., & Sarama, J. (2014). *Learning and teaching early math: The learning trajectories approach* (2nd
ed.). Routledge. doi: 10.4324/9780203520574

Clements, D., Sarama, J., Baroody, A., Joswick, C., & Wolfe, C. (2019). Evaluating the efficacy of a
learning trajectory for early shape composition. *American Educational Research Journal, 56*(6), 2509–
2530. doi: 10.3102/0002831219842788

Clements, D. H., Swaminathan, S., Hannibal, M. A. Z., & Samara, J. (1999). Young children's concepts of
shape. *Journal for Research in Mathematics Education, 30*(2), 192–212. doi: 10.2307/749610

Cornoldi, C., Venneri, A., Marconato, F., Molin, A., & Montinari, C. (2003). A rapid screening measure
for the identification of visuospatial learning disability in schools. *Journal of Learning Disabilities, 36*(4),
299–306. doi: 10.1177/00222194030360040201

Davis, B., Francis, K., & Drefs, M. (2015). A history of the current curriculum. In B. Davis & The Spatial
Reasoning Study Group (Eds.), *Spatial reasoning in the early years: Principles, assertions, and speculations*
(pp. 47–62). Routledge. doi: 10.4324/9781315762371

de Villiers, M. (2010, June). *Some reflections on the van Hiele theory.* Plenary at the Fourth Congress of
Teachers of Mathematics, Croatian Mathematical Society. Zagreb, Croatia.

Fuys, D., Geddes, D., & Tischler, R. (1988). The van Hiele model of thinking in geometry among
adolescents. *Journal for Research in Mathematics Education. Monograph, 3,* 1–196. doi: 10.2307/749957

Gagnier, K. M., Atit, K., Ormand, C. J., & Shipley, T. F. (2017). Comprehending 3D diagrams: Sketching
to support spatial reasoning. *Topics in Cognitive Science, 9*(4), 883–901. doi: 10.1111/tops.12233

Geary, D. C. (2003). Learning disabilities in arithmetic: Problem-solving differences and cognitive
deficits. In H. L. Swanson, K. R. Harris, & S. Graham (Eds.), *Handbook of learning disabilities* (pp. 199–
212). The Guilford Press.

Geer, E. A., Quinn, J. M., & Ganley, C. M. (2019). Relations between spatial skills and math performance
in elementary school children: A longitudinal investigation. *Developmental Psychology, 55*(3), 637–652.
doi: 10.1037/dev0000649

Gilligan, K. A., Hodgkiss, A., Thomas, M. S. C., & Farran, E. K. (2019). The developmental relations
between spatial cognition and mathematics in primary school children. *Developmental Science, 22*(4),
e12786. doi: 10.1111/desc.12786

Gilligan, K. A., Thomas, M. S. C., & Farran, E. K. (2020). First demonstration of effective spatial training
for near transfer to spatial performance and far transfer to a range of mathematics skills at 8 years.
Developmental Science, 23(4), e12909. doi: 10.1111/desc.12909

Gillum, J. (2014). Assessment with children who experience difficulty in mathematics. *Support for Learning, 29*(3), 275–291. doi: 10.1111/1467-9604.12061

Gravemeijer, K. P. (1998). From a different perspective: Building on students' informal knowledge. In R. Lehrer & D. Chazan (Eds.), *Designing learning environments for developing understanding of geometry and space* (pp. 45–66). Lawrence Erlbaum. doi: 10.4324/9780203053461-7

Grossner, K., & Janelle, D. G. (2014). Concepts and principles for spatial literacy. In D. G. Janelle, K. Grossner, & D. R. Montello (Eds.), *Space in mind: Concepts for spatial learning and education*. MIT Press. doi: 10.7551/mitpress/9811.003.0013

Gross-Tsur, V., Shalev, R. S., Manor, O., & Amir, N. (1995). Developmental right-hemisphere syndrome: Clinical spectrum of the nonverbal learning disability. *Journal of Learning Disabilities, 28*(2), 80–86. doi: 10.1177/00222194502800202

Gutiérrez, A., Jaime, A., & Fortuny, J. M. (1991). An alternative paradigm to evaluate the acquisition of the van Hiele levels. *Journal for Research in Mathematics Education, 22*(3), 237–252. doi: 10.5951/jresematheduc.22.3.0237

Halpern, D. F. (2000). *Sex differences in cognitive abilities* (3rd ed.). Psychology Press. doi: 10.4324/97814110605290

Harris, D., Lowrie, T., Logan, T., & Hegarty, M. (2020). Spatial reasoning, mathematics, and gender: Do spatial constructs differ in their contribution to performance? *British Journal of Educational Psychology*, e12371–e12371. doi: 10.1111/bjep.12371

Hawes, Z., LeFebre, J., Xu, C., & Bruce, C. D. (2015). Mental rotation with tangible three-dimensional objects: A new measure sensitive to developmental differences in 4- to 8-year-old-children. *Mind, Brain, and Education, 9*(1), 1–18. doi: 10.1111/mbe.12051

Heckman, J. J. (2006). Skill formation and the economics of investing in disadvantaged children. *Science, 312*(5782), 1900–1902. doi: 10.1126/science.1128898

Johnson, E. S., & Bouchard, T. J. (2003). The structure of human intelligence: It is verbal, perceptual, and image rotation (VRP), not fluid and crystallized. *Intelligence, 33*(4), 393–416. doi: 10.1016/j.intell.2004.12.002

Jordan, N., Resnick, I., Rodrigues, J., Hansen, N., & Dyson, N. (2017). Delaware longitudinal study of fraction learning: Implications for helping children with mathematics difficulties. *Journal of Learning Disabilities, 50*(6), 621–630. doi: 10.1177/0022219416662033

Kell, H. J., Lubinski, D., Benbow, C. P., & Steiger, J. H. (2013). Creativity and technical innovation spatial ability's unique role. *Psychological Science, 24*(9), 1831–1836. doi:10.1177/0956797613478615

Lehrer, R., Jacobson, C., Kemeny, V., & Strom, D. (1999). Building on children's intuitions to develop mathematical understanding of space. In E. Fennema & T. A. Romberg (Eds.), *Mathematics Classrooms that Promote Understanding*. Routledge. doi: 10.4324/9781410602619

Li, Y., & Geary, D. C. (2013). Developmental gains in visuospatial memory predict gains in mathematics achievement. *PLoS ONE, 8*(7): e70160. doi: 10.1371/journal.pone.0070160

Li, Y., & Geary, D. C. (2017). Children's visuospatial memory predicts mathematics achievement through early adolescence. *PLoS ONE, 12*(2), e0172046. doi: 10.1371/journal.pone.0172046

Liben, L. S., & Christensen, A. E. (2010). Spatial development: Evolving approaches to enduring questions. In U. Goswami (Ed.), *Blackwell Handbook of Childhood Cognitive Development* (pp. 446–472). Wiley-Blackwell. doi: 10.1002/9781444325485.ch17

Liu, M., Bryant, D., Kiru, E., & Nozari, M. (2021). Geometry interventions for students with learning disabilities: A research synthesis. *Learning Disability Quarterly, 44*(1), 23–34. doi: 10.1177/0731948719892021

Livy, S., Bobis, J., Downton, A., Hughes, S., McCormick, M., Russo, J., & Sullivan, P. (2019). Exploring spatial reasoning in the early years: Effective pedagogical approaches. *Australian Primary Mathematics Classroom, 24*(2), 26–32.

Lowrie, T., & Logan, T. (2018). The interaction between spatial reasoning constructs and mathematics understandings in elementary classrooms. In K. S. Mix & M. T. Battista (Eds.), *Visualizing mathematics: The role of spatial reasoning in mathematical thought* (pp. 253–276). Springer. doi: 10.1007/978-3-319-98767-5

Lowrie, T., Harris, D., Logan, T., & Hegarty, M. (2019). The impact of a spatial intervention program on students' spatial reasoning and mathematics performance. *The Journal of Experimental Education*, 1–19. doi: 10.1080/00220973.2019.1684869

Maier, P. H. (1994) *Räumliches Vorstellungsvermögen: Komponenten, geschlechtsspezifische Differenzen, Relevanz, Entwicklung, und Realisierung in der Realschule* [*Spatial presentation skills: Components, gender-specific differences, relevance, development, and implementation in the secondary school*]. Peter Lang.

Mason, M. M. (1998). The van Hiele levels of geometric understanding. In L. McDougal (Ed.), *The professional handbook for teachers: Geometry* (pp. 4–8). McDougal-Littell/Houghton-Mifflin.

Matthews, P., & Hubbard, E. (2017). Making space for spatial proportions. *Journal of Learning Disabilities, 50*(6), 644–647. doi: 10.1177/0022219416679133

McDuffie, A. R., & Eve, N. (2009). Break the area boundaries. *Teaching Children Mathematics, 16*, 18–27. doi: 10.5951/tcm.16.1.0018

Mix, K. S., & Cheng, Y.-L. (2012). The relation between space and math: Developmental and educational implications. *Advances in Child Development and Behavior, 42*, 197–243. doi: 10.1016/b978-0-12-394388-0.00006-x

Mix, K. S., Levine, S. C., Cheng, Y. L., Young, C., Hambrick, D. Z., Ping, R., & Konstantopoulos, S. (2016). Separate but correlated: The latent structure of space and mathematics across development. *Journal of Experimental Psychology, 145*(9), 1206–1227. doi: 10.1037/xge0000182

Möhring, W., Newcombe, N. S., Levine, S. C., & Frick, A. (2016). Spatial proportional reasoning is associated with formal knowledge about fractions. *Journal of Cognition and Development, 17*(1), 67–84. doi:10.1080/15248372.2014.996289

Moss, J., Bruce, C. D., Caswell, B., Flynn, T., & Hawes, Z. (2016). *Taking shape: Activities to develop geometric and spatial thinking grades K-2*. Pearson Canada.

Mullis, I. V. S., Martin, M. O., Foy, P., Kelly, D. L., & Fishbein, B. (2020). *TIMSS 2019 international results in mathematics and science*. Boston College, TIMSS & PIRLS International Study Center. https://timssandpirls.bc.edu/timss2019/international-results/

Nardi, D., Newcombe, N. S., & Shipley, T. F. (2013). Reorienting with terrain slope and landmarks. *Memory & Cognition, 41*, 214–228. doi: 10.3758/s13421-012-0254-9

National Center for Education Statistics (2017). *2017 NAEP Mathematics Assessments*. US Department of Education. www.nationsreportcard.gov

National Council of Teachers of Mathematics (1989, 2000). *Principles and standards for school mathematics*. www.nctm.org/Standards-and-Positions/Principles-and-Standards/

National Governors Association Center for Best Practices & Council of Chief State School Officers. (2010). *Common core state standards for mathematics*. www.corestandards.org/Math/

National Research Council. (2006). *Learning to think spatially: GIS as a support system in the K-12 curriculum*. National Academies Press. www.nap.edu/catalog/11019.html

Newcombe, N., & Huttenlocher, J. (2000). *Making space: The development of spatial representation and reasoning*. MIT Press. doi: 10.7551/mitpress/4395.001.0001

Ontario Ministry of Education (2014). *Paying attention to spatial reasoning*. Queen's Printer for Ontario. www.edu.gov.on.ca/eng/literacynumeracy/LNSPayingAttention.pdf

Owens, K. (2015). An ecocultural perspective on space, geometry, and measurement in education. In: *Visuospatial Reasoning. Mathematics Education Library, 111*, 291–308. Springer. doi: 10.1007/978-3-319-02463-9_10

Panorkou, N., & Greenstein, S. (2014, February). A learning trajectory for transformation-based reasoning in geometry. *Proceedings for the 42nd annual meeting of the Research Council on Mathematics Learning* (pp. 25–32). Las Vegas.

Passolunghi, M. C., & Mammarella, I. C. (2012). Selective spatial working memory impairment in a group of children with mathematics learning disabilities and poor problem-solving skills. *Journal of Learning Disabilities, 45*, 341–350. doi: 10.1177/0022219411400746

Peterson, E. G., Weinberger, A. B, Uttal, D. H., Kolvoord, B., & Green, A. E. (2020). Spatial activity participation in childhood and adolescence: Consistency and relations to spatial thinking in adolescence. *Cognitive Research: Principles and Implications, 5*(1), 43–43. doi: 10.1186/s41235-020-00239-0

Piaget, J. & Inhelder, B., & Szeminsak, A. (1948/1960). *The child's conception of geometry*. Routledge & Kegan Paul.

Pyke, A., Betts, S., Fincham, J. M., & Anderson, J. R. (2015). Visuospatial referents facilitate the learning and transfer of mathematical operations: Extending the role of the angular gyrus. *Cognitive, Affective, & Behavioral Neuroscience, 15*(1), 229–250. doi: 10.3758/s13415-014-0317-4

402 *Spatial and Geometric Reasoning*

Ramey, K. E., Stevens, R., & Uttal, D. H. (2020). In-FUSE-ing STEAM Learning with spatial reasoning: Distributed spatial sensemaking in school-based making activities. *Journal of Educational Psychology, 112*(3), 466–493. doi: 10.1037/edu0000422

Ramful, A., Lowrie, T., & Logan, T. (2017). Measurement of spatial ability: Construction and validation of the spatial reasoning instrument for middle school students. *Journal of Psychoeducational Assessment, 35*(7), 709–727. doi: 10.1177/0734282916659207

Resnick, I., Newcombe, N. S., & Jordan, N. C. (2019). The relation between spatial reasoning and mathematical achievement in children with mathematical learning difficulties. In A. Fritz, V. G. Haase, & P. Räsänen (Eds.), *International handbook of mathematical learning difficulties* (pp. 423–436). Springer. doi: 10.1077/978-3-319-97148-3-26

Reuhkala, M. (2001). Mathematical skills in ninth graders: Relationship with visuo-spatial abilities and working memory. *Educational Psychology, 21*(4), 387–399. 10.1080/01443410120090786

Rowan, T. E. (1990). The geometry standards in K-8 mathematics. *Arithmetic Teacher, 37*(6), 24–28. doi: 10.5951/at.37.6.0024

Rylant, C. (1998). *Scarecrow*. Houghton Mifflin Harcourt.

Satsangi, R. & Bouck, E. C. (2015). Using virtual manipulative instruction to teach the concepts of area and perimeter to secondary students with learning disabilities. *Learning Disability Quarterly, 38*(3), 174–186. doi: 10.1177/0731948714550101

Satsangi, R., Hammer, R., & Bouck, E. C. (2019). Using video modeling to teach geometry word problems: A strategy for students with learning disabilities. *Remedial and Special Education, 41*(5), 309–320. doi: 10.1177/0741932518824974

Shahbari, J. A., & Daher, W. (2020). Learning congruent triangles through ethnomathematics: The case of students with difficulties in mathematics. *Applied Sciences, 10*, 4950. doi: 10.3390/app10144950

Sidebotham, T. H. (2002). *The A to Z of mathematics: A basic guide*. John Wiley & Sons.

Sinclair, N., Moss, J., Hawes, Z., & Stephenson, C. (2018). Learning through and from drawing in early years geometry. In K. S. Mix & M. T. Battista (Eds.), *Visualizing mathematics: The role of spatial reasoning in mathematical thought* (pp. 229–252). Springer. doi: 10.1007/978-3-319-98767-5_11

Singer, B., & Shafer, K. G. (2018). Exploring spatial reasoning with SketchUp Make. *Teaching Children Mathematics, 25*(1), 46–52. doi: 10.5951/teacchilmath.25.1.0046

Smith, L., Ping, R. M., Matlen, B. J., Goldwater, M. B., Gentner, D., & Levine, S. (2014). Mechanisms of spatial learning: Teaching children geometric categories. In C. Freksa, B. Nebel, M. Hegarty, & T. Barkowsky (Eds.), *Spatial cognition IX: International conference proceedings* (pp. 325–337). Springer. doi: 10.1007/978-3-319-11215-2_23

Sorby, S., Veurink, N., & Streiner, S. (2018). Does spatial skills instruction improve STEM outcomes? The answer is 'yes'. *Learning and Individual Differences, 67*, 209–222. doi: 10.1016/j.lindif.2018.09.001

Sowell, E. J. (1989). Effects of manipulative materials in mathematics instruction. *Journal for Research in Mathematics Education, 20*(5), 498–505. doi: 10.2307/749423

Stylianides, G. J. (2007). Proof and proving in school mathematics. *Journal for Research in Mathematics Education, 38*, 289–321. doi: 10.2307/30034869

Swanson, H. L. (2012). Cognitive profiles of adolescents with math disabilities: Are the profiles different from those with reading disabilities? *Child Neuropsychology, 18*, 125–143. doi: 10.1080/09297049.2011.589377

Tian, J., & Siegler, R. (2017). Fractions learning in children with mathematics difficulties. *Journal of Learning Disabilities, 50*(6), 614–620. doi: 10.1177/0022219416662032

Unal, H., Jakubowski, E., & Corey, D. (2009). Differences in learning geometry among high and low spatial ability pre-service mathematics teachers. *International Journal of Mathematical Education in Science and Technology, 40*(8), 997–1012. doi: 10.1080/00207390902912852

Uttal, D. H., Meadow, N. G., Tipton, E., Hand, L. L., Alden, A. R., Warren, C., & Newcombe, N. S. (2013). The malleability of spatial skills: A meta-analysis of training studies. *Psychological Bulletin, 139*(2), 352–402. doi: 10.1037/a0028446

van Garderen, D. (2006). Spatial visualization, visual imagery, and mathematical problem solving of students with varying abilities. *Journal of Learning Disabilities, 39*, 496–506. doi: 10.1177/00222194060390060201 van Garderen, D., Scheuermann, A., & Jackson, C. (2012). Developing representation ability in mathematics for students with learning disabilities: A content analysis of grades 6 and 7 textbooks. *Learning Disability Quarterly*, *35*, 24–38. doi: 10.1177/0731948711429726

van Hiele, P. M. (1959/1984). The child's thought and geometry. In D. Fuys, D. Geddes, & R. Tishler (Eds.), *English translation of selected writings of Dina van Hiele-Geldof and Pierre M. van Hiele* (pp. 243–252). Brooklyn College School of Education. https://files.eric.ed.gov/fulltext/ED287697.pdf

Weiss, M., Herbst, P., & Chen, C. (2009). Teachers' perspectives on "authentic mathematics" and the two-column proof form. *Educational Studies in Mathematics*, *70*, 275–293. doi: 10.1007/s10649-008-9144-2

Wheatley, G. (1996). *Quick draw: Developing spatial sense in mathematics*. Mathematics Learning. www.mathematicslearning.org

Woolcott, G., Chamberlain, D., Hawes, Z., Drefs, M., Bruce, C. D., Davis, B., Francis, K., Hallowell, D., McGarvey, L., Moss, J., Mulligan, J., Okamoto, Y., Sinclair, N., & Whiteley, W. (2020). The central position of education in knowledge mobilization: Insights from network analyses of spatial reasoning research across disciplines. *Scientometrics*, *125*(3), 2323–. doi: 10.1007/s11192-020-03692-2

Xie, F., Zhang, L., Chen, X., & Xin, Z. (2020). Is spatial ability related to mathematical ability: A meta-analysis. *Educational Psychology Review*, *32*(1), 113–155. doi: 10.1007/s10648-019-09496-y

Yakimanskaya, I. S. (1980/1991). The development of spatial thinking in schoolchildren. In P. S. Wilson & E. J. Davis (Eds.), R. H. Silverman (Trans.), *Soviet Studies in Mathematics Education* (Volume 3). NCTM.

Yi, M., Flores, R., & Wang, J. (2020). Examining the influence of van Hiele theory-based instructional activities on elementary preservice teachers' geometry knowledge for teaching 2-D shapes. *Teaching and Teacher Education*, *91*, 103038–. doi: 10.1016/j.tate.2020.103038

Young, C., Levine, S. C., & Mix, K. S. (2018). What processes underlie the relation between spatial skill and mathematics? In K. S. Mix & M. T. Battista (Eds.), *Visualizing mathematics: The role of spatial reasoning in mathematical thought* (pp. 117–148). Springer. doi: 10.1007/978-3-319-98767-5_5

Zaslavsky, C. (1996). *The multicultural math classroom: Bringing in the world*. Heinemann.

Zhang, X., Räsänen, P., Koponen, T., Aunola, K., Lerkkanen, M., & Nurmi, J. (2020). Early cognitive precursors of children's mathematics learning disability and persistent low achievement: A 5-year longitudinal study. *Child Development*, *91*(1), 7–27. doi: 10.1111/cdev.13123

Strand E

Measurement, Data Analysis, and Probability

After studying this strand, the reader should be able to:

1. discuss common difficulties with measurement concepts and skills;
2. describe instructional approaches and interventions for measurement;
3. discuss common learning difficulties with data analysis and probability;
4. design interventions for statistical concepts and skills;
5. connect concepts within measurement and statistics to those in other mathematics domains.

Jess is a fifth-grade student at Pinetops Elementary School and has been doing well in his mathematics class with Ms. Wang until the current unit on geometry. Jess is fluent with rational-number operations, including fractions and decimals, and has strong multiplicative reasoning. He has developed a number of successful word-problem strategies as well. When Ms. Wang conducted a review session before beginning the geometry unit, Jess struggled to find the perimeter of simple rectangles using a ruler. Ms. Wang is exploring Jess's concept understanding further.

MS. WANG: Jess, you identified all your shapes correctly, even the most difficult shapes such as the rhombus, but your perimeter measurements are off. Can you tell me in your own words what a perimeter is?

JESS: Sure, that's the distance all the way around.

MS. WANG: Yes, that's right. Now I want you to determine the perimeter of this rectangle that has 4 inches on two sides and 6 inches in the other two.

JESS: That would be 8 + 12 so 20 total inches.

MS. WANG: Very good. You know your perimeters. Now, can you measure the perimeter of this rectangle using a centimeter ruler? [Shows Jess rectangle with two sides 5.3 cm and two sides 8.4 cm.]

JESS: Well, this side is just over six and the other side is nine and about a third.

DOI: 10.4324/9781003096733-13

MS.WANG: Jess, can you explain to me how you decided on those values?

JESS: I'm not great with centimeters, but I just read the numbers on the ruler and estimated.

Jess is not confident with measurement tools and using metric units. He is focused on the numbers of the ruler and is using fractional portions of centimeters. His otherwise strong concept understanding in geometry is hampered by weak measurement concepts.

Measurement and data analysis are strands of the mathematics curriculum that are developed from the earliest years and are closely connected with number and operation, geometry and spatial reasoning, and algebra. Measurement and data analysis involve critical concepts and skills for STEM (science, technology, engineering, mathematics) studies and careers, yet are often neglected in PreK–12 mathematics. The domains of measurement and data analysis provide excellent platforms for strengthening number concepts and problem-solving strategies from the earliest grade levels (Doabler, Clarke, Kosty, et al., 2019).

A measure "is a standard or mark against which we gauge or evaluate something" (Jones, 2013, p. 110). Measurement within mathematics typically refers to measuring physical properties, such as distance, volume, time, and energy. Data analysis involves collecting information (sometimes from measurements), analyzing sets of data, and representing those in various forms, including tables, graphs, narratives, and equations. Some diagnostic assessments in mathematics treat measurement and data analysis as merely applications of number and operations, geometry, and algebra. However, measurement and data analysis have broader concepts that develop over time and require careful instructional planning. This strand will describe some of the difficulties for students when engaged in measurement and data analysis and provide recommendations for interventions.

Measurement

The act of measuring or quantifying dates at least to 35,000 BCE with the Lembobo bone, a baboon's fibula inscribed with 29 notches (Mankiewicz, 2001). This bone is considered the oldest known mathematical artifact and is hypothesized to be a recording device for lunar phases because it is similar to calendar sticks used by a number of early and modern peoples. Whether time was being measured or some other measurement or quantification was made using this bone, it represents human's early attempt to measure.

Measurement is the mathematics domain that is most closely situated with everyday life. We measure time using clocks and calendars, distances using odometers, weight using scales, volume using measuring cups, and the temperature with a thermometer. It is difficult to traverse a day without measuring something. Measurement requires a coordination of spatial reasoning and number (quantitative) sense. Sometimes measurement can be an estimate, as with the approximate distance from your home to the grocery store. Other times measurement should be as precise as possible, as with the optics on the Hubble Space Telescope (NASA, 1990). When it was launched, the images it transmitted back to Earth were not in complete focus, which led to a finding that the primary mirror was ground and polished incorrectly. The curve at the perimeter was off by 2.2 micrometers (0.0022 mm). A very costly measurement error!

Conventional units of measure and their prefixes are governed by the International System of Units (*Système International d'Unités*, SI), the modern form of the metric system. US Customary System Units (USCS), commonly termed *standard units*, are derived from the British Imperial System (with some differences), which evolved from Roman, Anglo-Saxon, and Celtic traditions of using common references as measures, such as bushels, rods, and feet. Because both systems of measurement are used in the United States, mathematics standards challenge students to be proficient with both systems.

The National Council of Teachers of Mathematics (NCTM, 2015) recommended that the trajectory for developing measurement concepts should begin with developing concepts of attributes to be measured (e.g., length) by comparing and ordering. Next, students should use nonstandard units to compare and order objects (e.g., paper clips or tiles). Only then should students be introduced to standard systems of measure. This sequence has been adopted by most curricula, but there is evidence that elementary students can work with nonstandard and standard units simultaneously and some students may perform better with conventional tools than everyday objects (Smith et al., 2013). Students should be able to select appropriate units of measure, use tools for measurement, estimate, and reason proportionally to develop a sense of the relationship among units (e.g., inches to feet, mm to cm). Smith et al. (2013) noted that the *Common Core State Standards for Mathematics* (CCSSM, NGA, & CCSSO, 2010) failed to emphasize geometric measurement. The researchers found that in curricula for Grades K-3 only 3%–6% of lessons in kindergarten included linear measurement, with 4%–8% in first grade, 6%–10% in second grade, and 10%–13% in third. Further, lessons addressing linear measurement or applying those skills took place in the second half of the year in Grades K-2, although measurement concepts and skills support and strengthen number concepts, number-line understanding, spatial reasoning, fraction concepts, and multiplicative reasoning. Students have difficulty with linear measurement beyond third grade, often into middle and high school (Kamii, 2006; Tan-Sisman & Aksu, 2016), perhaps due to limited instructional exposure, focus on procedures over concept understanding, lack of appropriate representations and learning tasks, and teachers' weak understanding of concepts and student misconceptions.

Many researchers and curriculum developers have noted the risk of treating measurement as a separate topic within mathematics when it is so closely related to all other mathematics domains, especially geometry and statistics (e.g., Smith & Barrett, 2018). *Metric measurement* (assigning numerical values to continuous quantities) is somewhat different from descriptive geometry, but closely connected. For example, it is impossible to measure the volume of a container without reasoning about its geometric properties. Understanding the three-dimensional attributes of the container leads to making three linear measurements or visualizing and counting cubic units across space in rows and columns. Measurement serves as a conceptual bridge between number and spatial concepts, so explicit attention to measurement concepts and skills within mathematics instruction can support other domains and identify students' learning gaps.

Measurement affords many applications for the sciences and social sciences, as those are fields that also explore the world. Measurement concepts and skills are essential for all aspects of scientific inquiry and in the development of scientific concepts across the grade levels (National Research Council, 1996). Table E.1 illustrates physical and temporal aspects that are measured, with example measurement tools and units of measure. Within physics, there are seven base units of measure (SI units: meter, kilogram, second, ampere, kelvin, mole, and candela), from which other derived units are defined. Within social sciences there are four levels of measurement based on the type of data: nominal, ordinal, interval, and ratio scales. The data analysis applied will depend on the measurement scale.

Difficulties with Measurement

Measurement concepts, reasoning, and skills are assessed on national and state assessments. On the *National Assessment of Educational Progress* (NAEP, NCES, 2019), fourth-grade students on the whole scored below the proficient level, while students with disabilities scored just below basic (39 points lower than nondisabled students). Basic level indicates some evidence of understanding the concepts and procedures, while proficient-level students consistently apply integrated concepts and procedures to problem solving. Seventy-nine percent could identify a

Table E.1 Measurement Subjects, Tools, and Units

Category	Quantity	Example Tool	Example Unit
Space	Length	Meter stick	Centimeters
	Elevation	Altimeter	Meters
	Distance (length)	Odometer	Miles
	Angle	Protractor	Degrees
Matter	Volume	Graduated cylinder	Milliliters
	Mass	Scale	Grams
	Density	Scale, graduated cylinder	g/cm^{3*}
Time	Long periods	Calendar	Months
	Short periods	Clock	Hours
	Shorter periods	Stopwatch	Seconds
	Tempo	Metronome	Beats per minute*
Energy	Atmospheric	Barometer	Millibars*
	Electric current	Ammeter	Ampere*
	Temperature	Thermometer	Degrees Celsius
	Earthquake	Seismograph	Moment magnitude*
	Sound	Audiometer	Decibels*

* These units are actually derived units, from two or more measures. Scientists tend to focus on mass, length, and time for ratio units. For example: speed = distance/time, density = mass/volume

Vectors represent quantities with both magnitude and direction such as force, velocity, and acceleration.

tool for measuring temperature, but only 35% could convert feet to inches. For eighth graders, the scores overall were in the basic knowledge range and below basic for students with disabilities (42 points lower than nondisabled). Fifty-nine percent could reason amounts with pints, quarts, and gallons; 22% could use a given equation to convert Fahrenheit to Celsius; and only 21% could compute the volume of a cylinder. It may be that measurement tasks are too often taught in isolation or not taught with the reasoning, concept understanding, and connections with other topics needed to perform well on assessments. It is clear that most students in the US are not proficient with measurement concepts and skills, and that students with disabilities, on the whole, lack even basic understanding.

Most of the research on measurement misconceptions and difficulties has focused on linear measurement as that is the most emphasized within curricula. However, research on area, volume, angle, time, weight, and other physical quantities is increasing. Although measurement activities are common in everyday life, teachers should not assume that students understand quantities as measurable, the concept of unit, and the use of tools for measurement. Many students hold misconceptions beyond the elementary grades that will continue to affect their mathematics achievement. While student competencies will vary by age and experience, some of the most common issues include those listed below (Barmby et al., 2009; Clements, Barrett, & Sarama, 2017; Ren et al., 2019; Smith et al., 2013; Solomon et al., 2015; Tan-Sisman & Asksu, 2016).

- Lack of concept understanding about attributes of objects, such as length. The concept of length is difficult because it is one-dimensional. Length is not tangible; it is an abstract concept applied to physical objects. Before formal measurement, length concepts begin with comparisons, direct and indirect. Length understanding depends on identifying the beginning and ending points of a span or distance and can refer to curved or zig-zag distances. When students move on to perimeter, they have difficulty conceptualizing the length around an object as one-dimensional, because it is not a straight line. Other terms for length, such as width, height, depth, and distance, often confuse students. Length concepts also support area and volume reasoning, as well as number-line and coordinate-plane knowledge. Two- and three-dimensional (2D, 3D) objects have other measurable attributes that can be abstract unless developed using concrete and representational experiences: area, surface area, volume, edges, vertices.

- Misconceptions related to the concept of unit, an essential foundation for measurement.
    - Misconceptions about the iterations of a unit: Students may not understand that units (standard or nonstandard) must be the same and of equal length (or area or volume), they must be placed without gaps or overlaps, and the units are counted (a connection between numerical and spatial reasoning). The identical units must fill the quantity (tiling). If a tool isn't long enough to measure an object, some students may just keep counting rather than reposition the tool to continue the units (additive principle). Square units (tiles) are iterated for area and cubic units (cubes) iterated for volume with the same challenges.
    - Not generalizing the concept of composite unit: Composite unit, a concept important for many areas of mathematics, is the understanding that any unit (e.g., number, measure) is made up of multiple copies of a smaller unit. Repeated subtraction and counting by multiples require this understanding. A foot is a composite unit of 12 inches. The area of a rectangle is composed of multiple rows of unit squares. Volume of an object is composed of arrays of cubic units. Students' comprehension transitions from single units to composite units and then to multiplication, leading to formula understanding (Smith & Barrett, 2018).
    - Difficulty reasoning about parts of units, the hierarchical nature of units: Many students will assert "a half" for anything between two hash marks or part of a unit (subunit). If the unit is feet, a measure might be recorded as 5 feet, 3 inches or 5 ¼ feet. If the unit is meters, a measure could be 4.35 meters (4 meters, 30 centimeters, 5 millimeters). Related to number-line understanding and fractional and base-10 reasoning, partitioning a unit requires understanding that those parts must also be equal. Partitioning units adds precision to measurement.
    - Not understanding the principle of the inverse relationship between size of the unit and the resulting measure: When measuring the same quantity, larger units produce smaller measures. A pencil that is 15 cm long is also 150 millimeters long. A centimeter unit is longer than that of a millimeter, so it will have a smaller resulting value. Your desk might be 11 small paper clips wide and that is equivalent to 8 big paper clips wide. Proportional reasoning is involved in this understanding.
- Weak concepts and skills related to rulers. Rulers are broadly defined to include yard sticks, meter sticks, tape measures, and other customary tools for measuring length. Many students struggle to align the ruler, make iterations, find an accurate measurement using a ruler, and interpret portions of units. Some students count the hashmarks on a ruler rather than considering each unit as the space between hashmarks. Students who count the hashmarks instead of units between may even count the beginning and ending hash-marks.
- Misconceptions about starting point or zero point. Students often have difficulty with where to begin, whether using standard or nonstandard units. If provided a ruler, they may consider the "1" position the starting point. If asked to begin with another starting point, for example the "4" position, that is difficult to conceptualize. Other students may be too focused on the end point to consider the whole, and just read off the number corresponding with the end.
- Difficulty reasoning with standard units. As students begin using units in the USCS and SI systems, many students lack firm concepts of basic units when they are asked to reason about these units, convert between units, and partition units. For example, if your concept

of centimeter is vague, somewhere between little and not-so-little, then you would have difficulty choosing to measure using millimeters, centimeters, or meters for specific lengths or distances.

Some students have other misconceptions, such as length being the only attribute of an object, an assumption that their measurements are exact, and difficulty with considering continuous lengths as being composed of discrete units (Solomon et al., 2015). The continuous-length dilemma inhibits students' use of measurement tools—locating beginning points, identifying units of measure, partitioning units, and iterating units as with shifting a ruler to measure a longer object. Also related to students' growth in conceptual knowledge about measurement is their ability to *inhibit* less-mature strategies, such as reading hash marks, an executive function found to be a significant predictor of strategy improvement for linear measurement over Grades 1 to 3 (Ren et al., 2019).

When moving on to 2D objects, some of the same misconceptions and difficulties are apparent: lack of focus on appropriate attributes of objects, poor reasoning about units and portions of units, difficulties with measurement tools, weak reasoning with standard units, and problems with measuring continuous quantities with discrete objects. With two dimensions, students often confuse units of measure. They confuse the concept of perimeter with area and the unit of measure to use with each. Area is 2D (the product of two lengths), therefore we use square units such as square centimeters (cm^2) and square inches (in^2). Additional difficulties with measuring area include difficulty visualizing arrays, not applying multiplicative reasoning (array concepts), and difficulty applying formulas (symbolic level) before concrete and representational concepts are firm.

Students are often taught rote procedures (e.g., length × width = area) to determine area before area concepts and principles are understood. $A = l \times w$ is a formula applied to rectangles, not other polygons, and is often taught before students understand that area refers to the space within the shape, and that rectangles have two pairs of congruent sides (same length) even though only two sides might be labeled with their lengths. Area can also be computed for other shapes, including circles and irregular shapes, using other methods. Area concepts build upon multiplicative concepts of partitioning, unit iteration, conservation, and the structure of arrays. If a student is asked to find the area of a triangle, for example, she may be confused by using square units to measure a shape that isn't rectangular. She may attempt to count small triangles within the shape or leave gaps in the total area.

As students move on to 3D objects (three linear dimensions, e.g., length, width, height, base, depth), similar difficulties with reasoning about attributes, units, composite units, partitioning, and applying formulas may arise. Numerical, spatial, fractional, multiplicative, and proportional reasoning also apply. For 3D objects the unit is cubic. All three linear units should be expressed in the same unit, such as three measures in centimeters to find volume in cubic centimeters (except for some nonstandard units such as *acre feet* for measuring the capacity of reservoirs). Cubes, prisms, pyramids, cones, and other 3D objects have volume, defined as "the amount of three-dimensional space that a boundary contains" (Clements, Sarama, & Van Dine, 2017). Volume concepts require strong concepts and skills with length and area. Clements and colleagues (2017) emphasized that understanding volume involves *packing* (how many cubic units could be packed inside), *filling* (pouring liquid into a container, typically a linear measure), *building* (to create an object with volume), and *comparing* (objects and their volumes). The ability to perform these tasks develop simultaneously in the early stages of a learning trajectory for volume, up to the ability to structure 3D rows and columns mentally (between Grades 2 and 5), which precedes the ability to understand the abstract volume formula and mentally manipulate 3D arrays involving rectangular prisms, coordinating additive and multiplicative thinking (about 5th grade for many students).

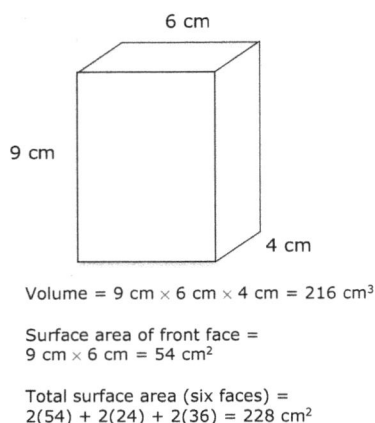

Volume = 9 cm × 6 cm × 4 cm = 216 cm³

Surface area of front face =
9 cm × 6 cm = 54 cm²

Total surface area (six faces) =
2(54) + 2(24) + 2(36) = 228 cm²

Figure E.1 Volume and Surface Area of Prism

Volume is conceptualized as both the *capacity* of the object and the *space* that is occupied. For example, a *cube* with *edges* that measure 5 cm (linear measure) has a volume of 125 cubic centimeters. It would hold 125 individual cm cubes. It takes up the space of 125 cm³. Each of its six sides is a 5 × 5 square, so the *surface area* of the entire cube is 25 × 6 = 150 cm². The liquid volume of this cube is 125 cubic milliliters (mL³) or 0.125 cubic liters (L³). Coordinating linear, area, and volume attributes and their numerical values for the same object is difficult for many students. Figure E.1 illustrates a *rectangular prism* with the dimensions 9 cm × 6 cm × 4 cm, and corresponding surface area and volume measures.

Students also have difficulty with the concepts and units of measure for angles and circles. As with using a ruler, students have difficulty using a protractor for measuring angles. They tend to see angles as having one horizontal ray, and if an angle is tilted it is more difficult to measure (Hansen, 2005). Reading the scale on the protractor in the appropriate direction involves some reasoning about *acute*, *obtuse*, and *reflex* angles . Some students find it difficult to understand that angles can have measures of 0°, 180°, or more than 180°. Students must reason about the unit of measure, *degrees* for geometric figures such as rectangles, angles, or circles (360°) in elementary and middle grades. In high school and college, *radians* are the unit (360 ° = 2π radians) because they are easier to use with trigonometry and calculus. Some misconceptions arise from poor language use about angles and their properties. Using "corner" or "point" instead of *vertex* can lead to miscommunication and misconceptions (Barmby et al., 2009). Circles and radians require abstract thinking about π (Pi), an irrational number that is the ratio of the circumference to the diameter. Many students memorize perimeter and area formulas for circles without understanding their meaning.

Other units difficult for students to comprehend include minutes and other units of time, mass compared with weight (amount of matter vs. pull of gravity), liquid compared with dry measures, and measures expressed as the ratio of two measurements (e.g., miles per hour). Many students hold on to misconceptions such as the smallest unit on the tool used is the smallest possible measure (there are no measures between the smallest hashmarks), length is the only measurable attribute, and only regular shapes and objects can be measured. Working with derived measurements, such as rate, requires proportional thinking in addition to other measurement concepts.

There is evidence that other factors contribute to students' measurement difficulties. These include insufficient teacher knowledge, weak curriculum design, insufficient time devoted to measurement, too much focus on numerical concepts that are not connected with other concepts such as measurement, and lack of dynamic representations 2D and 3D objects (Tan-Sisman & Aksu, 2012). In a review of preservice elementary teachers' knowledge, Browning et al. (2014) found gaps in general principles of measurement and how the number line connects measurement concepts with those of number. Preservice teachers may have only unconnected procedural knowledge about measurement, lacking conceptual understanding, struggling with the same challenges as their students. Smith and colleagues (2013) found, in a review of three US mathematics curricula (K-3), that textbooks and their teacher manuals tended to emphasize procedural knowledge (75% of all length content) over concept development. In some cases, procedures were taught prior to concepts. Further, some important concepts were rarely addressed, such as equal-size units, additive composition, ruler structures, and measuring complex or curved paths. None of the curricula addressed learning difficulties, concerning given the importance of this content. The researchers concluded that teachers should supplement provided curricula with research-based concept development through measurement tasks and address measurement standards earlier in the school year (K-3) to support other mathematics learning. Further, linear measurement concepts and tools should be reinforced in later grade levels.

Instruction in Measurement

Among the first measurements children make are comparisons of length. Who is taller? Who ran farther? Is my jump rope longer than yours? Young children should be involved with activities identifying attributes of objects and comparing those by attribute, learning comparison vocabulary. Children in first grade compare the lengths of objects and order them by length. They begin to measure length using same-size nonstandard units, such as tiles and rods. They should be encouraged to line up nonstandard units without gaps or overlaps, and to iterate the units as they count.

As students move from nonstandard units to standard ones, they should maintain the concept that each unit is the same size and there are no gaps between units. Students must develop the concepts of discrete and continuous units and the relation between number and spatial concepts. Further, they should develop the connection between number and measurement, realizing that different sized units can be used to measure the same length. "This table is 60 inches long; that's the same as five feet. I can measure this length in inches or feet, or with other units." As students move to second grade, they begin using standard units in both USCS (standard) and SI (metric) systems. They should gain an understanding of standard units so they can estimate before measuring and judge the reasonableness of their responses. When comparing lengths, second graders should determine not only the length of objects, but the difference in lengths, drawing on their number and operation concepts. How wide is this table from side to side? Will it fit through the door? Converting between units of measure (e.g., inches to feet) and between measurement systems (e.g., inches to centimeters) requires multiplicative and proportional reasoning as well as strong number line concepts for base-10 and fractional expressions.

Some instructional recommendations for young students (PreK-2) include:

- Develop the concept of length as the distance of a straight line between two points, the total distance along several straight sides, and the distance along a curved path.
- Plan engaging tasks that will develop critical concepts. Even young students can engage with interesting, realistic tasks involving length, area, and volume. For example, Sevinc

and Brady (2019) described model-eliciting activities (MEA) that were based on real-life problem stories. The *Proper Hop* story engaged students in thinking about unit iteration for length (frog hops as units) and the *Fussy Rug Bugs* story involved unit iteration for area (bugs as tiles). These MEAs encouraged students to reason about problems for the iterative process with teacher scaffolding.

- Have students compare the length of objects with each other (direct comparison) and with a third object (indirect comparison), something that must be reasoned about. Encourage the use of comparison terms (e.g., long, longer, longest).

- When measuring with nonstandard and standard units, emphasize the concept of unit. "Our unit is the blue rod. How many units will it take to cover the distance from this point to that point?" or "Our unit is a foot. How many feet tall is the doorway?" Emphasize the concepts of equal-size units, unit iteration without gaps or overlaps, the additive property (as measurement devices are shifted), and how to express part of a unit.

- Provide objects (discrete and continuous) for reasoning about and exploring linear measures. Students typically have experience with discrete quantities through their counting and base-ten operations, such as sets of linking cubes, number lines, and place-value representations. Continuous objects for measurement do not have measurement segments, so selecting a unit of measure and counting the number of units that cover that quantity provides the measurement value.

- When introducing rulers with numbers, teach their structure. Label the zero position and reinforce the concept that the first unit is the span between the 0 and 1. Each unit on the ruler is the same length. A good analogy is a child's birthday. If you are five years old, that means you have lived five years: birth to one, one to two, two to three, three to four, and four to five. We measure the spans or units between marks. Guide the transition between nonstandard and standard linear units by emphasizing the measurement of space, not the tally marks on a ruler. Teachers should use gestures across units to emphasize the space of a unit (Smith et al., 2013) or place unit rods or tiles next to a ruler with the same size units. Teachers should also provide "broken-ruler" or shifted-ruler tasks, where the beginning point on the ruler is not zero.

- Help students develop the inverse relationship between size of unit and number of units by using nonstandard units of different lengths (e.g., Cuisenaire rods). "Why did it take more red rods to measure this length than blue rods?" Then shift to standard units for tasks using the same reasoning. "Why are there more millimeters than centimeters when you measure the length of this box?"

- Guide and model accurate measurement language, such as longer and shorter for length, fewer and more for discrete units, specific unit names, and measurement result or measurement value.

- Use a research-based developmental trajectory, such as the one developed by Szilágyi et al. (2013) as a guide for sequencing learning activities and providing instructional scaffolding. See also the website *Learning & Teaching with Learning Trajectories* (Clements & Sarama, 2017/2019) for length, area, volume, angle, and other research-based trajectories (PreK-3; https://learningtrajectories.org).

By the end of second grade, students should have developed the essential measurement concepts related to unit and strategies for measuring length using rulers and standard linear units (e.g., inches and centimeters). Students in third grade begin estimating and measuring 2D and 3D objects, applying the concepts of area and volume, although many concepts are developed earlier (Clements, Barrett, & Sarama, 2017). Fourth graders are expected to convert among measurements (within a system). Elementary students' experiences with measurement are closely connected with geometry and number concepts (multiplicative reasoning, fraction concepts, and the decimal system) and can strengthen those understandings.

Often teachers jump right in with exploring attributes of objects and their measures, assuming that previous measurement concepts are firm. Students should also understand the correspondence between the unit of measure and the attribute. We use linear measurement units to measure the sides of a rectangle, but we use different units to measure the area of the rectangle (e.g., square centimeters and square inches). *The attribute of the object to be measured determines the unit of measure.* Some recommendations for teaching about measurement in Grades 3 through 5 include:

- Begin with review and informal assessment of students' measurement concepts, including their reasoning about unit iterations, estimations with standard units, manipulating measurement tools, and using accurate measurement language and symbols. A quick informal assessment task: Given a broken ruler (e.g., from 3 to 9), ask the student to measure the length of an object 4 inches long and another 71/4 inches long. Watch and listen to the student's reasoning about unit and ruler concepts. Provide targeted interventions for gaps in concept understanding and procedural knowledge.
- Build on prior knowledge of linear measurement to develop new concepts of perimeter, area, and volume and their attributes and units. Perimeter may be more challenging than area for some students who find it difficult to deal with "corners" when iterating units (Smith & Barrett, 2018). Make explicit connections by using concrete objects, representations on grids, and virtual tools. Introduce procedures and formulas only after concepts are firm, employing the concrete-representational-abstract (CRA) sequence for new concepts.
- Students should develop concepts of composite units for area and volume. Activities with composing, decomposing, and comparing objects promote understanding. "Can you create a polygon with area of 36 cm²? How many different polygons can we create? How do the perimeters of those polygons vary?"
- Teach about angles as *both* the "measure of divergence of two straight lines on a plane," a static view, and as a measure of turn or rotation, dynamic (Beneson et al., 2020, p. 11). The static view can be connected with 2D shapes, such as the angle of a triangle, or the degrees in a portion of a circle. The dynamic view can be related to the edges of scissors, the opening of a door, or the turning of a bicycle (Barmby et al., 2009). Only after students develop concepts through contextual tasks should abstract problems be introduced.
- Connect fraction and base-10 (decimal) concepts with those of unit partitioning for standard and metric units. Connect area and volume concepts with those of graphing and modeling. The coordinate plane has area shown by the x- and y-axes; the three-dimensional Cartesian coordinate system adds the z-axis.
- Support the transition from counting to multiplying for area and volume concepts. Bridging activities will support students' deeper understanding of formulas when those are introduced. For example, counting composite units of arrays (rows and columns) that, when stacked, comprise a rectangular prism can be connected to the measures of each dimension of the prism. That can lead to the formulas $V = l \times w \times h$ and $V = bh$.
- Teach students to use technology tools such as *Google SketchUp* (sketchup.com) for interacting with area and volume measurement concepts. "How many different shapes can I make for the same area? How does changing the surface area of the sides of a house affect its volume?"
- Plan engaging measurement tasks that promote measurement reasoning, active measuring, and communication about concepts and processes. For example: *Your task is to design the*

packaging for a new cereal. The box (rectangular prism) must hold 45 ounces of cereal. The nutrition label must appear on one side and cover 30 percent of that surface. Plan, measure, and cut out the net (pattern) for your cereal box and the nutrition label. You must show the weight of cereal in ounces and kilograms.

In the secondary grades, the CCSSM (NGA & CCSSO, 2010) incorporate measurement goals within geometry and data analysis. The NCTM (2000) measurement strand is developed PreK-12, but decreases in explicit content as geometry, data analysis, and algebra increase. Measurement becomes a tool for analyzing objects, understanding relationships, and solving problems. These connections are appropriate, but some students may still have gaps in basic measurement concepts and skills. Teachers should conduct screening for potential gaps and provide interventions where needed (Box E.1).

Box E.1   Intensive Intervention for Geometric Measurement

Target: Ten students in seventh-grade classrooms at Balsom Middle School were identified during pre-unit assessments with weak geometric measurement concepts. The students had been getting by with area, surface area, and volume formulas with rectangular objects, but lacked concept understanding necessary for new topics.

Intervention Schedule: Daily 30-minute lessons during the school's flex-lab times to supplement mathematics instruction. Approximately 4 weeks.

Nature of Intervention: Two five-student small groups will be engaged with explicit instruction on measurement concepts related to geometric topics using a CRA sequence (models and constructions) and high levels of discourse. Students are grouped by profiles on assessment. Topics follow the order:

1. Review of linear measurement, unit concepts, and ruler structure.
2. Development of linear concepts for 2D shapes (dimension, length, area, etc.) and perimeter, area of shapes (rectangles, quadrilaterals).
3. Development of linear concepts for 3D shapes (attributes, dimensions, etc.), surface area, volume of rectangular prisms.
4. Extend 2D concepts to triangles and other polygons. Connect with coordinate plane graphing and measuring angles.
5. Extend 3D concepts to other prisms.

Data Collection (Biweekly): Problem tasks assessed with rubric (measurement skills—US and metric—and concepts about units of measure for linear, area, and volume measurements and how to express those values).

Recommendations for the development of measurement competencies beyond fifth grade include:

- Have measurement tools at hand and re-teach their use as needed, including computer-based modeling.
- Middle-school students are still developing concepts related to perimeter, area, and volume, with increasingly complex shapes and objects. Plan real-world problems that are engaging and involve reasoning, communication, modeling, and tools. For example, Eskelson et al. (2021) posed a problem for maximizing the area of a fixed perimeter: planning a dog

pen for a new puppy using existing back-yard features (e.g., some existing fence, side of house) and 72 feet of new fencing. What shape pen will provide the best conditions for the puppy? Student groups used their related mathematical and pragmatic knowledge, attempted several approaches to the problem, made use of *GeoGebra Geometry* (geogebra. org) software, and discussed and justified their designs.

- Reinforce prior measurement vocabulary and concepts and build on those when transitioning to more complex topics such as properties of triangles and circles, making transformations, applying formulas, and reasoning about proofs.
- Make explicit measurement connections with other topics, such as coordinate-plane graphing, proportions, the decimal system, algebra, and other subject areas, such as science and geography. Include measurement situations in problem-solving tasks for these domains. High school topics involving measurement concepts and tools include polar coordinate systems, solid geometry, and pre-calculus.

Teachers at all grade levels can draw implications from the research of Project M^2 in the early grades on implementing geometry and measurement units with diverse groups of students (Gavin et al., 2013; gifted.uconn.edu/projectm2). Mathematics tasks in measurement and geometry should be complex and challenging, requiring students to reason at high levels. Students should be encouraged consistently to explain their reasoning verbally and in writing. Mathematics practices, such as reasoning, using tools and models, looking for structure, and problem solving, should be integrated closely with instruction. The researchers of Project M^2 noted that challenging content takes more planning and instructional time, but with students' deeper understanding of concepts, their time with measurement and geometry served as "value added" for other mathematics topics. Too often topics in measurement are treated as discrete skills or extras, when the tasks could be scaled up for deeper understanding and generalization.

Time and Money

Two everyday concepts we often take for granted are time and money. Some students struggle to develop the concepts and skills involved with these abstract concepts. However, both concepts are critical for managing one's daily life. Both involve symbolic interpretation, planning, measuring, and evaluating results. Teachers should evaluate students' informal knowledge of time and money and provide interventions for gaps in concept understanding.

Time

Time is an abstract concept that is even more elusive than most of us understand. Time can be measured in nanoseconds and centuries, minutes and days, and many other customary units. Tools for measuring time include clocks and calendars. Sometimes time telling should be precise and other times it can be estimated. With the popularity of digital clocks, many students find it difficult to read, interpret, and set analog clocks. In digital form time "appears," but in analog form time progresses for those watching. Mathematics standards call for fluency with both.

Students with learning difficulties often have problems with the general concept of time—estimating time needed or time elapsed. How long is a minute? An hour? How much time should you plan to spend on your mathematics homework? For next week's project, how many units of time will be needed each day in order to finish the project on time? Why does time seem to slow down when you're involved in something unpleasant, but it races by when you're having fun? Many students with specific mathematics disabilities

(MLD) have significant problems with planning their use of time and self-monitoring during task performance.

In a study of students in Grades 1 through 6, Burny et al. (2012) found that students with mathematics difficulties (MD) performed significantly worse than their typically achieving (TA) peers on clock-reading, writing, and transformation tasks with both digital and analog clocks. Some of the common errors were selective attention (attending to only one part of the clock), misinterpreting numbers, using a false reference point, miscounting, and confusing "to" and "past." TA students performed better with digital clocks, but students with MD did not, indicating the impact of underlying procedural and semantic cognitive deficits. Interestingly, using discriminate analysis, the researchers concluded that analog-clock reading was a predictor of overall mathematics difficulties. Prediction for mathematics difficulties based on analog-clock reading was 79% for first graders, 72% for second graders, and 70% for third graders. Clock-reading difficulties were not obvious in some students until 5-minute and 1-minute intervals were introduced, but difficulties with clock reading didn't decrease with these students until sixth grade.

Informal assessment of students' abilities to set clocks, read clocks, and interpret time can be conducted through routine activities and teacher-prepared tasks. Students who are English language learners may find it difficult to change from time-telling conventions such as saying "half six" to mean 5:30 or 6:30, depending on the culture, or the use of the 24-hour method for telling time. Other students with language deficits may struggle with relational vocabulary such as before, after, to, quarter 'till, o'clock, and half-past. Several phrases can mean the same time: three forty-five; quarter 'till four; fifteen minutes to four.

Instruction in telling time should include clock setting and clock reading, beginning with the larger units of time and moving to shorter segments—hour, half hour, quarter hour, five-minute increments, minutes—with real clocks. Tactics that help students discriminate between the large and small hands and the portions of the clock face include color-coding, clock overlays for specific portions, and prompting phrases such as, "Begin with the hour (small) hand and say the hour it is on or it just passed." Teachers should not assume that any student, regardless of other abilities, has mastered telling time because it can be easily disguised within the classroom. After mastering time telling, students should be challenged with time-based problems that involve comparing time, estimating time, and computing elapsed time, integrating their number-system concepts. "We will take a break in 35 minutes. What time will it be then?" "If your cousin was born on 1 May 2017, how old is he?" These problems require a lot of quantitative and spatial reasoning about abstract concepts.

Between Grades 2 and 4, students should develop time concepts and solve problems involving time. Problems involving comparing times (e.g., hours or years), finding a time difference, and finding a time in another time zone are especially challenging. For example, it is 6:00 pm in the eastern US (18:00 by the 24-hour clock). What time is it in Japan? Is it the same day? How do you reference Greenwich Mean Time to find the time in another place in the world? There are great connections with social studies in these investigations.

By sixth grade students are studying rates, computing, representing, and comparing *unit rates* such as miles per hour and meters per minute, using the distance/time formula: $r = \dfrac{d}{t}$. These students should have strong concepts of each measure within the rate as well as the concept of rate as a ratio. They will find any missing element within the rate formula and compare rates using proportions. Teachers should use realistic contexts and interesting problems for students to develop rate concepts. Tables and graphs become tools for problem solving with rate. Other secondary-school topics incorporate the unit-rate concept, including monetary concepts, physics, and statistics.

Money

Money refers to the physical coins and banknotes (bills) that are a medium for exchange. Without money representing a set value, we would have to barter for common goods and services. In the US, coins and paper money are legal tender (since 1933) and must be accepted at face value. Federal Reserve notes range from $1 to $100, with the most common $1, $5, $10, and $20. Coins include pennies, nickels, dimes, and quarters as well as less-common half-dollars and dollars.

Mathematics standards, such as those of NCTM and the CCSSM, include initial money concepts in second grade. After that, money is referenced only in connection with solving problems, as with percentage of cost problems. Some students, such as those with intellectual disabilities (ID), will need more explicit instruction over a longer period to be fluent with day-to-day money skills. Working with money includes identifying coins and bills by their values, computing costs, making change, and converting amounts. Students may also have problems with monetary notation, such as $0.35 and 35¢, and the concept of special decimal-place notation: $1.70 and not $1.7 or $1.702.

One of the most difficult skills, after students have learned the values of coins and combining amounts, is computing change. Teachers should conduct informal assessments of student understanding about coin and bill values, combining amounts, and making change from a given amount before beginning instruction. There are several approaches for computing change, but the most commonly used is the "counting up from the amount given" method. The counting-up method requires the student to count by 1s to the next number divisible by 5. Then a decision has to be made about using quarters, dimes, or nickels to reach the next even amount. Finally, counting on by dollars, 5 dollars, and 10 dollars if needed. For example, if an item costs 32 cents and a dollar was provided, then the student should count as follows: 33, 34, 35 (3 pennies); 40 (1 nickel); 50 (1 dime); 75, 1 dollar (2 quarters). Practice counting up by different denominations reinforces number sense and equivalence concepts. To check whether you've received the appropriate change, start with the total cost and count on using the change given to arrive at the amount given to the cashier. This forward and backward mental counting is difficult for many students because it requires additive, multiplicative, and inverse reasoning.

Price comparison skills have been identified as one of the more challenging, but important, money-related competencies (Weng & Bouck, 2016). Price comparison involves comparing the magnitudes of the prices in a grocery store (provided in numbers) and sometimes comparing the measures of the items. Students need to apply decimal-number understanding (value of each digit, mental number line), use number and comparison vocabulary, and one-to-one correspondence concepts. For students with ID, this task must be analyzed step-by-step and instruction supported by strategies for comparing decimal places one by one, picture or video modeling, simulated practice with real price tags, and reinforced practice in natural settings. Similar task analysis and support via augmented technology was used to teach middle-school students with ID to use ATM simulators independently (on iPads), for withdrawals and money transfer (Kang & Chang, 2019). Community-based instruction is best for generalization, but careful task analysis and practice with simulations can front-load skills needed for executing skills in real settings. Practice in actual settings often precludes allowing many errors or repeated practice with scaffolding. With the more common use of debit and credit cards, students will need more instruction in keeping track of balances, using ATM machines and financial apps, and managing budgets.

Students in life-skills mathematics should study other monetary topics, such as check-writing, taking out loans and the effects of interest over time, various savings options, and concepts for employment such as gross and net income, taxes, and other deductions. Many high-school students are motivated by these "adult" concepts, so a range of time and money topics can be integrated for interesting mathematics problem solving.

Data Analysis and Probability

Data are collections of information that are used to make decisions. Data can be collections of numbers, words, images, or other pieces of information. Data can be qualitative or quantitative, with quantitative forms the primary focus in school mathematics. Quantitative data are numeric and can be *discrete* (countable and finite) or *continuous* (measurable and infinite). Data analysis is a broad undertaking, from formulating questions and collecting data to making representations and interpretations of data sets. Probability is treated as a tool for data analysis in pre-college mathematics. The development of probability concepts progresses from informal discussions of "how likely or unlikely something is" in the early grades to the mathematical computations of probabilities in high school. This section will explore the data analysis and probability competencies expected for PreK-12 students, the common difficulties with these concepts and processes, and recommendations for interventions.

"Never have data and statistical literacy been more important" (Bargagliotti et al., 2020, p. 1). The *PreK-12 Guidelines for Assessment and Instruction in Statistics Education II* emphasized that "being able to reason statistically is essential in all disciplines in study and work" (Bargagliotti et al., 2020, p. 7). Further, these competencies are critical for navigating life (e.g., financial and health choices) and understanding governmental, agency, and media reports about topics such as extreme weather conditions, global pandemics, economic trends, and social issues. For example, an individual reading the news in the morning may study data related to weather, the stock market, and a favorite baseball team. This analysis will assist her in deciding what to wear, when to sell some stock, and whether to reserve seats for the next game. At work managing a large wholesale club, she will use sales data to predict the number of cashiers to schedule each day the following week and to plan special sales. She will prepare charts depicting annual sales trends for the next regional meeting. Even things that don't seem numerical in nature can be analyzed using data, such as opinions of consumers or the quality of products. The ability to reason statistically is critical for being a productive employee, informed citizen, discerning consumer, and healthy family member.

Statistics is "the science of learning from data, and of measuring, controlling, and communicating uncertainty; and it thereby provides the navigation essential for controlling the course of scientific and societal advances" (Davidian & Louis, 2012, p. 12). Over the past century, statistics—data analysis and probability—has gone from being fairly obscure to being studied at all grade levels (Jones & Tarr, 2010). The NCTM standards (2000) included a PreK-12 strand on data analysis and probability, beginning in the earliest grades with children collecting data about their own questions and creating simple graphs. The CCSSM (2010) integrated data analysis with measurement standards through Grade 5, then recommended a separate strand on statistics and probability for middle and high school.

The American Statistical Association (ASA) also recommended PreK-12 development of *statistical literacy* (Bargagliotti et al., 2020). In its curriculum framework, endorsed by NCTM, the ASA noted the key difference between statistics and mathematics—the focus of statistics is on *variability in data*. "In mathematics, context obscures structure. In data analysis, context provides meaning" (Cobb & Moore, 1997, p. 803). Statistics requires a different kind of thinking. Statistics becomes increasingly mathematical as the level increases, but data–collection design, exploration of data, and interpretation of results are heavily dependent on context and involve limited formal mathematics at the introductory level (Bargagliotti et al., 2020). Probability is a tool for statistics, as well as for mathematical modeling and applied mathematics. Probability assists us in reasoning about the likelihood of our data analysis findings being accurate.

The key concept of variability is difficult for students and takes years to develop. Children often believe that their answers are exact, that measurements are precise. It is human nature to attempt to categorize, to try make sense of and organize complexity. However, categorization leads to stereotyping and rigid boundaries that aren't reflections of reality and aren't productive for society. Students need to develop concepts of measurement variability, natural variability, induced variability, and sampling variability, and understand that variability provides strength and interest, as well as assistance for decision-making. Understanding the nature of variability can enable students to make more informed decisions and be more open-minded about their world.

Shafer and Özgün-Koca (2021) described variability as how stretched out or squeezed a group of data is. The authors described tasks appropriate for each grade-level band (PreK-2, 3–5, 6–8, 9–12) that ask students to make discoveries and develop conjectures by interpreting data using human dot plots for range concepts (PreK-2) and interactive *GeoGebra* applets for dot plots illustrating range, mean, mean absolute value (MAD), standard deviation, and characteristics of distributions. The human- and computer-based representations allowed students to reason about data distributions and the factors affecting variability.

NCTM standards (2000) recommended that study in data analysis and probability be generated by student questions. A question such as, "Which foods are most popular in our cafeteria?" can lead to the methods needed to gather, organize, analyze, and display or report results. Younger children are naturally curious and can design simple data-collection strategies, such as tallies or objects, and represent data on dot and bar graphs. They should understand that data representations provide information about their world. By the end of second grade, students should question inappropriate statements about data. In Grades 3 to 5 students are involved in investigations that produce data sets and use a wider range of representations, such as line graphs and computer-generated graphs. They study central tendencies, make predictions, and justify conclusions using data sets. They are introduced to more abstract concepts, such as population sampling.

By middle school, the emphasis should be on planning the most appropriate data collection and analysis methods and comparing those methods. Their tools and skills expand to include different graphing forms, statistical symbols, and the meanings of aggregated data and variability. Middle-school students connect concepts from algebra and geometry with those of data analysis, such as linear relationships, variables, and proportions. They should become more analytical about the benefits and limitations of different representations of data sets, the nature of relationships, and outcomes of predictions. By high school the range of applications increases as do the methods and tools. Students compare data sets (univariate and bivariate), study trends, use simulations, and apply more advanced concepts such as correlation and regression. They also explore related issues such as bias, randomization, error concepts, types of studies, and statistical predictions. They are more concerned with study design, selection of appropriate statistical methods, representing data relationships in the most valid and effective ways, and using data to justify predictions and conclusions.

The ASA curriculum framework (Bargagliotti et al., 2020) focused on statistical problem solving, an investigative process with four components: formulating questions, collecting data, analyzing data, and interpreting results. This process is similar to mathematics problem solving, scientific inquiry, and the engineering-design process. The framework also defined three developmental levels, a trajectory, for statistical problem solving for each of the four components.

- Level A involves initial learning and is more teacher driven. At this level students develop "data sense," that data are context-generated. They learn how to use graphical representations for data, describing features of distributions, and begin answering questions based on data. Beginning students also develop informal ideas about probability.

- Level B students have some statistical experience and are more self-directed. They begin to pose their own questions and are more aware of what makes a good investigative question and collect appropriate data, considering the possible types of data analysis for the situation. Students at this level use a range of representations and statistical tools and consider randomness, variability, and comparing distributions. They work with population samples, understanding the limitations and scope of this analysis.
- Level C students use questioning throughout the statistical problem-solving process. They recognize traditional and nontraditional data and select appropriate data analysis tools. These students engage in multivariable thinking and their questions expand to include causality and prediction (*inferential statistics*). Their interpretations are more insightful and sophisticated, integrating the context and objectives of the investigation to draw and support conclusions using statistical evidence.

The ASA framework document provides many examples of student competencies at each level as well as examples of assessment and learning activities to promote statistical literacy. In addition, the ASA provides free on-line journals, *Journal of Statistics Education* and *Statistics Teacher*, as well as investigations, modules, and lesson plans at all grade levels (www.amstat.org).

Statisticians categorize variables by one of four different levels of measurement that indicate the types of statistical analyses that can be performed. Nominal and ordinal data are considered for qualitative information, while ordinal (sometimes), interval, and ratio for quantitative. Interval and ratio data can be discrete (can take only specific values) or continuous (can take any value along an interval).

- *Nominal* data (from the Latin for *name*) are only labels or classifications and have no order. Nominal data can be grouped, and the groups compared but cannot be calculated. For example, gender, car manufacturer, zip codes, and ice cream flavors are nominal.
- *Ordinal* data have order, but the distance between measurements has no meaning. For example, opinion surveys often use categories such as "strongly agree, agree, disagree, and strongly disagree." The opinions are clearly in specific categories with an order, but the distance between categories is not clear and should not be calculated.
- *Interval* data have meaningful intervals but do not have a true zero or starting point, so some types of calculations cannot be performed. Temperature on the Celsius scale is interval data because the intervals between temperatures are consistent, but there is no starting point. Time is another example of an interval scale. We can add and subtract with interval data but not multiply or divide.
- *Ratio* data have the strongest properties for analysis. Their intervals have meaning and there is a beginning point so ratios computed also have meaning. Examples of ratio data are height and weight. With ratio data we can perform all four operations, including computations involving ratios, as well as central tendencies (e.g., mean) and measures of dispersion (e.g., standard deviation). Ratio data can have a value of zero on the scale of points with equivalent distances.

As PreK–12 students engage in data explorations they should develop increasingly sophisticated tools for data design, collection, representation, and analysis. They should develop the related problem-solving and decision-making abilities that rely on good data selection and analysis.

Difficulties with Data Analysis and Probability

Many students have difficulties with concepts and skills within data analysis and probability. For example, on the NAEP (NCES, 2019), fourth graders scored in the basic range, with students with disabilities scoring 27 points lower than nondisabled students. Eighth graders also scored in the basic range, with students with disabilities scoring 44 points lower, below basic levels. On the whole, students were not proficient with statistical competencies. For example, only 53% of fourth graders could interpret a data table and 51% could read and complete a pictograph. For eighth graders, 66% could complete a data table, but only 31% could determine the probability of an event presented in a word problem, and only 20% could interpret stem and leaf plots.

Data analysis and probability may not be getting the attention needed within a curriculum that is assessment driven. Skills may be taught in isolation or without concept understanding, when they could be connected with most other mathematics topics and other subject areas. Too much emphasis may be on exploratory data analysis and not enough on related variability and informal inference reasoning. There may be a lack of attention to a developmental trajectory for statistical reasoning, over the grade levels, with targeted interventions as needed.

Some of the most common difficulties with data analysis and probability concepts include:

- Problems formulating questions. Students with language disabilities, especially, have difficulty in this area. The question should lead to the type of data to collect. Questions will differ for *descriptive* (What is the average distance of paper planes thrown by our class?) and *inferential* statistics (What do you predict would be the average distance of paper planes thrown by all seventh graders?).
- Misinterpreting graphs. Students often have difficulty interpreting the scales on graphs or points not on grid lines. They also may assume that the starting place is zero when that is not always the case with data sets. Most students are familiar with graphs depicting change (relationship of independent and dependent variables), but are not aware of other depictions of data, such as frequency distributions and regression lines.
- Selecting the wrong graph for the data. Once students are comfortable with a type of graph, they tend to use that for other data sets, even when it is inappropriate. For example, using a simple bar graph (discrete categories) for numerical data showing trends. Graphs selected will depend on the method of analysis of the data and the types and number of variables (e.g., In & Lee, 2017).
- Interpreting graphs as pictures rather than representations of a data set (Ryan & Williams, 2007). For example, interpreting the slope of a line depicting the speed of a bicycle as a hill the bicycle is climbing.
- Misconceptions related to probability. Some students have difficulty making part to whole comparisons (e.g., 8 boys out of the total 15 children). Others are influenced by "recency effects," the most recent event in a pattern having a stronger impact on the next events. For example, if a coin toss resulted in H, T, H, H, H, then the student might assume the next toss would be tails, when it actually has the same probability on each toss. Other students find it difficult to reason about a total possible population and why the sample collected might be similar or different.
- Language issues. The study of statistics brings its own vocabulary and symbol system. Some familiar words have new and more precise meanings: mean, median, mode, significant, event, distribution, population. Some new terms have familiar parts: bivariate, percentile, and quartile. Some special symbols include: $\bar{x}$ for mean, n for number, r for correlation coefficient, μ for population mean, σ^2 for population variance, σ for population standard deviation, and Σ for summation. Some symbols have multiple meanings, such as r for rate, radian, and a correlation statistic. These symbols may vary by textbook or application.
- Problems with the abstractness of statistics. If concepts are not developed with adequate contexts and representations, students tend to use rote procedures, unable to generalize learning. They need a wide range of realistic problems and experiences over time, along with teacher guidance and peer interactions, to develop the reasoning skills required.

What does our data tell us if the mean and median are the same value? What effect does an outlier data point have on our results?

- Weak transnumerative thinking. "Transnumeration is a dynamic process of changing representations to engender understanding" (Shaughnessy & Pfannkuch, 2002, pp. 255–256). Students must capture and transform information from the real world, construct multiple statistical representations of that information, and then communicate what their representations suggest. Sometimes students' experiences bias information selection and predictions, so teachers use unfamiliar contexts on purpose (e.g., predicting the eruptions of Old Faithful).

- Less sophisticated thinking about data. As with whole- and rational-number concepts, some students have difficulty transitioning from additive to proportional reasoning (frequency counts rather than relative frequencies and percentages). Some students have difficulty integrating concepts and attending to and coordinating multiple aspects of data. For example, reasoning about sampling distributions requires the integration of center, shape, and variability (Noll & Shaughnessy, 2012).

- Little tolerance for uncertainty. Many students expect exact answers in mathematics. However, as variability is a cornerstone of statistics, students must develop an understanding, through context-rich experiences with data, that uncertainty is prevalent in life. Understanding how data can be collected and analyzed can provide a level of confidence in answers to questions. Even the measurement of the length of the same object by 20 different students will produce a range of values. Why do answers vary? How can we come close to the actual measurement value? It is impossible to collect all the data (entire population) for some investigations, such as finding the average retirement age for a specific country or region. How can we sample the population and have some confidence in our result?

- Gaps in learning. It is quite common to find gaps in student learning across mathematics. These might be concepts not fully developed within data analysis, such as the types of data or the range of possible analytic tools or representations of data. Other gaps are a result of weaknesses in other mathematics domains that have an impact on statistical reasoning, such as number understanding, spatial sense, and proportional reasoning.

Finally, if teachers are not confident with their content and pedagogical knowledge related to statistical concepts, they may teach these topics with limited examples and narrow procedures, neglecting the development of broader statistical reasoning and problem solving. The next section provides recommendations for instruction.

Instruction for Data Analysis and Probability

To promote statistical thinking, teachers should approach this domain from an exploratory, problem-solving perspective. The discrete methods and representations within statistics are tools for investigations. If students are experiencing difficulties with topics in data analysis and probability, teachers should conduct informal assessments that allow a view of students' statistical thinking, communication, making representations, and providing justifications. Posing a problem situation at the beginning of a unit and observing students approach the problem and apply previous concepts will provide information on prerequisite understanding. For example, a middle-school teacher posed the problem: *Detroit and Boston are on the same latitude. Which city is warmest in summer months? Which is coldest in winter months?* This problem led to discussions of what data to collect, how to analyze it, and how to represent the data using graphs. The teacher listened to student reasoning, vocabulary, and concepts about statistical methods.

Statistics is often compartmentalized into descriptive and inferential, with inferential the most difficult to master (Langrall et al., 2018). Inferential statistics encompasses a wider potential data field, beyond data at hand, therefore the conclusions are uncertain. Curriculum standards differ in their emphases with the NCTM (2000) requiring making inferences from the earliest grades. The CCSSM (2010) introduce making inferences about populations based on samples in seventh grade and include more formal inferential statistics in the high-school standards. If approached informally at first, younger students (K-2) can engage in inference through context-rich problems with teacher supports. Informal statistical inference involves a generalization (a claim), support of data-based evidence, and articulation that includes uncertainty (Makar & Rubin, 2009). These early experiences don't rely on probability distributions and formulas, but serve as a foundation for understanding population sampling, variability, distribution, and randomness.

Research on effective interventions for students with MD for data analysis is rare. One of the few projects focused on this domain addressed first- and second-grade measurement and data analysis standards integrated with grade-level science standards—the Tier 2 *Precision Mathematics* intervention (Doabler, Clarke, Kosty, et al., 2019; Doabler, Clarke, Firestone, et al., 2019). This intervention included eight modules (30-min small-group lessons, four days a week, eight weeks) that incorporated the four components of the investigative process (formulating questions, collecting data, analyzing data, interpreting results) plus language development and related word-problem subtypes. Lessons included technology-based and hands-on problem solving with explicit instruction and high levels of classroom discourse. An example of a first-grade module involving data analysis is sorting animal traits by attribute, solving compare word problems, and using a bar graph to answer questions about animal traits. Students in the intervention groups made significant gains on the measurement and data analysis assessment as compared with control students ($g = 0.45$), and, while not significant, positive effects on general-mathematics achievement and number identification. Further, differential gains in outcomes were greater for students with more intensive needs regarding initial understanding of whole numbers (and word reading skills). The researchers concluded that interventions for younger at-risk students beyond those of whole numbers, such as measurement and data analysis, have the potential to support overall mathematics learning and connections across domains.

Recommendations for developing data analysis and probability skills:

- Use real data sets, student questions, and real-life problems as the basis for introducing new concepts in data analysis. These explorations can be extremely motivating if connected with student interests. For example: We wanted a new boat ramp on a tidal creek near Charleston but had to demonstrate we were not damaging oyster beds. The regulations required data on the size and location of beds and number of oysters in each for the entire creek. Rather than counting all the oysters in 12 discrete beds to determine their density, we took a random sampling measurement (at low tide threw a frisbee 5 times and counted the oysters underneath) and compared that count and area of the frisbee with the area of each bed. We justified the argument that our ramp would not disturb established oyster beds between the ramp and river using these data.
- Practice developing investigation questions, then shifting those from descriptive to inferential forms, to go beyond populations in a narrow data set. For example, we want to explore the use of electric automobiles. From: How has the use of electric automobiles changed in our community over the past five years? To: What do we predict as trends for electric automobile use in our state?
- Provide hands-on (physical) objects with initial learning, along with computer-generated simulations. For example, students for a study of arm length, plants for a study of growth under different conditions, actual music selections for a school dance. Simulations can be made using *Fathom* (fathom.concord.org) and *TinkerPlots* (tinkerplots.com) applications.

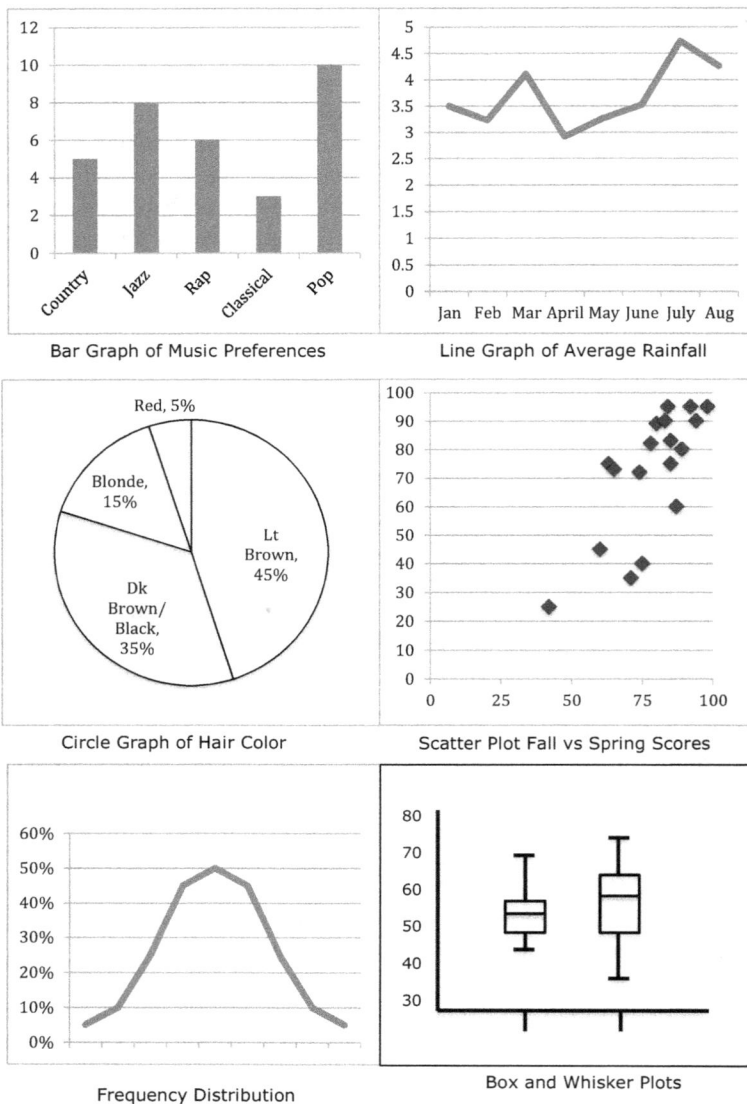

Bar Graph of Music Preferences

Line Graph of Average Rainfall

Circle Graph of Hair Color

Scatter Plot Fall vs Spring Scores

Frequency Distribution

Box and Whisker Plots

Figure E.2 Sampling of Graph Types

- Teachers should spend time exploring a problem situation or data set before using it with students. Without careful planning, the mean, median, and mode will all be the same! Good problems and data sets are those that contextualize the concepts being addressed within the limits of students' instructional levels.
- Provide students experience with a range of representations for interpretation, including graphs of different data types. Listen to student interpretations. See Figure E.2 for examples of graphs.
- Present nonroutine problems that assist students in developing concepts of ranking, where numbers are not immediately available. For example, Doerr and English (2003) presented "the sneakers problem," and other similar tasks, requiring students to develop and rank a

list of factors they considered important when buying sneakers. This task led to concepts such as frequency ranks, pairwise comparisons, and averaging, and eventually to a systems-view of data relationships.

- Use technology for its exploratory and efficiency properties. Students who have the fundamental concepts can ask "what if" questions of a data set and see immediate results without having to work through lengthy calculations.
- Use everyday objects to develop initial probability concepts, such as dice, spinners, and coins where the possible events are known. Then connect probability concepts to statistical results. Probability deals with the likelihood of an event or how certain you can be with your data-analysis results.
- Provide examples of how statistics can be misused, to demonstrate potential sources of bias. Younger students, especially, tend to believe numbers are correct, absolute, and unquestionable. But the premise of statistics is the variability of the world.

Measurement, data analysis, and probability concepts and skills should begin with the youngest students and be expanded and developed throughout the grade levels. These mathematics domains are closely connected with other topics in geometry, number systems, graphing, and algebra. Engagement in measurement and statistics learning activities promotes problem solving, using tools and models, and reasoning skills.

Multicultural Connection

Measurement tools have changed across millennia and cultures as needs and technologies have changed. The cubit, a length measure of the forearm (elbow to tip of middle finger—approximately 17 to 20 inches), was first mentioned in the *Epic of Gilgamesh* in Assyria (approx. 669-630 BCE), was also used in Ancient Egypt (about 7 hands), and mentioned in the Old Testament (e.g., the golden ark in *Exodus* of two cubits and a half; Stone, 2014). Leonardo da Vinci referred to nine units of measure in his comments on *Vitruvian Man* as fractions of the dimensions of six-foot man: yard; span; cubit; fathom; hand; foot; and Flemish, English, and French ell (Latin *ulna*, for forearm). The English also used body-referenced units, such as a foot, cubit (2 feet), yard (nose to thumb), and rod (12 feet), influencing today's measures.

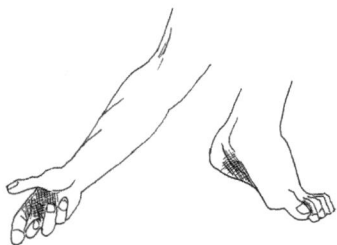

Some cultures still use nonstandard tools for length measurement. In studying measurement units with their contexts and vocabulary in Papua New Guinea, Owens and Kaleva (2007) surveyed and interviewed preservice teachers about practices in their home villages. Some teachers reported rope and paces as measures of garden lengths, with estimates of area based on those lengths. Other villages used a long cane of bamboo, sometimes with markings—to measure lengths of houses, bridges, gardens, and canoes—using concepts of unit iteration and composite units. Volume was measured in reference to the size of a pig-cooking pit. The researchers concluded that these teachers should use the informal measures and familiar contexts to bridge concept understanding to standard measurement units required by the curriculum with their future students.

References

Bargagliotti, A., Franklin, C., Arnold, P. Gould, R., Johnson, S., Perez, L., & Spangler, D. A. (2020). *The PreK-12 guidelines for assessment and instruction in statistics education II (GAISE II)*. American Statistical Association. www.amstat.org/asa/education/Guidelines-for-Assessment-and-Instruction-in-Statistics-Education-Reports.aspx

Barmby, P., Bilsborough, L., Harries, T., & Higgins, S. (2009). *Primary mathematics: Teaching for understanding*. Open University Press. www.mheducation.co.uk/

Benenson, W., Harris, J. W., Stocker, H., & Lutz, H. (2020). *Handbook of physics*. Springer-Verlang. www.springer.com

Browning, C., Edson, A. J., Kimani, P. M., & Aslan-Tutak, F. (2014). Mathematical content knowledge for teaching elementary mathematics: A focus on geometry and measurement. *The Mathematics Enthusiast, 11*, 333–384. https://scholarworks.umt.edu/tme/vol11/iss2/7/

Burny, E., Valcke, M., & Desoete, A. (2012). Clock reading: An underestimated topic in children with mathematics difficulties. *Journal of Learning Disabilities, 45*(4), 351–360. doi: 10.1177/0022219411407773

Clements, D. H., Barrett, J. E., & Sarama, J. (2017). Measurement in early and elementary education. In J. E. Barrett, D. H. Clements, & J. Sarama (Eds.), *Children's measurement: A longitudinal study of children's knowledge and learning of length, area, and volume* (JRME Monograph No. 16; pp. 3–21). NCTM.

Clements, D. H., & Sarama, J. (2017/2019). *Learning and teaching with learning trajectories [LT]²*. Marsico Institute, Morgridge College of Education, University of Denver. www.learningtrajectories.org/early-math/birth-to-grade-3

Clements, D. H., Sarama, J., & Van Dine, D. W. (2017). Volume. In J. E. Barrett, D. H. Clements, & J. Sarama (Eds.), *Children's measurement: A longitudinal study of children's knowledge and learning of length, area, and volume* (JRME Monograph No. 16; pp. 151–157). NCTM.

Cobb, G. W., & Moore, D. S. (1997). Mathematics, statistics, and teaching. *The American Mathematical Monthly, 104*(9), 801–823. doi: 10.2307/2975286

Davidian, M., & Louis, T. A. (2012). Why statistics? *Science, 336 (6077)*, 12. doi: 10.1126/science.1218685

Doabler, C. T., Clarke, B., Firestone, A. R., Turtura, J. E., Jungjohann, K. J., Brafford, T. L., Sutherland, M., Nelson, N. J., & Fien, H. (2019). Applying the curriculum research framework in the design and development of a technology-based tier 2 mathematics intervention. *Journal of Special Education Technology, 34*(3), 176–189. doi: 10.1177/0162643418812051

Doabler, C., Clarke, B., Kosty, D., Turtura, J., Firestone, A., Smolkowski, K., Jungjohann, K., Brafford, T., Nelson, N., Sutherland, M., Fien, H., & Maddox, S. (2019). Efficacy of a first-grade mathematics intervention on measurement and data analysis. *Exceptional Children, 86*(1), 77–94. doi: 10.1177/0014402919857993

Doerr, H. M., & English, L. D. (2003). A modeling perspective on students' mathematical reasoning about data. *Journal for Research in Mathematics Education, 34*(2), 110–136. doi: 10.2307/30034902

Eskelson, S. L., Townsend, B., E., & Hughes, E. K. (2021). Maximizing area and perimeter problems. *Mathematics Teacher: Learning & Teaching PK-12, 114*(1), 41–46. doi: 10.5951/MTLT.2019.0310

Gavin, M. K., Casa, T. M., Adelson, J. L., & Firmender, J. M. (2013). The impact of challenging geometry and measurement units on the achievement of grade 2 students. *Journal for Research in Mathematics Education, 44*(3), 478–509. doi: 10.5951/jresematheduc.44.3.0478

Hansen, A. (2005). *Children's errors in mathematics*. Sage Publishing.

In, J., & Lee, S. (2017). Statistical data presentation. *Korean Journal of Anesthesiology, 70*(3), 267–276. doi: 10.4097/kjae.2017.70.3.267

Jones, B. E. (2013). Measurement: Past, present and future: Part 1 Measurement history and fundamentals. *Measurement and Control, 46*(4), 108–114. doi: 10.1177/0020294013485673

Jones, D., & Tarr, J. E. (2010). Recommendations for statistics and probability in school mathematics over the past century. In B. J. Reys & R. E. Reys (Eds.), *Mathematics curriculum issues, trends, and future directions: 72nd yearbook* (pp. 65–75). NCTM.

Kang, Y., & Chang, Y. (2019). Using an augmented reality game to teach three junior high school students with intellectual disabilities to improve ATM use. *Journal of Applied Research in Intellectual Disabilities, 33*(3), 409–419. https://doi.org/10.1111/jar.12683

Kamii, C. (2006). Measurement of length: How can we teach it better? *Teaching Children Mathematics*, *13*(3), 154–158. doi: 10.5951/tcm.13.3.0154

Langrall, C. W., Makar, K., Nilsson, P., & Shaughnessy, J. M. (2018). Teaching and learning probability and statistics: An integrated perspective. In J. Cai (Ed.), *Compendium for research in mathematics education* (pp. 490–525). NCTM.

Makar, K., & Rubin, A. (2009). A framework for thinking about informal statistical inference. *Statistics Education Research Journal*, *8*(1), 82–105. www.stat.auckland.ac.nz/~iase/serj/SERJ8(1)_Makar_Rubin.pdf

Mankiewicz, R. (2001). *The story of mathematics*. Princeton University Press.

NASA (1990). *The Hubble space telescope optical systems failure report* (Document No. 19910003124). http://ntrs.nasa.gov/

National Center for Education Statistics (2019). *2019 NAEP Mathematics Assessments*. US Department of Education. www.nationsreportcard.gov

National Council of Teachers of Mathematics (2000). *Principles and standards for school mathematics*. www.nctm.org/Standards-and-Positions/Principles-and-Standards/

National Council of Teachers of Mathematics (2015). *The metric system: A position of the National Council of Teachers of Mathematics*. www.nctm.org

National Governors Association Center for Best Practices & Council of Chief State School Officers. (2010). *Common core state standards for mathematics*. www.corestandards.org/Math/

National Research Council (1996). *National science education standards*. National Academy Press. www.nap.edu/

Noll, J., & Shaughnessy, M. (2012). Aspects of students' reasoning about variation in empirical sampling distributions. *Journal for Research in Mathematics Education*, *43*(5), 509–556. doi: 10.5951/jresematheduc.43.5.0509

Owens, K., & Kaleva, W. (2007). Changing our perspective on measurement: A cultural case study. *Mathematics: Essential Research, Essential Practice*, *2*, 571–580. www.academia.edu/21269371

Ren, K., Lin, Y., & Gunderson, E. (2019). The role of inhibitory control in strategy change: The case of linear measurement. *Developmental Psychology*, *55*(7), 1389–1399. doi: 10.1037/dev0000739

Ryan, J., & Williams, J. (2007). *Children's mathematics 4–15: Learning from errors and misconceptions*. Open University Press. www.mheducation.co.uk

Sevinc, S., & Brady, C. (2019). Kindergarteners' and first-graders' development of numbers representing length and area: Stories of measurement. In *Mathematical Learning and Cognition in Early Childhood* (pp. 115–137). Springer. doi: 10.1007/978-3-030-12895-1_8

Shafer, K., & Özgün-Koca, S. A. (2021). GPS: Investigating variability. *Mathematics Teacher: Learning & Teaching PK-12*, *114*(1), 78–82. doi: 10.5951/MTLT.2020.0176

Shaughnessy, J. M., & Pfannkuch, M. (2002). How faithful is Old Faithful? Statistical thinking: A story of variation and prediction. *The Mathematics Teacher*, *95*, 252–259.

Smith, J. P., & Barrett, J. E. (2018). Learning and teaching measurement: Coordinating quantity and number. In J. Cai (Ed.), *Compendium for research in mathematics education* (pp. 355–385). NCTM.

Smith, J., Males, L., Dietiker, L., Lee, K., & Mosier, A. (2013). Curricular treatments of length measurement in the united states: Do they address known learning challenges? *Cognition and Instruction*, *31*(4), 388–433. doi: 10.1080/07370008.2013.828728

Solomon, T. L., Vasilyeva, M., Huttenlocher, J., & Levine, S. C. (2015). Minding the gap: Children's difficulty conceptualizing spatial intervals as linear measurement units. *Developmental Psychology*, *51*(11), 1564–1573. doi: 10.1037/a0039707

Stone, M. H. (2014). The cubit: A history and measurement commentary. *Journal of Anthropology*, *2014*, 1–11. doi: 10.1155/2014/489757

Szilágyi, J., Clements, D. H., & Sarama, J. (2013). Young children's understandings of length measurement: Evaluating a learning trajectory. *Journal for Research in Mathematics Education*, *44*(3), 581–620. doi: 10.5951/jresematheduc.44.3.0581

Tan-Sisman, G., & Aksu, M. (2012). The length measurement in the Turkish mathematics curriculum: Its potential to contribute to students' learning. *International Journal of Science and Mathematics Education*, *10*(2), 363–385. doi: 10.1007/s10763-011-9304-1

Tan-Sisman, G., & Aksu, M. (2016). A study on sixth grade students' misconceptions and errors in spatial measurement: Length, area, and volume. *International Journal of Science and Mathematics Education*, *14*(7), 1293–1319. doi: 10.1007/s10763-015-9642-5

Weng, P., & Bouck, E. (2016). A toolbox for teaching price comparison to students with disabilities. *Teaching Exceptional Children, 49*(5), 347–354. doi: 10.1177/0040059916671136

Strand F
Algebraic Reasoning

After studying this strand, the reader should be able to:

1. describe the scope of algebraic reasoning across the grade levels;
2. discuss early learning experiences that contribute to algebraic reasoning;
3. identify common misconceptions of students with learning difficulties in algebra;
4. plan targeted instruction for challenging topics within pre-algebra and algebra.

Sara is a new sixth-grade student at Balsam Middle School. Ms. Smith is assisting her mathematics teacher in conducting informal assessments of Sara's mathematics skills and concept knowledge.

MS. SMITH: Sara, have you seen an equation like this one before? [$2x + 5 = 11$]. Can you solve this?

SARA: I think so. Two times five is ten so it should be 10, not 11.

MS. SMITH: Why did you multiply to find that answer?

SARA: Because of the multiply sign right after the 2.

MS. SMITH: OK, can you solve for what is missing in this problem? [$7 = 15 + ?$]

SARA: Well, the answer is seven, but it should be on the other side.

MS. SMITH: Can you write what you mean by that?

SARA: [Writes $15 + ? = 7$]. That's the way it should look.

MS. SMITH: OK, can you solve for the question mark?

SARA: I think it is 22 but I'm not sure. I haven't had this kind of addition in a long time.

Sara is faced with algebraic reasoning before she is firm with foundational concepts such as equivalence, mathematical notation, multiplicative reasoning, arithmetic with negative integers, and fraction representations. It is likely that Sara stumbled with mathematics notations and syntax and didn't develop deeper understandings of equations and reasoning with arithmetic processes and properties in earlier grades. Even in fifth grade, Sara's teachers may have taught numerical expressions with symbols and graphing ordered pairs as a sequence of rules and procedures without developing understanding.

This strand addresses the development of algebraic reasoning, from patterns and properties of numbers in elementary school, through expressions and equations in middle school, to the functions and modeling in high-school mathematics, through the concepts of linear functions and equations. It is no coincidence this strand is last in a series of content strands that cross the grade levels. The development of algebraic reasoning and the study of algebra draw on all previous mathematics concepts and provide a means for connecting those in increasingly abstract and varied ways. By the end of high school, students should be able to think mathematically through the use of multiple representations including equations, graphs, and other models.

DOI: 10.4324/9781003096733-14

Stop five people on the street and ask, "What is algebra?" You will hear five different responses, such as "That's when we solved equations," and "Algebra is just arithmetic with letters," or "Algebra is when math gets abstract and when I started not understanding anything." Even dictionaries and other trusted sources offer poor definitions such as, "Algebra is a branch of mathematics in which letters are used to represent numbers," or "Algebra is solving for the unknown." In fact, algebra is a very broad term that includes *school algebra*, for middle- and high-school settings, and *abstract algebra*, which is the abstract study of number systems and their operations (e.g., groups, rings; Renze & Weisstein, n.d.). The word *algebra* is also used for a type of algebraic structure (a vector space over a field with multiplication). This strand will focus on the first type, the algebra studied in school that includes the solution of polynomial equations with one or more variables and basic properties of functions and graphs in the secondary grades, as well as algebraic thinking and pre-algebra topics developed in earlier grades. Algebra requires abstract thinking about mathematical relationships that is developed across the grade levels.

Foundations of Algebra

The report of the 2008 National Mathematics Advisory Panel emphasized the importance of students succeeding in mathematics through at least Algebra II (or equivalent integrated course) for better college and career options. Algebra, the report noted, is a gateway to later achievement in mathematics and completion of Algebra II correlates significantly with college graduation and employment earnings. By 2019, most states required three or four mathematics course credits for high-school graduation, except California and Montana (2 credits) and five other states with locally determined requirements (Macdonald et al., 2019). Of the states with credit requirements, 32 required Algebra I or an integrated geometry and algebra course (e.g., Math I) and 18 also specified Algebra II or the integrated equivalent.

Many states and districts recommend Algebra I in eighth grade for students with strong prerequisites, followed by three or four more mathematics courses in high school. However, many students are not ready for Algebra I even by the end of eighth grade. Data from the *National Assessment of Educational Progress* (National Center for Education Statistics, 2019) on the algebra scale showed that fourth graders scored just below proficient and students with disabilities scored 28 points lower than those without disabilities, at the lower end of the basic level. Eighth graders also scored just below proficient while students with disabilities scored 40 points lower than their peers, below basic. By 12th grade, students with disabilities were still well below basic and 35 points below their peers. Released items from the 2017 assessments showed that many fourth graders had difficulties with number patterns (45% correct) and extending number patterns in a table (22%). While 49% of eighth graders could determine the intercept of a line on the x and y axes, only 36% could interpret the meaning of a linear equation in context, and only 15% could make and explain a conclusion about a linear relationship in context. Most students are struggling with algebraic thinking in elementary school and are not prepared for Algebra I in eighth grade.

Development of Algebraic Reasoning

The National Council of Teachers of Mathematics considered algebra a "strand that unfolds across the PreK-12 curriculum" (NCTM, 2014). In order to access the more formal and abstract algebra in secondary grades, students must develop ways of thinking and have opportunities to generalize, model, and analyze situations that are purely mathematical from the earliest grades. Students entering algebra often face an abrupt ending to arithmetic and beginning for algebra as if there were no connection (Cai & Moyer, 2008). Younger students can develop algebraic thinking along with arithmetic competencies if teachers understand that this is not just pushing secondary topics earlier, but a developmental trajectory that connects content and provides a transition to more abstract mathematics.

Mathematics educators have outlined the subdomains or strands of algebra, helpful for considering foundational concepts and learning trajectories. Kaput (2008) identified three strands for the core aspects of algebra:

1. the study of structures and systems abstracted and generalized from arithmetic and quantitative reasoning;
2. the study of functions, relations, and joint variation (generalization of variability); and
3. the application of a cluster of modeling languages (inside and outside mathematics), including notation systems, visual representations, and computer modeling.

The actions of abstracting and generalizing are central to algebraic thinking (Hackenberg & Lee, 2015). *Abstracting* is the process of viewing a particular example in a more general way. *Generalizing* involves broadening the range of applications, extending reasoning beyond specific examples. For example, the student who measures the sides of a right triangle as 3, 4, and 5 inches begins to generalize when she wonders if a triangle 6, 8, and 10 inches will have a right angle as well. Then abstract thinking allows her to consider the original 3, 4, and 5 as variable values that are related in some way and could be represented with letters as a way of abstracting the values even more: a, b, and c. Dynamic software assists this student in manipulating the right triangle and examining the properties and relationships of its angles and sides, before the Pythagorean Theorem is introduced.

The NCTM offered four areas of emphasis within algebra across the grade levels, similar to those of Kaput: 1) understand patterns, relations, and functions; 2) represent and analyze mathematical situations and structures using algebraic symbols; 3) use mathematical models to represent and understand quantitative relationships; and 4) analyze change in various contexts (NCTM, 2000, pp. 37–40). How will these areas be addressed in elementary and secondary grades to promote algebraic thinking and connections with arithmetic?

Patterns, relations, and functions are closely related to early concepts of attributes, classification, comparison, and the fundamental concept of change. Identifying the attributes of objects as well as grouping and ordering those objects typically take place in preschool settings. Teachers can present ordered patterns (seriations) and encourage children to continue the sequence (e.g., yellow, red, red, yellow, red, …). In the early grades, numbers are used to create patterns and are symbols for values. Elementary students also have early experiences with functions using function tables for equations such as $a + 2 = b$. Inserting specific values for a will result in a range of specific values for b. Hines et al. (2001) summarized the levels of function conceptualization as: 1) Learner notices no link between two variables; 2) Learner establishes underlying pattern in a recursive (forward and backward), sequential manner; and 3) Learner generalizes understanding of relationships to systematic co-variation through verbal and symbolic means.

As students progress through the upper-elementary and middle grades, their repertoire of function types expands and they should use multiple representations for functional relationships: tables, graphs, equations, diagrams, and verbal descriptions. Figure F.1 illustrates various representations for the same function. In this function, when x changes in value, y will change according to the relation. Function concepts are also developed within other domains. In geometry the area of a polygon is a function of the measures of the sides. In measurement, the length of an object is a function of iterated units. Fractions, ratios, and proportions are symbolic representations of the relations of values. For data analysis, graphs and other representations allow us to consider the relationships among data, whether there are functional

Equation Form: $y = 5x + 1$

Table Form: Graph Form:

x	y
-3	-14
-2	-9
-1	-4
0	1
1	6
2	11
3	16

Diagram Form:

Insert number for x

( )5 $+ 1$ changes y to

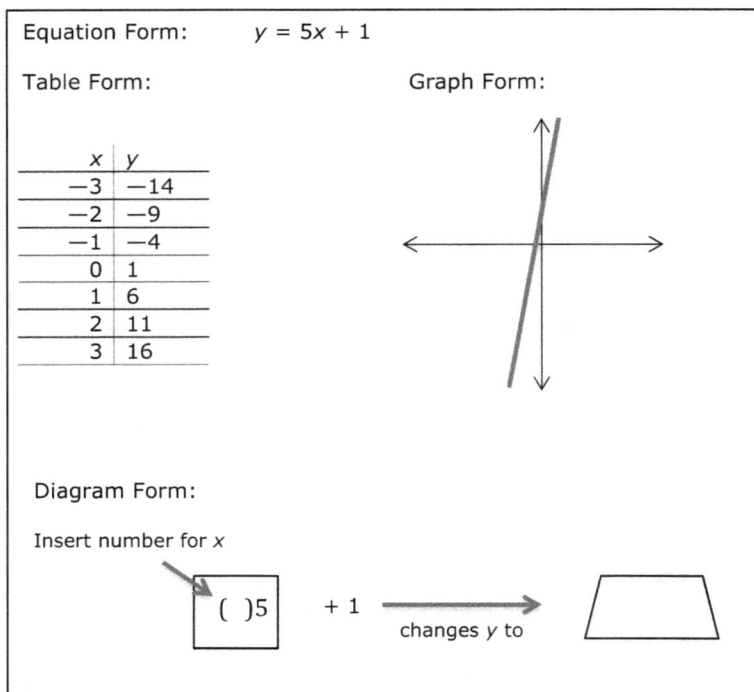

Figure F.1 Various Representations for the Same Function

relationships where some variables depend on the values of others. Functional relationships also are central to other areas of study, such as physics and social sciences.

Representing and analyzing mathematical situations and structures using algebraic symbols evolves from understanding the properties and operations of numbers within arithmetic. In fact, some mathematicians describe algebra as a generalization of arithmetic (Kilpatrick & Izsák, 2008). It is not as simple as using letters for unknowns within arithmetic equations, however. Kieran (2007) suggested deliberate shifts in thinking while doing arithmetic to promote algebraic thinking. These included focusing on relations among quantities, not just the numerical calculation; operations and their inverses (doing and undoing); both representing and solving problems (not just finding an answer); and the meaning of the equal sign as a symbol that denotes an equivalent relationship between quantities. Students should also develop deep understanding of properties of numbers and operations. For example, what is actually happening with the distributive property of multiplication over addition? What are the effects of multiplying by 10 or 100? The concept of variable should also evolve and deepen from a simple value substitution to more complex one-variable ($4x + 2 = 15$) and two-variable equations ($y = 3x + 7$). Students should begin to use algebraic symbols consistently and use conventional notations and syntax to simplify expressions.

Using mathematical models to represent and understand quantitative relationships also evolves across the grade levels. Young children represent situations with concrete objects and pictures. Elementary students begin to represent situations with symbolic notations. For example, the class softball team wants to keep some player statistics, such as batting averages. A person's batting average is the number of hits divided by the number of at bats. Sally had 10 hits for her 35 at bats, so her batting average is 10/35 or 0.286 (decimal notation is the convention). In order to generalize this method for finding batting averages, the class creates an algebraic equation: $a = h/b$. Entering the hits (h) and at bats (b) for any player will yield the batting average (a).

Other mathematical models include graphs, diagrams, tables, and narratives. Students should have practice with modeling situations within other mathematics domains. They should also use technologies to model, taking advantage of quick analysis and computation so that students can focus on making predictions, adjusting variables, and drawing conclusions. Students should be encouraged to represent the situations presented in word problems through diagrams, equations, and other models. Through these problem-solving exercises, as much emphasis is placed on how to represent the problem situation as on the solution. Students can consider the relationships of the quantities and/or objects in the problem, whether the relationships are static or dynamic (changing), and how best to represent those relationships.

Finally, *analyzing change in various contexts* permeates the PreK-12 mathematics curriculum. Young children describe change qualitatively and quantitatively, as with explaining the growth of plants. Elementary students use graphs and tables to notice and describe change, as with trends in outdoor air temperature or growth in populations. In middle school, rates of change are described using tables and graphs of equations. Other classes of functions are examined in high school (e.g., quadratic, exponential, logarithmic). Change is encountered in graphing equations—when parts of the equation change so does the graph. Change is observed in arithmetic operations and the manipulation of geometric and graphic models. We want students to notice and analyze change and function situations within contexts, such as word problems and learning tasks.

Concepts and Skills Foundational for Algebra

For developing algebraic reasoning during the elementary grades and with pre-algebraic lessons during middle school, it is important that algebraic topics and notations be connected with students' prior knowledge and not treated as a new, discrete area of mathematics. If topics are taught with concept connections, multiple strategies and examples, varied representations, and increasing generalization and abstraction during the earlier grades, students will be prepared to work with structure in symbolic expressions and reason with equations, inequalities, and systems in the secondary grades. If students have gaps in foundational concepts and skills, interventions should be provided based on individual student profiles. Table F.1 provides an overview of concepts and skills developed across Grades PreK-8 that are essential prior to formal algebra in high school.

It is not sufficient to teach and reteach the concepts and skills depicted in Table F.1 through procedures and practice. Teachers must engage students with well-designed tasks that promote concept understanding and reasoning through scaffolding, questioning, and dialog about number and object relationships. Dougherty and colleagues (2015) warned about simply reteaching the same procedures and skills using modeling and practice items. Often the proficiency level for students, especially those with learning difficulties, is short-lived because retention of isolated skills and memorized steps is not robust. The authors recommended a framework of three types of questions to support the acquisition of concepts leading to algebraic thinking: reversibility, flexibility, and generalizations. For example, a standard item for arithmetic with integers is $-4 + -7$. Rather than practice a page full of similar items, ask a reversibility question: What are two integers whose sum is -11? A flexibility question asks students to describe how the following problems are alike: $-4 + (-7)$, $-5 + (-7)$, and $-6 + (-7)$. For generalization, the authors proposed questions such as: What are two negative integers whose sum is negative? What are a positive integer and a negative integer whose sum is negative? By engaging students with these types of questions, teachers promote deeper understanding of algebraic concepts and generalizations.

Table F.1 Algebraic Concepts and Skills PreK-8

Concepts/Skills	Examples
Number relationships and patterns	Addition reverses subtraction, Multiples of 2, 5, 10
Arithmetic operations	Dividing 36 by 4, Multiplying by 2, Subtracting by 5s
Properties of operations	Distributive: $2(3 + 4) = 6 + 8$
	Associative: $(5 - 2) + 7 = 5 + (-2 + 7)$
Equivalence, relational meaning of equal sign	$5 + y = 12$ $15 - 7 = ? + 5$
Positive and negative integers	$5 + (-7) = z - 8 - 4 = ?$
Fraction notation	$\dfrac{9}{5} = 1\dfrac{4}{5}$
Fraction meaning	Equivalent fractions: $\dfrac{2}{5} = \dfrac{6}{15}$ $4 = \dfrac{36}{6}$
Factoring	Factors of 12: 1, 2, 3, 4, 6, 12 $(4 + 6) = 2(2 + 3)$
Solving word problems	Expressing WP as equation, solving for the unknown
Whole number exponents	$6^3 = 6 \cdot 6 \cdot 6 = 216$
Applying formulas	$A = bh$ $P = 2l + 2w$
Coordinate plane graphing	Graph and connect points: $(-4, 3), (-4, 1), (3, 1), (3, -3)$.
Decimal fractions	$\dfrac{4}{10}, \dfrac{37}{100}$
Inequalities	$7 + x > 15$
Decimal forms of rational numbers	0.4 5.25 $\dfrac{4.5}{12x}$
Ratios, rates, and proportions	$\dfrac{6}{30} = \dfrac{x}{40}$ $d = 65t$
Functional relationships	$a = 3b$; if $b = 5$, what is the value of a? $f(x) = 9(3x - 5)$
Order of operations	Brackets, Multiply/Divide, Add/Subtract
Expressions (with terms)	$5 + 6x$ (an expression with two terms)
Equations (in one variable)	$y - 9 = 25$ $32 - 7 = x + 14$
Mathematics processes	Reasoning and proof, Modeling and representing

Students taking Algebra I in eighth grade may need prerequisite concepts and skills related to other types of exponents (negative and fractional), other types of functions, square and cube roots, linear equations in two variables, and solving simultaneous equations, depending on the local curriculum.

Carraher and Schliemann (2018) referred to *algebraic thinking* as "reasoning that expresses itself as statements or other representations denoting relations among sets of elements, typically numbers or quantities" (p. 131). Algebraic expression may take on various forms, including linguistic, tabular, graphical, and diagrammatic, before algebraic notation is introduced. The researchers encouraged teachers to approach algebraic thinking with younger students from the perspective of relations, especially functional relations. For example, in the *Candy Boxes Problem* below for third graders, students can state verbally, draw, or even assign random values to boxes. Through class discussion of student efforts and refining those efforts, students can represent the unknown amounts using letters. Further tasks allow students to explore other numerical relationships and represent those in various ways.

> *John and Mary each have a box of candies. The two boxes have exactly the same number of candies. Mary has three extra candies on top of her box. Draw or write something that shows how many candies John and Mary have* (Carraher & Schliemann, 2018, p. 114).

Function-oriented tasks emphasizing the *structure* of number and operations were investigated by Kieran (2018). From working with additive and multiplicative structures of numbers to creating patterns with figures, applying properties of arithmetic, and identifying the terms of expressions, students need to be engaged in composing, decomposing, and seeking patterns and structures within numerical tasks. An example task, *Five Steps to Zero*, asked 12-year-old students to use their calculators to start with the number 144 (or 151, 172, etc.) and write down as many ways as possible to bring it to zero with the fewest steps possible, using only the whole numbers 1 to 9 and the four operations. Students gained a deeper understanding of divisibility, multiples, and the effects of adding and subtracting to achieve a more divisible number (Kieran, 2018). These tasks could be scaled down for younger students by changing the range of numbers, operations, and steps.

Researchers have demonstrated that young children (ages 5 and 6) have the potential to generalize and reason with structure and relationships. Children can be guided to explore relationships between quantities, spatial skills, and repeated patterns through teacher-guided story contexts, block play, and representational tasks (e.g., Blanton, Brizuela, et al., 2015; Rittle-Johnson et al., 2019). For example, Mulligan et al. (2020) implemented a year-long intervention (four hours a week) with kindergarteners on patterns and structures that included 11 components such as repeated patterns, grid structure, base-10 structure, multiplicative structure, and growing patterns. Each component included 10 to 20 sequenced subcomponents, each with a succession of tasks. Each mathematical process of the model was presented in sequence: modeling and representing, visualizing and generalizing, and sustaining. For example, in the component "growing patterns" a subcomponent was "pattern of squares" and a task was prompting, "Use the squares to show the pattern 1, 4, 9, 16…then the 10th. Tell me about it." The intervention had a significant, positive impact on students' development of awareness of mathematical pattern and structure, with differences maintained a year after intervention. Students a year later were able to represent (through drawings and objects) and explain their representations, making connections between mathematical ideas. Even less-able students showed impressive growth, with half moving from pre-structural to at least emergent. The most capable students demonstrated emerging generalizations at the structural and advanced structural levels, supporting early algebraic thinking, well beyond curricular expectations.

Young students can understand symbols that show how quantities are related. The most compelling research concerns the equal sign—children with a relational understanding of the equal sign can more successfully solve equations and produce equivalent quantities though substitution (e.g., Jones & Pratt, 2012). Matthews and Fuchs (2020) found that second graders' equal sign knowledge (open equations with missing numbers, equivalence of other representations) predicted fourth-grade algebraic reasoning (solving equations with variables and function tables) and was the strongest predictor over IQ, attention, arithmetic skill, and SES factors. Second graders in this study were only 50% accurate with equal-sign items, even with numbers of low quantities, putting half the students at risk for not developing algebraic concepts needed for middle-school topics.

As early as first grade, students can represent generalizations using variable notation in meaningful ways. Withholding variable notation until students are "ready" may actually disadvantage students' reasoning abilities. For example, Blanton et al. (2017) implemented a series of teaching episodes (30–40 min lessons, twice a week in two, four-week cycles) in two first-grade, diverse classrooms. Variable notation was introduced intentionally in the context of representing rules for functional relationships (e.g., someone's height and their height wearing a one-foot hat) for the function types $y = mx$, $y = x + b$, and $y = mx + b$ in everyday contexts (using their own reasoning, language, and letters such as S, R, or V to represent their understanding). The researchers found a possible developmental progression for students' thinking about variable notation: pre-variable/pre-symbolic, pre-variable/letters as labels, letters representing variables with fixed values, letters representing variables with fixed but arbitrary values, letters representing variables that are varying unknowns, and letters representing variables as mathematical objects. The researchers observed that the thinking of these first-graders was not unlike that of some adolescents about what letters represent, delayed from a lack of experiences needed to mathematize situations.

Blanton, Stroud, et al. (2019) implemented a sequence of early algebraic lessons (Project LEAP, www.didax.com/math/leap-books.html) with third, fourth, and fifth graders in 18 one-hour lessons for each grade. The lessons focused on the big ideas of equivalence, expressions, equations, inequalities; generalized arithmetic; and functional thinking. Each of the big ideas included learning goals of generalizing, representing, justifying, and reasoning with mathematical structure and relationships. Students in the experimental classrooms significantly outperformed control students on researcher-developed assessments at the end of each grade, although there was a plateau in rate of growth for experimental students in Grade 4. Experimental students were more successful in noticing and representing generalized quantities and operations on them and generalizing and representing mathematical relationships. They were more skilled in using variable notation and meaningful ways to represent arithmetic properties, expressions, and equations. Both experimental and control students were better able to represent functional relationships with variable notation than with words. The very significant growth in third-grade students' learning regarding generalization and representing mathematical structure and relationships (from 0% to 60%) underscores the need to introduce these concepts earlier.

This body of research is compelling for a trajectory, beginning in the earliest grades, for developing algebraic thinking. More research is needed on the details of that trajectory for a range of student profiles (Blanton et al., 2017; Chimoni et al., 2018; Stephens et al., 2018). We need to learn more about how functional thinking transitions from elementary to middle grades with more complex contexts (covariance and rates of change). There are also reasoning challenges related to the shift to non-whole number domains (e.g., fractions, decimals) and discrete vs continuous quantity relationships in the elementary grades. Research is also needed on which types of instruction facilitate the progression from early algebra to more advanced. Very little is known about the algebraic-reasoning transitions across grade levels for students with mathematics learning difficulties and disabilities.

Assessment of Algebraic Concepts and Skills

Screenings and progress monitoring of algebraic reasoning should be conducted across the grade levels, with algebraic reasoning and foundational skills the focus in the elementary grades and foundational gaps as well as algebraic concepts and skills in secondary. Screening in elementary grades tends to focus on number knowledge and operations, with less attention on other important topics, such as algebraic reasoning. Assessment instruments should be checked for algebraic reasoning content, such as understanding equivalence (equal sign), solving open sentences, understanding and using variables in expressions, using function rules, and modeling problem situations with representations and equations. For example, Blanton, Stroud et al. (2019) provided example assessment items for Project LEAP (Grades 3–5 early algebra), such as $7 + 3 = \_\_ + 4$, explaining a rule about odd numbers, and completing a function table. Other predictors for algebra success by eighth grade are fraction number-line estimation (Booth et al., 2014), nonsymbolic ratio processing (Matthews et al., 2016), fact retrieval, fraction competence, and proportional reasoning (Cirino et al., 2019).

More algebraic screening and progress-monitoring instruments are available for middle- and high-school settings. Powell et al. (2020) described the use of data-based individualization for middle-school algebra readiness (Grades 6–8, Project STAIR). Teachers in the study used *Algebra Readiness Progress Monitoring* (ARPM, Ketterlin-Geller et al., 2015) as pretest, weekly progress monitoring, and posttest over 12–15 weeks. Instructional coaches worked with the teachers to interpret assessment data and adapt instruction based on those data for targeted students with mathematics learning difficulties (MD) in their classrooms. Pre- to post-test differences were significant for ARPM and one other measure of algebra-readiness knowledge and skills, but not for a distal measure. Some teachers in this study did not use the assessment technology consistently, however.

Accardo and Kuder (2017) illustrated a formative-assessment strategy for middle-school classrooms co-taught by mathematics and special educators, also supported by professional development and coaching. The formative-assessment methods included breaking problems into steps for error analysis, using data-collection charts to identify student-response patterns, providing multiple probes to assess student understanding, and embedding one key question (exit ticket) into formative assessments for closer analysis. The "exit ticket" for one unit on writing equations for word problems involved breaking down a word problem into four visible steps: write a basic expression, write an expression with parentheses, combine the expressions into one equation, and solve the equation. Charting each step of this problem for each student showed the teachers exactly where students lacked understanding so they could target additional instruction.

High-school screening measures for procedures and concepts in algebra were described by Genareo et al. (2021). Researchers in the *Algebra Screening and Progress Monitoring* project (ASPM, 2017) developed three procedural measures and three conceptual measures for screening and progress monitoring in initial algebra learning. The three procedural measures had high levels of reliability ($r = .72–.99$) and moderate concurrent and predictive validity ($r = .36–.64$) and are suitable for use as screening measures and diagnostic testing. The conceptual measures demonstrated moderate to low validity ($r = .10–.44$; with course grades, state test, select NAEP items) and can be used for instructional or diagnostic purposes with permission. The researchers emphasized that procedures are easier to target and measure than concepts, but concept understanding is essential. Teachers using screening and progress-monitoring tools should compare their content and response expectations with local and state standards and outcome measures. These tools are critical for identifying students at risk for poor performance and targeting instructional interventions.

Difficulties with Algebraic Reasoning

Before we consider instructional approaches for algebraic reasoning, this section provides an overview of the most common difficulties for students. These problem areas can be targeted with preinstruction and intervention strategies as well as more carefully designed and sequenced core instruction. Teachers can identify student misconceptions through error analysis and interviews or discussions. Misconceptions about new learning may be a result of incomplete or misunderstood prior knowledge, when the new learning is incompatible with previous knowledge, or when previous learning is isolated, a set of meaningless disconnected procedures (Jankvist & Niss, 2018). Misconceptions also can be the result of instruction that does not adequately connect prior knowledge and intuitive thinking to more generalized thinking through an emphasis on concept understanding, instruction that does not consider the developmental progressions students move through from arithmetic thinking to algebraic.

The most common difficulty with algebraic reasoning cited in the literature is the attempt by many students to maintain arithmetic thinking when moving into algebraic situations. Too often during the development of arithmetic concepts and procedures, students were encouraged to find the answer. Not enough attention was given to reasoning about the relationships of quantities in problem situations. Stacey and MacGregor termed these arithmetic intrusions a "compulsion to calculate" (1999, p. 149). Students often have well-entrenched arithmetic processes that have served them well in the past and some students have difficulty inhibiting those processes when an algebraic strategy is required. In a study of 14-year-olds solving algebraic word problems, Khng and Lee (2009) found that 23% of variance in arithmetic intrusions could be explained by inhibitory control problems. Although they had been taught algebraic strategies for solving word problems, students attempted arithmetic strategies first. The researchers recommended that, rather than ask students to abandon previously learned strategies, a more effective approach would build new knowledge using prior knowledge, connecting arithmetic concepts with those in algebra, generalizing their arithmetic knowledge.

Another arithmetic intrusion is misconception of the equal sign, or equivalency. Many elementary students view the equal sign as a unidirectional symbol, indicating that the answer comes next. The equal sign should be taught as a relational symbol, signifying that the numbers or expressions on each side are equivalent. McNeil (2014) suggested that young children be presented with nonarithmetic equivalencies first (e.g., $8 = 8$, dozen $= 12$) before introducing arithmetic equations (e.g., $3 + 8 = 15$ and $12 = 4 + 8$). Nonstandard equations should be introduced simultaneously (e.g., $3 +$ ___ $= 15$ and $2 + 3 + 1 =$ ___ $+ 4$). However, in a study of eight major curricula across Grades K to 5, Powell (2012) found that only one of eight programs presented nonstandard equations with frequency and none offered no-operation equations. Powell also found that the equal sign was rarely mentioned in these curricula, not introduced when it was first used, and not reviewed when used again.

Prior arithmetic learning plays a role when students move to algebra and lack strong concept understanding of operations and properties and how algebraic thinking is involved. Students in elementary school are engaged in algebraic thinking when they notice and can represent the structures, patterns, and functions involved in arithmetic operations. For example, algebraic thinking requires students to consider reciprocal and reverse operations. For the equation $4x - 5 = 7$, why are we able to add 5 to each side without changing the equivalence of the sides? This action would change the equation to $4x = 12$. What properties of numbers and operations allow us to multiply each side by $\frac{1}{4}$? Students with poor concept understanding may apply the wrong property (e.g., a commutative property for subtraction as with $5z - 4$ is the same as $4 - 5z$) or apply an incomplete property (e.g., distributing partially as with $3(2x + 8)$ to achieve $6x + 8$). Deep understanding of the properties illustrated in Strand B (Table B.3) is essential for algebraic reasoning. Additional properties for real numbers involve negative numbers, exponents, and fractions, as are illustrated in Table F.2. It is beyond the scope of this book to address properties of complex numbers, vectors, and matrices.

Another major category of difficulties involves forming accurate algebraic concepts. The concept of *variable* is abstract and challenging to develop. In algebra, students must understand that letters have referents, representing a range of values. Students often view letters as labels instead, a concrete-level misconception. For example, a student may interpret the equation $3c + 5 = b$ as "three cubes plus five more equals a new number" rather than "the value of c multiplied by 3, then added to 5, will be equivalent to the value b; as c changes, so will b." Some students persist in believing that variables cannot represent more than one value. Students may also fail to distinguish among ways letters are used within equations. For example, a student may attempt to combine unlike terms ($9y + 2 = 11y$) or manipulate and eliminate variables without regard to standard procedures (e.g., simplifying $8 = 7 + 4p - 3p$ by adding 3 to each side $11 = 7 + 4p$). Some students also have difficulty understanding the implied 1 in

Table F.2 Select Algebraic Properties

Property	Rule	Example
Additive Inverse	$a + (-a) = 0$	$5 + (-5) = 0$
Multiplicative Inverse	Multiplication of number by reciprocal (nonzero real number) equals 1	$8 \cdot \dfrac{1}{8} = 1$
Distributive Property of Multiplication	$a(b + c) = ab + ac$ $a(b - c) = ab - ac$ is the same as $a(b + (-c)) = ab + (-ac)$	$4(x + 9) = 4x + 36$ $y(6 - z) = 6y - yz$
Properties of Negation	$(-1)a = -a$ $-(-a) = a$ $(-a)b = -(ab) = a(-b)$ $(-a)(-b) = ab$ $-(a + b) = (-a) + (-b)$	$(-1)4y = -4y$ $-(-12) = 12$ $(-6)z = -6z = 6(-z)$ $(-5)(-3) = 15$ $-(6 + x) = -6 - x$
Properties of Equality	If $a = b$, then $a + c = b + c$ If $a = b$, then $ac = bc$	If $5 + 2 = x$, then $5 + 2 + (-2) = x + (-2)$ If $x = 7$, then $3x = (3)7$
Rules of Signs for Fractions	The negative sign can be placed anywhere in the fraction. Two negatives result in a positive.	$-\dfrac{2}{3} = \dfrac{-2}{3} = \dfrac{2}{-3}$ $\dfrac{-1}{-4} = \dfrac{1}{4}$
Rules for Exponents	Multiplication, add exponents (of same variable) Division, subtract exponents Negative exponent is reciprocal Powers of a power Exponent of 1 Exponent of 0	$x^2(x^3) = x^{2+3} = x^5$ $y^4/y^2 = y^{4-2} = y^2$ $z^{-3} = \dfrac{1}{z^3}$ $(x^4)^3 = x^{12}$ $3^1 = 3$ $5^0 = 1$

See additional properties in Strand B, Figure B.3.

expressions such as $-(2x + 5) + 7$ [the -1 would be distributed across $2x + 5$ for $-2x - 5 + 7 = -2x + 2$] and $xy - 4xy + y$ [the $1xy$ added to a negative $4xy$ would result in $-3xy + y$].

The negative sign is also a challenging concept for many students. In a study of misconceptions of Algebra I students across a year, Booth et al. (2014) found that negative sign errors were the most prominent type of misconception and persisted across six key topics in the course (order of operation, one-step equations, multistep equations, systems of equations, inequalities, and polynomials). Examples of negative sign errors include maintaining the sign when a positive number results (e.g., $6 - (-2) = -8$) and not changing the sign with the corresponding equation manipulation (e.g., for $3 + 5y = 1 - 2y$ the student adds $5y$ to each side, instead of adding $-5y$). Students may lack strong understanding of negative signs because they tend to be taught via rules to memorize and are difficult to represent concretely. Recommendations for teaching about negation can be found in Strand A.

A concept often taught in isolation with surface understanding is factoring. The process of factoring depends on multiplicative thinking, to break a product into two or more factors, reversing the multiplication process. The number 16 can be factored into 2(8) and 4(4) and even 2(2)(2)(2). In algebra, factoring requires breaking an expression into factors, to write the expression as a product of factors. It is not always obvious what factors might be useful for making another representation. The expression $2x^2 + 4x + 6$ can be factored as $2(x^2 + 2x + 3)$. This type of factoring requires understanding coefficients as well as multiplicative relationships and the distributive property of multiplication. Understanding that the quadratic expression $x^2 + 5x + 6$ can be factored as $(x + 3)(x + 2)$ requires even more noticing and applying properties and operations in reverse. If the numbers in these expressions are negative, fractions, or decimals, the complexity of this thinking increases.

It is often stated that strong fraction understanding is critical for algebraic reasoning. Algebraic expressions and equations use fraction notation rather than division symbols and involve variables, negative signs, exponents, roots, and other symbols. How can a student deal with a fraction such as $\dfrac{5x + 4y}{4}$? How are the terms within the fraction related? Students need grounding in the meanings and manipulations of the range of fraction types, such as simplifying $\dfrac{3}{6}$ or understanding the meanings of $5\dfrac{1}{4}$, $-\dfrac{1}{2}$, and $\dfrac{8}{8}$, requiring strong multiplicative reasoning. They need a lot of experience with the concept of *terms* and the meaning of each part of the term. For example, $\dfrac{4}{5x}$ is not equivalent to $\dfrac{4}{5}x$ (hint: try substituting a 2 for the x). Fractions within expressions and equations represent multiplicative relationships, but they also can be part of expressions that have additive elements (e.g., $\dfrac{5z}{2} + 4z - 7$). Zielinski and Glazner (2019) identified errors involving fractions among the most common misconceptions, including random cancellation errors ($\dfrac{6 + 2x}{2x} \neq 6$), rules of fraction errors ($\dfrac{3}{a} + \dfrac{3}{b} \neq \dfrac{6}{a + b}$), and rules of negative exponent errors ($(3x)^{-2} \neq \dfrac{-9}{x^2}$).

Hackenberg and Lee (2015) found that seventh- and eighth-grade students' abilities with equation writing and solving could be differentiated by their multiplicative concept of understanding related to fractions. Students who could not use fractions as multipliers unknowns (e.g., $\dfrac{1}{3}x$) also had difficulty with algebraic notations with whole numbers as multipliers. Students who had developed iterative-fraction schemes (multiples of unit fractions) and could apply reversibility and reciprocal concepts were able to use whole numbers and fractions as multipliers of unknowns. A student with strong multiplicative sense should be able to add $\dfrac{1}{2} + \dfrac{3}{4}$ by visualizing $\dfrac{1}{2}$ as $\dfrac{2}{4}$ and adding the fractions, rather than adding a sequence of fourths.

Students also have difficulty understanding algebraic expressions, those mathematical statements that don't require solving, such as $x + 8$ or $9 - 5x$ or even $12 - 4 + 3 \div 4$. Expressions include at least two terms (composed of numbers and/or variables) and at least one math operation (addition/subtraction). Expressions can sometimes be simplified using rules of operations and their properties. Students may have difficulty understanding that expressions aren't solved and disentangling the order in which expressions can be understood and processed (Jupri et al., 2014). Many students have difficulty evaluating the two expressions that comprise an equation, noticing the similarities and differences in the expressions. For example, in the equation $20 - 7x = 6x - 6$, both expressions have whole numbers (constants) as terms as well as terms with coefficients and the variable

x (like terms). Students often confuse expressions with equations, the latter connecting two expressions by an equal sign.

A related aspect of algebraic reasoning that is often taught by rule rather than through concept understanding is the order of operations and use of brackets (parentheses and other grouping symbols) to signify the sequence. Brackets can be used to set off an additive relationship (4 + 5) or indicate a multiplicative one such as 3(7). Students often fail to see the usefulness of brackets for clarifying parts of expressions and indicating which parts to combine or operate on first. For example, are the expressions $3 \times 3 + 2$ and $3 \times (3 + 2)$ equivalent? Along with the order of operations, other algebraic conventions are difficult for many students. When they first examine an expression, they have difficulty interpreting its structure (e.g., $4xy \neq 4 + x + y$). They often struggle to perform valid transformations for solving multi-step equations. For the equation $4z - 6 = 8z + \frac{2}{3}$, what operation can you apply to each side, under what valid properties of operations, to achieve a simpler equation? What steps are needed to isolate the z on one side of the equation? To simplify algebraic expressions and equations, students are often required to work completely in the abstract, manipulating symbols without meaning. For students who need concrete representations to learn new concepts, this is like playing a nonsense game without knowing the rules or the purpose.

Understanding function concepts as both static and dynamic relationships and represented by equations, tables, graphs, and other means is also challenging for many students. How can the equation $4y = x$ be the same as a line on a graph? Common difficulties for students include not moving flexibly among various representations, viewing functions as the same as equations, not considering a constant function (e.g., $y = 5$) as a function, not understanding that an input value could be an expression, not understanding the relations of input to output values (only one y value for each inputted x), and misinterpretations of graphs of functions. These difficulties typically arise from instruction that focuses on algebraic procedures rather than concept understanding (Carlson & Oehrtman, 2005). In the secondary grades, functions begin taking on special notation that can be confusing, with the symbols f, g, and h often used. For example, the linear function $y = 3x + 7$ can be written as $f(x) = 3x + 7$ (read f of x is equal to 3 times x plus 7). The f and x are not multiplied; $f(x)$ means the function whose input is x. The set of x values (inputs) is the domain of f; the set of y values (outputs) the range of f.

Related to most other issues are difficulties solving word problems using algebraic thinking. Students often struggle to represent parts of the problem with expressions, represent known quantities relationally and unknowns with variables, simplify or combine expressions as needed, and carry out the solution strategy by choosing from a range of representational options (e.g., solving equation, table, graph, diagram). For example: *A total of 240 tickets were sold for the school play on Thursday night. The number of student tickets was three times as many as the number of adult tickets. How many adult tickets were sold?* How many unknowns? What is known? How can students represent the parts of this word problem with equations? How can students represent the results? Pramesti and Retnawati (2019), in studying the algebraic word-problem solving of 65 middle schoolers (13–15 years), found that only 37% could select an equation that would solve a given problem (understanding how the information and situation within the problem should be represented) and only 19% could explain what the variable represented. Forty-two percent could reason with an area problem using variables but only 23% could create a word problem given an equation ($7x - 11 = 24$). Using algebraic reasoning and skills to solve word problems poses many difficulties for students.

Another category of difficulties, especially for students with mathematics learning disabilities (MLD), involves cognitive processes, such as various aspects of memory, attention, and language. Memory problems can be related to retrieving information from long-term storage, holding important information in working memory while it is used (verbal and visual-spatial), and encoding new information for later retrieval and use. A student may see a polynomial such as $2x^2 + 3x - 4 - x^2 + x + 9$ and not recall how to combine terms and sequence the result. Another student has difficulty recalling algebraic methods when working on a geometric application such as $a^2 + b^2 = c^2$. Working memory (WM), in particular, contributes to student success in mathematics, especially as tasks become more complex and abstract. Trezise and Reeve (2014) found that low working-memory ability added to anxiety about algebra predicted poor algebraic problem-solving performance among 14-year-old female students, even controlling for other factors. Students with moderate WM did not differ in algebraic performance with level of anxiety.

Attention issues can affect students' focus during instruction and their focus on important parts of word problems, equations, and graphs. Students may not be able to block out extraneous information or prior knowledge that does not contribute to new learning. Language issues arise with new symbol systems and the syntax of algebra. Many students attempt to turn word problems into algebraic equations through direct translation, rather than using reasoning and algebraic structures. The classic example is "Write an equation using the variables s and p to represent the following statement: There are six times as many students as professors at this university." The problem was posed by Clement (1982) to freshmen engineering students, with only 63% responding with the correct equation: $s = 6p$. Many students translated the sentence word-by-word into the algebraic notation: $6 \times s = p$, resulting a lot of professors! Algebraic word problems are particularly challenging for students who may have difficulties in both abstract reasoning and language interpretation. They often don't know where to begin and typically fall back on arithmetic reasoning and a sequence of calculations (Walkington et al., 2012).

Language difficulties often affect algebraic thinking, not only for interpreting word problems, but for learning and applying new vocabulary, following directions, collaborating, and justifying reasoning. Stegall and Malloy (2019) described a tenth-grade class of students who failed Algebra I in ninth grade. The students could define only six vocabulary words for algebra in 15 minutes (e.g., variable, exponent, coefficient). The educators implemented a vocabulary intervention over 15 weeks, with 20 minutes of supplemental instruction on one critical term each week. Students were taught a modified Frayer diagram for drawing a quadrant with the word in the middle and student-worded definition, example and non-example, tool for recalling definition, and algebraic application in each section (see Figure F.2). Activities followed that required terminology understanding with discussions revealing misconceptions of algebraic concepts that could be targeted for reteaching.

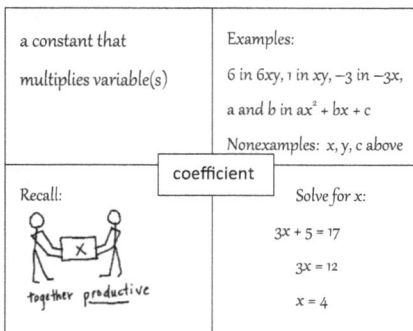

Figure F.2 Diagram for Algebra Vocabulary

Students with MLD also have problems with metacognitive processes such as self-monitoring, self-regulation, and organizational skills required for lengthy or multistep processes (Impecoven-Lind & Foegen, 2010). They often begin having difficulties when faced with multiplicative reasoning and increasing abstractions. These areas include reasoning about properties and operations in generalized ways, understanding relationships depicted by fractions and proportions, and relating concepts within geometry and measurement to other domains. When they reach algebra, many students with MLD struggle with the expanded symbols and syntactic systems and how to work with those at the abstract level or moving among representations. How does a graph of a straight line with the line crossing the y-axis at 3 relate to the equation $y = -2x + 3$? Can you visualize the line's slope without graphing coordinate pairs?

In more general terms, students with difficulties in algebraic reasoning have not developed the algebraic processes of noticing, reasoning, generalizing, representing, and justifying with mathematical structures across the grade levels (Chimoni et al., 2018). Unfortunately, typical arithmetic-based mathematics curricula and instruction in the elementary grades does little to prepare students for success with more formal algebra in later grades (Blanton, Stephens, et al., 2015). Often students' misconceptions and unproductive beliefs about mathematics, developed over years, will affect their algebra learning in the secondary grades, requiring individually focused interventions. An intervention based on misconceptions of high-school students in algebra courses is depicted in Box F.1, with elements adapted from Zielinski & Glazner (2019). The intervention allows individualization by focusing on persistent errors of each student.

Box F.1   Intensive Intervention for Algebra Misconceptions

Target: Algebra I or II students with persistent misconceptions such as rules for exponents, distribution, fractions, and negatives.

Intervention Schedule: Daily, 8 minutes over 8 weeks. Intervention can be used during the first quarter while previous algebra concepts are reviewed, or during the second quarter, after misconceptions are identified.

Nature of Intervention: Based on conceptual learning (activating prior knowledge and refuting misconceptions), spaced and interleaved practice, process mnemonics (for remembering rules and procedures, existing or created) and learning to mastery.

1. Students take 28-item pretest with four questions each in seven misconception categories (or categories identified in classroom).
2. For each misconception category with errors, students view an online module (3–5 minutes) and take a practice quiz with instant feedback.
3. Every day at the beginning of class, students complete practice on paper practice slips for each category completed online. To earn mastery, students must correctly complete three different slips for a category on three days.
4. Upon earning mastery for all seven categories, students unlock the eighth, a mixed review which interleaves all categories of items.
5. Posttest confirms mastery of concepts. Students needing more practice return to select modules.

Data Collection: Individual student graphs of categories, items completed to mastery, and categories mastered.

Instructional Approaches for Challenging Concepts

Algebra is a method for solving problems that requires logical reasoning and abstract thinking, not the same as arithmetic with letters. The notation system serves as an aid to this abstract thinking. Students (and their teachers) often attempt to maintain arithmetic thinking when approaching algebra, which will only hold for the simplest equations. Devlin (2011) offers the example of writing macros for spreadsheets to explain the differences. The spreadsheet can perform simple arithmetic calculations with an algorithm such as = A7 + B9 (operation of addition of the values from two cells). But a macro, a set of programming instructions written in code, can simplify repetitive tasks, such as complex calculations, communicating with data sources, or formatting. For example, you can create a macro for entering a fraction into a cell on the spreadsheet and reduce the fraction to simplest terms or convert it into a decimal number with a specific number of decimal places. You are thinking algebraically because you are working at the symbolic level, thinking in general terms rather than about a specific number, and reasoning about the relationships of numbers and their properties.

Over the years there have been several approaches and emphases for algebra. For most of the 19th and early 20th centuries, algebra textbooks extended arithmetic, focused on processes (factoring, roots, operations), and emphasized manipulating algebraic equations (Kilpatrick & Izsák, 2008). Algebra was taught as a purely mathematical, abstract discipline with little practical application. Despite the European emphasis on functions of the early 20th century, North American schools continued to view the equation as central for algebra. Enrollments in algebra declined through the 1950s as it was viewed as having little value for average students. During the "new math" period (1950s to 1970s), algebra was conceptualized from an axiomatic approach, with a focus on properties, proofs, and set theory. Algebra remained an abstract study without critical connections and applications. The reforms of national standards in the 1990s and 2000s resulted in a functions approach with emphases on problem solving and modeling. Reasons for this shift included the increasing capabilities of computers, the need to integrate mathematics domains for applications in other fields, and the importance of concept understanding and reasoning. This shift included the PreK-12 strand on algebra within the NCTM (2000) standards to promote earlier development of foundational concepts. Today's textbooks and curricula for algebra (and pre-algebraic topics) tend to focus on developing algebraic thinking through concept understanding and multiple representations over many grade levels, connecting prior arithmetic concepts to new algebraic reasoning. However, they may differ in their emphasis on symbol systems and structure, patterns and functions, the use of models, and even how word problems are incorporated. Most research on effective teaching for typical students focuses on two aspects critical for learning: careful design and sequencing of tasks and well-conceived questioning and guidance by teachers (Kieran, 2014).

The sections that follow explore some of the most critical foundational work for algebra concepts—early algebraic thinking, algebraic notation and variables, simplifying expressions and solving one-variable equations, coordinate plane graphing, and linear functions—along with teaching strategies for these concepts across the elementary and secondary curriculum. It is beyond the scope of this text to explore other functions and algebraic systems. The strand concludes with research on interventions with students with difficulties in mathematics.

Early Algebraic Thinking

From the earliest grades, teachers should promote algebraic thinking by providing tasks and problems that foster reasoning about patterns, numbers, and operations in more general ways. Rather than modeling a procedure for computing with dozens of similar practice items, teachers should consider posing a problem that requires reasoning and the application of math concepts. For example, Chimoni et al. (2018) posed the problem: If ¤ + ¤ = 4, then ¤ + ¤ + 6 =? It requires students to notice similarities and differences in the two equations, conjecture about the relationship, and justify their responses based on properties of equalities. There are several ways to reason about the problem, including substituting numbers for the ¤s and substituting the 4 for the two ¤s in the second equation.

Another example for generalizing common math activities is after counting by 5s or 10s using a hundred chart, ask students to make a generalization about their counts, such as "when counting by 5s, every other count in the ones place is the same." Teachers can also pose word problems that require reasoning and generalization: *A frog is at the bottom of a 20-meter well. Each day he makes a 3-meter leap up the well. But each night, very tired, he slips 2 meters back down. How many days will it take him to get out of the well?* This problem requires reasoning, one of several possible solutions strategies, and justification. Other word problems require developing or selecting solution expressions and justifying those: *A tailor is making school uniforms and bought 5 bolts of cloth. She has finished using all the bolts, but still needs 20 yards of cloth to finish the uniforms. Write an expression to describe the situation of starting with 5 bolts and needing more* (adapted from Pramesti & Retnawati, 2019). This problem requires reasoning, creating an expression with a variable, and justifying the expression.

As early as first grade, students should represent generalizations with variables accurately. For example, students listen to the story *Ten Monkey Jamboree* (Ochiltree, 2002) and enjoy hearing the rhymes that explore all the combinations that will add up to ten. Then the teacher challenges students to create one-variable equations using addition describing the events in the story for peers to solve, with *m* for a number of monkeys (e.g., 2 + *m* = 10, 4 + *m* + 5 = 10, 10 = *m* + 10). An example for upper-elementary students is taken from *Anno's Magic Seeds* (Anno, 1992) about a young man named Jack who was given two seeds—one to bake and eat and one to grow into two more seeds in a year. Students are asked to represent the growth in seeds when Jack realizes he can skip a year and have one to eat and three to plant, continuing this pattern each year. This problem promotes reasoning about powers of number and exponential growth, powerful concepts even with simple numbers. Students can create equations and represent those on graphs. Students in elementary grades learn to generalize and represent mathematical structure and relationships if provided robust tasks and teacher guidance. Engaging in early algebraic thinking improves students' algebra readiness for middle school (Blanton, Isler-Bayka, et al., 2019). Additional recommendations for younger students are included in some of the following sections.

Algebraic Notation

Unlike the alphabetic symbols for reading that represent phonemes, mathematics symbols and combinations of symbols represent concepts. To make matters more confusing, sometimes several symbols can represent the same concept (e.g., for division 7/3, 7 ÷ 3, $3\overline{)7}$, and $\frac{7}{3}$). Sometimes the same symbol can have different meanings (semantics) based on the context (syntax). For example, consider the role of the number 2 in the following expressions: $2x$, x^2, $2 + x$, and x_2. Sometimes the insertion of symbols changes an expression (syntax), while other times it does not: $3y$ is the same as $3 \cdot y$, the same as $3(y)$, and the same as $(3)(y)$, but not the same as $3 + y$ or y^3. Some similar symbols have different meanings: the division (⌐) and square root ($\sqrt{\ }$) symbols, the angle (∠) and less than (<) symbols, and the ∈ (element of set) and Σ (the sum of) symbols. Further, punctuation symbols in English have different meanings in mathematics (e.g., : for ratio, ! for factorials, – for negative numbers or subtraction, [ ] for matrices). Chalouh and Herscovics (1998) identified "concatenation of symbols" as a common error, where students confuse place-value concepts from arithmetic with symbolic notation. The value of the 4 in the following expressions may appear to mean the same by some students: $43, 4\frac{1}{3}, 4x$, and $34x$.

The semantics and syntax of mathematical symbols also govern the interpretation of longer expressions, much like word meaning, morphemes, and grammatical rules govern the interpretation of sentences. For example, at first glance the following expressions seem to represent the same value:

$$(2 + 6)5 - 4^2 = A \qquad (2 + (6 \cdot 5) - 4^2) = B$$

But algebraic rules called the *order of operations* require the following sequence:

1. simplifying within grouping symbols such as ( ), [ ], and { };
2. applying exponents;
3. performing all multiplicative (multiplication and division) operations from left to right;
4. performing all additive (addition and subtraction) operations from left to right.

Some students use the mnemonic device *Please Excuse My Dear Aunt Sally* to remember this sequence (parentheses, exponents, multiplication/division, addition/subtraction). Following the order of operations for the expressions above, the value for A is 24 and the value for B is 16. Some students may be misled by the PEMDAS acronym (or BEDMAS in Canada, BODMAS in the UK, using brackets) and think that multiplication precedes division and addition precedes subtraction when those occur at the same time (as reverse operations). The order may not be as simple with multiple brackets, so students should be taught to work "inside outward" and to consider whether terms can be simplified before beginning, such as the 4^2 in the example above. Some grouping symbols are also operations, the fraction bar $\frac{4+x}{5-2}$, absolute value symbol $|x - 4 + x|$, and the square root symbol (radical) $\sqrt{3+6}$, so simplifying

within those is performed first. One could argue that division symbols do not appear within expressions and that subtraction is addition of the opposite, so addition and multiplication are the primary operations within expressions, and both can be manipulated using the commutative property regardless of the order of the terms. For example, $6(7) + 3\sqrt{48+1} - 2$ could be simplified intuitively as $42 + (-2) + 3(7)$ or $40 + 21 = 61$.

For students in algebra coursework (typically eighth grade and above), Taff (2017) recommended first ensuring students understand the critical vocabulary *term* and *factor*, often misused. Then students should be encouraged to examine the structure of terms (separated by addition and subtraction signs) and factors (joined by multiplication) in an expression first (iTAFF—identify terms and factors first), viewing exponents as a list of identical factors. Then students learn to identify the nested structure of terms and factors. Next, they look for plusses and minuses not within grouping symbols, to perform those last (and using a highlighter on those as a reminder). As soon as factors are known those can be multiplied. Students work on each term individually until the only tasks left are adding and subtracting. For example, the expression $24 - x(2x + 3y - 4) - x^2 + 4xy - 12 + 2x^2$ has six terms. Changing minus signs to 'plus negative' can assist student understanding of the separate terms and not lose the negative when distributing, with parentheses used to clarify. Next, work within grouping symbols and multiply factors. You have $24 + (-2x^2) + (-3xy) + 4x + (-x^2) + 4xy + (-12) + 2x^2$; and now all like terms can be combined and reordered: $-x^2 + xy + 4x + 12$. Natural extensions of this approach were expressions with square roots, absolute value signs, and fraction bars.

Recommendations regarding symbol and syntax instruction:

- From the earliest grades, teach that the equal sign is a relational symbol, showing that the two sides of the equation are equivalent. Use "is equivalent to" and "is the same as" instead of "makes" or "results in."
- Post symbols on wall charts for easy reference. Students may even want to keep a symbol page at the beginning of their notebook.
- Ask students to read expressions that contain symbols and communicate their meaning. Prompt students to "think algebraically" when working with algebraic notation.
- Teach bracket concepts earlier, with the properties of operations. For example, examine why the equation $3 \times 3 + 2 = 15$ needs brackets (parentheses). Parentheses have three functions within expressions: clarifying, grouping, and signaling multiplication. Ask frequently, "What is the purpose of the parentheses in this expression?"
- Have students create "challenge expressions and equations" for each other to solve using the order of operations.
- Use clear hand printing or an equation editor when preparing overhead projections, charts, and worksheets. Encourage students to use lined or grid paper to keep mathematics notation clear. Students may write the 't' as a plus and 's' as a 5, so encourage a hook (t), loop, or cursive for those letters.
- Avoid stringing out equations (e.g., $4(2 + x) - 10 = 8 + 4x - 10 = 4x - 2$); instead teach students to write equations in steps on each line, with each equal sign lined up below the previous, to reinforce the concept of balance.
- Encourage the use of conventional syntax when writing expressions and equations. We write it in this form so it will be clear to others and we can work with it easier. It is like learning the rules for driving; someone created those so we all drive safely and get where we are going.

Variables are another source of confusion within algebraic and other notion systems. In mathematics, some letters are from the English alphabet (originally from the Latin, Etruscan, and Greek) while others (e.g., π, θ, α) are from the Greek alphabet. But not all letters within formulas and equations are variables. The symbols used to represent numbers with fixed values are called constants. In the following equation, the a is a constant, it has only one value: $8 + a = 15$. In this equation the 8 and the 15 are also constants. As students progress through the grade levels, significant numbers are represented by specific symbols. For example, the π represents a specific value and is a constant, as is φ for the golden ratio.

Variables are symbols for changing values. Letters that represent variables are, by convention, in lower-case form from the end of the alphabet, and most often x, y, z, t, and u. Frequently letters from the beginning of the alphabet (a, b, c) as well as capital letters (C, K, A) are used to represent constants. Sometimes specific letters are used with certain values, such as f for functions (not a variable), x and y for ordered pairs, d for derivative, p for probability, m for slope, n for number, r for correlation coefficient or radius, t for time, and d for distance or diameter. It is good practice to avoid using letters whose shapes can be confused with other mathematical symbols (o, i, l, j, u, v, w) unless there is a specific use as with set, velocity, or imaginary number. Within expressions and equations, variables are typically listed in alphabetical order unless exponents are involved (e.g., $x + 2y + z$). If students are still working with the symbol $\times$ for multiplication, in second and third grades, try to avoid x as a variable. Make an explicit transition to other ways to represent multiplication.

Recommendations for teaching about variables:

*   In early grades, use a __,?, or □ for an unknown value. Example: $5 +$ ___ $= 2 + 7$.
*   Younger students may prefer to use their initials or other significant letters as variables. For example, Anna wanted to use an A for her name and B for her friend Betty's name in the equation $5A + 3B = 30$ This is fine as long as the A and B are variables (unknowns), such as the cost of each of Anna's 5 pencils and the cost of each of Betty's 3 pens. However, it is easy to confuse these letters with labels, confusing the structure of the equation. Letters should not be used as labels because that will not hold up for multiplicative reasoning (e.g., P1 + P2 = T for Part 1 plus Part 2 = total). Spell out labels for clarity (e.g., 4 pencils + 5 pens = 9 writing instruments), especially in proportions.
*   Discuss explicitly the multiple roles of letters, as unknown constants ($5y = 10$), as placeholders for measurements ($A = b{\cdot}h$), or as values that vary ($y = 2x - 5$).
*   Some students interpret abbreviations as variables, so they need to write out labels when using measures or other labels (e.g., 12m for 12 meters, 5h for 5 hours).
*   Another common mistake is viewing a variable as representing only one number or value, such as the letter b for base and assuming b still represents base in $a + b = c$. Differentiate formulas from equations and provide many examples of letters as variables. Some students confuse the use of the same variable in one equation as two different values (e.g., $2x + 10 + x = 40$).
*   Assist students making the transition from arithmetic and algebraic forms of equations. In the equation $x + 4 = 10$, x is an unknown constant (sometimes termed one-variable equation). There is only one possible value for x. But in the algebraic equation $x + y = 10$, there are limitless values for the two variables, with a dependent relationship.
*   Develop a sound understanding of the concepts *like* and *unlike terms*, especially at the beginning of algebra courses. For example, $3x$ and $4x$ are like terms, while $3xy$ and $4x^2$ are unlike those. Some teachers use color coding for like terms on white boards within group activities for combining and justifying. Others have assigned terms via name tags (e.g., $-14, 7y^2, xy$) to each student and asked students to group themselves by like terms (Stegall & Malloy, 2019).
*   Teach students how to translate word problems into algebraic equations without the syntactic and semantic errors that are so common, by translating into a diagram or table first. Students may interpret some variables derived from word problems as labels rather than varying values. For example, if sodas cost s and cookies cost c, write an equation that gives the total cost t for 12 sodas and 36 cookies. ($t = 12s + 36c$) Ask students to describe the values of s and c. Have students practice creating equations from word problems and vice versa.

Monomials and *polynomials* are also fundamental concepts for understanding algebraic expressions and equations. Monomials are single terms while polynomials have several terms, connected with addition or subtraction symbols. A *term* can be a real number, a variable, or the product of a real number and one or more variables. The following terms are all monomials:

$$4 \quad 4x \quad \frac{4x}{y} \quad 4xy \quad 4xyz \quad 4x^2yz^3$$

The degree of a monomial is the sum of the exponents on the variable. For example, the last monomial above has a degree of 6. Polynomials are the sum and/or difference of monomials and do not include negative exponents. A polynomial with two terms is called binomial, three terms trinomial, and all others polynomial. The degree of a polynomial is the greatest exponent. A second-degree polynomial is called a quadratic ($3x^2 + 24x + 2$), a third-degree term a cubic ($3x^3 + 4xy + y^2$), and so forth. Polynomials are usually arranged in descending or ascending order by the degree of each term. Similar terms (same variables and exponents) can be combined (added or subtracted):

descending: $3x^2 + xy + 8$;
ascending: $4 + 3xy − z^3$;
addition of these trinomials results in the polynomial $12 + 4xy + 3x^2 − z^3$.

The word *coefficient* refers to the number that multiplies the variable(s) in a term. In the term $5x$, the coefficient is 5. In the term xy, the coefficient is 1. The multiplier coefficient can also be a fraction: $\frac{2}{3}xy$, also written $\frac{2xy}{3}$. The most confusing concepts about monomials and polynomials are the words used (term, degree, coefficient), the use of a minus symbol or the plus symbol and a negative, and the rules for simplification of these expressions. Students are often asked to manipulate these symbols in the abstract, without concrete or graphic representations to promote understanding.

Suggestions for teaching polynomial concepts:

* Stress correct vocabulary use through word walls, charts, and frequent verbal communication of mathematics expressions. Make clear distinctions among words such as *expression* and *equation* and the directions on many tests such as simplify, solve, transform, and interpret. Be aware of differences in mathematics vocabulary used by different textbooks. Send home vocabulary with definitions and examples so that parents can use the same terminology with their children.
* Teach a line-drawing strategy for separating the terms in a polynomial for clear understanding of each symbol and for easier simplification (M. Del Mastro, personal communication, 28 May 2005). Some teachers have students cut apart terms before each addition or subtraction symbol for rearranging (see Figure F.3).
* Instead of teaching that the expression $2x− (−3)$ should be changed into $2x + (+3)$, where the negatives turn into positives, teach that the $− (−3)$ term is a negative-negative (double negative), therefore a positive ($+3$).
* Have students create long polynomials for rearranging and simplifying. Teach students the conventions of arrangement—that the terms are usually listed in order of degree of exponents for easier understanding and use. Practice reading simplified polynomials and naming their degrees.
* Some students aren't comfortable with a result such as $3n + 2$. Practice simplifying terms and discussing why unlike terms cannot be combined. An answer is not always an integer.

Simplifying Algebraic Expressions and Solving Equations

A process that depends on understanding the concepts of algebraic notation is that of simplifying algebraic expressions. Simplification, including isolating and combining terms, working

$$4 \Big/ + 3xy \Big/ - (-z^2) \Big/ -y$$

Figure F.3 Line Strategy for Isolating Terms

with negative integers and variables, and using the order of operations, is required for solving algebraic equations, which also builds upon the elementary concepts of equivalence and arithmetic properties. However, simplifying expressions and solving equations are not goals of mathematics learning, they become tools for solving problems.

Textbooks devote many pages to exercises in simplifying terms and solving one-variable equations. But early experiences with "keeping an equation in balance" should establish fundamental concepts for these skills. Elementary students should have experiences manipulating equations using properties of arithmetic. For example, in the identity equation $5 + 3 = 2 + 6$, students can add and subtract the same amounts from each side and keep the equation in balance. In the equation $8 - 6 + 4 = x - 2 + 5$, students can explore adding and subtracting different values while keeping the equation in balance and eventually end up with a value for x. They can even flip the two sides and maintain equivalence (and the x on the left).

The following sequence for attacking an expression or equation is recommended in many textbooks:

1. Substitute an equivalent expression for any expression in the equation, sometimes combined with Step 2 because it is a simplification of like terms. For $x - 4 = 9 - (-5)$, the number 14 can be substituted for the right side.
2. Simplify each side of the equation by combining like terms, using the distributive property, and removing unnecessary parentheses, using the *order of operations*.
   For the expression: $2x + 4(3 + 7) - 5x$, the $3 + 7$ becomes 10, then the expression becomes $2x + 40 - 5x$ by multiplying, then $40 - 3x$ from combining like terms.

From this point *the order of operations is suspended* and a new order is followed. Basically, for an expression within an equation, follow the order of operations but to solve for one variable, start with additive combining. The approach for solving is to *isolate the variable* on one side of the equation.

3. Add (or subtract—add the reciprocal) the same real number to each side of the equation.
4. Multiply (or divide—multiply the reciprocal) the same nonzero real number to each side.
5. If the final transformation is a false statement ($3 = 5$), then the equation has no solution ($\emptyset$). If the final result is an equivalence ($x = x$), then the equation is true for all real numbers.

Consider what actions should be applied to each side of the equations in Table F.3 to result in a solution for x. Your mission here is to end up with the x on one side of the = and everything else, in simplest and neatest form, on the other.

Recommendations for teaching equation solving and simplification:

- Use procedural tasks such as the evaluation of arithmetic expressions to enhance students' awareness of structure and to assess their structure sense (Banerjee & Subramaniam, 2012). For example, evaluating $19 - 3 + 6$ or $7 + 3 \times 5$ can help students focus on relationships and structure.
- Teach the concepts of *value, operators, variable, term, expression, factor, equality, equation, combining terms, like terms, coefficient, constant, base, exponent, simplest form,* and *distributive property* explicitly and with frequent use in teacher and student language.
- Practice identifying equal expressions without computing. For example, judge these two expressions for equivalence: $18 - 15 + 4 \times 6$ and $3 \times 6 + 15 - 12$.
- Use a balance-scale analogy for demonstrating the two sides of the equation, especially with younger students. What's with the x that we must remove using our rules for inverse properties?

Table F.3 Solving for *x* Using Properties of Algebra

Solve for x by isolating it on one side of the equation.

---

a.	$x - 3 = -8$	Hint: add 3 to both sides
b.	$x + 5 = 2(6 + 2)$	Hint: distribute the 2 first and combine terms on the right, then add -5 to both sides
c.	$-\dfrac{x}{4} = -10.5$	Hint: multiply both sides by -4
d.	$6x = -96$	Hint: divide both sides by 6 (or multiply by 1/6)
e.	$1 = \dfrac{x}{12} + 5$	Hint: add -5 to both sides, multiply both by 12
f.	$\dfrac{x+2}{9} = -5$	Hint: multiply both sides by 9, then add -2 to both
g.	$\dfrac{2x}{3} + \dfrac{x}{2} = 7$	Hint: find like terms for the fractions and combine ($\dfrac{7x}{6}$), then multiply both sides by $\dfrac{6}{7}$.

- Review work with fractions, including properties and operations with fractions (with negative and positive terms) and expressing fractions in different forms. Connect previous fraction learning with algebraic forms for simplifying terms.
- Use real-life data and personal experiences to introduce and create equations for solving. The goal for learning these skills, after all, is to solve problems. Students can also create real-life problems for one-degree equations with one and two variables. Situations for algebraic problems include sports, video games, part-time jobs, cell phones, and other applications that use time, costs, changing rates, and data analysis.
- Use concrete objects, such as Algebra Tiles, Algeblocks, or Lab Gear (see Chapter 7), to model equations, but explore their representations and limits before using them with students. The red sides of Algebra Tiles represent negative values and help to reinforce reciprocal concepts. Each time these objects are used, it is critical each shape be defined for the situation. Sometimes the small square represents a numerical value (1), and sometimes it represents a variable (*y*). Monitor whether these shifts in value are too confusing for some students. These manipulatives can also be placed on a Cartesian plane to make the geometric connection (lengths, widths, area) with algebraic expressions.
- When teaching algebraic problem solving, support students' understanding of the relationships in the problem by teaching students to translate the information into a table, graph, or diagram before they translate into equation form (Moseley & Brenner, 2009).

Coordinate Plane Graphing

The Cartesian coordinate system is named after René Descartes who invented the grid system in the early 17th century, calling the *x*-axis the East axis (horizontal) and the *y*-axis the North axis (vertical; Sidebotham, 2002). Points, two-dimensional figures, and equations can be plotted on this four-quadrant plane (see Figure F.4). This concept of plotting points on a plane has been generalized to various mapping and graphing applications such as using latitude and longitude in geography and constructing a line graph to depict census data or heart rates over time.

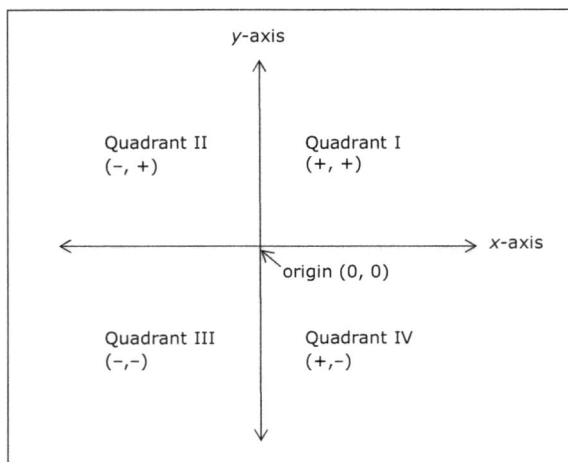

Figure F.4 Cartesian Coordinate Plane

In the mathematics curriculum, formal point plotting in the first quadrant (both the x and y values are positive) is typically introduced in the fifth grade. However, kindergarten and first-grade students should be involved in data collection and the construction and interpretation of simple object, picture, and bar graphs of meaningful data. Students in Grades 3 through 5 gain experience with line plots and graphs and begin exploring integers less than 0 on the bi-directional number line. Middle-school students should use four-quadrant graphs for representing data, geometric figures, and linear equations with some explorations into inequalities and nonlinear equations. In high school, graphing applications include a range of function classes and vectors and extend learning to other coordinate systems such as spherical and polar. Coordinate graphing truly connects the mathematics curriculum: number theory (integers); operations (ratio and proportions); algebra (equation graphing); geometry (figure graphing); measurement (area and perimeter on a grid); and data analysis (depicting data in graph form). Graphing is a powerful tool for representing many mathematical concepts.

Recommendations for teaching graphing skills:

- Start with meaningful data graphed in more representational forms—such as picture and bar graphs. For example, first graders could use actual buttons to create a graph of the number of buttons on each child's clothing. Second graders can create a bar graph depicting the length of each student's arm span.
- Use graphing terminology, such as *origin* (0, 0), *intercept*, *point*, *axis*, and *line*, even with younger students.
- When more challenging terms are introduced in the middle grades, mnemonic devices should be used to assist learning: *a* of *abscissa* comes before *o* of *ordinate*, *x* comes before *y*, and *domain* (input) comes before *range* (output). We always start with the abscissa or x-axis when graphing. Don't avoid these correct terms; teach and reinforce their use.
- Teachers in the early grades should take care to use graphing conventions that will hold up throughout the grade levels. For example, number lines and graphs should have arrows at the ends of axes, depicting infinity. Number lines for students in the earlier grades should always have the zero-position indicated (unless the number line depicts a sequence of numbers for a different purpose, such as a range from 100 to 150). Graphs may have

differing units depicted (count by 2, count by 10, count by 0.5). The axes may use different scales, depending on the nature of values. For graphing functional relationships, the x-axis value should depict the independent variable and the y-axis value the dependent variable, connected to the equation form of $y = f(x)$. Time should be depicted on the x-axis. Each axis should be labeled, and the overall graph given a descriptive title. As students gain experience with various types of graphs, they should be asked to consider the type of data (e.g., nominal, ratio) and the best type of graph and scale to represent those data.

- Have students view a coordinate pair and state in which quadrant its point will fall before starting with the x value along the horizontal line in a positive or negative direction. For example, the point $(-2, 3)$ will fall into the second quadrant.

- Build flexible graphing skills by using a variety of materials and tools—computer spreadsheet software, graphing calculators, whiteboards, and charts. Graph paper with appropriate-sized grids should be available for student use, with the teacher modeling how to draw and mark off each axis. Graphs from newspapers and magazines should be collected for interpretation and display. Students with spatial deficits need more concrete, and perhaps longer, practice.

- Have students predict graphs from equations or from changes in equations. They can check results on a graphing calculator (e.g., desmos.com/calculator). For example, the graph for $y = 2x + 5$ will be a straight line with the line crossing the y-axis at 5. The graph for $y = 3x^2$ will be a narrow parabola with an upward turn from the origin $(0,0)$. Teach students to rearrange and simplify equations into more standard formats before graphing.

- *Slope* (also called gradient) is another key concept for many areas of mathematics and can be linked to students' prior knowledge of a skateboard ramp, a roofline, or a wheelchair ramp. The formula for a slope is expressed as a fraction: rise/run, with the common symbol m. A positive m is depicted with a line in the direction / (increasing from left to right) and a negative m in the direction \ (decreasing from left to right; Sidebotham, 2002). In high-school algebra, students will explore gradients of curves, which must change along the curve, so that tangents are used to find gradients at specific points along the curve.

- Communication skills are critical for students to use when making the connections among representations, such as between data sets and graphs, tables and graphs, or graphs and equations. Students should describe and interpret graphs for others, using consistent terms. They should answer questions about their graphs such as, "What do the values on the x-axis mean?", "Where does this line intercept the y-axis?", "What would happen to this graph if the equation were changed?"

Linear Functions

Linear functions are relationships that when graphed form a straight line. Two variables (independent and dependent) are represented in equation form so the solution for one is always in terms of the other. If a value is known for x, the equation can be solved for y. If 0 is substituted for x, that is the point where the line will cross the y-axis, giving the value for y. Likewise, if 0 is substituted for y, the x-intercept is found. A linear function, in practical applications, represents a constant rate or change in one direction.

There are three conventional forms for writing linear equations (Muschla & Muschla, 1995):

- Slope-intercept: $y = mx + b$ (m is the slope and b is the y-intercept).
- Standard form: $Ax + By = C$ (A and B cannot be 0).
- Point-slope form: $(y-y_1) = m(x-x_1)$ (the point (x_1, y_1) lies on the line, m is the slope).

$y = 2x - 3$

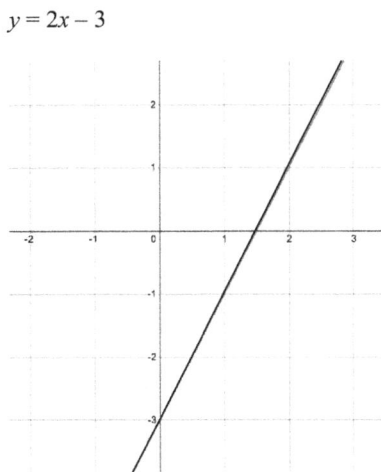

Figure F.5 Linear Function

Note that the degree of linear equations (greatest exponent used) is one and the independent variable (x) may not be the exponent (exponential function) in the denominator (rational function) or under the radical (radical function), as those functions produce nonlinear graphs. Figure F.5 depicts a linear function in graph form with the corresponding equation. Linear functions can be represented in many forms including equations, graphs, tables, diagrams, and words.

Recommendations for teaching about linear functions:

- Provide meaningful and symbolic experiences with patterning with younger children. They should not only be asked to predict the next item in a pattern sequence but provide the controlling rule for the sequence (function). For example, in the number pattern 2, 5, 8, 11, ___, 17, the missing number is 14 and the controlling rule for the sequence is +3. Patterning activities can range from simple to complex.

- Begin with meaningful contexts, not abstract definitions and rules. Definitions based on set theory are particularly opaque for students engaged in initial learning. Research has demonstrated that rule-bound instruction at the symbolic level is related to student (and teacher) misunderstanding of overall function concepts and problems with flexibility in moving among representations of the same function (Tall, 1992). Moreover, good problem contexts add meaning to how functions can be used in a variety of applications.

- Use realistic situations to illustrate the properties of linear functions, such as computer games, driving speed, purchases, and TV ratings, and guide the transition to abstract notations. Even in units rich in situations, students can interpret linear equations as static, "storage containers" for inserting values rather than as a relationship between the x and y values (Lobato & Ellis, 2002). Teachers should continue making connections with the situation when moving to the abstract.

- Ensure students understand that not all functions are linear by introducing the concept of nonlinear functions (e.g., quadratics). Also show that a linear inequality is not a function. Technology can assist with representing models of functions if teachers guide the technology use so that students make connections about changes to the dependent and independent variables (concept of co-variation).

- Use tables to facilitate student understanding of the relationships and patterns within functions. Use a realistic problem for which students can generate data into table form, such as testing physical models with mechanical devices or exploring the effects of compounded interest in a bank account (Hines et al., 2001).
- Use a balance-scale model for teaching the concept of equality for linear equations. This model can be deceptively complex, so teachers should explore drawn, physical, and computer-based representations before instruction (Otten et al., 2019). Otten and colleagues reviewed 34 articles on using the balance scale for linear-equation instruction and found that positive effects were more frequent for younger students when they are introduced to the concepts involved, such as equality and strategies for maintaining balance. Some limitations in later grades included representing negative numbers and some equations detached from the concrete model (e.g., $-6x = 24$), which caused misconceptions (e.g., Vlassis, 2002). Students need to understand the concepts of *term* and *coefficient* as well as review properties before using this model.
- Use technology to display instant graphs and changes in graphs so that the characteristics of functions can be analyzed without the tedium of drafting and erasing. Take care to link graphs to equations (and original data) and teach students how to graph data manually. The graphing process helps students understand functions as relationships.

The What Works Clearinghouse practice guide on teaching strategies for algebra recommended three primary research-based approaches (Star et al., 2015, 2019). The first was to use solved problems (worked examples) to engage students in analyzing algebraic reasoning and strategies. Guided discussions and small-group work on solved or partially solved problems (correct or incorrect) can highlight algebraic structures, properties of numbers and operations, and connections among concepts. Riccomini and Morano (2019) described how to use worked solutions with students with MLD for multi-step algebra problems. In addition to worked, step-by-step examples, the researchers used self-explanation, strategic use of solutions, scaffolding understanding, partner practice, and embedded worked solutions within practice. The authors concluded that the use of worked solutions can decrease cognitive load (WM) and encourage self-explanation and reasoning skills. The second recommendation was to teach students to use the structure underlying algebraic representations for greater understanding. Teachers should model precise mathematics language when examining the structure of an expression, equation, graph, word problem, or other representation. When examining an expression, what are the terms and how are those connected? When examining a linear function, do I recognize the form? What are the variables?

Finally, teach students "to intentionally choose from alternative algebraic strategies when solving problems" (Star et al., 2015, 2019, p. 26). This recommendation does not mean that students become fluent in all possible strategies but have options when faced with problems to solve. Teachers should introduce one or two at a time and have students articulate their reasoning when using the strategies. Special education students may need explicit instruction in alternative strategies and when and why a specific strategy is useful for specific problems. Lynch and Star (2014) investigated the use of multiple strategies with six struggling students over the course of a year-long Algebra I course (https://scholar.harvard.edu/contrastingcases). The intervention included worked examples (typically two); carefully selected routine examples for comparison; engaging cartoon characters for context; explicit, step-by-step instruction; and a structured implementation model (compare, contrast, discuss). Student reports were generally positive, with students commenting on the advantage of having more than one strategy for solving a problem, finding a method that worked best, and reducing anxiety.

Research on Algebra Instruction for Students with Mathematics Difficulties

Students with MD and MLD may need a more carefully sequenced foundation for algebraic reasoning (PreK to 7) than typically achieving students. It is critical this reasoning be screened at *each* grade level, interventions provided where needed, and those documented for future IEP committees. Students with difficulties may also need more explicit connections with concrete representations or virtual manipulatives and a longer period of time to develop essential concepts for high-school algebra. It is likely that nine-week supplemental interventions will not work for algebraic concepts unless those are individualized and target specific misconceptions. Some students will require courses that allow additional time for concept development and differentiation of instruction. For example, the *Promoting Student Success in High School Project* (Knudson & Sorensen, 2017) investigated a supplementary learning opportunity for students struggling in Algebra I—double-dose algebra. A special companion course provided students a second period of algebra to offer more instructional time in grade-level content. Research concluded that instruction in this model should be targeted and intentional (*not* more of the same), high-quality practices are essential (e.g., real-world connections, formative assessment), the right teachers should be selected (believe all students can learn), and individual student needs must be identified and targeted. These courses can be developed to integrate with any high-school scheduling format.

Regardless of their performance in elementary mathematics, some students with MLD may be at a disadvantage when they reach high school due to improper mathematics placement in middle school. In an analysis of longitudinal data on students, teachers, and schools from kindergarten through Grade 8, Faulkner et al. (2013) found that students with IEPs who demonstrated inconsistently high performance in mathematics (70th-84th percentile) by fifth grade had only one-fifth the placement rate in algebra by eighth grade as compared with their peers (also inconsistently high performance). Teachers' placement predictions (perceptions) were not as related to future math placements as actual performance, especially for low-performing students. However, for higher-achieving students, teacher predictions had stronger correlations. "Having an IEP and low teacher perception, … was virtually prohibitive of placement in algebra, even in the presence of high math performance" (Faulkner et al., 2013, p. 338). If students do not have the chance for success in a first course in algebra by eighth or ninth grade, their options for high-school coursework and college diminish substantially, also an effect on careers and other life options.

In a review of effective intervention research for algebra and students with MLD, Watt et al. (2014) identified only 15 studies since 1980. The most common successful intervention was the concrete-representational-abstract (CRA) sequence, with an average effect size of 0.53 on student achievement. The CRA approach involved concrete and pictorial representations prior to working in the abstract, but afforded students the opportunity to move back and forth among representations. Strickland and Maccini (2012) termed this adjustment CRA-I instruction, with the "I" an integration of concrete manipulatives and sketches with abstract notation in their studies on multiplication of linear expressions and factoring quadratic expressions. These lessons also incorporated review to connect prior knowledge, teacher modeling and guided practice, and visual organizers. Cognitive-strategy instruction was the approach with the greatest effects ($g = 0.83$), especially with algebraic word problems. Building on arithmetic problem solving, algebraic strategies assisted students in understanding the problem, representing the problem with diagrams and equations, following problem-solving steps, and evaluating solutions. One research group focused on enhanced-anchored instruction (EAI), using real-life situations to provide a problem context (Bottge et al., 2007). Students in EAI groups outperformed traditional-instruction groups for solving word problems involving distance, rate, and time ($g = 0.80$). In addition to these approaches, most studies in the literature review also employed explicit instruction, the use of visual representations, formative-assessment data, and the use of strategies. In all 15 studies the mathematics content was below the grade level of the study participants, highlighting a need for research on more complex algebraic skills.

Lee and colleagues (2020) conducted a systematic review of interventions for algebraic concepts and skills for secondary students with MLD, identifying 12 studies. The studies included the effects of six instructional components: multiple representations, a sequence or range of examples, explicit instruction, student verbalizations, the use of "heuristics" (cognitive and meta-cognitive strategy instruction), and the use of real-world problems. All of these studies demonstrated large or very large intervention effects and most combined two or more effective components. The authors emphasized that teachers should choose age-appropriate visualizations for representing word problems in algebra, such as virtual manipulatives, number lines, and drawings, and that teachers need to be knowledgeable about algebra concepts and skills in order to plan effective interventions.

While the components of effective instruction have been confirmed for secondary topics such as algebra, studies with struggling students tended to focus on a narrow range of topics—representing and solving word problems involving integers, proportion and ratio problems, one-variable equations, multi-step equations and inequalities, linear equations, and representing quadratic expressions. Most of the early studies on algebra and students with MD tended to focus on procedures and skills, often below students' grade levels. However, since more challenging state and national standards for mathematics have been implemented, more studies include concept development and connecting concepts with skills as well as address more challenging content. More research is needed on the application of rational number knowledge in algebra, instruction in combined algebra and geometry concepts, students' understanding and use of functions to model relationships between quantities, and how progress monitoring can inform algebraic reasoning and algebra instruction in all tiers of interventions.

Algebraic reasoning connects all other mathematics topics across the grade levels. There must be a deliberate and consistent program of instruction PreK-12, with screening and progress monitoring, for students to be successful in the more advanced coursework in high school. Early experiences with reasoning about the attributes of objects and the properties of numbers and operations are formative. Discussing and justifying their reasoning for arithmetic problems and thinking about problems in generalized ways will support algebraic reasoning. Developing multiplicative and proportional reasoning by the middle grades will assist the transition. Implementing a coherent curriculum that balances concept understanding with study of structure and procedure, along with effective instructional strategies supported by appropriate and guided representations and verbalizations, has the most potential for improving student achievement.

Multicultural Connection

Functions are found everywhere in daily life from interpersonal relationships to inflating a tire and baking bread. They are also relevant for leisure-time activities. People in most cultures enjoy board games and have developed quite a number of varieties. In most board games, a given move is a function of a previous move. Participants try to predict future moves to plan their next moves. Many board games around the world are based on three-in-a-row objects, including the common tick-tack-toe. For example, Native Americans of New Mexico played a "Square Game" called *pitarilla* (little stones) that is thought to have been adopted from the European Spanish (Culin, 1975), and most likely from Egypt originally but developed

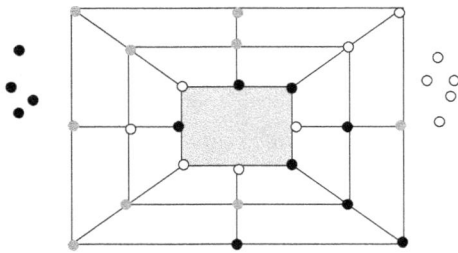

Figure F.6 Native American Board Game

simultaneously in Asia (Zaslavsky, 1996). In most of these games, two players take turns moving objects along lines to specific points on the board, attempting to achieve three objects in a straight line while blocking the other player from getting three-in-a-row first (see Figure F.6). Each move becomes a function of the locations available on the board as a result of previous moves.

References

Accardo, A. L., & Kuder, S. J. (2017). Monitoring student learning in algebra. *Mathematics Teaching in the Middle School, 22*(6), 352–359. doi: 10.5951/mathteacmiddscho.22.6.0352

Algebra Screening Progress Monitoring (2017). *About the project.* https://education.iastate.edu/soe-outreach/algebra-screening-and-progress-monitoring/

Anno, M. (1992). *Anno's magic seeds.* Philomel Books.

Banerjee, R., & Subramaniam, K. (2012). Evolution of a teaching approach for beginning algebra. *Educational Studies in Mathematics, 80*(3), 351–367. doi: 10.1007/s10649-011-9353-y

Blanton, M., Brizuela, B., Gardiner, A., Sawrey, K., & Newman-Owens, A. (2015). A learning trajectory in 6-year-olds' thinking about generalizing functional relationships. *Journal for Research in Mathematics Education, 46*(5), 511–558. doi: 10.5951/jresematheduc.46.5.0511

Blanton, M., Brizuela, B. M., Gardiner, A. M., Sawrey, K., & Newman-Owens, A. (2017). A progression in first-grade children's thinking about variable and variable notation in functional relationships. *Educational Studies in Mathematics, 95*(2), 181–202. doi: 10.1007/s10649-016-9745-0

Blanton, M., Isler-Baykal, I., Stroud, R., Stephens, A., Knuth, E., & Gardiner, A. (2019). Growth in children's understanding of generalizing and representing mathematical structure and relationships. *Educational Studies in Mathematics, 102*(2), 193–219. doi: 10.1007/s10649-019-09894-7

Blanton, M., Stephens, A., Knuth, E., Gardiner, A. M., Isler, I., & Kim, J.-S. (2015). The development of children's algebraic thinking: The impact of a comprehensive early algebra intervention in third grade. *Journal for Research in Mathematics Education, 46*(1), 39–87. doi: 10.5951/jresematheduc.46.1.0039

Blanton, M., Stroud, R., Stephens, A., Gardiner, A., Stylianou, D., Knuth, E., Isler-Baykal, I., & Strachota, S. (2019). Does early algebra matter? The effectiveness of an early algebra intervention in grades 3 to 5. *American Educational Research Journal, 56*(5), 1930–1972. doi: 10.3102/0002831219832301

Booth, J. L., Barbieri, C., Eyer, F., & Paré-Blagoev, E. J. (2014). Persistent and pernicious errors in algebraic problem solving. *Journal of Problem Solving, 7*(1), 10–23. doi: 10.7771/1932-6246.1161

Bottge, B. A., Rueda, E., Serlin, R. C., Hung, Y. H., & Kwon, J. M. (2007). Shrinking achievement differences with anchored math problems: Challenges and possibilities. *Journal of Special Education, 41*(1), 31–49. doi: 10.1177/00224669070410010301

Cai, J., & Moyer, J. (2008). Developing algebraic thinking in earlier grades: Some insights from international perspectives. In C. E. Greenes & R. Rubenstein (Eds.), *Algebra and algebraic thinking in school mathematics* (70th Yearbook, pp. 169–180). NCTM.

Carlson, M., & Oehrtman, M. (2005). *Key aspects of knowing and learning the concept of function* [Research Sampler 9]. Mathematical Association of America. www.maa.org

Carraher, D. W., & Schliemann, A. D. (2018). Cultivating early algebraic thinking. In C. Kieran (Ed.), *Teaching and learning algebraic thinking with 5- to 12-year-olds*, ICME-13 Monographs (pp. 107–138). Springer. doi: 10.1007/978-3-319-68351-5_5

Chalouh, L., & Herscovics, N. (1998). Teaching algebraic expressions in a meaningful way. In A. F. Coxford (Ed.), *The ideas of algebra, K-12* (1988 Yearbook) (pp. 33–42). NCTM.

Chimoni, M., Pitta-Pantazi, D., & Christou, C. (2018). Examining early algebraic thinking: Insights from empirical data. *Educational Studies in Mathematics*, *98*(1), 57–76. doi: 10.1007/s10649-018-9803-x

Cirino, P., Tolar, T., & Fuchs, L. (2019). Longitudinal algebra prediction for early versus later takers. *The Journal of Educational Research*, *112*(2), 179–191. doi: 10.1080/00220671.2018.1486279

Clement, J. (1982). Algebra word problem solutions: Thought processes underlying a common misconception. *Journal for Research in Mathematics Education*, *13*(1), 16–30. doi: 10.5951/jresematheduc.13.1.0016

Culin, S. (1975). *Games of North American Indians*. Dover Publications.

Devlin, K. (2011, 20 November). *What is algebra?* https://profkeithdevlin.org/2011/11/20/what-is-algebra/

Dougherty, B., Bryant, D. P., Bryant, B. R., Darrough, R. L., & Pfannenstiel, K. H. (2015). Developing concepts and generalizations to build algebraic thinking: The reversibility, flexibility, and generalization approach. *Intervention in School and Clinic*, *50*(5), 273–281. doi: 10.1177/1053451214560892

Faulkner, V. N., Crossland, C. L., & Stiff, L. V. (2013). Predicting eighth-grade algebra placement for students with individualized education programs. *Exceptional Children*, *79*, 329–345.

Genareo, V. R., Foegen, A., Dougherty, B. J., DeLeeuw, W. W., Olson, J., & Dundar, R. K. (2021). Technical adequacy of procedural and conceptual algebra screening measures in high school algebra. *Assessment for Effective Intervention*, *46*(2), 121–131. doi: 10.1177/1534508419862025

Hackenberg, A. J., & Lee, M. Y. (2015). Relationships between students' fractional knowledge and equation writing. *Journal for Research in Mathematics Education*, *46*(2), 196–243. doi: 10.5951/jresematheduc.46.2.0196

Hines, E., Klanderman, D. B., & Khoury, H. (2001). The tabular mode: Not just another way to represent a function. *School Science and Mathematics*, *101*(7), 362–371. doi: 10.1111/j.1949-8594.2001.tb17970.x

Impecoven-Lind, L. S., & Foegen, A. (2010). Teaching algebra to students with learning disabilities. *Intervention in School and Clinic*, *46*(1), 31–37. doi: 10.1177/1053451210369520

Jankvist, U. T., & Niss, M. (2018). Counteracting destructive student misconceptions of mathematics. *Education Sciences*, *8*(2), 53. doi: 10.3390/educsci8020053

Jones, I. & Pratt, D. (2012). A substituting meaning for the equals sign in arithmetic notating tasks. *Journal for Research in Mathematics Education*, *43*(1), 2–33. doi: 10.5951/jresematheduc.43.1.0002

Jupri, A., Drijvers, P., & Van den Heuvel-Panhuizen, M. (2014). Difficulties in initial algebra learning in Indonesia. *Mathematics Education Research Journal*, *26*, 683–710. doi: 10.1007/s13394-013-0097-0

Kaput, J. J. (2008). What is algebra? What is algebraic reasoning? In J. L. Kaput, D. W. Carraher, & M. L. Blanton (Eds.), *Algebra in the early grades* (pp. 5–17). Routledge. doi: 10.4324/9781315097435-2

Ketterlin-Geller, L. R., Gifford, D., & Perry, L. (2015). Measuring middle school students' algebra readiness: Examining validity evidence for experimental measures. *Assessment for Effective Intervention*, *41*(1), 28–40. doi: 10.1177/1534508415586545.

Khng, K. H., & Lee, K. (2009). Inhibiting inference from prior knowledge: Arithmetic intrusions in algebra problem solving. *Learning and Individual Differences*, *19*(2), 262–268. doi: 10.1016/j.lindif.2009.01.004

Kieran, C. (2007). Learning and teaching algebra at the middle school through college levels. In F. K. Lester (Ed.), *Second handbook of research on mathematics teaching and learning*. New Age Publishing, NCTM.

Kieran, C. (2014). Algebra teaching and learning. In S. Lerman (Ed.), *Encyclopedia of Mathematics Education* (pp. 27–32). Springer. doi: 10.1007/978-94-007-4978-8

Kieran, C. (2018). Seeking, using, and expressing structure in numbers and numerical operations: A fundamental path to developing early algebraic thinking. In C. Kieran (Ed.), *Teaching and learning algebraic thinking with 5- to 12-year-olds*, ICME-13 Monographs (pp. 79–105). Springer. doi: 10.1007/978-3-319-68351-5_4

Kilpatrick, J., & Izsák, A. (2008). Historical perspectives on algebra in the curriculum. In C. E. Greenes & R. Rubenstein (Eds.), *Algebra and algebraic thinking in school mathematics* (70th Yearbook, pp. 3–18). NCTM.

Knudson, J., & Sorensen, N. (2017). Double-dose algebra: Profile of practice brief. *Promoting Success in Algebra I*. American Institutes for Research. www2.ed.gov/programs/dropout/doubledoseprofile.pdf

Lee, J., Bryant, D., Ok, M., & Shin, M. (2020). A systematic review of interventions for algebraic concepts and skills of secondary students with learning disabilities. *Learning Disabilities Research and Practice*, *35*(2), 89–99. doi: 10.1111/ldrp.12217

Lobato, J., & Ellis, A. B. (2002). The teacher's role in supporting students' connections between realistic situations and conventional symbol systems. *Mathematics Education Research Journal*, *14*(2), 99–120. doi: 10.1007/bf03217356

Lynch, K. & Star, J. R. (2014). Views of struggling students on instruction incorporating multiple strategies in Algebra I: An exploratory study. *Journal for Research in Mathematics Education*, *45*(1), 6–18. doi: 10.5951/jresematheduc.45.1.0006

Macdonald, H., Zinth, J. D., & Pompelia, S. (2019). *50-state comparison: High-school graduation requirements*. Education Commission of the States. www.ecs.org/high-school-graduation-requirements/

Matthews, P. G., & Fuchs, L. S. (2020). Keys to the gate? Equal sign knowledge at second grade predicts fourth-grade algebra competence. *Child Development*, *91*(1), e14–e28. doi: 10.1111/cdev.13144

Matthews, P. G., Lewis, M. R., & Hubbard, E. M. (2016). Individual differences in nonsymbolic ratio processing predict symbolic math performance. *Psychological Science*, *27*(2), 191–202. doi: 10.1177/0956797615617799

McNeil, N. M. (2014). A change-resistance account of children's difficulties understanding mathematical equivalence. *Child Development Perspectives*, *8*(1), 42–47. doi: 10.1111/cdep.12062

Moseley, B., & Brenner, M. E. (2009). A comparison of curricular effects on the integration of arithmetic and algebraic schemata in pre-algebra students. *Instructional Science*, *37*(1), 1–20. doi: 10.1007/s11251-008-9057-6

Mulligan, J., Oslington, G., & English, L. (2020). Supporting early mathematical development through a 'pattern and structure' intervention program. *ZDM*, *52*, 663–676. doi: 10.1007/s11858-020-01147-9

Muschla, J. A., & Muschla, G. R. (1995). *The math teacher's book of lists*. Prentice Hall.

National Center for Education Statistics (2019). *NAEP report card: 2019 NAEP mathematics assessment*. www.nationsreportcard.gov

National Council of Teachers of Mathematics (2000). *Principles and standards for school mathematics*. www.nctm.org/Standards-and-Positions/Principles-and-Standards/

National Council of Teachers of Mathematics (2014). *Algebra as a strand of school mathematics for all students* [Position Statement]. NCTM. www.nctm.org

National Mathematics Advisory Panel (2008). *Foundations for success: The final report of the National Mathematics Advisory Panel*. US Department of Education. www2.ed.gov/about/bdscomm/list/mathpanel/report/final-report.pdf

Ochiltree, D. (2002). *Ten monkey jamboree*. Simon & Schuster.

Otten, M., Van den Heuvel-Panhuizen, M., & Veldhuis, M. (2019). The balance model for teaching linear equations: A systematic literature review. *International Journal of STEM Education*, *6*(1), 1–21. doi: 10.1186/s40594-019-0183-2

Powell, S. R. (2012). Equations and the equal sign in elementary mathematics textbooks. *The Elementary School Journal*, *112*(4), 627–648. doi: 10.1086/665009

Powell, S., Lembke, E., Ketterlin-Geller, L., Petscher, Y., Hwang, J., Bos, S., Cox, T., Hirt, S., Mason, E., Pruitt-Britton, T., Thomas, E., & Hopkins, S. (2020). Data-based individualization in mathematics to support middle school teachers and their students with mathematics learning difficulty. *Studies in Educational Evaluation*, 100897–. doi: 10.1016/j.stueduc.2020.100897

Pramesti, T. I., & Retnawati, H. (2019). Difficulties in learning algebra: An analysis of students' errors. *Journal of Physics: Conference Series*, *1320*(1), 12061. doi: 10.1088/1742-6596/1320/1/012061

Renze, J., & Weisstein, E. W. (n.d.). *Algebra*. [From *MathWorld*—A Wolfram web resource.]http://mathworld.wolfram.com

Riccomini, P., & Morano, S. (2019). Guided practice for complex, multistep procedures in algebra: scaffolding through worked solutions. *Teaching Exceptional Children*, *51*(6), 445–454. doi: 10.1177/0040059919848737

Rittle-Johnson, B., Zippert, E. L., & Boice, K. L. (2019). The roles of patterning and spatial skills in early mathematics development. *Early Childhood Research Quarterly*, *46*, 166–178. doi: 10.1016/j.ecresq.2018.03.006

Sidebotham, T. H. (2002). *The A to Z of mathematics: A basic guide.* John Wiley & Sons.

Stacey, K., & MacGregor, M. (1999). Learning the algebraic method of solving problems. *Journal of Mathematical Behavior, 18*(2), 149–167. doi: 10.1016/s0732-3123(99)00026-7

Star, J. R., Caronongan, P., Foegen, A., Furgeson, J., Keating, B., Larson, M. R., …Zbiek, R. M. (2015, 2019). *Teaching strategies for improving algebra knowledge in middle- and high-school students (NCEE 2015-4010).* National Center for Education Evaluation and Regional Assistance, Institute of Education Sciences, US Department of Education. http://whatworks.ed.gov

Stegall, J. B. & Malloy, J. A. (2019). Addressing misconceptions in Algebra 1. *The Mathematics Teacher, 112*(6), 450–454. doi: 10.5951/mathteacher.112.6.0450

Stephens, A. C., Ellis, A. B., Blanton, M., & Brizuela, B. M. (2018). Algebraic thinking in the elementary and middle grades. In J. Cai (Ed.) *Compendium for research in mathematics education* (pp. 386–420). NCTM.

Strickland, T. K., & Maccini, P. (2012). The effects of the concrete-representational-abstract integration strategy on the ability of students with learning disabilities to multiply linear expressions within area problems. *Remedial and Special Education, 34*(3), 142–153. doi: 10.1177/0741932512441712

Taff, J. (2017). Rethinking the order of operations (or what is the matter with Dear Aunt Sally?). *The Mathematics Teacher, 111*(2), 126–132. doi: 10.5951/mathteacher.111.2.0126

Tall, D. (1992). The transition to advanced mathematical thinking: Functions, limits, infinity, and proof. In D. A. Grouws (Ed.), *Handbook of research on mathematics teaching and learning* (pp. 495–511). Macmillan.

Trezise, K., & Reeve, R. A. (2014). Working memory, worry, and algebraic ability. *Journal of Experimental Child Psychology, 121*, 120–136. doi: 10.1016/j.jecp.2013.12.001

Vlassis, J. (2002). The balance model: Hindrance or support for the solving of linear equations with one unknown. *Educational Studies in Mathematics, 49*(3), 341–359. doi: 10.1023/A:1020229023965

Walkington, C., Sherman, M., & Petrosino, A. (2012). "Playing the game" of story problems: Coordinating situation-based reasoning with algebraic representation. *Journal of Mathematical Behavior, 31*, 174–195. doi: 10.1016/j.jmathb.2001.12.009

Watt, S. J., Watkins, J. R., & Abbitt, J. (2014). Teaching algebra to students with learning disabilities: Where have we come and where should we go? *Journal of Learning Disabilities, 49*(4), 437–447. doi: 10.1177/0022219414564220

Zaslavsky, C. (1996). *The multicultural math classroom: Bringing in the world.* Heinemann.

Zielinski, S. F. & Glazner, M. (2019). The no-to-yes project: Conquering common algebraic mistakes that drive us and our students crazy. *The Mathematics Teacher, 112*(6), 432–439. doi: 10.5951/mathteacher.112.6.0432

Appendix I: Mathematics Activities to Promote Positive Dispositions

Activity 1: Take a walk around your neighborhood (or campus) and make a list of all the mathematics applications you notice (e.g., numbers, comparisons, shapes, and measurements). Alternatively, keep a list of mathematics-related tasks you perform during one day. Think about the mathematics behind these everyday examples.

Activity 2: Locate a set of dominoes (or cut out a set from a website) and follow the rules below for a simple game. A set has 28 pieces. The pieces are sometimes called bones; the dots are called pips. Any group of bones with a common end is called a suit and bones with identical ends are called doublets. Consider the mathematics involved. Create new rules or look up more complex rules online.

- Place all pieces face down and shuffle them around.
- Draw for the lead and reshuffle. Each player then draws seven bones.
- The lead player usually plays his highest domino first. The object of this game is to have the fewest pips left in hand.
- The second player must play a piece that matches one end of a piece on the table. Doublets are played crosswise. Players who cannot make a match must draw another bone.
- The player who matches all bones in his hand calls "Domino" and earns the number of points as there are pips in his opponent's hand.
  Note that this activity requires new vocabulary and rules. It can promote subitizing, unit construction and coordination, and strategy development (MacDonald et al., 2020).

Activity 3: Using a copy of the hundred chart (1 to 100) in Appendix II, explore the concepts of multiples and factors (prime and composite numbers):

- Circle 2 and strike through all multiples of 2.
- Circle 3 and strike through all multiples of 3.
- Continue with 5, 7, 11, 13, 17 (the number 1 is neither prime nor composite).
- Circle all numbers not crossed out. The circled numbers are prime numbers and don't have factors other than one and themselves. This process is called the "Sieve of Eratosthenes" after the Greek mathematician (275–195 BCE) given credit for this simple method for identifying prime numbers (Wolfram, 2002).

Activity 4: Fibonacci numbers are numbers of the following sequence: 1, 1, 2, 3, 5, 8, 13, 21 ... where successive numbers are the sum of the two preceding numbers. This sequence of numbers or patterns formed by this sequence appears naturally in the

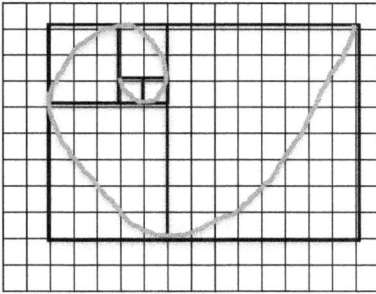

Figure I.1 Fibonacci Pattern

environment. Find a leaf, stem, pinecone, or seashell with the spiral pattern shown here. Look for this pattern elsewhere in nature or art.

Draw your own spiral on a piece of graph paper following these directions:

1. Select one square near the center and trace it.
2. Trace in one square to the right.
3. Above the two previous squares, trace in a square that is 2 blocks by 2 blocks.
4. Trace in a square that is 3 by 3 to the left.
5. Trace in a square below that is 5 by 5.
6. To the right, trace in the next square 8 by 8.
7. Above, trace in a square 13 by 13.
8. Continue as long as you have room on the paper.
9. Beginning in the first square, connect the corners of each square in the same sequence with a looping spiral, like that of a nautilus shell. That spiral is a Fibonacci pattern!

References

MacDonald, B. L., Moss, D. L., & Hunt, J. H. (2020). Dominoes: Promoting units construction and coordination. *Mathematics Teacher, 113*(7), 551–557. doi: 10.5951/MTLT.2019.0237

Wolfram, S. (2002). *A new kind of science.* Wolfram Media. wolframscience.com

Appendix II: Resources

0	1	2	3	4	5	6	7	8	9
10	11	12	13	14	15	16	17	18	19
20	21	22	23	24	25	26	27	28	29
30	31	32	33	34	35	36	37	38	39
40	41	42	43	44	45	46	47	48	49
50	51	52	53	54	55	56	57	58	59
60	61	62	63	64	65	66	67	68	69
70	71	72	73	74	75	76	77	78	79
80	81	82	83	84	85	86	87	88	89
90	91	92	93	94	95	96	97	98	99

0 to 99 Number Chart

1	2	3	4	5	6	7	8	9	10
11	12	13	14	15	16	17	18	19	20
21	22	23	24	25	26	27	28	29	30
31	32	33	34	35	36	37	38	39	40
41	42	43	44	45	46	47	48	49	50
51	52	53	54	55	56	57	58	59	60
61	62	63	64	65	66	67	68	69	70
71	72	73	74	75	76	77	78	79	80
81	82	83	84	85	86	87	88	89	90
91	92	93	94	95	96	97	98	99	100

1 to 100 Number Chart

10	9	8	7	6	5	4	3	2	1
20	19	18	17	16	15	14	13	12	11
30	29	28	27	26	25	24	23	22	21
40	39	38	37	36	35	34	33	32	31
50	49	48	47	46	45	44	43	42	41
60	59	58	57	56	55	54	53	52	51
70	69	68	67	66	65	64	63	62	61
80	79	78	77	76	75	74	73	72	71
90	89	88	87	86	85	84	83	82	81
100	99	98	97	96	95	94	93	92	91

100 Chart (Right to Left)

thousands	hundreds	tens	ones

ones		tenths	hundredths	thousandths

Place-Value Mats

1 whole							

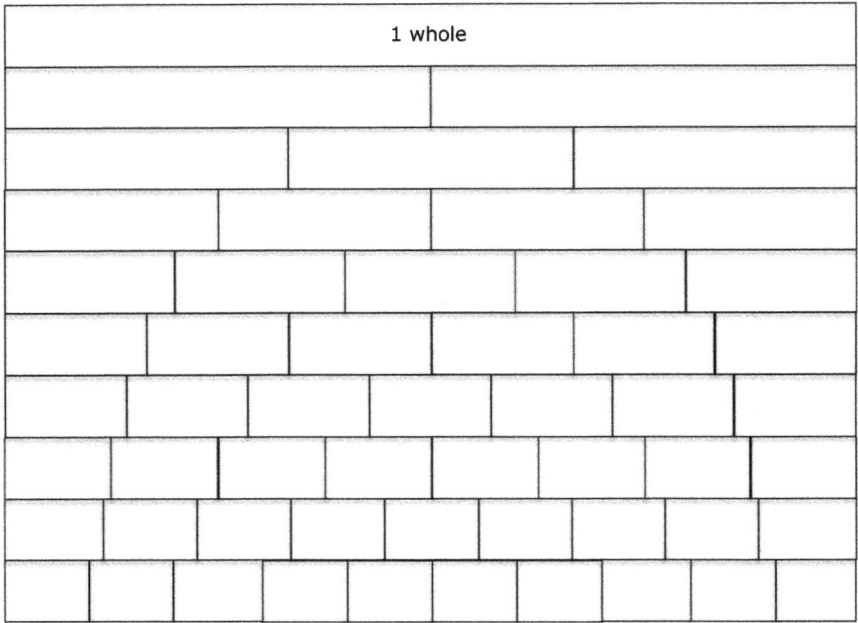

Fraction Strips

1	2	3	4	5	6	7	8
9	10	0	1	2	3	4	5
6	7	8	9	10	=	=	+
+	−	−	X	X	/	/	0
(	)	?	$	>	<	.	0

Number and Symbol Tiles

Glossary of Select Terms

504 plan Derives from Section 504 of the Rehabilitation Act of 1973, which prohibits discrimination, such as barriers to an education, based on disability. 504 plans are developed for students who have an identified disability and need accommodations but are not eligible for services under the Individuals with Disabilities Education Act (2004).

Algorithm A step-by-step procedure for accomplishing a task or for solving a problem.

Anchored instruction A technology-based learning approach that places learning within an authentic, problem-solving context.

Attribution The process of inferring the causes of events or behaviors.

Axiom A statement or proposition that is commonly accepted. In mathematics, the term *postulate* is more often used for these truths that don't require proof. An example from Euclid: all right angles are congruent.

Cardinal (form of number) Numbers used when counting. The total quantity achieved when counting.

Comorbidity The simultaneous presence of two or more conditions.

Data (plural) Information. Data can be collected in many forms, such as numbers, observations, facts, and measurements.

Decimal fraction A fraction or mixed number in which the denominator is a power of ten, expressed using a decimal point, such as 1.25 or .067.

Deductive reasoning Logical thinking that begins with general statements to reach a specific conclusion.

Digit A single whole number 0 to 9 (see number).

Dyscalculia A term coined in the mid-20th century to describe a neurological disorder that now refers to difficulty acquiring basic arithmetic skills.

Effect Size A quantitative measure of the magnitude of the experimental effect. The larger the effect size, the stronger the relationship between two variables. Effect sizes cited in this text include *Cohen's d* (comparison of two means), *Pearson r correlation* (strength of bivariate relationship, between -1 and $+1$), *Hedges' g* (modified Cohen's d for sample sizes < 20), η^2 (uses several explanatory variables to predict response variable), and *Tau-U* (used in single-case designs to correct for trends).

Equal Identical.

Equivalent Having the same value. Twelve (12) is equivalent to a dozen.

Euclidian geometry The study of plane and solid figures based on Greek mathematician Euclid's axioms and theorems, taught in secondary schools.

Factors Numbers that can be multiplied together to get another number. For example, 4 and 5 are factors of 20 (as are 2 and 10).

Fewer Refers to a comparison that can be counted (e.g., marbles). Often confused with *less*. However, the mathematical terminology for $<$ is *less than*, even when numbers are used.

Function A relation from a set of inputs to a set of possible outputs where each input is related to exactly one output.

Hippocampus Brain structures deep in the temporal lobe of each cerebral cortex. Important for regulating learning, memory encoding, memory consolidation, and spatial navigation (Dutta, 2019).

Inductive reasoning To make broad generalizations based on specific observations.

Inequality One mathematics quantity or expression is not equal to another. For example, $5y > 20$ ($5y$ is greater than 20).

Inferential statistics A collection of procedures that allow using random samples from a population to make inferences about the entire population.

Irrational numbers Real numbers that cannot be written as a simple fraction (not rational). Examples include π and $\sqrt{2}$.

Java applet A small program (little application) written in Java code and delivered to users in the form of bytecode. An applet is typically embedded in a web page and can be an animation, calculation, or other simple task.

Learning trajectory "Descriptions of children's thinking and learning in a specific mathematical domain, and a related conjectured route through a set of instructional tasks designed to engender those mental processes or actions hypothesized to move children through a developmental progression of levels of thinking" (Clements & Sarama, 2004, p. 83).

Less Refers to a comparative amount that cannot be counted (e.g., amount of paint in a bucket). Often confused with fewer.

Myelination Formation of a protective, insulating sheath (of lipids and lipoproteins) around axons of neurons, from the prenatal period through adolescence.

Net (for geometric object) A shape that is formed by unfolding a three-dimensional object.

Number A general term referring to a quantity. Numerals are symbols to represent numbers.

Ordinal (form of number) An adjective that describes a numerical position, such as *third* or *7th*.

Percent (adverb or noun) Latin for "of each hundred," typically refers to a unit and is used with numbers. Only three *percent* of students had their books today.

Percentage (noun) Part of a whole expressed in hundredths. Refers to the concept, not used with numbers. What *percentage* of students had their books?

Percentile A percentage position in a range of data. The *n*th percentile of a set of data is the number such that *n*% of the data is less than that number. A score at the 95th percentile is higher than 95% of the other scores.

Plasticity (of brain) Malleability, ability to change and adapt as a result of experience.

Prism Three-dimensional shape with two identical sides facing each other (bases). For example, cube, triangular prism, rectangular prism.

Second-order meta-analysis A meta-analysis (systematic, quantitative study of the results of previous research) of a number of statistically independent and methodologically comparable first order meta-analyses examining the same relationship in different contexts.

Subitize To perceive at a glance the number of items presented (objects or markings on paper) without counting.

Topological geometry, or topology A study of the properties that are preserved through deformations, twistings, and stretching of objects. In topology, a circle is equivalent to an ellipse (Weisstein, n.d.).

Triangle Plane figure with three straight sides and three angles. Types include *right* (one 90° angle, *acute* (largest angle < 90°), *obtuse* (largest angle > 90°), *equilateral* (all sides congruent), *isosceles* (two congruent sides), and *scalene* (no sides congruent).

Virtual manipulatives Computer-based, interactive "objects" or manipulatives, typically in Java applets.

References

Clements, D. H., & Sarama, J. (2004). Learning trajectories in mathematics education. *Mathematical Thinking and Learning, 6*(2), 81–89.

Dutta, S. S. (2019). *Hippocampus functions*. News-Medical.Net. www.news-medical.net/health/Hippocampus-Functions.aspx

Individuals with Disabilities Education Improvement Act (2004). 20 USC § 1400 et seq.

Rehabilitation Act (1973). 29 USC § 701 et seq.

Weisstein, E. W. (n.d.). *Topology*. Mathworld: A Wolfram web resource. http://mathworld.wolfram.com

Index

For Product Safety Concerns and Information please contact our EU
representative GPSR@taylorandfrancis.com
Taylor & Francis Verlag GmbH, Kaufingerstraße 24, 80331 München, Germany

www.ingramcontent.com/pod-product-compliance
Lightning Source LLC
Chambersburg PA
CBHW081220220326
41598CB00037B/6846